Boston Studies in the Philosophy and History of Science

Volume 338

The series *Boston Studies in the Philosophy and History of Science* was conceived in the broadest framework of interdisciplinary and international concerns. Natural scientists, mathematicians, social scientists and philosophers have contributed to the series, as have historians and sociologists of science, linguists, psychologists, physicians, and literary critics.

The series has been able to include works by authors from many other countries around the world.

The editors believe that the history and philosophy of science should itself be scientific, self-consciously critical, humane as well as rational, sceptical and undogmatic while also receptive to discussion of first principles. One of the aims of Boston Studies, therefore, is to develop collaboration among scientists, historians and philosophers.

Boston Studies in the Philosophy and History of Science looks into and reflects on interactions between epistemological and historical dimensions in an effort to understand the scientific enterprise from every viewpoint.

More information about this series at http://www.springer.com/series/5710

Adam LaCaze • Barbara Osimani
Editors

Uncertainty in Pharmacology

Epistemology, Methods, and Decisions

Springer

Editors
Adam LaCaze
School of Pharmacy
The University of Queensland
St Lucia, QLD, Australia

Barbara Osimani
Department of Biomedical Sciences
and Public Health
Polytechnic University of the Marche
Ancona, Italy

Munich Center for Mathematical
Philosophy
Munchen, LMU
Munich, Germany

ISSN 0068-0346 ISSN 2214-7942 (electronic)
Boston Studies in the Philosophy and History of Science
ISBN 978-3-030-29178-5 ISBN 978-3-030-29179-2 (eBook)
https://doi.org/10.1007/978-3-030-29179-2

This Springer imprint is published by the registered company Springer Nature Switzerland AG.
The registered company address is: Gewerbestrasse 11, 6330 Cham, Switzerland

Preface

Pharmacology is the science of drug action. Combined with other biomedical sciences, it underpins clinical drug development, drug regulation and clinical use of drugs and informs broader interactions between drugs and society. Many deep and practically important philosophical questions arise in this domain: How should pharmacological theories be understood? How should diverse lines of evidence in the biological, medical and clinical sciences be amalgamated to inform the care of individuals? How should drugs be evaluated and regulated? What is the appropriate role of the pharmaceutical industry? to name just a few. Work from the philosophy of biology, epistemology, philosophy of science, philosophy of medicine and ethics is relevant to these questions. In recent years, the quantity of work that engages with the practice and specific problems that arise in pharmacology and the use of drugs has increased. This work examines the foundations, methods and values of clinical and experimental pharmacology and contributes to our understanding of the science and its implementation in clinical and regulatory decision-making. This volume seeks to showcase the depth and breadth of work broadly characterised as philosophy of pharmacology and to present it within a single volume for the first time.

The contributions are diverse in terms of the disciplinary backgrounds of the authors, topics and methodological approaches. The volume contains work from philosophers, clinicians, regulators and statisticians. Some tackle practical issues as they arise within their discipline, while others take a more theoretical focus. The topics range from the appropriate way to think about evidence of mechanisms in clinical medicine to the normative foundations of pharmaceutical regulation, taking in inference patterns when evaluating the effects of drugs, the role of values in pharmacology and the history of masking treatment allocation in clinical trials along the way. Contributions are presented according to the following topic themes: EPISTEMOLOGY; METHODS; and DECISIONS. The methodological approach of many of the contributions fits under the broad umbrella of analytical philosophy. Some of the methods employed include the in-depth exploration of a case (Bueter and Jukola, Chap. 20), historical analysis (Holman, Chap. 17) and formal epistemology (Poellinger and Landes, Chaps. 5 and 11).

The role of mechanisms and evidence of mechanisms in medicine is an important topic in the philosophy of pharmacology and one that many of the contributions in this volume touch on. The first four chapters in the EPISTEMOLOGY section explore this topic. Aronson provides a clinical pharmacologist's perspective on the definition of "mechanism" and then differentiates two types of relation between mechanisms and evidence: evidence of mechanisms, which he labels "evidence-based mechanism", and the use of mechanisms as evidence, "mechanism-based evidence". The distinction highlights the different ways that mechanisms are discussed and used in pharmacology and clarifies some aspects of the debate regarding the role of mechanisms in evidence-based decision-making. Rocca, Anjum and Mumford take a different perspective, focusing on how mechanistic knowledge can develop in cases of *causal failure*. Causal failure occurs when the treatment fails to deliver the expected outcome. They also argue that instances of causal failure provide an opportunity to learn more about the context-sensitive, intrinsic, tendential and complex mechanisms by which drugs produce effects.

The next two chapters explore the role that mechanisms play in extrapolation. Parkkinen and Williamson compare strategies for extrapolating causal claims from model organisms to humans using atherosclerosis research as a case study. They argue that evidence of mechanisms plays an important role in the key strategies employed when extrapolating from model organisms: comparative process tracing, phylogenetic reasoning and robustness analysis. The importance of evidence of mechanisms to these strategies, they argue, provides further support for the Russo-Williamson thesis—the epistemological claim that evidence of mechanisms and evidence of difference-making are typically needed in order to establish causal claims. Belzung, Billette de Villemeur and Lemoine provide an analysis of drug development in psychiatry to argue for a different set of conclusions regarding extrapolation from animal models. Belzung and colleagues argue that the use of previously successful animal models in the development of psychopharmacology frequently relies on a form of *statistical extrapolation* and that the received view of extrapolation tends to underestimate the importance of this kind of extrapolation.

The remaining chapters in the EPISTEMOLOGY section examine a range of theoretical and practical topics regarding pharmacological knowledge. Poellinger seeks to illustrate and distinguish analogy-based inference patterns in the causal assessment of drug harms. Boem, Malagrinò and Bertolaso examine the role of in silico clinical trials in addressing complexity in pharmacology. Sözüdoğru and Clarke examine the use of Lipinski's rule of five as a heuristic strategy within drug discovery and explore what this means for standard mechanistic and capacity accounts of causation. And finally, Ruthenberg completes the section with a discussion of underdetermination in chemistry and pharmacology.

Contributions within the METHODS section examine the role and limits of randomised trials and the challenge of reliable biomarkers and consider a range of approaches to amalgamating pharmacological evidence produced by different methods. The first two chapters in this section explore the logic of randomised trials. Festa, Tambolo and Cevolani suggest that the principles governing the assessment of efficacy hypotheses, and, more generally, hypotheses of statistical causality, are provided by an appropriate statistical version of the method of difference put forward by John Stuart Mill. Annoni and Boniolo examine assumptions regarding specific and non-specific treatment effects and question the commonly held view that these effects are additive and easily separated for the purposes of efficacy assessment.

The contributions from Landes and LaCaze and Winckel look at the application of evidence hierarchies. Landes approaches the problem of ranking drugs for a particular outcome using diverse lines of evidence as a multi-criteria decision problem. This chapter employs the formal machinery of decision sciences to provide a ranking of drugs based on the available evidence. Using metformin-associated lactic acidosis as a case study, LaCaze and Winckel argue that a causal approach to drug safety assessment along the lines of the Russo-Williamson thesis is superior to the method-focussed approach advocated within evidence-based medicine.

The next two chapters explore methods to reduce uncertainty in clinical trials. Keyser and Sarry analyse the methodological use of clinical biomarkers in pharmacological measurement. Their contribution presents a systematic methodology for assessing the reliability of multiple biomarkers underpinned by robustness analysis. Mansmann and Boulesteix continue this theme by examining four standard statistical methods to reduce and model uncertainty about individual response to treatments. Mansmann and Boulesteix propose ways to improve the analysis and presentation of heterogeneous treatment outcomes. Osimani's contribution closes the second part of the volume and introduces the last one by addressing (implicit) strategic dimensions related to how evidence is gathered, evaluated and used. The chapter analyses the dissent around evidence standards in medicine and pharmacology as a result of distinct ways to address epistemic losses and to conceptualise reliability in distinct scientific paradigms and philosophical schools of thought.

The final section, DECISIONS, considers the relations between science, policy and regulatory decision-making. Hansson begins the section with a consideration of the appropriate role of non-epistemic values in pharmacological science. The next four chapters approach pharmaceutical regulation, pharmaceutical markets and the influence of industry and society more broadly. Holman discusses the development of standards for assessing therapeutic claims and the contribution of the American Medical Association's Council on Pharmacy and Chemistry on a growing pharmaceutical market prior to the establishment of the US Food and Drug Administration. Teira examines two normative principles that underpin pharmaceutical regulation: a liberal argument for protecting pharmaceutical markets in terms of quality control and a paternalistic argument for protecting consumers of medicines. Teira argues for a paternalistic approach to pharmaceutical regulation based on the need for *impartial* information regarding the benefits and harms of medicines. Solomon's contribution considers the issue of industry bias and offers both qualitative and quantitative proposals to reduce the impact of this bias. Bueter and Jukola close out the volume with an analysis of the FDA's decision to approve flibanserin as a treatment for a "hypoactive sexual desire disorder". Bueter and Jukola suggest that criticising the approval of flibanserin primarily in terms of medicalisation is problematic and argue in favour of institutional rather than conceptual safeguards against unnecessary pharmaceutical medicalisation.

We would like to thank the contributors for submitting their work, the reviewers for their contribution to the volume and the editorial team at Springer for their assistance. It has been a pleasure to put together this volume. We believe it well illustrates the breadth of work in the emerging subdiscipline of philosophy of pharmacology.

Funding for the research and editorial work by Barbara Osimani has been provided through the ERC funded project: "Philosophy of Pharmacology: Safety, Statistical Standards, and Evidence Amalgamation" – PhilPharm GA 639276.

Brisbane, Australia
Ancona, Italy
May 2019

Adam LaCaze
Barbara Osimani

Contents

List of Contributors

Rani Lill Anjum Norwegian University of Life Sciences (NMBU), Ås, Norway

Marco Annoni National Research Council, Rome, Italy
National Institute of Biomedical Technologies, Milan, Italy
Fondazione Umberto Veronesi, Milan, Italy

Jeffrey K. Aronson Centre for Evidence Based Medicine, Nuffield Department of Primary Care Health Sciences, Radcliffe Observatory Quarter, Oxford, UK

Catherine Belzung University of Tours, INSERM U930, Tours, France

Marta Bertolaso Università Campus Bio-Medico di Roma, Rome, Italy

Etienne Billette de Villemeur University of Lille, LEM (UMR CNRS 9221), Lille, France

Federico Boem University of Milan, Milan, Italy

Giovanni Boniolo Dipartimento di Scienze Biomediche e Chirurgico Specialistiche, Università di Ferrara, Ferrara, Italy
Institute for Advanced Study, Technische Universität München, Munich, Germany

Anne-Laure Boulesteix Institute for Medical Informatics, Biometry and Epidemiology, Ludwig-Maximilians Universität München, Munich, Germany

Anke Bueter Leibniz University Hannover, Hannover, Germany

Gustavo Cevolani IMT School for Advanced Studies Lucca, Lucca, Italy
Center for Logic, Language, and Cognition, University of Turin, Turin, Italy

Brendan Clarke Department of Science and Technology Studies, UCL, London, UK

Roberto Festa Department of Humanistic Studies, University of Trieste, Trieste, Italy

Sven Ove Hansson Division of Philosophy, Royal Institute of Technology (KTH), Stockholm, Sweden
Department of Learning, Informatics, Management and Ethics, Karolinska Institutet, Solna, Sweden

Bennett Holman Underwood International College, Yonsei University, Seoul, South Korea
Faculty of Humanities, University of Johannesburg, Johannesburg, South Africa

Saana Jukola Bielefeld University, Bielefeld, Germany

Vadim Keyser California State University, Fresno, CA, USA

Adam LaCaze School of Pharmacy, The University of Queensland, Brisbane, QLD, Australia

Jürgen Landes Munich Center for Mathematical Philosophy, LMU Munich, Munich, Germany

Maël Lemoine University of Bordeaux, Immunoconcept, Bordeaux, France

Ilaria Malagrinò Università Campus Bio-Medico di Roma, Rome, Italy

Stephen Mumford Durham University, Durham, UK

Ulrich Mansmann Institute for Medical Informatics, Biometry and Epidemiology, Ludwig-Maximilians Universität München, Munich, Germany

Barbara Osimani Department of Biomedical Sciences and Public Health, Polytechnic University of the Marche, Ancona, Italy
Munich Center for Mathematical Philosophy, Munchen, LMU, Munich, Germany

Veli-Pekka Parkkinen Department of Philosophy, University of Bergen, Bergen, Norway

Roland Poellinger Department of Arts and Culture of the City of Munich, Munich, Germany

Elena Rocca Norwegian University of Life Sciences (NMBU), Ås, Norway

Klaus Ruthenberg Coburg University of Applied Sciences and Arts, Coburg, Germany

Louis Sarry California State University, Fresno, CA, USA

Miriam Solomon Temple University, Philadelphia, PA, USA

Erman Sözüdoğru Department of Science and Technology Studies, UCL, London, UK

Luca Tambolo Department of Humanistic Studies, University of Trieste, Trieste, Italy

David Teira Universidad Nacional de Educación, Madrid, Spain

Jon Williamson Centre for Reasoning, University of Kent, Canterbury, UK

Karl Winckel School of Pharmacy, The University of Queensland, Brisbane, QLD, Australia
Princess Alexandra Hospital, Brisbane, QLD, Australia

Part I
Epistemology

Chapter 1
Defining Aspects of Mechanisms: Evidence-Based Mechanism (Evidence *for* a Mechanism), Mechanism-Based Evidence (Evidence *from* a Mechanism), and Mechanistic Reasoning

Jeffrey K. Aronson

Abstract Evidence derived from mechanisms and evidence derived from probabilistic studies are complementary.

There are six key components of mechanisms: entities (structures), activities (functions), organization, potential interactions, phenomena (outcomes), and the internal milieu in which the mechanism is situated, which can affect outcomes. These features suggest the following definition of "mechanism":

> **mechanism,** n. /ˈmɛkənɪz(ə)m/ One or more entities and activities spatially and temporally organized and potentially interacting, depending on the internal milieu, in such a way as to be wholly or partly responsible for a phenomenon or phenomena [ancient Greek μηχανή a contrivance]

The relation between mechanisms and evidence implies two separate concepts:

(a) Evidence-based mechanism (evidence *for* a mechanism). This describes two types of evidence that demonstrate the existence of mechanisms (ontic evidence) and recognition of how they operate (anagnoristic evidence).
(b) Mechanism-based evidence (evidence *from* a mechanism; polychrestic evidence). This describes how biological mechanisms can be used in many ways: to analyse how adverse drug reactions occur and show how they can be prevented; to explain outcomes; to extrapolate from clinical trials to clinical practice; to suggest practical procedures (praxis); to inform drug discovery; and to generate testable hypotheses. However, clinical interventions based on predictive mechanisms alone should generally not, with a few exceptions, be used unless they have been probabilistically tested in high quality trials.

J. K. Aronson (✉)
Centre for Evidence Based Medicine, Nuffield Department of Primary Care Health Sciences, Radcliffe Observatory Quarter, Oxford, UK
e-mail: jeffrey.aronson@phc.ox.ac.uk

A. LaCaze, B. Osimani (eds.), *Uncertainty in Pharmacology*, Boston Studies in the Philosophy and History of Science 338,
https://doi.org/10.1007/978-3-030-29179-2_1

These two terms are more specific than the ambiguous terms "evidence of mechanism" and "mechanistic evidence".

Mechanistic reasoning combines these two approaches with the results of probabilistic clinical observations.

Keywords Mechanisms · Evidence · Evidence-based mechanism · Mechanism-based evidence · Mechanistic reasoning

1.1 Introduction

> It has not escaped our notice that the specific pairing we have postulated immediately suggests a possible copying mechanism for the genetic material. [Watson JD Crick FH. Nature 1953; 171(4356): 737-8]

The famous final sentence in the paper in which Jim Watson and Francis Crick described a proposed structure of DNA, quoted above, reminds us that even when scientists are primarily occupied with other matters, they are still cognizant of the primacy of mechanisms. Indeed, they have been at least since the time of Descartes (see below) and possibly even earlier.

I want to make it clear from the outset that I view mechanisms in bioscience, and specifically in physiology and pharmacology, with which I am primarily concerned, as being important and indeed necessary in many ways, which I shall outline. This view is only slightly vitiated by the observation that there are occasions on which mechanistic reasoning may be misleading.

Interest in scientific mechanisms has increased among philosophers since the 1960s, in three phases, as shown by publications listed in the Philosophical Index. There were two early phases in 1965–1985 and in 1985–1999 (Fig. 1.1, upper panel), although the percentages of all publications (Fig. 1.1, lower panel) did not increase during those periods, since the total numbers of publications also increased. However, in the third phase, since 2000, the number of publications on mechanisms has increased both in absolute terms and in proportion to all publications. This large increase in interest is generally regarded as having been initiated, despite earlier work by such as Wimsatt, Richardson, Glennan, and Cartwright (see below), by the publication in 2000 of what is known as the MDC paper, from the initials of its authors, Machamer, Darden, and Craver (2000).

In contrast, the numbers of publications listed in the general bioscience database PubMed in which mechanisms were mentioned has increased steadily since 1970, both in absolute terms and in relation to the total number of publications listed (Fig. 1.2).

In this chapter I have explored definitions of aspects of the term "mechanism" in different ways. First, by seeking an intensional definition, using a fourfold linguistic approach, which involves exploring etymology, usage (deduced from definitions and examples given in major dictionaries), previous definitions that others have proposed, and aspects of the definiendum that show how the processes involved actually operate (Aronson 2009). Synthesis of these has generally, in my experience,

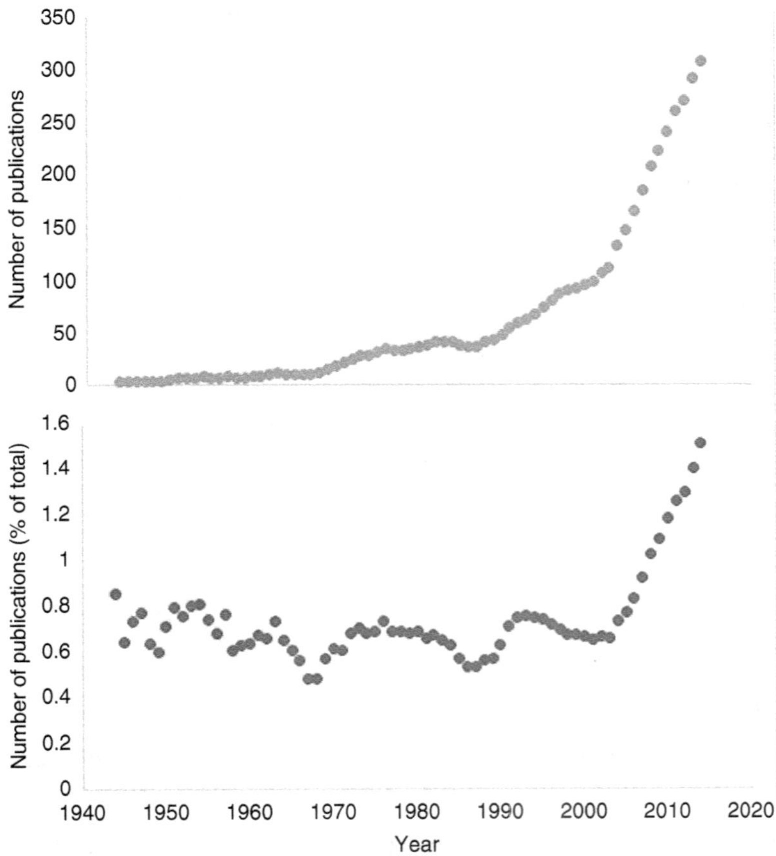

Fig. 1.1 Numbers of publications in the Philosopher's Index in which mechanisms were mentioned (five-year running averages); upper panel—absolute numbers of publications; lower panel—percentage of all publications

yielded satisfactory definitions, as evidenced by international adoption of some of them (compare, for example, Ferner and Aronson (2000) with European Medicines Agency (2018) and Hauben and Aronson (2009) with Council for International Organizations of Medical Sciences (2010)). Secondly, I have conducted an extensional exploration of the different ways in which mechanisms operate. Finally, I have considered how my proposed intensional and extensional definitions affect the ways in which mechanistic reasoning is used in clinical medicine.

In formulating their definitions some have aspired to cover all kinds of mechanisms, including mechanisms in areas as disparate as physics and social science. For example, Illari and Williamson (2012) claimed that they had formulated "a characterization that gives an understanding of what is common to mechanisms in *all* fields" [their emphasis], "widely applicable across the sciences", and that others "should have no serious objections to [their] account". Since the intensional definition that I have proposed is different in only small, although potentially

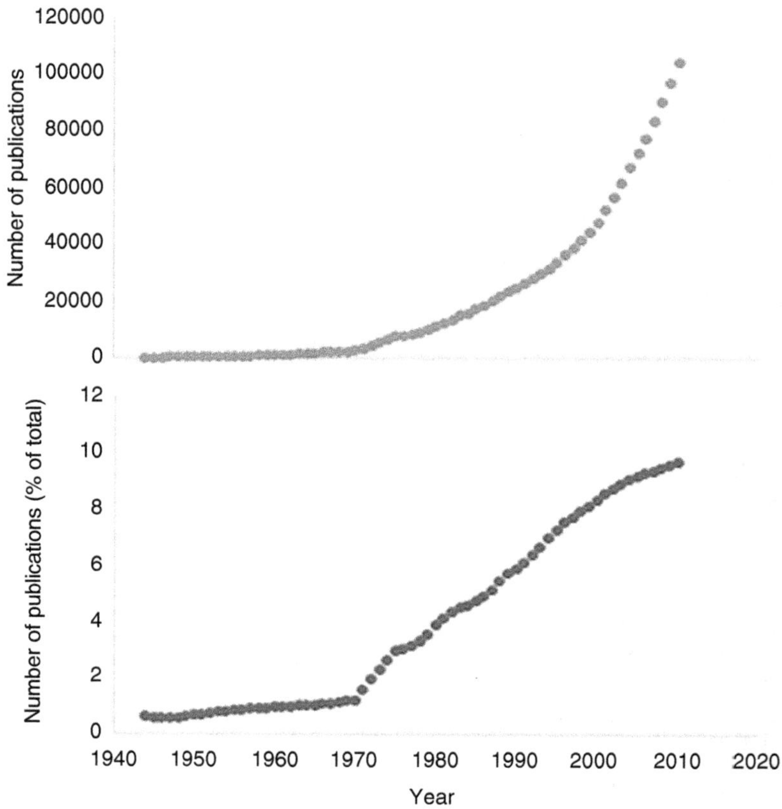

Fig. 1.2 Numbers of publications listed in PubMed in which mechanisms were mentioned (five-year running averages); upper panel—absolute numbers of publications; lower panel—percentage of all publications

important, ways from theirs, it may be that my analysis can be similarly regarded. However, I make no such claim, having considered only biological mechanisms, and particularly physiological and pharmacological ones. This is also reflected in the multiplicity of mechanistic subdivisions in the extensional lists of the modes, enumerated below, in which physiological and pharmacological mechanisms can operate.

1.2 Defining "Mechanism"

The Greek word for a definition was ὅρισμός, from ὁρίζειν, to draw a boundary around something. In English, "orismology" means a collection of technical terms or the explanation of such terms; in a word, terminology. The word "define" also etymologically implies circumscribing a boundary (Latin finis). This implies the inclusion of all those things that are essential to the definiendum, its essence or quiddity (Aristotle τό τί ἦν εἶναι), and the exclusion of all those things that are not essential.

For etymology, I generally begin with the earliest identifiable linguistic root. In English, that is usually a theoretical construct in the system known as Indo-European, which hypothesizes that a large range of languages, including Albanian, Armenian, Balto-Slavic, Celtic, Germanic, Hellenic, Indo-Iranian, Italic, and some extinct languages, have descended from a single parent language, known as proto-Indo-European (Aronson 2016). In some cases other proto-languages, such as proto-Semitic, are relevant to English (Agmon 2010).

For usage (i.e. "senses relating to habit, custom or practice"), I consult the *Oxford English Dictionary* (OED), which gives the earliest identified written occurrences of specific meanings; I chart these on time lines and extract the relevant meanings.

For previous definitions, I undertake systematic searches of the usual databases (such as PubMed, Embase, and the Cochrane Library). I combine and synthesize these different pieces of information, in conjunction, when appropriate, with an approach that recognizes that the meaning of a term is implied by the relevant scientific theory in which it is framed, including all the assertions that the theory makes about the term, adding, in the case of clinical terms, an understanding of the practical aspects of the relevant theory (Winnie 1967).[1]

1.2.1 Etymology

Etymology can be informative in reaching an understanding about what a word or phrase means and in starting to frame an appropriate definition and it gives insights into the history of the development of ideas surrounding the definiendum. It therefore points to items that should not be included as well as those that might be. For example, in this case the fact that "mechanism" is etymologically related to "machine" points to a debate about the difference. On the other hand, it is important to avoid the etymological fallacy, assuming that what a word once meant is what it must mean now.

The Indo-European root of "mechanism" is MAGH, power, and its e-grade form[2] is MEGH, great. Words from Sanskrit that reflect this begin with mah(a)-: maharaja

[1]See also http://www.jimpryor.net/teaching/courses/mind/notes/ramseylewis.html

[2]The basic form of Indo-European (IE) roots is the so-called e-grade form, which contains a short *e* or a long *ē*. That can change into the so-called o-grade form, with a short *o* or a long *ō*. The vowel can also disappear or be replaced by a neutral vowel sound called a schwa, after the Hebrew vowel of that name, the so-called zero-grade form. The schwa, represented in print by an inverted *e* (ə), typically occurs in weakly stressed syllables. This variation in vowel sounds is called ablaut. In addition, each ablaut form can have suffixes and prefixes added. Other changes include shortening, nasalization by the insertion of the letter n, reduplication, or compounding with other forms. Each resultant form can by itself give rise to several different words. For further information see Beekes RSP. *Comparative Indo-European Linguistics: An Introduction*. Amsterdam: John Benjamins, 1995.

and maharani, maharishi, mahatma, mahout (from mahāmātra, an elephant-keeper), and the Mahabharata, one of the two major ancient Hindu epics, traditionally ascribed to the poet Vyāsa, which tells the story of the Bharata dynasty and its war and contains the religious text the Bhagavadgita.

Related Greek words included μάχη (battle), μέγας (large), and μηχανή (a contrivance). The kind of contrivance the Greeks originally had in mind was a crane for lifting heavy weights or an engine of war for throwing them. Corresponding English words include might and main, magus and magic (from Latin via Persian), magnify, and a couple of dozen words ending in -machy, denoting fighting, such as logomachy (disputation about words) and sphaeromachy (tennis). And of course machine and mechanism.

1.2.2 Usage

The word "mechanism" appeared first in English in the late seventeenth century. Various usages of the word, as cited in the OED, are shown in Fig. 1.3. The meanings it accrued at that time all focused on machine-like properties. This is not surprising, in view of the etymology and the rise of mechanical philosophy during the seventeenth century, as propounded and applied not only by mathematicians and physicists, such as Descartes in his *Principia Philosophiae* (1644) and Newton in his *Philosophiæ Naturalis Principia Mathematica* (1687), a text that dealt with what he called "Rational Mechanics", but also by those who took a mechanical approach to biological problems. These included William Harvey, who described the circulation in mechanical terms in *Exercitatio Anatomica de Motu Cordis et Sanguinis in Animalibus* (1628), and Robert Hooke, who described the use of the microscope in his studies of small insects and plants in *Micrographia* (1665) (Lawson 2015). John Floyer's invention of the pulse-watch is a later example (1707) (Gibbs 1971). Hooke, for example, used "mechanism" to mean a system of mutually adapted parts working together in a machine or in a manner analogous to that of a machine. The extent to which such scientists may have taken non-mechanistic approaches to some matters does not invalidate their mechanistic approaches to others.

These usages reveal five essential features that are typical of machines: structures, functions, organization, and potential interactions, yielding outcomes. In modern times we find the same elements being invoked time and again, although the terminology has changed, partly in order to avoid confusion with meanings that have been applied to some of these words in other contexts: structures, the anatomy of biological mechanisms, are now generally called "entities" (often referred to as "parts"); functions, the physiology and pathophysiology of biological mechanisms, are called "activities" (or sometimes "capacities"); and outcomes are called "phenomena" (or sometimes "behavior"). Here, for the purposes of definition, I use the terms "entities", "activities", and "phenomena", rather than "structures", "functions", and "outcomes". However, I continue to use the term "functions" when referring to clinical phenomena (e.g. liver function).

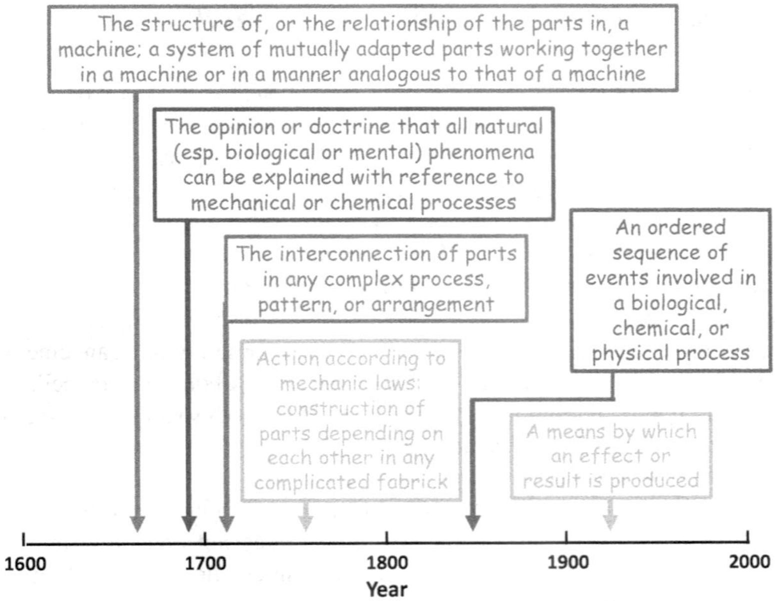

Fig. 1.3 Usages of "mechanism" from definitions listed in the OED

That mechanisms are not necessarily linearly organized is a feature that is rarely if ever mentioned explicitly in definitions, although the terms "interaction" and "organized in such a way", used by some, may in some cases imply non-linearity.

In passing, it is worth noting that Samuel Johnson's definition of mechanism in his dictionary of 1755 includes a reference to "action according to mechanic laws". However, Cartwright's (1999) deconstruction of the idea of natural laws, which she claims to be true only ceteris paribus, not universally, suggests that invariant laws should not be included as a feature of mechanisms. Indeed, our current understanding of the ways in which many biological mechanisms work suggests that laws, even locally applicable laws, should not be invoked in descriptions of mechanisms, at least of biological, and specifically physiological and pharmacological ones, with which I am specifically concerned here. For example, a pharmacological outcome mediated by a mechanism may be altered when physiological conditions change or when pathology supervenes. Furthermore, such mechanisms may themselves change during long-term exposure to an intervention (e.g. upregulation of receptors in response to long-term exposure to a pharmacological antagonist). This point will emerge most forcefully when we come to consider both the extensional range of mechanistic operations and the effects of the internal milieu on mechanisms. I do not intend to imply that this stricture should necessarily apply to other types of mechanism (e.g. physical or social ones).

1.2.3 Modern Intensional Definitions of "Mechanism"

An intensional definition is one that provides the meaning of a word or expression by specifying necessary and sufficient conditions for its appropriate use (Cook 2009). In exploring modern intensional definitions of "mechanism" I have taken into account the five key features identified above: entities, activities, organization, potential interactions, and phenomena, and have added a sixth—internal milieu, as discussed below. I have also excluded certain terms relating to laws, linearity, regularity, and invariance, which appear in some definitions.

I believe that entities and activities need no further discussion here.

Organization is defined in the OED as "systematic ordering or arrangement" and "the development or coordination of parts (of the body, a body system, cell, etc.) in order to carry out vital function". I take this to include both spatial organization and temporal organization, i.e. the anatomical arrangement of parts of the mechanism and the order in which different parts are effectuated.

Interaction is defined as "reciprocal action; action or influence of things on each other". In other words a simple train of effects does not constitute interaction. However, even a single feedback loop introduces interaction. This is discussed in relation to the examples below.

Milieu is defined in the OED as "environment", which is defined as "a particular set of surroundings or conditions which something ... interacts with". I stress here that by "milieu" I mean the *milieu intérieur*, as first described by Claude Bernard, the concept that Walter B Cannon used to inform the notion of homoeostasis (Cannon 1926). This is in contrast to the *external* physical environment, which in the face of strong homoeostatic mechanisms plays little or no part in modulating physiological mechanisms, except in extreme circumstances. I include under the heading of *milieu intérieur* not only the physiological state of a cell (e.g. its internal pH) and of the body as a whole (e.g. the degree of sympathetic nervous system tone), but also epigenetic effects that are not explained by changes in DNA (e.g. modifications of chromatin causing altered gene expression) but which nevertheless result in changes in inheritable gene activity. Although others have impliedly recognized the importance to mechanisms of internal milieu, in referring, for example, to background conditions and context, none has including this concept in any definition. Indeed, in some cases, where reference is made to such conditions, it turns out to be nothing to do with the internal milieu; for example, when Craver and Darden (2013) refer to "modulating conditions" in studies of learning behaviours in mice, it turns out that they are referring to factors such as the number of learning trials, the grouping of the trials, and the nature of the stimuli used, rather than the internal milieu.

Table 1.1 shows how each of the definitions I have explored has taken these features into account.

Bechtel and Richardson (1993) defined a machine as "a composite of interrelated parts, each performing its own functions, that are combined in such a way that each contributes to producing a behavior of the system". To this definition they added the

Table 1.1 Features covered by various definitions of "mechanism"

Author(s)	Key features that should be included						Features that should be excluded	
	Entities (structures)	Activities (functions)	Organization	Interactions	Outcomes	Milieu	Laws	Linearity/regularity/ invariance
Bechtel and Richardson (1993)	+	+	+	±	+			
Glennan (1996)	+		±	+	+		+	
Cartwright (1999)	+	+	+		+	+	+	+
Machamer, Darden, and Craver (2000)	+	+	+		+			+
Glennan (2002)	+		+	+	+		+	+
Bechtel and Abrahamsen (2005)	+	+	+		+			
Illari and Williamson (2012)	+	+	+	±	+			
Glennan (2015)	+	+	+	+	+			
This paper	+	+	+	+	+	+		

+ = feature explicitly mentioned
± = feature possibly implied

codicil that a mechanistic explanation "identifies these parts and their organization, showing how the behavior of the machine is a consequence of the parts and their organization." Insofar as these two statements can be taken together as a definition of mechanism, they incorporate four, possibly five, of the six identified features.

Glennan (1996) defined "a mechanism underlying a behaviour" as "a complex system which produces that behavior by the interaction of a number of parts according to direct causal laws". This definition, read explicitly, incorporates only three of the six essential features identified above, entities, potential interactions, and phenomena, although the term "system" implies organization; it also invokes laws. Glennan added that "[a] complex system has many mechanisms underlying its different behaviors", making it unclear what exactly he meant by "a complex system" and whether a complex mechanism might be composed of simpler submechanisms. However, according to Illari and Williamson (2012), Glennan subsequently (in 2009) withdrew the word "complex" from this definition.

Cartwright (1999) defined a mechanism, which she called "a nomological machine", as "a fixed (enough) arrangement of components, or factors, with stable (enough) capacities that in the right sort of stable (enough) environment will, with repeated operation, give rise to the kind of regular behaviour that we describe in our scientific laws". This definition incorporates five of the six essential features identified above, entities ("components"), functions ("capacities"), organization ("arrangement"), and outcomes ("behaviour"), and milieu ("environment"), but omits interactions. Her use of the term "regular [behavior]" excludes mechanisms that do not act regularly, if by that she means "in a steady, predictable, or uniform manner" or "repeatedly, without interruption" (OED).

Machamer et al. (2000) defined mechanisms as "entities and activities organized such that they are productive of regular changes from start or set-up to finish or termination conditions". This definition incorporates four of the six essential features identified above, entities, functions, organization, and outcomes, but omits interactions and milieu. Furthermore, the definition refers to "regular [changes]", and the statement that mechanisms always have a start and a finish implies linearity and ignores mechanisms that involve, for example, feedback loops.

Glennan (2002) then proposed the following definition: "A complex system that produces [a] behavior by the interaction of a number of parts, where the interaction between parts can be characterized by direct, invariant, change-relating generalizations". This definition incorporates four of the six essential features identified above, entities, organization ("system"), interactions, and outcomes, but omits functions, at least on an explicit reading, and milieu. Furthermore, "direct, invariant, change-relating generalizations" implies laws, as in Glennan's earlier formulation, and specifies invariance, although, as Glennan (2002) points out, the original formulation by Woodward (2000) was intended to resolve the tension in the view that most generalizations in the special sciences fail to conform to the standard criteria on which laws are based. However, in my view, Woodward's definition of an invariant generalization, which specifies that it is stable or robust when challenged by an intervention, does not chime with the ways in which many mechanisms in physiology and pharmacology behave.

A later definition by Bechtel and Abrahamsen (2005) is "a structure performing a function in virtue of its component parts, component operations, and their organization", with a codicil, "the orchestrated functioning of the mechanism is responsible for one or more phenomena". This definition incorporates four of the six essential features identified above—entities, activities, organization, and outcomes—and introduces the useful notion that mechanisms can have more than one outcome. However, it omits interactions, unless that is what is meant by "orchestrated", and also omits milieu.

Illari and Williamson (2012) then proposed the following: "A mechanism for a phenomenon consists of entities and activities organized in such a way that they are responsible for the phenomenon". This is an excellent description, but it omits the idea of interaction (although that is perhaps implied by "in such a way") and the possible plurality of outcomes, deficiencies that were later partly remedied by Glennan (2015): "A mechanism for a phenomenon consists of entities (or parts) whose activities and interactions are organized in such a way that they produce the phenomenon". Both definitions omit milieu.

We are almost there, but problems remain, since these descriptions do not take account of milieu (apart from Cartwright's) and other facets of mechanisms: (a) both entities and phenomena can be single or multiple; (b) not all of them are wholly responsible for observed phenomena—they may for example initiate phenomena but not maintain them or maintain phenomena that have been initiated by other mechanisms; (c) the milieu can affect a mechanism—pharmacological mechanisms are not only capable of producing variable outputs depending on the internal milieu, but even diametrically opposite outcomes can occur, depending on the milieu in which the mechanism is situated (Smith et al. 2012): xamoterol, for example, is predominantly a beta-adrenoceptor agonist when sympathetic nervous system activity is low (e.g. in mild heart failure), but an antagonist when it is high (e.g. in severe heart failure) (Cruickshank 1993), and tamoxifen can act as an estrogen receptor agonist or antagonist depending on the tissue in which it is acting (Gallo and Kaufman 1997). The internal milieu may also encompass factors that render an individual susceptible to a particular beneficial or adverse outcome (Aronson and Ferner 2003), for example a genetic variant (Tangamornsuksan et al. 2015) or a cation deficiency (Aronson and Hardman 1992).

Finally, these descriptions are not strictly speaking definitions of the sort that one would expect to find in a dictionary.

I have therefore modified the above descriptions, have incorporated the missing components, and have cast them as a intensional dictionary definition of the count noun "mechanism", including pronunciation according to the IPA system[3] and a brief etymology:

> **mechanism,** n. /ˈmɛkənɪz(ə)m/ One or more entities and activities spatially and temporally organized and potentially interacting, depending on the internal milieu, in such a way as to

[3]For the International Phonetic Alphabet, see: http://www.antimoon.com/how/pronunc-soundsipa.htm

be wholly or partly responsible for a phenomenon or phenomena [ancient Greek μηχανή a contrivance]

This formulation recognizes the plurality of entities and phenomena and adds the concept of internal milieu. It also recognizes the idea that some aspects of mechanisms may not be wholly responsible for the outcomes. For example (Table 1.2) when a compartment syndrome is initiated by trauma, the trauma does not maintain the syndrome once established. The syndrome does not occur without the trauma, but is maintained in its absence; trauma is in that sense only a partial mediator of the outcome.

1.2.4 Some Examples

I give here four examples that demonstrate features of this definition. Each case uses a physiological or pharmacological construct to illustrate an aspect or aspects of the theoretical claims about the essential components of the proposed definition of mechanisms.

Table 1.2 Extensional lists of different categories of the modes of operation of biological mechanisms

Feature	Examples
1. Roles of the mechanism	
(a) The mechanism stabilizes an outcome	(a) Homoeostatic mechanisms
(b) The mechanism initiates an outcome that terminates spontaneously	(b) Sneezing and other reflexes
(c) The mechanism initiates an outcome or outcomes that are terminated by another mechanism or mechanisms	(c) Self-limiting conditions, such as infections that need no intervention and some forms of headache
(d) The mechanism initiates the outcome but does not maintain it	(d) Chemical carcinogenesis; compartment syndrome (initiated by trauma, maintained by bleeding)[a]
(e) The mechanism maintains the outcome but does not initiate it	(e) Muscle clonus (maintained by failure of inhibition of contraction after initiation by stretching)
(f) The mechanism both initiates and maintains the outcome	(f) The Na/K-ATPase (the sodium-potassium pump) maintaining transmembrane cation gradients
(g) The mechanism prevents the outcome (inhibition)	(g) Receptor antagonists; enzyme inhibitors; channel blockers; physiological negative feedback systems
(h) The mechanism is doubly preventive (inhibition of an inhibitor)	(h) Inhibition of antithrombin III by heparin
(i) A defective or absent mechanism produces the outcome	(i) Many pathological mechanisms (e.g. enzyme deficiencies)

(continued)

Table 1.2 (continued)

Feature	Examples
2. Temporality of activities (functions)	
(a) Mechanisms that function once only	(a) Fertilization of an ovum by a sperm
(b) Mechanisms that function once, are exhausted, and cannot be reactivated	(b) Platelet aggregation, coagulation (new platelets and coagulation factors are required)
(c) Mechanisms that function once and are exhausted, but can be reactivated	(c) Nerve reflexes
(d) Mechanisms that function repeatedly and regularly, including oscillations	(d) The cardiac cycle; diurnal rhythms
(e) Mechanisms that function repeatedly and irregularly	(e) The ventricular response in atrial fibrillation
(f) Mechanisms that function continuously	(f) Physiological homoeostatic mechanisms
3. Types and temporality of outputs	
(a) Mechanisms with no apparent outcomes	(a) Na^+/Na^+ transmembrane exchange in normal cells
(b) Mechanisms that produce different outcomes depending on the internal milieu	(b) Xamoterol in mild and severe heart failure; tamoxifen affecting estrogen receptors in breast and uterus
(c) Mechanisms that produce constant outcomes	(c) Renal glomerular filtration; some pharmacological effects at steady state
(d) Mechanisms that produce intermittent outcomes	(d) Pulsatile secretion of hormones such as growth hormone, insulin
(e) Mechanisms that produce fluctuating outcomes	(e) Mechanisms with diurnal variation (e.g. temperature control, respiratory function)
4. Concentration-responsiveness	
(a) Mechanisms that are monotonically concentration-responsive	(a) Almost all pharmacological mechanisms, within either an individual or a population
(b) Mechanisms that are biphasic (hormesis)	(b) Secretion of prolactin in response to dopamine

[a]I am grateful to Brendan Clarke for this example

The first example (Fig. 1.4) shows the physiological control of corticosteroid secretion by the adrenal glands. Stimuli from higher centres cause the hypothalamus to secrete corticotropin releasing factor (CRF), which stimulates the pituitary to secrete adrenocorticotropic hormone (ACTH), which in turn stimulates the adrenal cortex to secrete cortisol. Cortisol then inhibits this sequence by negative feedback at glucocorticoid receptors (GRs) and mineralocorticoid receptors (MRs) in the areas shown. This is a relatively simple mechanism, involving at its core a linear chain of events, moderated by negative feedback loops. Pharmacological intervention by the administration of a potent glucocorticoid, such as prednisolone, changes the internal milieu, activates the negative feedback loops, and chronically reduces the stimulatory pathway, resulting in reduced endogenous secretion of cortisol and other corticosteroids and consequent atrophy of the adrenal glands. This example illustrates all the features of the definition, showing the organization and interactions of entities and activities, and demonstrating the importance of feedback mechanisms

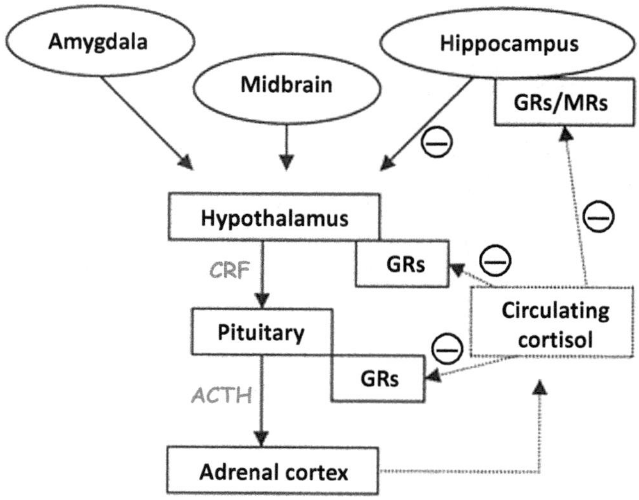

Fig. 1.4 Physiological control of corticosteroid secretion

to the maintenance of the internal milieu. It also implies that changes in the internal milieu may affect the responses of the system; for example, the addition of a histone deacetylase inhibitor would be expected to modify the effects of excess amounts of endogenous corticosteroid (Lee et al. 2018).

The second example (Fig. 1.5) shows the physiological control of the function of a neurotransmitter (5-hydroxytryptamine) in neurons. The details are described in the legend. This is also a relatively simple mechanism, involving at its core a linear chain of events, moderated at the end of the chain by negative feedback loops. An important point about this mechanism is that inhibitors of 5-hydroxytryptamine reuptake (selective serotonin reuptake inhibitors, SSRIs) can in certain circumstances upregulate components of the 5HT system (Yamada et al. 2013). This is in contrast to Woodward's (2002) proposition that the components of a modular mechanism should be independent, in the sense that it should be possible to change or interfere with the behaviour of one component without interfering with the behaviour of others. Furthermore, it gives an example of interaction between different mechanisms, since SSRIs also cause downregulation of beta-adrenoptors (Koe and Lebel 1995). At a higher level, interactions of this sort across various mechanisms can lead to emergent properties (Margineanu 2016), which encourages a systems approach to drug discovery (Green and Aronson 2012).

The third example (Fig. 1.6) shows the physiological response to reduced blood volume and reduced blood pressure following haemorrhage. The renin–angiotensin system is switched on, causing the formation of angiotensin II, which has two separate effects, vasoconstriction and salt and water retention, the effects of which cause negative feedback. This is a series of branching mechanisms with two outcomes and a negative feedback loop.

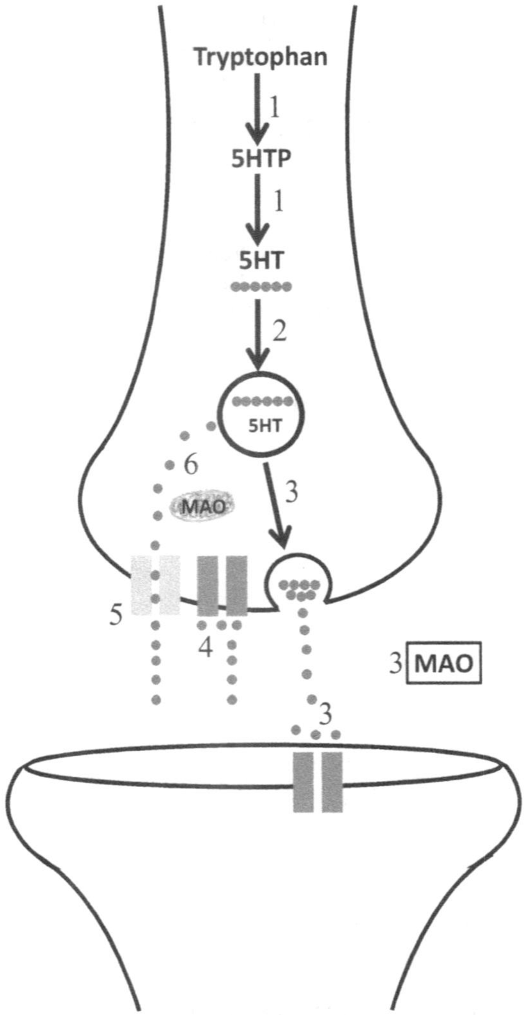

Fig. 1.5 Regulation of 5-hydroxytryptamine (serotonin, 5HT) neurotransmission
1. In the presynaptic neuron, tryptophan is converted to 5-hydroxytryptophan (5HTP) by tryptophan hydroxylase and then to 5-hydroxytryptamine by aromatic L-amino acid decarboxylase.
2. 5HT is stored in presynaptic vesicles.
3. Stored 5HT is released by exocytosis and acts on 5HT receptors on postsynaptic neurons, transmitting the signal; it can be metabolized by monoamine oxidase (MAO) in the synaptic cleft.
4. 5HT also acts on presynaptic 5HT receptors (autoreceptors), inhibiting further release.
5. 5HT is also taken up into the presynaptic terminal.
6. It can be metabolized by mitochondrial MAO or stored again in vesicles

The fourth example (Fig. 1.7) shows that the same outcome can arise from more than one mechanism. Gastric acid secretion by the parietal cells of the stomach is stimulated by the actions of histamine and acetylcholine, via G protein coupled receptors, with negative feedback loops; secretion is inhibited by somatostatin, also via G protein coupled receptors, which has a positive feedback loop; the final common pathway is the secretion of hydrogen ions into the gastric lumen by the H^+/K^+-ATPase. The model also shows interactions between the different mechanisms; for example, somatostatin can inhibit the release of acetylcholine.

Biological mechanisms described in this way are models of what is supposed to happen in vivo, structurally, functionally, and sometimes statistically. In some cases, such as the foregoing examples, there may be a great deal of evidence

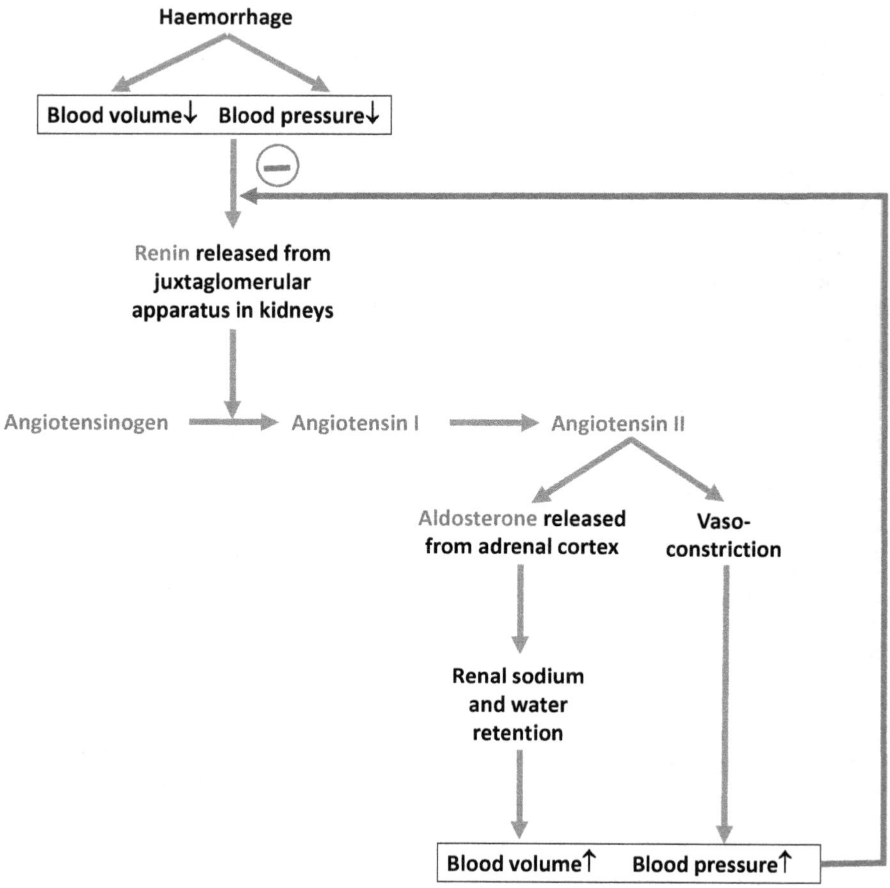

Fig. 1.6 Two outcomes of activation of the renin–angiotensin system after haemorrhage

that the mechanisms are constructed and act in the way that the model states. But even if models may be useful (Box 1979) they are simplified constructs, and even mechanisms whose existence and modes of action, thus modelled, are well supported by evidence do not necessarily predict specific outcomes when the mechanisms are evoked in practice, as discussed below.

1.2.5 *Definitions of "Mechanism" in Terms of Different Forms of Operation*

The intensional definition of a [biological] mechanism proposed above encompasses what Aristotle, in τα τόπικα, called τό τί ἦν εἶναι, essential features: entities

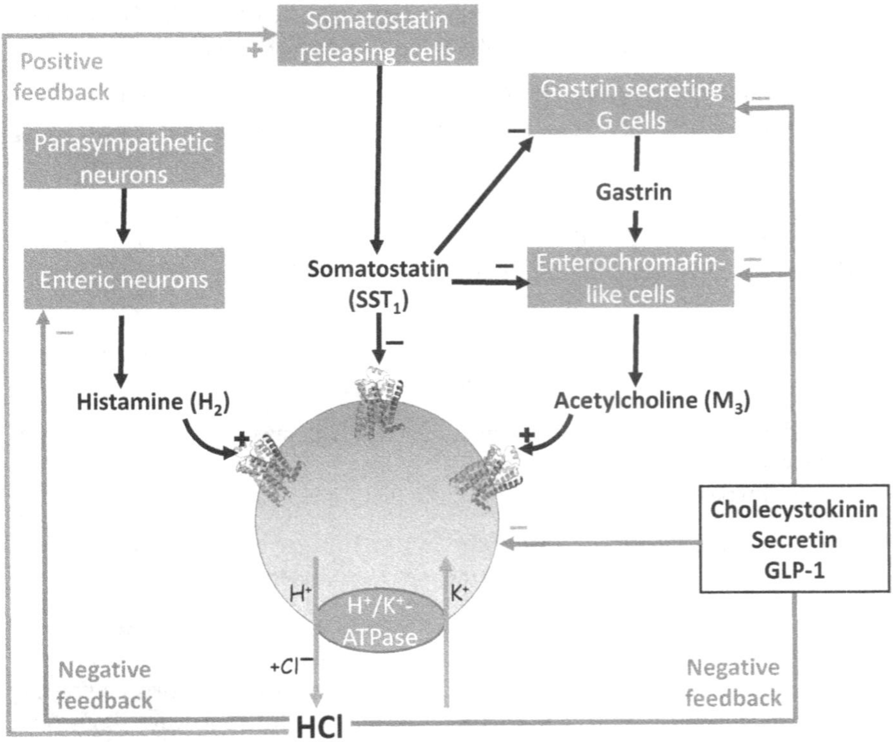

Fig. 1.7 More than one mechanism controlling the same outcome, gastric acid secretion, with positive and negative feedback loops

(structures), activities (functions), organization, and potential interactions, yielding phenomena (outcomes), depending on the internal milieu. However, it does not describe the various ways in which mechanisms can operate. A breakdown of those is given in Table 1.2 above with examples. The diversity of mechanisms identified in this way demonstrates the potential problem in generalizing different definitions to mechanisms in different fields.

1.3 "Evidence of Mechanism": The Problem of "of"

The term "evidence of mechanism" is often used in discussions about mechanisms, but it is ambiguous. The ambiguity arises from the use of the form "evidence of [noun]", which can mean one of two things, and sometimes both simultaneously. This is not surprising, considering that the OED lists 79 separate meanings of the word "of", of which 47 are in current use, 32 being labelled archaic or obsolete.

The ambiguity can be illustrated using examples taken from the OED, which contains over 1400 unique instances of the form "evidence of [noun]".

Example 1 (Pusey WA. *Principles and Practice of Dermatology*, 3rd ed. 1917: 575): Palpable adenopathy is present in many spare persons and is not, alone, strong evidence of syphilis.

In this example, adenopathy provides evidence of the existence of syphilis. Of all the instances in the OED this is the most common variety by far (about 74%).

The second most common meaning of "evidence of" (about 15% of cases) is evidence that something is happening or has happened.

Example 2 (*Strand Magazine*, December 1896: 699): All the Arctic equipment bore evidence of having been severely tested in actual use.

Here the evidence shows that something has happened—the equipment was severely tested.

The third type of usage relates to evidence that someone or something gives.

Example 3 (Journal of African Law 1958; 2: 10): The above recited evidence of the woman in the case 'seems contra to the views expressed by the Assessors.'

Here the evidence is not evidence of the existence of the woman, but evidence that, as it were, belongs to, is "of", her—evidence that she gives. This is not merely a legal use, as other instances in other contexts show. Here are two contrasting instances of the same formulation being used in two different ways, both from etymological notes in the OED.

Example 4: However, the suggestion of an Indian origin is no longer accepted, as the only evidence of the Sanskrit word comes from a modern glossary

Example 5: ... an original sense 'to rub' has been suggested for the Indo-European base[d] on the evidence of Sanskrit

Example 4, which is of the same form as Example 1, refers to the existence of a Sanskrit word, while Example 5, which is cognate with Example 3, deals with evidence that Sanskrit gives of an original sense of the word "milk".

Example 6 (Quinton A. *Hume*. Routledge, 1997: 52): That well-loved intellectual device of the eighteenth century, which infers the existence of God from evidence of order and of the adaptation of means to ends in nature.

In Example 6, Quinton refers to evidence of the existence of order (cf evidence of syphilis in Example 1), and he takes order (cf the woman in Example 3) as evidence of the existence of God. Both meanings in the same sentence.

Thus, the phrase "evidence of mechanism" should, in my view, be avoided. The term "mechanistic evidence" bears a similar ambiguous burden; it could mean evidence of the existence of a mechanism or the use of a mechanism as evidence to support some supposed outcome. An analysis of the relations between evidence and mechanisms suggests alternative terms, "evidence-based mechanism" (or evidence *for* a mechanism) and "mechanism-based evidence" (or evidence *from* a mechanism).

1.4 Evidence-Based Mechanism (Evidence *for* a Mechanism) and Mechanism-Based Evidence (Evidence *from* a Mechanism)

The relation between mechanisms and evidence is complicated and has two facets, shown in Fig. 1.8, derived with pharmacological mechanisms primarily in mind.

1.4.1 Evidence-Based Mechanism (Evidence for a Mechanism)

Two forms of evidence together constitute *evidence-based mechanism* (Fig. 1.8, top half).

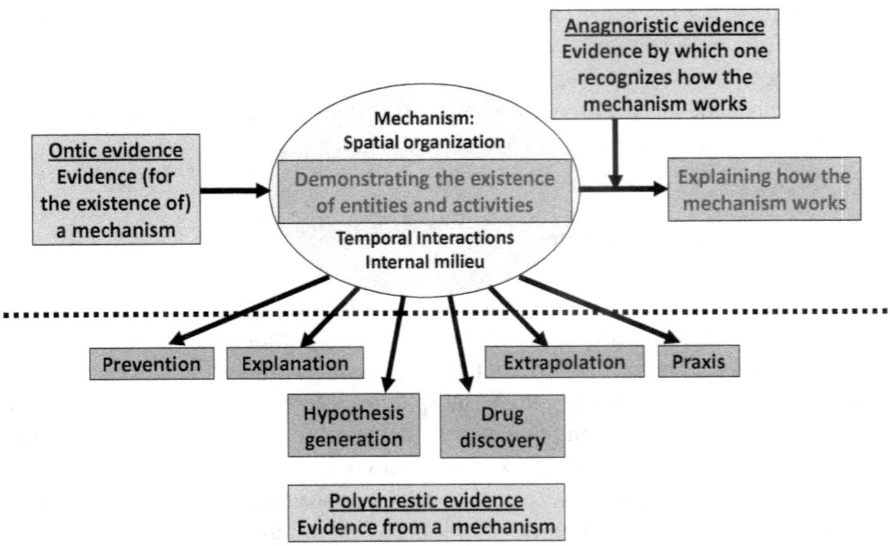

Fig. 1.8 *Evidence-based mechanism* (evidence *for* a mechanism) describes evidence that demonstrates the existence of mechanisms or from which the existence of a mechanism can be inferred (evidence for the existence of a mechanism—ontic evidence; top left), and recognition of how it works (anagnoristic evidence; top right). *Mechanism-based evidence* (interpretative evidence *from* a mechanism—polychrestic evidence; bottom) describes how pharmacological mechanisms can be used as evidence to inform analysis and prevention of adverse drug reactions, to explain outcomes, to extrapolate from clinical trials to clinical practice, to suggest practical procedures (praxis), to inform drug discovery, and to generate testable hypotheses

1.4.1.1 Ontic Evidence

First, there is evidence that a mechanism exists, or evidence *for* a mechanism, which I have called ontic evidence (Greek, *τά ὄντα, things that actually exist).*

Although the term "evidence-based medicine" has come to be associated primarily with randomized controlled trials or systematic reviews of such trials, the original idea that such evidence is at the top of a hierarchy of evidence has been strongly criticized and modified, for example in the GRADE recommendations[4]. Admissible evidence can come from sources other than randomized trials, even in medicine, and certainly in other fields (e.g. "evidence-based policy"). Pharmacological ontic evidence, for example, is any evidence that convincingly implies the existence of a pharmacological mechanism. Such evidence may come from experiments showing that a drug produces an outcome in vitro, ex vivo, or in vivo. Furthermore, it is possible to infer the existence of a mechanism without being able to envisage how it works. For example, we may observe that digoxin consistently increases cardiac output without being able to infer how it does that at a molecular, cellular, or tissue level (respectively inhibition of an enzyme-based pump, a change in intracellular cation concentrations, and altered contractility of the myocardium). But how can there not be a mechanism, knowing that there is an outcome, even though we can come to no conclusions about the entities or activities involved, or their organization or interactions, or the effects of the internal milieu? If we also observe under similar circumstances that another drug reduces cardiac output, we may similarly not know what the mechanism is, nor even whether it is wholly or even partly mediated through a mechanism that is the converse of the first.

1.4.1.2 Anagnoristic Evidence

Evidence that results in recognition of how the mechanism works I have called anagnoristic evidence (Greek, *ἀναγνώρισις, recognition, dénouement*). Such evidence can be obtained at different levels, from the molecular level, through cellular, tissue, and organ physiology and pharmacology, to evidence at the level of the whole person or even in populations. Just as ontic evidence may be obtained from any types of studies, in vitro, ex vivo, or in vivo, so can anagnoristic evidence.

Although different types of experiment may be used to elicit information about physiological and pharmacological mechanisms at different levels, the clinician needs to understand how mechanisms work at all levels. An example is shown in Fig. 1.9.

The pharmacological receptor for cardiac glycosides, such as digoxin, is a cell membrane-bound protein that contains an enzymatic site and pumping sites for sodium and potassium. As an enzyme, it is called the sodium-potassium adenosine triphosphatase (Na/K-ATPase). This enzyme hydrolyses ATP to ADP and inorganic

[4]See http://www.gradeworkinggroup.org

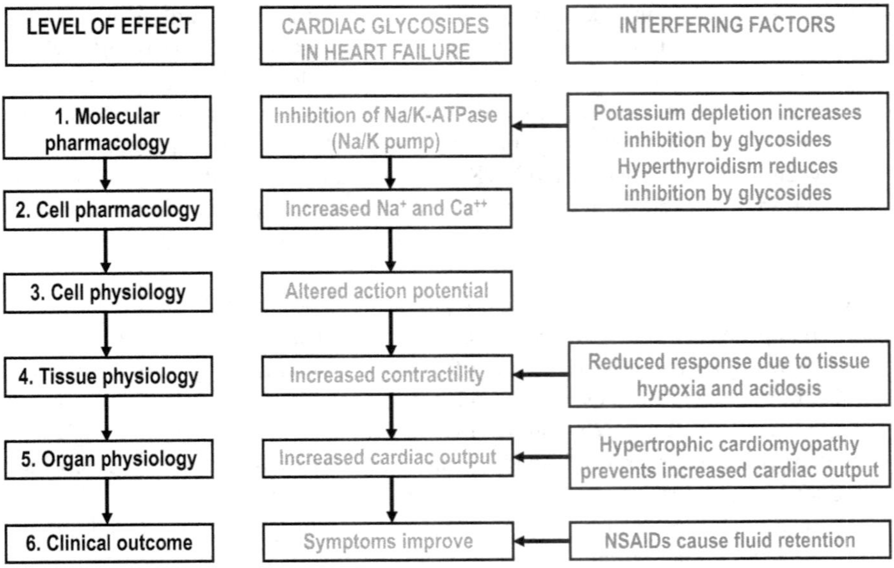

Fig. 1.9 The chain of events linking the pharmacological effects of cardiac glycosides to their therapeutic effects and factors that can interfere

phosphate, providing energy to work the pump, the Na/K pump, which binds potassium and sodium and transports potassium into cells and sodium out, maintaining the transmembrane gradients of these cations and hence the transmembrane potential. These actions are important for various cellular functions in different cells, including hormone secretion, nerve impulses, and muscle contractility. Normally, cardiac glycosides act as follows in the treatment of cardiac failure:

1. Inhibition of the activity of the Na/K pump by competing with potassium for its binding site; this is a pharmacological effect at the molecular level.
2. This inhibition of transmembrane sodium and potassium transport causes an increase in the intracellular concentration of sodium, which in turn alters the intracellular disposition of calcium; this is a pharmacological effect at the cellular level.
3. The altered disposition of intracellular calcium alters the action potential of cardiac muscle; this is a physiological effect at the cellular level.
4. As a result the rate of contractility of the cardiac muscle fibres increases; this is a physiological effect at the tissue level.
5. Cardiac output increases; this is a physiological effect at the level of the whole organ.
6. If this chain of events occurs without interruption, the clinical result will be relief, partial or complete, of the signs and symptoms of heart failure.

However, there are drug–disease and drug–drug interactions that can alter the therapeutic outcome at different stages in the mechanistic chain:

- At stage 1 potassium depletion enhances the binding of cardiac glycosides to the pump, and there is increased inhibition of sodium transport, which can in turn lead to drug toxicity. In this case a lower dose may produce a satisfactory therapeutic response, although it is better to remedy the potassium deficiency instead. Conversely, in hyperthyroidism the nature of the interaction between cardiac glycosides and Na/K-ATPase is altered, resulting in resistance to inhibition; in this case, increasing the dose in order to try to obtain a therapeutic effect can result in drug toxicity without ever producing a therapeutic effect in the heart.
- At stage 4, in patients with chronic cor pulmonale, cardiac glycosides may inhibit the pump and even cause toxic effects, such as cardiac arrhythmias, without ever causing an increase in the rate of myocardial contractility. The reasons for this are not clear, but they may be related to tissue hypoxia and acidosis in cor pulmonale. Tissue hypoxia and acidosis in patients with cardiac failure after an acute myocardial infarction can also contribute to an increased risk of glycoside-induced cardiac arrhythmias without therapeutic benefit.
- At stage 5 the increased rate of myocardial contractility is not translated into an increase in cardiac output in patients with hypertrophic cardiomyopathy, since there is a fixed obstruction in the left ventricular outflow tract; instead cardiac output falls, a paradoxical effect.
- At stage 6, fluid retention due, for example, to a non-steroidal anti-inflammatory drug (NSAID), such as diclofenac, can prevent the therapeutic response to cardiac glycosides.

For a complete understanding of how to use cardiac glycosides clinically, it is necessary to appreciate all these aspects of their mechanism of action and the factors that can interfere.

1.4.2 Mechanism-Based Evidence (Evidence from a Mechanism)

Next, there are ways of using evidence-based mechanisms to make evidential inferences. This is *mechanism-based evidence*, or evidence *from* a mechanism (Fig. 1.8, bottom half), which I have called polychrestic evidence (Greek, πολύχρηστος, *useful for many purposes*). Biological mechanisms can be used as evidence in several ways:

- to analyse how adverse drug reactions can be caused (predictively or post hoc) and how they can be prevented;
- to explain outcomes;
- in extrapolation;
- in praxis;
- in drug discovery.

I shall deal with each of these in the following discussion. I shall not discuss causality.

1.4.2.1 Analysing How Adverse Drug Reactions Can Be Caused and Prevented (Aronson and Ferner 2010)

A 50-year-old woman was given what would normally have been a therapeutic dose of tramadol, but died after 4 days. Post-mortem blood sampling showed an excess of tramadol, suggesting that she had died of tramadol toxicity, but this had not been diagnosed, partly because her pupils were dilated. Since tramadol is an opioid analgesic, constricted pupils would have been expected in the presence of toxicity. However, mechanistic analysis led to an explanation. Tramadol is metabolized to an active metabolite, which has potent opioid effects, whereas tramadol itself is both an opioid and an inhibitor of serotonin reuptake in nerve cells. In poor metabolizers of tramadol, an uncommon genetic variant, there is an excess of tramadol over its metabolite, and this causes a mixture of opioid toxicity and the serotonin syndrome, explaining death from opioid toxicity in the presence of paradoxically dilated pupils.

Knowing that acetylcholinesterase inhibitors in myasthenia gravis increase acetylcholine action at nicotinic receptors, you can prevent or reduce the adverse effects of acetylcholine at muscarinic receptors by using a muscarinic antagonist at the same time (Szathmáry 1981). Other examples are given in Table 1.3.

1.4.2.2 Explanation of Outcomes

Paradoxical and bidirectional drug effects can sometimes be explained mechanistically (Smith et al. 2012). For example, amiodarone can cause either hypo- or hyperthyroidism because of multiple competing effects, which include blockade of thyroid hormone entry into cells, inhibition of type 1 and type 2 5′-deiodinase, reduced T3 binding to its receptors, the Jod–Basedow effect, failure to escape from the Wolff–Chaikoff effect, thyroid cytotoxicity, and potentiation of thyroid autoimmunity. However, in attempting to explain newly described drug–event associations, whether paradoxical or not, it is better to invoke mechanisms that have been previously well demonstrated, rather than creating new explanations to fit—the proposed mechanism may deceive.

1.4.2.3 Extrapolation

A mechanism that predicts an outcome in the individuals in a trial may be used to extrapolate that outcome to others, tackling the problem of external validity (Howick et al. 2013). For example, people with the HLA B*5701 genotype are at risk of rashes from abacavir in clinical studies; this knowledge has been used to prevent

Table 1.3 Preventability of adverse drug reactions according to mechanism (Aronson and Ferner 2010)

Mechanistic factor	General strategy	Types of strategies	Examples
Extrinsic species (the drug, a metabolite, a breakdown product, a contaminant, an adulterant)	Prevention at the level of the medicinal product	(a) Knowledge of cross-reactivity	(a) Avoidance of beta-lactam antibacterials in penicillin hypersensitivity
		(b) Eliminating excipients, adulterants, or contaminants	(b) Eosinophilia-myalgia syndrome (due to a tryptophan adulterant)
		(c) Replacing the parent compound with an active metabolite	(c) Fexofenadine replacing terfenadine avoids adverse reactions
		(d) Defining storage conditions (e.g. limiting shelf-life)	(d) Avoiding tetracycline-induced Fanconi syndrome
		(e) Inactivating the extrinsic species	(e) Prevention by mesna of haemorrhagic cystitis due to acrolein formed from cyclophosphamide and ifosphamide
Intrinsic species (the pharmacological target)	Prevention at the level of the target of the adverse effect (e.g. the receptor)	(a) Direct interaction at the site of the adverse effect while preserving other actions	(a) Atropinic drugs to block the muscarinic actions of acetylcholinesterase inhibitors (e.g. neostigmine)
		(b) Physiological counteraction	(b) Atropinic drugs to relieve Parkinsonism due to dopamine receptor antagonists
Distribution	Prevention by preventing or avoiding co-distribution of the extrinsic and intrinsic species	(a) Administration at a distance from the site of the adverse effect	(a) Local administration to the skin, eyes, ears; inhaled drugs; regional analgesia
		(b) Reducing access of the drug to the site of the adverse effect	(b) Levodopa + a dopa decarboxylase inhibitor; H_1 receptor antagonists (antihistamines) with poor brain penetration

rashes in other patients during routine practice (Tangamornsuksan et al. 2015). Other types of mechanism of this kind are listed in Table 1.4.

1.4.2.4 Praxis

Mechanisms can suggest practical procedures. For example, the knowledge that blood flows from the periphery towards the heart, controlled by venous valves,

Table 1.4 A taxonomy of the types of mechanisms that can inform whether to extrapolate from trial populations to other populations [42]

Type of mechanism	Examples	Comments
Genetic	(a) Colorectal tumours with the wild-type KRAS genotype respond better to monoclonal antibodies that target epidermal growth factor receptors (EGFR)	In both cases, knowing the genetic components of the mechanisms responsible for therapeutic benefit (colorectal tumours) or an adverse drug reaction (abacavir) helps maximize benefit and reduce harm from these therapies; in these cases trial data can be extrapolated to individual patients
	(b) The adverse cutaneous effects of abacavir can be avoided by screening for human leukocyte antigen (HLA) B*5701	
Age	(a) Benoxaprofen proved efficacious in trials, but killed some elderly patients in routine practice	In all three cases extrapolation was dangerous because of age-related differences: (a) Benoxaprofen: Benoxaprofen is metabolized in the liver; its pharmacokinetics are different in elderly patients with reduced liver function from the patients who were studied in trials
	(b) The dosage of growth hormone for adults with growth hormone deficiency is 25% that of children, in spite of greater body mass	(b) Growth hormone: The degree of growth required is much greater in children
	(c) Antihypertensive drugs reduce total mortality in middle-aged patients but may not in elderly ones, or not to the same extent	(c) Antihypertensive drugs: Many physiological mechanisms change with age, altering responses to these drugs
Sex	The risk of drug-induced lupus-like syndrome is greater in women than in men	The mechanisms that cause idiopathic systemic lupus erythematosus, which is more common in women, make them more susceptible to the drug-induced form; extrapolation from men to women may be dangerous
Physiological variants	Obese patients require different weight-related drug dosages	The dosage of a drug may be affected by its distribution into body fat; knowing which drugs are thus affected informs extrapolation from lean to obese individuals

(continued)

Table 1.4 (continued)

Type of mechanism	Examples	Comments
Co-morbidities	(a) Renal impairment, hepatic impairment	(a) Knowledge of pharmacokinetic mechanisms can predict outcomes for drugs eliminated by the kidneys or liver; extrapolation from individuals with normal renal and hepatic function to those with impaired function is possible with the requisite pharmacokinetic knowledge
	(b) Digoxin normally relieves heart failure, but paradoxically worsens it if there is also hypertrophic cardiomyopathy (HCM)	(b) Increased cardiac contraction leads to reduced cardiac output in the face of obstruction to flow caused by hypertrophic cardiomyopathy; one cannot extrapolate from, say, ischaemic heart disease to HCM when treating heart failure
Drug–drug interactions	Parkinsonism due to antipsychotic drugs should be treated with anticholinergic drugs, not levodopa	Antipsychotic drugs cause parkinsonism by inhibiting the action of dopamine at its receptors, making levodopa ineffective; one cannot extrapolate from idiopathic parkinsonism to the drug-induced form, because the mechanisms are different
Sociological factors	Integrated Nutrition Programs succeeded in Tamil Nadu, but failed when implemented in Bangladesh	The mechanisms for improving the access of malnourished children to nutrition were not the same in the two countries; extrapolation failed

predicts that it will be easier to sample blood from small veins by inserting the needle in a retrograde direction (Dorrington and Aronson 2014).

1.4.2.5 Drug Discovery

Drug discovery often starts with a mechanism that leads to the synthesis of a drug to affect that mechanism (Vandenbroucke 2008), although success does not always come from targeting a single mechanism (Green and Aronson 2012).

1.4.2.6 Hypothesis Generation

Mechanisms can generate hypotheses, but such hypotheses must be formally tested. It is a capital mistake to use mechanisms to justify a course of action without first demonstrating that the hypothesized outcome really does occur. There are several cases in which mechanisms have been inappropriately used to justify methods of treatment without first testing them in a trial (Howick 2011). Dr Benjamin Spock, for example, hypothesized that if a baby lying on its back vomited it might inhale the vomit and die. His advice to put babies to sleep prone was followed, until studies

Table 1.5 Examples of conditions in which outcomes predicted by mechanisms either did not occur or were the opposite of predicted

Intervention	Expected outcome based on mechanism	Actual outcome
Antiarrhythmic drugs to prevent sudden death after myocardial infarction	Frequency of sudden cardiac death reduced	Frequency of sudden cardiac death increased
Human growth hormone in critically ill patients	Reduced morbidity and mortality	Increased morbidity and mortality
Hormone replacement therapy at the menopause	Cardiac disease reduced	Cardiac disease increased
Radical mastectomy for breast cancer	Reduced mortality	Increased mortality
Nimodipine after stroke	Beneficial effects	No beneficial effects
Evening primrose oil for eczema		
Centoxin for Gram –ve septicaemia		
Arthroscopic surgery for osteoarthritis		
Fluoride for postmenopausal fractures		

showed that the risk of cot death is increased by following this advice, and that babies should in fact be put "back to sleep" (Gilbert et al. 2005).

Other examples are listed in Table 1.5.

1.5 Mechanistic Reasoning

Mechanistic reasoning in clinical medicine is the use of evidence-based mechanisms (evidence *for* mechanisms) and mechanism-based evidence (evidence *from* mechanisms), taken in conjunction with the methods of evidence-based medicine and evidence-based health care, in analysing practical problems.

However, mechanistic reasoning is not itself a form of evidence. To take an analogy: the inclusion of "expert opinion" as the lowest form of evidence in early forms of the evidence hierarchy in evidence-based medicine was a category error. Expert opinion that is based on high-quality evidence is a high-quality opinion; expert opinion that is based on low-quality evidence is a low-quality opinion. Similarly, mechanistic reasoning may be of high or low quality depending on the quality of the evidence adduced to support it. The inclusion of mechanistic reasoning in the Oxford Centre for Evidence Based Medicine (OCEBM) levels of evidence at the lowest step (Level 5) is, in my view, a similar category error (OCEBM Working Group 2011). The higher the quality of the evidence, the more useful mechanistic reasoning becomes (Howick et al. 2010). Mechanistic reasoning, like expert opinion, should be placed outside schemas of levels of evidence. In the OCEBM levels of evidence it could be omitted or replaced by an item such as "Non-evidential information incorrectly treated as if it were evidence". This would include mechanisms and expert opinion as well as, for example, formal consensus, reports of expert committees, and basic principles.

A model of the processes of mechanistic reasoning is presented in Fig. 1.10. The top part of the figure reproduces, with slight variation, what is shown in Fig. 1.8. Evidence that supports the existence of a mechanism (ontic evidence) and evidence that shows how the mechanism works (anagnoristic evidence) together lead by direct inference to the outcomes described above: prevention, explanation, praxis, and drug discovery. In addition they can generate hypotheses. If a hypothesis is supported by a strong chain of evidence, it may be provable by recourse to case reports, particularly the definitive reports that have been called "between-the-eyes" reports (Aronson

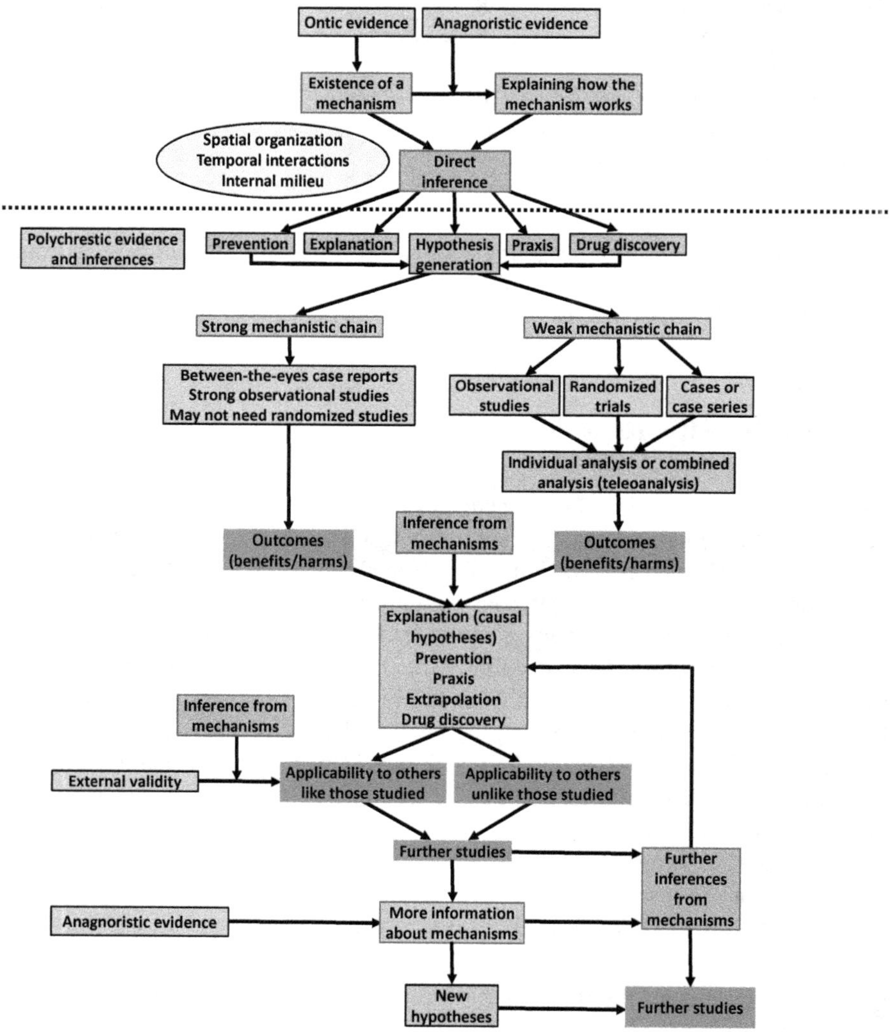

Fig. 1.10 A definition of mechanistic reasoning derived from the relevant operations

and Hauben 2006), or high-quality observational studies; randomized controlled studies may not be necessary to demonstrate benefits (Glasziou et al. 2007) or harms (Aronson and Hauben 2006). If, however, the evidence chain is of poor quality, a randomized study will probably be necessary, in order to test the hypothesis. Since it is not always easy to know how strong the chain of evidence is, randomized studies will often be required if hypotheses are to be properly tested. Combined analysis (teleoanalysis, see Wald and Morris 2003) of the results of randomized studies, observational studies, and case reports or case series may be of additional value. When the outcomes of such studies are known, further inferences may be drawn from known mechanisms, which can lead to further explanations, including causal hypotheses that can be further tested, preventive strategies, praxis, and extrapolation to other individuals who have not been studied. This in turn may lead to more information about mechanisms and new hypotheses.

As a practical example, consider the development and use of non-steroidal anti-inflammatory drugs (NSAIDs), of which the following is an abbreviated account (Fig. 1.11).

The knowledge that willow bark has an antipyretic effect in people with fevers, clearly established in the eighteenth century, was ontic evidence of an antipyretic mechanism. Other evidence later emerged that aspirin also has analgesic and anti-inflammatory properties.

Various types of anagnoristic evidence enabled the mechanism to be understood:

- the active component of willow bark was shown to be a salicylate;
- acetylsalicylic acid was synthesized and later rediscovered as a component of the meadowsweet plant (*Spiraea*), and given the brand name Aspirin, which is now generally adopted as the compound's common name;
- aspirin was shown to interfere with the actions of a substance or substances that turned out to be prostanoids;
- this effect was mediated by inhibition of prostanoid synthesis by an enzyme called prostaglandin synthetase, later called cyclo-oxygenase.

This knowledge led to the view that compounds other than aspirin that inhibited cyclo-oxygenase (now called non-steroidal anti-inflammatory drugs or NSAIDs) would probably have similar pharmacological properties, i.e. would be antipyretic, analgesic, and anti-inflammatory. This turned out to be the case. The pharmacology of the COX enzyme suggested that this result, demonstrated in clinical trials and observational studies, would be widely applicable (external validity).

Further refinements in understanding of the mechanism of action of aspirin and NSAIDs subsequently emerged, principal among them being the discovery of at least two types (isoforms) of cyclo-oxygenase, COX-1 and COX-2, the former being constitutive, i.e. responsible for various normal cellular functions, and the latter formed in response to inflammation. This led to the hypothesis that medicines that selectively inhibit COX-2 (called coxibs) would have analgesic, antipyretic, and anti-inflammatory actions, like non-selective NSAIDs, but would lack the major adverse effect that the non-selective drugs cause, by inhibition of COX-1, namely

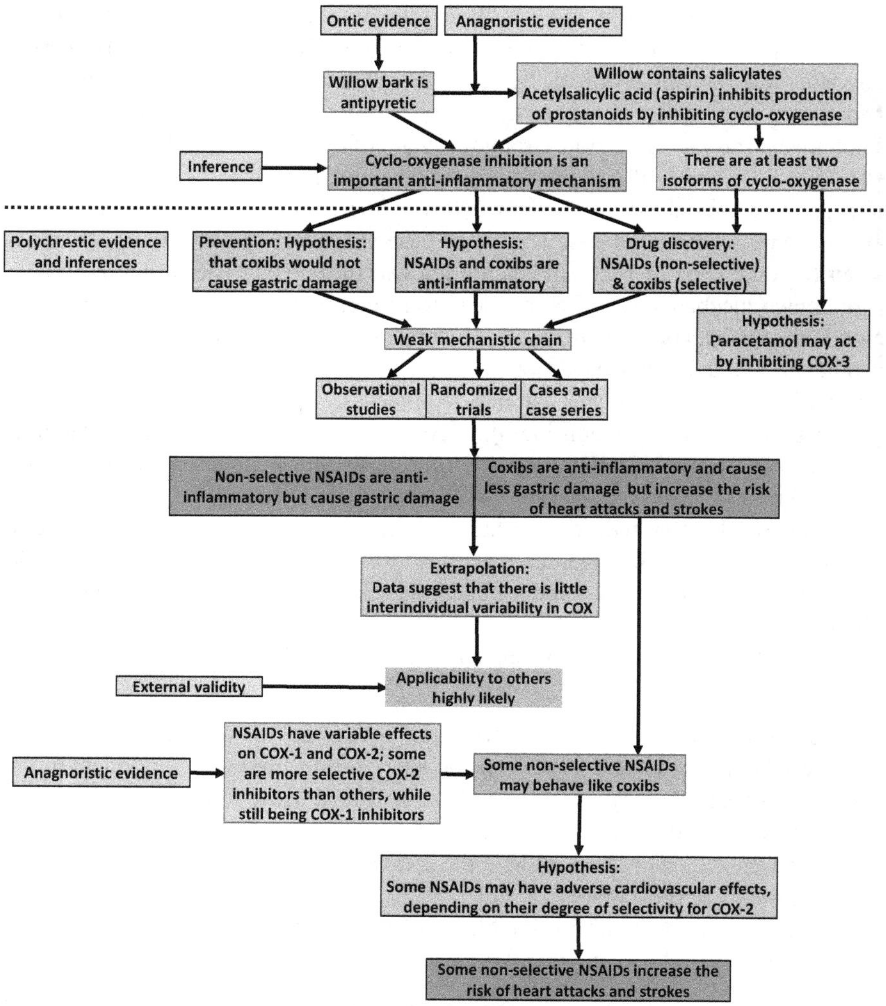

Fig. 1.11 The example of NSAIDs and coxibs using the schema depicted in Fig. 1.10. Other examples would result in different modifications to the general schema

gastric damage, causing erosions, ulcers, and bleeding. New molecules with COX-2 selectivity were therefore synthesized.

Clinical trials of the new medicines showed that the prediction of reduced gastric damage was fulfilled, but revealed a different set of adverse effects, principally an increased risk of cardiovascular adverse outcomes (heart attacks and strokes) and less commonly damage to the large intestine.

These observations led to the view that some of the NSAIDs that had been considered to be non-selective, but whose selectivity varied from drug to drug, might also cause an increased risk of cardiovascular damage; that has proved to be so.

This has led to a reappraisal of the use of these drugs in the long-term management of diseases of joints, modified by the concurrent development of monoclonal antibodies with different mechanisms of action, primarily mediated by actions on inflammatory cytokines, such as TNFalfa.

This short account of the development and use of NSAIDs and coxibs illustrates how the schema shown in Fig. 1.10 can be used, in this case in modified fashion, to chart the progress of a pharmacological intervention mechanistically, guiding the use of mechanistic reasoning in tackling clinical problems.

1.6 Mechanisms Without Trials and Trials Without Mechanisms

Ideally mechanistic evidence and probabilistic evidence should be used together. I have detailed above the several ways in which the former can be used to enhance the latter. There is, however, one problematic area, the use of mechanisms to predict clinical outcomes and hence to support the use of interventions that have not been supported by probabilistic studies.

Some take the view, supported by the existence of studies such as are listed in Table 1.5, that only high-quality evidence from randomized clinical trials and systematic reviews of such trials can be used to support interventions intended to produce specific outcomes, and that when such outcomes are not as predicted by mechanistic reasoning the trial result trumps the reasoning. Others claim that mechanisms can be used predictively without the need for formal confirmation in a clinical trial. Both of these claims need to be scrutinized.

1.6.1 Mechanisms Without Trials

In some well described cases it is possible to use mechanistic reasoning to interpret outcomes, or likely outcomes, in clinical practice without the need for formal confirmation in a randomized trial or even an observational study.

The classic, somewhat whimsical, example of this in relation to the benefit of an intervention is the obvious benefit conferred by a parachute for those who choose to jump from an aeroplane, by harnessing air pressure to counteract gravity (Smith and Pell 2003). This does not obviate the possibility of randomized trials of different types of parachute, investigating the effects of different designs and different materials.

Another good example is the mother's kiss (Glasziou et al. 2007). A mother brings her young child to the Accident and Emergency Department of a hospital; he has inhaled a bead, which has stuck in his nose. The doctor instructs the mother to open the child's mouth and blow hard into it. The bead flies out. The mechanism in this case is the application of air pressure behind the obstruction. No clinical trial is needed to prove the efficacy of the manoeuvre.

The case cited above of the mechanistic prediction that blood will be more easily obtained from a small vein on the back of the hand if the needle is inserted in a retrograde direction, because venous valves determine the direction of flow of the blood (Dorrington and Aronson 2014), provides another example, in which the mechanism is highly predictive of efficacy, and for which simple observation on one or a few occasions provides definitive evidence, based in this case on anatomical mechanistic reasoning.

Pharmacological examples illustrate that some harms can be predicted mechanistically and demonstrated by single observations, without the need for randomized controlled trials (Aronson and Hauben 2006) For example, if an intravenous cannula has been used to inject a range of substances without harm, but an injection of a new substance is immediately followed by severe pain, it is reasonable to suppose that that substance has caused the pain, for example by causing phlebitis; clinical evidence of phlebitis in that case would be confirmatory and it would not take many, if any, more instances of such an effect to provide convincing evidence of a cause-and-effect relation.

The common feature of all these examples is that in each case the mechanism is obvious, rendering a clinical trial or observational study unnecessary. The statistician Joseph Berkson reportedly called this the interocular traumatic test—the conclusion hits you between the eyes (Edwards et al. 1963).

1.6.2 Trials Without Mechanisms

The claim that a well-designed randomized controlled clinical trial can give an answer to a clinical question without the need for a priori mechanistic reasoning is clearly correct; there are many examples. However, it is not the case that the results of clinical trials always trump the predictions of mechanistic reasoning. When the outcome of a trial and the mechanistic prediction are discordant, both must be considered.

Take such a case:

- Mechanism M gives prediction P and an outcome O(M).
- Clinical trial T gives an outcome O(T).
- O(T) is not the same as O(M).

The possible explanations of this discordance are that:

(a) M is wrong generally;
(b) M is wrong in this case, because of some interfering factor, but not generally;
(c) T is wrong.

If T is a well-designed trial, in which the risks of bias and confounding are low, and in which the effect size is large and of high probability, one would conclude either (a) or (b), although more information would be needed to distinguish these two possibilities. However, if T is a poorly designed trial with high risks of bias and/or confounding, one would be uncertain. Even if T was well-designed, but if the effect size was small and of low probability, one could not be sure if M was right, or wrong (a), or wrong (b). Nor would one be sure how much credence to give to O(T), given the relevant mechanistic reasoning. A strong mechanism M could vitiate the strength of the outcome O(T) of a small trial.

Conversely, strong evidence that there is *no* mechanism (Illari 2011) (e.g. the postulated memory of water to explain homoeopathic effects, which is strongly contradicted by everything we know about the statistical mechanics of fluids) would also trump the outcomes of trials; extraordinary mechanistic claims require extraordinary probabilistic evidence.

Post hoc mechanistic reasoning is also often used when explaining outcomes in clinical trials, particularly if those outcomes were unexpected. However, it is [almost] always possible to invent a plausible mechanistic explanation for any outcome, even when there is no evidence to support such an interpretation. Such explanations should always be regarded as new hypotheses, needing to be confirmed by further studies. When no alternative mechanistic explanation is forthcoming it may be either that one does not know about an explanation, even though it may exist, or that one does not have the imagination to think one up, even one that may not exist.

1.7 Conclusions

Evidence derived from mechanisms and evidence derived from probabilistic studies are complementary.

Philosophers' interest in mechanisms has burgeoned in recent years, following the realization that scientists spend a lot of time studying mechanisms. A survey of historical usages and recent definitions of "mechanism" reveals five key components of mechanisms: structures (or entities), functions (or activities), interactions, and organization (spatial and temporal), yielding outcomes (or phenomena). A sixth feature, based primarily on physiological and pharmacological mechanisms, is the milieu in which the mechanism is situated, which can affect the outcome. Taking account of these features yields the following intensional definition of the count noun "mechanism", as it would appear in a conventional dictionary:

> **mechanism,** n. /ˈmɛkənɪz(ə)m/ One or more entities and activities spatially and temporally organized and potentially interacting, depending on the milieu, in such a way as to be wholly or partly responsible for a phenomenon or phenomena [ancient Greek μηχανή a contrivance]

Extensional operational definitions of mechanisms describe: the role of the mechanism; temporality of functions (activities); types and temporality of outputs; and dose-responsiveness.

The relation between mechanisms and evidence leads to two separate concepts:

(a) Evidence-based mechanism (evidence *for* a mechanism), which describes two types of evidence that demonstrate the existence of mechanisms (ontic evidence) and recognition of how they produce their outputs (anagnoristic evidence).
(b) Mechanism-based evidence (evidence *from* a mechanism; polychrestic evidence), which describes how biological mechanisms can be used as evidence in many different ways: to analyse how adverse drug reactions occur and how they can be prevented; to explain outcomes; to extrapolate from clinical trials to clinical practice; to suggest practical procedures (praxis); to inform drug discovery; and to generate testable hypotheses. However, clinical interventions based on predictive mechanisms alone should generally not, with a few exceptions, be used unless they have been probabilistically tested in high quality trials.

These two terms are more specific than the ambiguous terms "evidence of mechanism" and "mechanistic evidence", which, in my view, should be abandoned.

Mechanistic reasoning combines these two approaches with the results of probabilistic clinical observations.

Acknowledgements I am grateful to Stuart Glennan, Adam LaCaze, and Barbara Osimani for pertinent and helpful comments on an earlier version of this chapter.

References

Agmon, N. (2010). Materials and language: pre-semitic root structure change concomitant with transition to agriculture. *Brill's Annual of Afroasiatic Languages and Linguistics, 2*(1), 23–79. https://doi.org/10.1163/187666310x12688137960669.

Aronson, J. K. (2009). Medication errors: Definitions and classification. *British Journal of Clinical Pharmacology, 67*(6), 599–604. https://doi.org/10.1111/j.1365-2125.2009.03415.x.

Aronson, J. K. (2016). When I use an evidence-based word . . . 2. The roots of language. http://www.cebm.net/use-evidence-based-word. Accessed 25 Feb 2018.

Aronson, J. K., & Ferner, R. E. (2003). Joining the DoTS: New approach to classifying adverse drug reactions. *BMJ, 327*(7425), 1222–1225. https://doi.org/10.1136/bmj.327.7425.1222.

Aronson, J. K., & Ferner, R. E. (2010). Preventability of drug-related harms – Part II: proposed criteria, based on frameworks that classify adverse drug reactions. *Drug Safety, 33*(11), 995–1002. https://doi.org/10.2165/11538280-000000000-00000.

Aronson, J. K., & Hardman, M. (1992). ABC of monitoring drug therapy. Digoxin. *BMJ, 305*(6862), 1149–1152. https://doi.org/10.1136/bmj.305.6862.1149.

Aronson, J. K., & Hauben, M. (2006). Anecdotes that provide definitive evidence. *BMJ, 333*(7581), 1267–1269. https://doi.org/10.1136/bmj.39036.666389.94.

Bechtel, W., & Abrahamsen, A. (2005). Explanation: a mechanist alternative. *Studies in History and Philosophy of Science Part C: Studies in History and Philosophy of Biological and Biomedical Sciences, 36*(2), 421–441. https://doi.org/10.1016/.shpsc.2005.03.010.

Bechtel, W., & Richardson, R. C. (1993). *Discovering complexity. Decomposition and localization as strategies in scientific research*. Princeton: Princeton University Press.

Box, G. E. P. (1979). Robustness in the strategy of scientific model building. In *Robustness in statistics* (pp. 201–236). Elsevier. https://doi.org/10.1016/b978-0-12-438150-6.50018-2.
Cannon, W. B. (1926). In Pettit A (Ed), *A Charles Richet: ses amis, ses collègues, ses élèves* (p. 91). Paris: Editions médicales.
Cartwright, N.. (1999). *The Dappled World. A Study of the Boundaries of Science*. New York: Cambridge University Press. https://doi.org/10.1017/cbo9781139167093.
Cook, R. T. (2009). In tensional definition. In: *Dictionary of philosophical logic* (p. 155). Edinburg: Edinburgh University Press.
Council for International Organizations of Medical Sciences. (2010). *Practical aspects of signal detection in pharmacovigilance: Report of CIOMS Working Group VIII* (Technical report). https://cioms.ch/working_groups/working-group-viii/
Craver, C. F., & Darden, L. (2013). *Search of mechanisms* (p. 58). Chicago: The University of Chicago Press.
Cruickshank, J. M. (1993). The xamoterol experience in the treatment of heart failure. *The American Journal of Cardiology, 71*(9), C61–C64. https://doi.org/10.1016/0002-9149(93)90088-t.
Dorrington, K. L., & Aronson, J. K. (2014). Failed phlebotomy? Think William Harvey. *BMJ, 349*, g5232. https://doi.org/10.1136/bmj.g5232.
Edwards, W., Lindman, H., & Savage, L. J. (1963). Bayesian statistical inference for psychological research. *Psychological Review, 70*(3), 193–242. https://doi.org/10.1037/h0044139.
European Medicines Agency. (2018). *Medication errors*. http://www.ema.europa.eu/ema/index.jsp?curl=pages/special_topics/general/general_content_000570.jsp. Accessed 25 Feb 2018.
Ferner, R. E., & Aronson, J. K. (2000). Medication errors, worse than a crime. *The Lancet, 355*(9208), 947–948. https://doi.org/10.1016/s0140-6736(00)99025-1.
Gallo, M. A., & Kaufman, D. (1997). Antagonistic and agonistic effects of tamoxifen: Significance in human cancer. *Seminars in Oncology, 24*(1 Suppl 1), S1-71–S1-80.
Gibbs, D. D. (1971). The physician's pulse watch. *Medical History, 15*(2), 187–190. https://doi.org/10.1017/s0025727300016409.
Gilbert, R., Salanti, G., Harden, M., & See, S. (2005). Infant sleeping position and the sudden infant death syndrome: Systematic review of observational studies and historical review of recommendations from 1940 to 2002. *International Journal of Epidemiology, 34*(4), 874–887. https://doi.org/10.1093/ije/dyi088.
Glasziou, P., Chalmers, I., Rawlins, M., & McCulloch, P. (2007). When are randomised trials unnecessary? Picking signal from noise. *BMJ, 334*(7589), 349–351. https://doi.org/10.1136/bmj.39070.527986.68.
Glennan, S. (1996). Mechanisms and the nature of causation. *Erkenntnis, 44*(1). https://doi.org/10.1007/bf00172853.
Glennan, S. (2002). Rethinking mechanistic explanation. *Philosophy of Science, 69*(S3), S342–S353. https://doi.org/10.1086/341857.
Glennan, S.. (2015). Mechanisms and mechanical philosophy. In P. Humphreys (Ed.), *The Oxford handbook of philosophy of science* (Online). Oxford: Oxford University Press. https://doi.org/10.1093/oxfordhb/9780199368815.013.39.
Green, A. R., & Aronson, J. K. (2012). An agenda for UK clinical pharmacology. From basic to clinical neuropharmacology: targetophilia or pharmacodynamics? *British Journal of Clinical Pharmacology, 73*(6), 959–967. https://doi.org/10.1111/j.1365-2125.2012.04246.x.
Hauben, M., & Aronson, J. K. (2009). Defining 'signal' and its subtypes in pharmacovigilance based on a systematic review of previous definitions. *Drug Safety, 32*(2), 99–110. https://doi.org/10.2165/00002018-200932020-00003.
Howick, J. (2011). A qualified defence of the EBM stance on mechanistic reasoning. In *The Philosophy of Evidence-Based Medicine* (pp. 122–157). Oxford: Wiley-Blackwell. https://doi.org/10.1002/9781444342673.ch10.
Howick, J., Glasziou, P., & Aronson, J. K. (2010). Evidence-based mechanistic reasoning. *Journal of the Royal Society of Medicine, 103*(11), 433–441. https://doi.org/10.1258/jrsm.2010.100146.

Howick, J., Glasziou, P., & Aronson, J. K. (2013). Can understanding mechanisms solve the problem of extrapolating from study to target populations (the problem of 'external validity')? *Journal of the Royal Society of Medicine, 106*(3), 81–86. https://doi.org/10.1177/0141076813476498.

Illari, P. M. (2011). Mechanistic evidence: Disambiguating the Russo–Williamson thesis. *International Studies in the Philosophy of Science, 25*(2), 139–157.

Illari, P. M., & Williamson, J. (2012). What is a mechanism? Thinking about mechanisms *across* the sciences. *European Journal for Philosophy of Science, 2*(1), 119–135. https://doi.org/10.1007/s13194-011-0038-2.

Koe, B. K., & Lebel, L. A. (1995). Effects of serotoninergic agents on downregulation of beta-adrenoceptors by the selective serotonin reuptake inhibitor sertraline. *Archives internationales de pharmacodynamie et de therapie, 329*(2), 231–244.

Lawson, I. (2015). Crafting the microworld: How Robert Hooke constructed knowledge about small things. *Notes and Records: The Royal Society Journal of the History of Science*. https://doi.org/10.1098/rsnr.2015.0057.

Lee, H.-A., Kang, S.-H., Kim, M., Lee, E., Cho, H.-M., Moon, E.-K., & Kim, I. (2018). Histone deacetylase inhibition ameliorates hypertension and hyperglycemia in a model of Cushing's syndrome. *American Journal of Physiology. Endocrinology and Metabolism, 314*(1), E39–E52. https://doi.org/10.1152/ajpendo.00267.2017.

Machamer, P., Darden, L., & Craver, C. F. (2000). Thinking about mechanisms. *Philosophy of Science, 67*(1), 1–25. https://doi.org/10.1086/392759.

Margineanu, D. G. (2016). Neuropharmacology beyond reductionism – likely prospect. *Biosystems, 141*, 1–9. https://doi.org/10.1016/j.biosystems.2015.11.010.

OCEBM Levels of Evidence Working Group. (2011). *The Oxford Levels of Evidence*. https://www.cebm.net/index.aspx?o=5653. Accessed 25 Feb 2018.

Smith, G. C. S., & Pell, J. P. (2003). Parachute use to prevent death and major trauma related to gravitational challenge. Systematic review of randomised controlled trials. *BMJ, 327*(7429), 1459–1461. https://doi.org/10.1136/bmj.327.7429.1459.

Smith, S. W., Hauben, M., & Aronson, J. K. (2012). Paradoxical and bidirectional drug effects. *Drug Safety, 35*(3), 173–89. https://doi.org/10.2165/11597710-000000000-00000.

Szathmáry, I., Magyar, P., & Szobor, A. (1981). Myasthenia gravis: protective effect of ipratropium bromide (Atrovent®) on airways obstruction caused by edrophonium chloride (Tensilon®). *European Neurology, 20*(1), 56–61. https://doi.org/10.1159/000115206.

Tangamornsuksan, W., Lohitnavy, O., Kongkaew, C., Chaiyakunapruk, N., Reisfeld, B., Scholfield, N. C., & Lohitnavy, M. (2015). Association of HLA-B*5701 genotypes and abacavir-induced hypersensitivity reaction: a systematic review and meta-analysis. *Journal of Pharmacy & Pharmaceutical Sciences, 18*(1), 68–76. https://doi.org/10.18433/j39s3t.

Vandenbroucke, J. P. (2008). Observational research, randomised trials, and two views of medical science. *PLoS Medicine, 5*(3), e67. https://doi.org/10.1371/journal.pmed.0050067.

Wald, N. J. & Morris, J. K. (2003). Teleoanalysis: combining data from different types of study. *BMJ, 327*(7415), 616–618. https://doi.org/10.1136/bmj.327.7415.616.

Winnie, J. A. (1967). The implicit definition of theoretical terms. *The British Journal for the Philosophy of Science, 18*(3), 223–229. https://doi.org/10.1093/bjps/18.3.223.

Woodward, J. (2000). Explanation and invariance in the special sciences. *The British Journal for the Philosophy of Science, 51*(2), 197–254. https://doi.org/10.1093/bjps/51.2.197.

Woodward, J. (2002). What is a mechanism? A counterfactual account. *Philosophy of Science, 69*(S3), S366–S377. https://doi.org/10.1086/341859.

Yamada, M., Kawahara, Y., Kaneko, F., Kishikawa, Y., Sotogaku, N., Poppinga, W. J., Folgering, J. H. A., Dremencov, E., Kawahara, H., & Nishi, A. (2013). Upregulation of the dorsal raphe nucleus-prefrontal cortex serotonin system by chronic treatment with escitalopram in hyposerotonergic Wistar-Kyoto rats. *Neuropharmacology, 72*, 169–178. https://doi.org/10.1016/j.neuropharm.2013.04.044.

Chapter 2
Causal Insights from Failure: Post-marketing Risk Assessment of Drugs as a Way to Uncover Causal Mechanisms

Elena Rocca, Rani Lill Anjum, and Stephen Mumford

Abstract In order to make predictions about potential harms and benefits of treatments for the single patient, pharmacology needs to gain *deep causal knowledge*. That is, one needs to understand the causal mechanism underlying a certain outcome. Here, we base our argument on a particular philosophical framework, causal dispositionalism, which urges that causes should be understood as complex, intrinsic, tendential and context-sensitive. Applied to pharmacology, dispositionalism suggests that it is by digging into contextual influence, and not by eliminating it, that we can learn more about the how and why the intervention does its causal work. When a treatment fails to give the expected outcome, therefore, it offers an opportunity to investigate the local context of failure and identify possible interferers. Potentially, this helps uncovering more of the causal nexus by which the outcome is produced. Both pre-clinical research and clinical experimentation alone are poorly fit for uncovering causal mechanisms through failure, since they are based on screening off, or disregard, of interferers. This leaves the post-marketing drug monitoring as the best scenario for systematically generating mechanistic hypotheses about a treatment through studying instances of causal failure of treatment in individuals. We suggest an integrated framework in which post market monitoring, through the study of treatment failure, feeds pre-clinical and clinical research with mechanistic hypothesis.

Keywords Pharmacovigilance · Causal mechanism · Causal dispositionalism · Treatment failure · Clinical predictions

E. Rocca (✉) · R. L. Anjum
Norwegian University of Life Sciences (NMBU), Ås, Norway
e-mail: elena.rocca@nmbu.no; rani.anjum@nmbu.no

S. Mumford
Durham University, Durham, UK
e-mail: stephen.mumford@durhman.ac.uk

A. LaCaze, B. Osimani (eds.), *Uncertainty in Pharmacology*, Boston Studies in the Philosophy and History of Science 338,
https://doi.org/10.1007/978-3-030-29179-2_2

2.1 Deep Causal Knowledge for Pharmacology

Pharmacology is as much about understanding causation as it is about establishing it. Accordingly, the discovery that a certain drug causes a positive outcome is no more than half the task. One needs also to understand the potential harms that might accompany the benefits. We will be considerably aided in this aim if we have, what we will call, deep causal knowledge. Deep causal knowledge consists in more than simply knowing *that* one thing caused another: it involves also knowledge of *how*.

To justify our stance, we use the ontological framework of causal dispositionalism (developed in detail in Mumford and Anjum 2011) to motivate an explorative approach to causation: one that develops mechanistic knowledge concerning how a pharmacological intervention works. The framework explains why cases of causal failure, if properly acknowledged, provide opportunities for new insights, paving the way for deep causal knowledge. The features of causation emphasised in causal dispositionalism also highlight the importance of post-marketing risk assessment of drugs. This is in part because post-marketing research can contribute to evidence of mechanisms, but also because of the nature of causation itself. A trial might show us that A causes B, but in a context which is often artificial. The context-sensitive, intrinsic, tendential and complex nature of causation explains why we should expect that new information comes to light in the post-marketing phase, especially about off-target effects. Accordingly, we argue in support of a better integration of post-marketing monitoring with the other phases of pharmacological research.

2.2 What Is This Thing Called Causal Science?

It will help if we set out some of our basic assumptions and define our main terms, starting with the idea of causal science (with an acknowledgment to Chalmers 1999). We call it causal science because it is the part of science that is concerned with causes: for instance, the discovery of what causes what, or the application of causal knowledge in deciding a medical intervention or developing a new technology. Given that there are multiple ways in which we can discover causal connections, and multiple uses to which we can put that knowledge, it might be better to speak of the causal sciences, plural. Randomised controlled trials (RCTs), for instance, would be a part of causal science, if one interprets this method's purpose as the discovery of effects of a trial intervention. It also seems plausible that causal science is the largest part of science. It might not be the whole of it, but science is at least partly a tool that enables us not just to understand but also to manipulate the world. Central to any such endeavour is the acquisition of causal knowledge. Once we know what tends to follow from what, we are in a position to decide courses of action.

Pharmacology is primarily a causal science. What explains its utility is that it can tell us the effects of drugs, with the ultimate aim of using them to produce benefit and reduce suffering, while minimising harms. We can also, then, use this science to illustrate another important distinction: between deep causal knowledge and surface causal knowledge. It is one thing to know *that* A causes B, for instance. Evidence-based medicine professes to have made significant progress on the correct methods for finding causal knowledge of this kind (Howick 2011). But, on our classification, to know that an intervention is causally linked to a specific outcome, without knowing anything about how it does so, would count as surface causal knowledge only. Deep causal knowledge, however, requires both knowing-that and knowing-how A causes B, in our terminology. Knowledge-that could for instance be indicated statistically, through covariance or increased relative risk, without the backing of a causal theory. Knowledge-how, on the other hand, would require a theoretical approach. Such an approach would include a range of knowing-thats, but it would not end there.

Clearly knowing *how* A causes B entails knowing *that* A causes B. But the entailment does not automatically run the opposite way. Knowing-that would be a necessary, but not sufficient condition for knowing-how. This shows that deep causal knowledge is more than surface causal knowledge. Now, of course, knowing-how an effect is produced may involve some further knowledge-that; for example, that there is such-and-such a mechanism at work. But this knowledge is deeper than our original knowledge simply that A causes B: the knowing-how consists in knowing all sorts of relevant causal facts and how they relate in order to produce the effect. So such knowledge is deeper knowledge in the sense that to know how A causes B is to know something about the mechanisms that take the causal process from A to B. In the case of pharmacology, this will require an understanding, among other things, of the molecular interactions involved. Typically, this will concern the interaction between the drug and the individuals in the target group, including their genetics and biochemical milieu.

Knowing in one case how the cause led to the effect amounts to knowledge of the mechanism involved, as we think mechanisms should be understood. Conceptions of mechanism vary, of course (see Glennan 1996, 2002; Machamer et al. 2000; Machamer 2004; Bechtel and Abrahamsen 2005; Bechtel 2011; Psillos 2011; Illari and Russo 2014: ch. 12). Some understand mechanisms on a reductive basis; that is, they seek explanations of wholes in terms of the activity of their parts. There is also a necessitarian strand to the understanding of mechanisms: that, if functioning correctly and under some specific conditions, they bring about their effects as a matter of necessity. We reject a number of these connotations, as should become clearer in what follows. If we simply say that the mechanism is the process leading from A to B, then we leave open the possibility that explanations can come from a higher level than that which they explain and not just a lower level. This will be important in social explanation, for instance, but it also allows for emergent, top-down causation generally. An obvious candidate for top-down causation in medicine is the placebo effect.

An exclusively reductive explanation could bring a loss of information, since it would fail to account for top-down causation, but also for the interaction of the parts. For instance, two drugs taken together could interact in a synergistic or an antagonistic way, where their effect in combination is not simply the sum of their individual effects, had they been taken separately. In this sense, the whole is not simply the sum of the parts.

2.3 The Nature of Causation

We have set out some of our basic concepts that will be useful in what follows. We explained what we meant by causal science and by deep causal knowledge, which amounts to knowledge of mechanisms. We will be arguing for what we call causal failure as a significant opportunity for the acquisition of deep causal knowledge: knowledge showing us the *how*, and not just the knowledge *that*, of causation.

We have further groundwork to lay out, however, for our point is to be made within a particular understanding of causation. What we say is built upon the main claims of causal dispositionalism and its application to scientific method (Mumford and Anjum 2011). Such a view accepts an ontology of dispositions, or real causal powers (developed in Harré and Madden 1975; Mumford 1998; Molnar 2003; Cartwright 1989; Martin 2008). It is not necessary to embrace every detail of this metaphysics in order to see the appeal of the features of causation to which we attend. In any case, those who support the ontology often disagree over the details. However, that causation works in a tendential and context-sensitive way, as we argue here, can be motivated by reasons that are independent of the powers ontology, even though we think such an ontology is the best way to understand how all these features are united in causation. We will explain them briefly and then, in later sections, say how they relate to discovery of deep causal knowledge through causal failure in post-marketing pharmacological settings.

(a) *Intrinsicality*

We understand causes as powerful properties of things, which we sometimes call dispositions, and these manifest themselves in their effects. Crucially, this is a non-relational view of causation that stands in opposition to accounts that make it extrinsic. In which accounts would it be extrinsic? A regularity theory of causation (Hume 1739; Psillos 2002), for instance, claims that particular event *a* causes particular event *b* only if there is a general pattern in which every instance like *a* is followed by an instance like *b*. In other words, there could not be a unique case of causation. The defender of intrinsicality rejects this as counterintuitive, and instead supports a form of causal singularism. For a singularist, the fact of whether *a* causes *b* is not ontologically dependent on what happens at other times and places. There could, then, be a case of causation that only happened once in the history of the universe. If a person takes a pill and then experiences an allergic episode, there is no ontological requirement, therefore, in order to say that the pill caused the episode, that other people took the same kind of pill and had a similar allergic reaction. What happens to other people, at other times and places, is instead a matter of the epistemology of causation.

Taking causation as intrinsic and singular is important in the medical context. David Hume, in his extrinsic account of causation, was committed to the principle that ‘like objects, plac’d in like circumstances, will always produce like effects’ (Hume 1739: I, iii, 8, p. 105). The singularist is not logically committed to Hume’s principle. Under singularism, it does not matter whether there are any such ‘like circumstances’ in order to claim causation. Arguably, there are no ‘like circumstances’ in biology, since no two individuals are identical. When it comes to pharmacological cases, there seems to be an acknowledgment that this is prima facie the case; namely that there can be individual variation and even medical uniqueness (Smith et al. 2012). Indeed, that the same pharmacological intervention can have different effects in different people is acknowledged within the structure of trial design. It is the reason why RCTs are conducted on at least two randomised large groups. There is an acceptance that not everyone in each group will have identical responses but, instead, with large-enough groups we can gather trend data on incidences of outcome.

(b) *Tendencies*

A singularist still has to account for why there is a degree of order and stability in the world. Unaccompanied, singularism is consistent with a cause having any kind of effect at all: that is, anything could follow (be caused by) anything. This does not seem to describe our world, which is regular enough and predictable enough for science to be a worthwhile enterprise. But in rejecting Hume’s principle, above, that like objects will always produce like effects, we should not find ourselves defending the equally absurd principle that like objects could produce *any* effect.

Causal dispositionalism tells us that a cause tends towards a particular type of effect, where *tendency* is to be considered an irreducible modality that is less than necessity but more than pure contingency (Mumford and Anjum 2011: ch. 8). A cause need not result in the same effect always, therefore, but will tend to produce some types of effect more than others. We can see that pharmacological interventions prima facie work this way. For example, SSRIs tend to improve depression but they don’t in every case. The taking of an SSRI bears a more than contingent connection with a lessening of depression: it will tend to produce that outcome. Such a tendency can be adequate grounds for prescribing such an intervention even if there is no guarantee of the desired outcome.

Tendencies come in different degrees of strength. For example, oral contraception is causally linked both to prevention of ovulation (strongly) and producing venous thromboembolism (weakly). The former happens in >99% of cases, while the latter in around 15 in 100,000 cases. A weak tendency can still be a very important matter if it is towards a drastic outcome; for example, we should aim to thwart even a weak tendency towards death.

There are reasons why causes do not necessitate their effects. The simplest is that causes can be counteracted and prevented from manifesting their effects by other causal factors. For instance, oral fexofenadine tends to an antihistaminic effect, but this tendency is counteracted by grapefruit juice, which reduces its bioavailability (Bailey 2010). A deeper reason is that causes simply work in this dispositional kind of way. A radioactive radium atom has a tendency to decay by a certain time, specified by its half-life. But it only has a disposition to do so. There is no necessity that it does. Indeed, the specification is in the form of a half-life because that is the point at which there is a 50/50 chance of decay having occurred. If it fails to decay, it is not because anything stopped it, under at least one plausible interpretation. It is simply that its tendency did not manifest. A similar example from biology could be the disposition of ion channels to be either open or closed. Whether an individual channel is open or not is dependent on the presence or absence of a drug but only in a way that can be described in a dose/response curve for opening/closing. Whether or not it does is not, for all we know, a matter of necessity.[1]

The ontological commitment to tendencies has some empirical basis. It is the way we experience the world as working. Indeed, in our own actions, we know that efforts to achieve a certain outcome are never guaranteed to succeed. But it is still worth the while because our actions will at least tend towards those outcomes. The commitment also fills any gap left when we reject a regularity theory of causation in favour of singularism. Singularism does not have to mean that anything goes.

(c) *Complexity*

Dispositionalism accepts complexity as essential to the proper understanding of causation. A complexity view is also a rejection of monocausality: the idea that an effect has just one cause. Instead, we argue that an effect typically occurs when a number of different causes form a mutual manifestation partnership (see Martin 2008), which produces something together than none of the component causes could have produced alone. Ampicillin tends to cure staphylococcal infections, for instance, but only in interaction with a non-resistant bacterium and proper conditions for being absorbed by the patient's body and being distributed to the site of the infection. The drug in isolation can do no causal work. Another example is how some hallucinogenic drugs (e.g. N,N-dimethyltryptamine) are inactive when ingested, but once combined with a trigger (e.g. beta-carbolines) they become bioavailable. On a dispositionalist account, both these substances should count as causes of the hallucinogenic effect.

Despite the implausibility of monocausality, philosophical theories often begin with an attempt at understanding what it is to be a cause (singular), abstracted from the world's complexity. Such an abstraction is relatively harmless if complex causes are nothing but the additions of single causes. However, this is often not the case; for instance, in cases of synergistic and antagonistic interactions. In contrast, philosophical analysis frequently works by abstracting a cause away from the messy reality in which it is located, to see how a 'pure' or 'perfect' instance works, and using that insight to understand the regular, impure cases.

This is a dubious approach in the case of causation, however. For one thing, we have already explained how causes can prevent each other from manifesting. Unless we see that a number of different factors are usually in play, we may fail

[1] Thanks to Jeffrey Aronson for this example.

to see that causes only tend towards their effects rather than guarantee them. But there is a further and even more important reason why we have to acknowledge complexity, which, again, we have encountered. Effects do not simply add together. They often combine in nonlinear and sometimes emergent ways. Emergence occurs where causes that are brought together interact, change each other, and in so doing form a differently empowered whole (for the details on our account of emergence, see Anjum and Mumford 2017). Chemical bonding typically works in nonlinear and emergent ways with respect to causal powers, where the molecule can have very different properties and powers from each element taken separately. For instance, the power of water to put out fires is an emergent power, which is formed when its parts undergo a chemical bonding. None of the parts – hydrogen and oxygen atoms – individually has a power to put out fires. Indeed, they would do the opposite. A pharmacological counterpart to this example is benorylate, which is a chemical compound of aspirin and paracetamol. But unlike aspirin, this compound does not cause gastric erosion, because the chemical bond remains stable while in the gut (Wright 1976).

Now it might be thought that the idea that powers can change each other when they interact looks incompatible with a commitment to intrinsic tendencies. But these are not inconsistent. There is nothing to say that intrinsic features cannot change through their causal interactions. Nor should the facts of emergence lead one to doubt that those intrinsic tendencies are there in the components. They all contribute something of their own towards the interaction, which might not have occurred if any of the contributors were absent.

Causal complexity, in particular nonlinear interactions (such as synergisms and antagonisms), is a crucial feature in pharmacology, and, more generally, in chemistry. In this sense, causal dispositionalism grounds the view that knowledge of a substance's full causal potential can never be exhausted, since it depends on the near-infinite combinations of other entities that it might encounter (Ruthenberg 2016). Accordingly, complexity also represents a well-known challenge concerning risk, in the so-called cocktail effect. The method of RCT tests one intervention at a time and pronounces on its singular effect. However, when one or more drugs work in combination with a particular individual, new causal powers or side effects could emerge. These effects might only come to light in the post-marketing phase, however, once drugs are used in combinations that have never before been tested.

(d) *Context-sensitivity*

The final feature of causation to which we draw attention is context-sensitivity, which might be considered the natural consequence of the three previous features, already outlined. The current point tells us that a causal intervention is capable of producing different effects according to the different mutual manifestation partnerships into which it enters. In the case of pharmacology, we can say that different contexts will contribute to bring out different causal powers of a drug.

For example, antiarrhythmic drugs can calm irregular heartbeat (by fast sodium channel inhibition) but in some patients, in a different context, they can worsen it (by potassium channel inhibition) (Smith et al. 2012). Context-sensitivity might be taken as the single notion encapsulating the idea that causes enter into complex partnerships, mutually manifesting their intrinsic tendencies or, in some cases, their joint tendency. The upshot is that we cannot expect the same effect always to come from the same cause. For each person with whom the drug interacts, different manifestation partners are involved, many of which are unique to that person.

Acceptance of the context-sensitivity of causation is a rejection of the epistemic basis of the regularity theory. Hume talks as if we infer causal connections from constant conjunctions of cause and effect. And this assumption is accepted by many of Hume's opponents, who think that in addition to the constant conjunction of cause and effect, causation also involved the necessitation of the effect by the cause. We can note that this necessitarian claim is built on, and indeed requires, the principle of *same cause, same effect*. But regularity is not our experience, as most empiricists recognise. Instead of moving to a more tendential account that acknowledges causation's context-sensitivity, however, there are usually attempts to bypass it. One strategy is to think of the cause as larger, including also the background conditions under which the effect would be produced. We call this strategy 'causal expansion'. Another strategy is to shield off the cause from possible interferers, mainly by abstraction. We call this strategy 'causal isolation'. Both these are attempts to get a situation where we can assume that the effect is guaranteed to follow the cause under otherwise like circumstances. Indeed, we saw with Hume's statement of the *same cause, same effect* principle, that it was supposed to apply only to causes in the same circumstances. Presumably, then, causes in exactly the same situation would result in the same.

There is a problem with the assumption that causes are conjoined with their effects in this way, however. The only way to demonstrate it empirically would be to create identical circumstances and check whether the same cause would produce the same effect. But this is not practically possible, even in lab settings. Acknowledging that each patient is different, one will never meet the requirements of identical conditions in medicine or pharmacology, since the patient is part of the causal interaction (as a mutual manifestation partner). What is left, then, for the causal necessitarian, is to make a thought experiment or a theoretical model in which one did have identical conditions. For instance, one might infer that if the patient had an identical twin, with the same diet, lifestyle, history, and so on, then if both got the same intervention, the same effect would be guaranteed. But since this is a theoretical presumption, and not something that can be demonstrated empirically, we can only conclude that causal necessitation is an assumption about the nature of causation. Such necessity is thus philosophically motivated and based on an abstraction from our experience, rather than an accurate report of it. Causal necessitarianism – the view that the way in which a cause produces its effect is by necessitating it – has a long tradition in philosophy. It is supported, for instance, by Aristotle (*Metaphysics* Θ 5), Spinoza (1677: I, axiom III), Kant (1781: II, ii, 3, second analogy), Bergson (1889: 199), Ducasse (1924: 55) and many others (for a detailed discussion, see Mumford and Anjum 2011: chapter 3.) It can be found even in Mackie's (1980: 62) INUS-condition account, which might otherwise be thought to allow much of what we want. Even here, however, an effect is produced only when a collection of factors in jointly *sufficient* for the effect, which is to say necessary.

There are thus two consequences from assuming a tendential view of causation, as we promote here. One is that the same cause does not have to produce the same effect under some alleged like circumstances. This is what it means to say that the cause *tends* and no more than tends towards its effect. The second conclusion is that the same effect does not require that the circumstances were alike. We see this in pharmacology, where the same effect can occasionally be produced in very different patients. Accordingly, there is not a one-to-one relationship between cause and effect in dispositionalism, even if the cause includes a large set of background conditions.

2.4 The Dispositional Nature of Mechanisms

We said that mechanistic understanding is a requirement for what we called deep causal knowledge. But what is a mechanism? One could say that a causal mechanism is what takes the cause to the effect: a process between the intervention and the outcome. In its simplest form, therefore, one could understand a mechanism as something that is added between the intervention and the outcome; such as when the drug D causes its effect E via the mechanism M:

$$\mathrm{D}\ (\rightarrow \mathrm{M}) \rightarrow \mathrm{E}$$

This would be both causal knowledge-*that* and causal knowledge-*how*. For instance, ibuprofen reduces inflammation by inhibiting the COX2 enzyme and synthesis of inflammatory prostaglandins:

$$\text{Ibuprofen}\ (\rightarrow \text{COX2} \rightarrow \text{inh prostaglandins}) \rightarrow \text{inh Inflammation}$$

However, if we think of mechanisms in the *dispositional* sense – as tendential, complex and context-sensitive – then this cannot be the full causal picture. Understood in this way, the arrows are just tendencies. This means that they can be counteracted or amplified by contextual interferers. In addition to the basic mechanism, we thus need to understand also which factors could interact with the causal process, including how the patient contributes to the outcome. For instance, we would want to know *how* ibuprofen could provoke problems with the cardiovascular system. Moreover, higher-level factors such as stress, anxiety and expectations can affect any pharmacological treatment. Because of the many potential mutual manifestation partnerships, a drug is capable of far more variety of action and different causal interactions than what is described by the basic model of mechanism.

From the dispositionalist perspective, we cannot learn everything about mechanisms simply by many repetitions of the effect. There is a diminishing return on positive evidence: a new corroboration can increase our confidence in the hypothesis further, but beyond a certain point, the extent to which confidence is increased is less than the previous instance. In contrast, cases of causal failure, when something unexpected happens – or there is a discrepancy – could show us where to look to expand our causal understanding. Potentially, causal failure tells us that there is something about a mechanism that we did not understand, that it was capable of another causal pathway, partnered within a certain context, towards a different result than those previously known. We might also uncover more of the causal mechanism of the drug itself: some previously ignored aspect of the molecule, for example, or reaction to the molecule.

Pre-clinical research and clinical experimentation are poorly fit for uncovering causal mechanisms through failure, since they are based on isolation from causal interferers. This leaves the post-marketing drug monitoring as a valuable source of systematically generated mechanistic hypotheses through studying instances of failure of treatment in individuals. Furthermore, discovery through post-marketing causal failure has empirical considerations in its favour, as we will now go on to show. The approach that we present here, although starting from an ontology of causal dispositionalism, is in line with work by those who promote the epistemological value of single case reports for pharmacovigilance (see for instance Aronson 2003, 2005; Hauben and Aronson 2006; Karimi et al. 2014; Härmark et al. 2016).

2.5 The Knowledge Potential of Post-marketing Monitoring

Recently, voices have been raised that vindicate the importance of causal irregularities and failures for the process of knowledge-building in the fields of pharmacological research (London and Kimmelman 2015; Ioannidis 2016). These authors encourage the scientific community to abandon the traditional designation of 'negative results' in favour of the more constructive concept of 'unrecognised opportunities' (London and Kimmelman 2015: 28). Such a shift, they argue, will result in the advance of mechanistic understanding, basic research, translation and ultimately healthcare in general. Interestingly, the argument starts from considerations that strike us as close to the dispositional view:

> . . . drugs alone are not therapeutic agents. Drugs on their own are substances [. . .] that have the *capacity* to interact with particular physiological systems. When they are used in the absence of the knowledge and capacity for applying them properly, they are likely to be toxic. (London and Kimmelman 2015: 29, our emphasis)

Capacities, or dispositions, are intrinsic properties that make something causally powerful. Drugs are effective precisely because they have a causal power toward a curative outcome. However, the therapeutic effect is complex: it emerges from the mutual manifestation of such dispositions within a specific context, namely the patient. A thorough knowledge of the context of a drug's action, therefore, maximises benefits and minimises harms of the treatment. Confirmatory research about the targeted effect of the drug does not alone fulfil this need. An analysis of the patient's context might sometimes start from confirmatory research, for instance by analysis of covariates statistically associated to a certain outcome (see Mannsmann, this volume). However, this approach alone cannot tackle the issue thoroughly, since confirmatory studies work with stringent inclusion criteria and therefore offer information only about some precise patient groups (Rothwell 2005).

After almost 60 years under the shadow of the thalidomide disaster, the link between clinical monitoring of drugs and basic medical understanding is undeniable. Before thalidomide was proved able to interfere with normal foetal development, the mechanisms by which the drug could pass the placental barrier were not accounted for or even considered (McBride 1977). Yet, basic medical knowledge is still largely associated with experimental isolation. When basic researchers investigate real-world complexity instead of lab isolation, however, their efforts do not go unnoticed. An example of this is a study that joined data from the post-marketing monitoring of a large number of marketed drugs, and used the information to gain understanding of the mechanism by which drugs interact with human molecules (Campillos et al. 2008). In particular, they grouped drugs according to similarity in their spectra of undesired effects, and used this comparison to detect common molecular targets. Through this approach, 20 possible unexpected drug-drug relations were identified, between drugs with dissimilar structures and different therapeutic indications. The authors emphasised that their hypotheses could be formulated only by observing the complexity of interactions between drugs and the human body, that is, outside experimental isolation. Their approach, they suggested, hinted at 'new uses of marketed drugs' (Campillos et al. 2008: abstract).

Once we accept that information from clinical failures is valuable for basic research, more questions follow. Are all types of information equally useful for the purpose of deep causal knowledge? No, we argue. As suggested above, evidence of failure, unlike repeated corroboration of positive effects, can lead to the *how* of causation, as long as it also provides detailed information about the specific causal processes, and of how they can be counteracted.

This view reflects well the reality of post-marketing monitoring of drugs. Large observational studies are useful to give an overview of a drug's impact on the population. Certainly, such surface causal knowledge can potentially improve clinical use of some drugs, for example by awareness of risk factors (or susceptibility factors), and by identifying the reference class to which the patient belongs. Optimising the impact of a treatment for the single patient, however, requires the deepening of mechanistic understanding. Susceptibility to rare adverse reactions, for instance, is hard to predict with population data. This is why genetic tests are being developed in order to make precise individual predictions (see for instance the detection of HLA $B_{*}5701$ polymorphism as a predictor of allergic reactions to abacavir in the treatment of HIV/AIDS). Clearly, genetic tests require that we understand the molecular underpinning of the adverse reaction. A further reflection is that the mere knowledge of patient's risk factors might inform us about the cases in which a treatment should be avoided, but does not easily serve the purpose of optimising it. Some vaccination programs, for instance, fail to work in obese patients

(Wiedermann et al. 2016). However, without the understanding of the molecular processes underlying this failure, it is hard to find a way to overcome the problem.

While evidence of mechanisms can come from a variety of sources, it is easier to formulate informed hypotheses from details of specific clinical cases than from frequentist data alone. We can illustrate this with an example.

Valproic acid was marketed in the late 70s as an anti-epileptic drug. Soon after, its use was broadened to the treatment of migraine, bipolar affective disorder, alcohol withdrawal and more (Sitarz et al. 2014). Although this drug is usually well tolerated, it can cause a range of important adverse reactions, including rare but potentially deadly liver injury. In 1984, a first report on liver damage associated with valproic acid counted 42 cases with fatal outcomes, of which most were paediatric cases and cases of polypharmacy (Powell-Jackson et al. 1984). The same report hinted at hypotheses for the mechanism of harm, together with some evidence in lab animals, but the total evidence of mechanism at the time was inconclusive. Two years later, the incidence of hepatic fatalities had decreased nearly fivefold, despite an overall increase in the use of the drug (Dreifuss 1989). This improvement was probably due to awareness among prescribers, and decreased prescription in high-risk patients, such as young children, because of the availability of population data. Despite this improvement, fatal liver injury persisted. In 1996, 29 additional cases were reported (Bryant and Dreifuss 1996). Since the causes of such reactions were still mysterious, it was extremely difficult to predict a fatal outcome in an otherwise healthy patient.

A new development in the mechanistic knowledge occurred when detailed evidence about valproic acid's victims began to be gathered. Histological analyses showed some specific modifications, suggesting that the principal cause of harm started from interference with mitochondrial processes (Fromenty and Pessayre 1995). A number of observations confirmed the fact that valproic acid affects mitochondrial functions. For instance, a detailed analysis of urine composition in treated children, compared with untreated children, revealed that the drug influences mitochondrial metabolism (Price et al. 2011). Anecdotal reports pointed out that valproic acid-induced liver failure was fulminant in patients with genetic mitochondrial diseases (Krähenbühl et al. 2000). Ultimately, detailed molecular characterization of valproic acid's victims revealed that the risk of liver toxicity is determined by point mutations in the gene POLG, which codes for a mitochondrial protein (DNA polymerase γ) (Stewart et al. 2010).

We see, then, that different contexts of interaction brought out new causal powers of the drug. Even if the mutual manifestation partnerings for some of those effects are rare, they are nevertheless valuable for understanding how the drug does its causal works. Such a precise determination of the mechanism of harm is valuable for two reasons. First, the patient's tendency to be harmed by valproic acid might be revealed in advance through genetic screening. Secondly, the discovery triggered a new line of research aimed at a deeper causal understanding: why do patient with POLG mutations have a tendency to develop valproic acid-induced liver injury? This question was addressed, for instance, by culturing cells from patients carrying a mutation of the POLG gene, and observing how these cells react to valproic acid administration (Sitarz et al. 2014).

The example illustrates well the knowledge potential of post-marketing monitoring. In pharmacology, patterns of harm detection are intermingled with patterns of causal discovery. Such mingling is facilitated by evidence of causal failure.

2.6 How Much Failure Do We Need?

Having more data is usually thought to provide better evidence of causation, which is why quantitative research is given epistemological priority over for instance case studies. The question is what we mean by more data. More of what sort of data? And how much is enough?

Kimmelman et al. (2014) argue for the necessity of distinguishing two phases of pre-clinical research: exploratory and confirmatory. The first is focused on generating hypotheses, and the second on testing them through robust evidence. The authors point out that the two stages should use different and complementary types of study, and urge the need of a 'greater volume of confirmatory investigation'.

We maintain that this is a useful distinction, and that the process of corroborating a hypothesis needs a large amount of experimental confirmation. On the other hand, the exploratory phase might benefit more from failure than from confirmation. Think for instance about the role of serendipitous discoveries in the history of science (Rocca et al. 2019)[2]. The failure of reproducing the expected result can generate new hypotheses, and expand knowledge far beyond what a mere confirmation would have allowed. However, if we need a reasonable number of confirmations in order to establish knowledge, how much failure do we need in order to explore it? For cases of serendipity, the success of a discovery seems to rely greatly on 'catching' the unexpected as soon as it is manifested. The key for discovery is thus the quality, rather than the frequency, of an unexpected observation. This is a general truth for all causal science, and pharmacology is no exception.

Unexpected effects of drugs might be useful for research, even if they do not happen frequently. Since causation is intrinsic and singular, as described above, we cannot deny causation only because it happens too rarely. If the drug has causal powers that almost never manifest themselves, because the mutual manifestation partner is rare, it is nevertheless a power of the drug that ought to be considered a real risk. For example, Zolpidem, a GABAergic hypnotic used to treat insomnia, has provided the medical literature with a number of surprising anecdotes (Rocca 2017). Patients under its effect were reported to sleepwalk, sleep-cook, sleep-eat and sleep-drive. But also, some patients suddenly recovered after stroke, re-gained mobility after brain injury or improved from post-traumatic states shortly after taking the drug (Hoque and Chesson 2009). Although being only circumstantial,

[2]Thanks to Samantha Copeland for bringing this point to our attention.

and unsuitable for a frequentist analysis, these cases were not ignored by basic research. Quite the contrary. They led to the hypothesis that the brain might have some redundant patterns of self-recovery, which Zolpidem could unmask.

Unexpected effects of drugs can thus function as a springboard for basic research, even when they are very rare (or maybe even because of this). We will now explain why we say that 'more data' ought to mean more data about the *single* cases of causal failure, and then propose a practical way forward for systematically collecting and using such data.

2.7 Reporting Suspected Causal Failure: A Community Effort

The problem of improving knowledge translation has gained increasing attention in the medical literature over the last years (Ioannidis 2016; Edwards and Hugman 1997). Following the inefficiency of pre-clinical and clinical research for the ultimate improvement of healthcare, legitimate questions have been asked about mistakes in the information flow from research to clinical practice. This step in the process is often called bench-to-bedside, or knowledge-to-action translation.

However, these concerns cover only half the problem. Indeed, a community that acknowledges the central role of real-world complexity for the advance of medical research must make sure that clinical observations feed research with hypotheses in the most fruitful ways. In other words, not only knowledge-to-action, but also action-to-knowledge translation must be optimised for improvement of the process as a whole. In the specific case of harm detection, how and when should clinical observation of off-target effects occur? And how should it be reported and communicated to best fulfil its potential for advancing knowledge?

In the view of causal dispositionalism, deep causal understanding happens by analysing the individual causal process in a detailed way. In the context of harm detection, this view results in a central focus on the clinical observation of side effects. Recognition of the epistemological value of the individual clinical observation has to be seen as an advantage in the context of post-marketing drug monitoring. Indeed, what can be identified as the systematic trigger of pharmacovigilance, if not clinical assessment of suspected side effects? Note that with this we are not denying that large population studies have an important role in monitoring harm. However, we agree with Edwards when he points out that 'any consideration of causality based only on epidemiological approaches is not logically defensible and certainly not socially acceptable' (Edwards 2012: 51). A population study needs sound hypotheses and a good experimental design in order to be fully reliable. Historical examples show that, in the field of pharmacovigilance, population data should not be produced or interpreted in isolation from other types of evidence, including single suspected clinical diagnoses (Edwards 2012: 42 and 51).

The type and amount of information available about a suspected effect determines which hypotheses can be made about its biological underpinning. In other words, mechanistic hypotheses depend on the quality of the side effect reports. In order to gain a deep causal understanding of harm, beyond the assessment that a certain effect occurred, also including the understanding of how it occurred, it is useful to consider the side effect report from a dispositionalist perspective. That is, the main input of the report is a set of conditions (context) that in combination with the drug provoked a certain outcome (Rocca 2017).

So although some effects are rare, the causal processes leading up to them must nevertheless be considered and understood. In this sense, drug agencies would benefit most from receiving as many details as possible, to facilitate the detection of signals of harm, the performance of comparative analyses, and ultimately the formulation of causal hypotheses (see also Aronson 2011; Gagnier et al. 2014).

Consider 'classic' case reports, which marked the beginning of important lines of medical research. For instance, we can recall the first case-series describing patients affected by AIDS (Gmerk 1990). Clearly, the way in which doctors reported their observations influenced the immediate research hypotheses. Clinicians selected some common features of the patients that they suspected to be somehow causally connected with their condition (e.g. the fact that they were all young and homosexual). However, they did not discard other details, although they had no explanatory backing for including them (e.g. cytomegalovirus infection, pneumonia, use of inhalant drugs like 'poppers', acquaintances). With such details at hand, researchers moved immediately in several directions in order to verify causal relations and discard red herrings. This feature, of containing unfiltered details that at the time of the observation are not organised in a causal picture, is an important epistemological value of clinical reports (for a complete analysis of this case, see Ankeny 2011).

Is this expectation feasible in the reality of clinical life? After all, the doctor's task is to provide healthcare, not to write medical literature. Doctors spend an increasing amount of time with administrative duties, and in the last decades studies confirmed that they dedicate less than half their working day to patients, resulting in a decline in overall productivity (Røhme and Kjekshus 2001; Sinsky et al. 2016). Not all side effects, therefore, can be reported in detail. In fact, under-reporting is a factor that delays pharmacovigilance (Hazell and Shakir 2006). Studies show that health practitioners tend to select the suspected side effects also by novelty, hindering the possibility of getting a realistic overview of the impact of a treatment on the population (Haramburu et al. 1997). A side effect or failure, indeed, might not be novel in itself, but the description of the context of failure – that is, the particular process – might contain information that is important for a deep causal understanding.

How, then, should clinicians prioritise reporting in a way that is feasible with their routine, if not only based on novelty? This problem becomes manageable once we consider that it does not pertain to practitioners in isolation, but rather to a community of practitioners, carers, patients and researchers. From this perspective, science ought to provide clinicians with the criteria they need for prioritising the cases that at a specific time or place are worth more effort in reporting. The priority should not be determined by strict or permanent rules, but should be adaptable over time and in relation to different practices. It should be adjusted based on the current state of knowledge and uncertainties about the specific pharmacological treatments.

We thus suggest an improvement of *action-to-knowledge* translation based on an integrated system of close collaboration between post-marketing monitoring and all phases of research, as described in Rocca (2017). Pharmacological developments should not be seen as straight patterns starting from basic research and ending with the clinical treatment, as depicted in the standard model of clinical translation. Rather, information should flow in a continuous loop among all phases of research and clinical practice. Some authors have recently exhorted this development (London and Kimmelman 2015). Their motivation, however, remains improvement of *knowledge-to-action* translation. We extend these considerations to the whole process of information flow: from the lab bench to the patient's side, *and back*, in a continuous loop. Reports of suspected side effect should be written with the intent of feeding back to basic research with hypotheses for deep causal knowledge. For this, statistical analyses of correlations is only a part of the process. The reports should also be analysed by detailed comparisons, with the aim of providing basic research with mechanistic hypotheses of harm (for more about this information loop, see Rocca 2017: figure 1).

If the report of suspected harm should not be produced by a clinician in isolation, but by a clinician as part of an integrated system, two practical considerations follow. First, the production of high quality reports must be promoted actively by the system. A strategy for this could start from medical education, with a shift to a type of communication that is focused on patients and their context. Models for medical interview based on bio-psychosocial, patient-centred communication have already been developed and are being taught in some medical schools (Larivaara et al. 2001). Second, the side effect report can be seen as a common effort of one or more practitioners together with the patient. It could be the result of an integrated network of clinical data collected about the patient in several instances and by different practitioners (general practitioner, specialist, physiotherapist, pharmacist, surgeon, et cetera). The utility of data from health registries for statistical population studies is at the moment still under debate (Grimes 2015; Norén et al. 2010). Some see it as a goldmine for statistical analysis, while others are sceptical of using data that were not collected following a specific design and research question. However, a network of registries can surely give more thorough and detailed information about *single* cases, and be highly valuable for exploratory research. Confirmation belongs to another stage of the loop.

2.8 Conclusion

While arguing that we need better systems in place to learn from causal failure, we do not, of course, encourage failure itself. In pharmacological cases, such failures can be catastrophic. But when they do occur, we must acknowledge them and adjust our theories accordingly. There seems to be no point in protesting that current approaches to drug safety, based upon the large-scale statistical techniques of evidence based medicine, are adequate as they stand. Adverse drug reactions constitute, on some accounts, the fourth biggest cause of death in the United States (FDA Report 2016). Not all of these deaths are down to basic errors, such as dosage mistakes. Rather, it seems that something fundamental needs improving in how we assess and understand the effects of drugs.

We have provided a philosophical framework that explains some of the difficulties: causes should be understood as complex, intrinsic, tendential and context-sensitive. Applied to drug use, this means that the post-marketing phase represents the reality better than the methods seeking to isolate causes in artificial conditions. Real-life contexts of causal failures – unexpected results – is where new hypotheses can be developed and, with due consideration, deep mechanistic understanding can grow.

Acknowledgements Thanks to Jeffrey Aronson, Ralph Edwards and Barbara Osimani for useful comments and feedback on the paper. Some of the examples used were suggested to us by Jeffrey Aronson. The research in this paper was conducted with the support of the Norwegian Council for Research (NFR) FRIPRO scheme.

References

Anjum, R. L., & Mumford, S. (2017). Emergence and demergence. In M. Paoletti & F. Orilia (Eds.), *Philosophical and scientific perspectives on downward causation* (pp. 92–109). London: Routledge.

Ankeny, R. (2011). Using cases to establish novel diagnoses: Creating generic facts by making particular facts travel together. In P. Howlett & M. Morgan (Eds.), *How well do facts travel? The dissemination of reliable knowledge* (pp. 252–272). Cambridge: Cambridge University Press.

Aronson, J. K. (2003). Anecdotes as evidence. *British Medical Journal, 326*, 1346.

Aronson, J. K. (2005). Unity from diversity: The evidential use of anecdotal reports of adverse drug reactions and interactions. *Journal of Evaluation in Clinical Practice, 11*, 195–208.

Aronson, J. K. (2011). 'Guidelines and a checklist for reporting suspected adverse drug reactions anecdotally in journals. In J. Talbot & J. K. Aronson (Eds.), *Stephens' detection and evaluation of adverse drug reactions: Principles and practice* (6th ed., pp. 699–708). Oxford: Wiley-Blackwell.

Bailey, D. G. (2010). Fruit juice inhibition of uptake transport: A new type of food-drug interaction. *British Journal of Clinical Pharmacology, 70*, 645–655.

Bechtel, W. (2011). Mechanism and biological explanation. *Philosophy of Science, 78*, 533–557.

Bechtel, W., & Abrahamsen, A. (2005). Explanation: A mechanist alternative. *Studies in History and Philosophy of Science Part C: Studies in History and Philosophy of Biological and Biomedical Sciences, 36*, 421–441.

Bergson, H. (1889). *Time and free will: An essay on the immediate data of consciousness*, F. L. Pogson (trans.). New York: Dover Publications Inc., 2001.

Bryant, A. E., & Dreifuss, F. E. (1996). Valproic acid hepatic fatalities. III. US experience since 1986. *Neurology, 46*, 465–469.

Campillos, M., Kuhn, M., Gavin, A. C., Jensen, L. J., & Bork, P. (2008). Drug target identification using side-effect similarity. *Science, 321*, 263–266.

Cartwright, N. (1989). *Nature's capacities and their measurement*. Oxford: Oxford University Press.

Chalmers, A. (1999). *What is this thing called science?* (3rd ed.). Maidenhead: Open University Press.

Dreifuss, F. E., Langer, D. H., Moline, K. A., & Maxwell, J. E. (1989). Valproic acid hepatic fatalities. II. US experience since 1984. *Neurology, 39*, 201–206.

Ducasse, C. J. (1924). *Causation and the types of necessity*. New York: Dover, 1969.

Edwards, I. R. (2012). Considerations on causality in pharmacovigilance. *International Journal of Risk & Safety in Medicine, 24*, 41–54.

Edwards, I. R., & Hugman, B. (1997). The challenge of effectively communicating risk-benefit information. *Drug Safety, 17*, 216–227.

FDA Report (2016). *Preventable adverse drug reactions: A focus on drug interactions*. Online: http://www.fda.gov/Drugs/DevelopmentApprovalProcess/DevelopmentResources/DrugInteractionsLabeling/ucm110632.htm. Accessed 8 Feb 2017.

Fromenty, B., & Pessayre, D. (1995). Inhibition of mitochondrial beta-oxidation as a mechanism of hepatotoxicity. *Pharmacology & Therapeutics, 67*, 101–154.

Gagnier, J. J., Kienle, G., Altman, D. G., Moher, D., Sox, H., Riley, D., & the CARE Group. (2014). 'The CARE guidelines: Consensus-based clinical case report guideline development. *Journal of Clinical Epidemiology, 67*, 46–51.

Glennan, S. (1996). Mechanisms and the nature of causation. *Erkenntnis, 44*, 49–71.

Glennan, S. (2002). Rethinking mechanistic explanation. *Philosophy of Science, 69*, S342–S353.

Gmerk, M. D. (1990). *History of AIDS: Emergence and origin of a modern pandemic*, R. C. Maulitz and J. Duffin (trans.). Princeton: Princeton University Press.

Grimes, D. A. (2015). Epidemiologic research with administrative databases: Red herrings, false alarms and pseudo-epidemics. *Human Reproduction, 30*, 1749–1752.

Haramburu, F., Bégaud, B., & Moride, Y. (1997). Temporal trends in spontaneous reporting of unlabelled adverse drug reactions. *British Journal of Clinical Pharmacology, 44*, 299–301.

Härmark, L., Raine, J., Leufkens, H., Edwards, R. I., Moretti, U., Sarinic, V. M., & Kant, A. (2016). Patient-reported safety information: A renaissance of pharmacovigilance? *Drug Safety, 39*, 883–890.

Harré, R., & Madden, E. H. (1975). *Causal powers: A theory of natural necessity*. Oxford: Blackwell.

Hauben, M., & Aronson, J. K. (2006). Gold standards in pharmacovigilance: the use of definitive anecdotal adverse drug reactions as pure gold and high grade ore. *Drug Safety, 30*, 545–555.

Hazell, L., & Shakir, S. A. (2006). 'Under-reporting of adverse drug reactions: A systematic review. *Drug Safety, 29*, 385–396.

Hoque, R., & Chesson, A. L. (2009). Zolpidem-induced sleep walking, sleep related eating disorder, and sleep-driving: Fluorine-18-fluorodeoxyglucose positron emission tomography analysis, and a literature review of other unexpected clinical effects of zolpidem. *Journal of Clinical Sleep Medicine, 5*, 471–476.

Howick, J. (2011). *The philosophy of evidence-based medicine*. Chichester: Wiley-Blackwell.

Hume, D. (1739). *A treatise of human nature*, L. A. Selby-Bigge (Ed.). Oxford: Clarendon Press, 1888.

Illari, P., & Russo, F. (2014). *Causality*. Oxford: Oxford University Press.

Ioannidis, J. P. (2016). Why most clinical research is not useful. *PLoS Medicine, 13*, e1002049.

Kant, I. (1781). *Critique of pure reason*, N. Kemp Smith (trans.). London: MacMillan, 1929.

Karimi, G., Star, K., Lindquist, M., & Edwards, I. R. (2014). Clinical stories are necessary for drug safety. *Clinical Medicine, 14*, 326–327.

Kimmelman, J., Mogil, J. S., & Dirnagl, U. (2014). Distinguishing between exploratory and confirmatory preclinical research will improve translation. *PLoS Biol, 12*, e1001863.

Krähenbühl, S., Brandner, S., Kleinle, S., Liechti, S., & Straumann, D. (2000). Mitochondrial diseases represent a risk factor for valproate-induced fulminant liver failure. *Liver, 20*, 346–348.

Larivaara, P., Kiuttu, J., & Taanila, A. (2001). The patient-centred interview: The key to biopsychosocial diagnosis and treatment. *Scandinavian Journal of Primary Health Care, 19*, 8–13.

London, A. J., & Kimmelman, J. (2015). Why clinical translation cannot succeed without failure. *Elife, 4*, e12844.
Machamer, P. (2004). Activities and causation: The metaphysics and epistemology of mechanisms. *International Studies in the Philosophy of Science, 18*, 27–39.
Machamer, P., Darden, L., & Craver, C. F. (2000). Thinking about mechanisms. *Philosophy of Science, 67*, 1–25.
Mackie, J. L. (1980). *The cement of the universe*. Oxford: Oxford University Press.
Martin, C. B. (2008). *The mind in nature*. Oxford: Oxford University Press.
McBride, W. G. (1977). Thalidomide embryopathy. *Teratology, 16*, 79–82.
Molnar, G. (2003). *Powers: A study in metaphysics*, S. Mumford (Ed.). Oxford: Oxford University Press.
Mumford, S. (1998). *Dispositions*. Oxford: Oxford University Press.
Mumford, S., & Anjum, R. L. (2011). *Getting causes from powers*. Oxford: Oxford University Press.
Norén, G. N., Hopstadius, J., Bate, A., Star, K., & Edwards, I. R. (2010). Temporal pattern discovery in longitudinal electronic patient records. *Data Mining and Knowledge Discovery, 20*, 361–387.
Powell-Jackson, P. R., Tredger, J. M., & Williams, R. (1984). Hepatotoxicity to sodium valproate: A review. *Gut, 25*, 673–681.
Price, K. E., R. E. Pearce, U. C. Garg, B. A. Heese, L. D. Smith, J. E. Sullivan, M. J. Kennedy, J. F. Bale Jr, R. M. Ward, T. K. H. Chang, F. S. Abbott and J. S. Leeder. 2011. 'Effects of valproic acid on organic acid metabolism in children: A metabolic profiling study', *Clinical Pharmacology & Therapeutics*, 89: 867–874.
Psillos, S. (2002). *Causation and Explanation*. Chesham: Acumen.
Psillos, S. (2011). The idea of mechanism. In P. M. Illari, F. Russo, & J. Williamson (Eds.), *Causality in the sciences* (pp. 771–788). Oxford: Oxford University Press.
Rocca, E. (2017). Bridging the boundaries between scientists and clinicians: Mechanistic hypotheses and patient stories in risk assessment of drugs. *Journal of Evaluation in Clinical Practice*. https://doi.org/10.1111/jep.12622.
Rocca, E., Copeland, S., & Edwards, R. (2019). Pharmacovigilance as scientific discovery: An argument for trans-disciplinarity. *Drug Safety, 42*, 1115–1124.
Røhme, K., & Kjekshus, L. E. (2001). When time counts: Physicians' work assignments and use of TIME. *Tidsskriftet den Norske Lægeforening, 121*, 1458–1461.
Rothwell, P. M. (2005). External validity of randomised controlled trials: "To whom do the results of this trial apply?". *The Lancet, 365*(9453), 82–93.
Ruthenberg, K. (2016). About the futile dream of an entirely riskless and fully effective remedy: Thalidomide. *HYLE–International Journal for Philosophy of Chemistry, 22*, 55–77.
Sinsky, C., Colligan, L., Li, L., Prgomet, M., Reynolds, S., Goeders, L., Westbrook, J., Tutty, M., & Blike, G. (2016). Allocation of physician time in ambulatory practice: A time and motion study in 4 specialties. *Annals of Internal Medicine, 65*, 753–760.
Sitarz, K. S., Elliott, H. R., Karaman, B. S., Relton, C., Chinnery, P. F., & Horvath, R. (2014). Valproic acid triggers increased mitochondrial biogenesis in POLG-deficient fibroblasts. *Molecular Genetics and Metabolism, 112*, 57–63.
Smith, S. W., Hauben, M., & Aronson, J. K. (2012). Paradoxical and bidirectional drug effects. *Drug Safety, 35*, 173–189.
Spinoza, B. (1677). *The ethics*, R. H. M. Elwes (Ed.). New York: Dover, 1955.
Stewart, J. D., Horvath, R., Baruffini, E., Ferrero, I., Bulst, S., Watkins, P. B., Fontana, R. J., Day, C. P., & Chinnery, P. F. (2010). Polymerase γ gene POLG determines the risk of sodium valproate-induced liver toxicity. *Hepatology, 52*, 1791–1796.
Wiedermann, U., Garner-Spitzer, E., & Wagner, A. (2016). Primary vaccine failure to routine vaccines: Why and what to do? *Human Vaccines & Immunotherapeutics, 12*, 239–243.
Wright, V. (1976). A review of benorylate. A new antirheumatic drug. *Scandinavian Journal of Rheumatology, 4*(sup13), 5–8.

Chapter 3
Extrapolating from Model Organisms in Pharmacology

Veli-Pekka Parkkinen and Jon Williamson

Abstract In this chapter we explore the process of extrapolating causal claims from model organisms to humans in pharmacology. We describe and compare four strategies of extrapolation: enumerative induction, comparative process tracing, phylogenetic reasoning, and robustness reasoning. We argue that evidence of mechanisms plays a crucial role in several strategies for extrapolation and in the underlying logic of extrapolation: the more directly a strategy establishes mechanistic similarities between a model and humans, the more reliable the extrapolation. We present case studies from the research on atherosclerosis and the development of statins, that illustrate these strategies and the role of mechanistic evidence in extrapolation.

3.1 Introduction

How does extrapolation work in pharmacology? In pharmacology, the causal claims of primary interest are those which assert the efficacy (or lack of efficacy) of a drug and those that assert that a drug causes a particular harm (or, alternatively, that it is safe, i.e., that it causes no significant harm). Causal claims will normally be tested on model organisms, and the results of these tests are used as evidence for corresponding causal claims on humans. This chapter aims to shed some light on this process of extrapolating causal claims from model organisms to humans in pharmacology.

The ultimate goal of pharmacological inquiry is usually to *establish* the relevant causal claim, i.e., to secure evidence which both confirms the causal claim to a

V.-P. Parkkinen (✉)
Department of Philosophy, University of Bergen, Bergen, Norway
e-mail: veli-pekka.parkkinen@uib.no

J. Williamson
Centre for Reasoning, University of Kent, Canterbury, UK
e-mail: j.williamson@kent.ac.uk

A. LaCaze, B. Osimani (eds.), *Uncertainty in Pharmacology*, Boston Studies in the Philosophy and History of Science 338,
https://doi.org/10.1007/978-3-030-29179-2_3

sufficiently high degree and makes it sufficiently likely that further evidence will not significantly lower this degree of confirmation. Once established, a causal claim can be added to our stock of claims and treated as evidence for further claims. For instance, establishing a causal claim in a model organism allows it to be taken as evidence for the corresponding claim in humans. Establishing is very demanding, however, and lower standards of surety are sometimes also of use: a drug approval committee may only need a reasonable suspicion of harm in humans in order to reject an application for approval; if the cost of treatment is sufficiently low or the cost of failing to treat is sufficiently high, a drug may be approved for use in certain cases even where efficacy or safety in humans is not conclusively established. For example, during the 2014 Ebola outbreak, the World Health Organization recommended the use of certain yet unregistered treatments based on just the results of model organism studies, as the cost of failing to treat patients was certain to be catastrophic (World Health Organization 2014). Such cases are rather exceptional, however. Normally, for a new drug to be approved, it must be established to be efficacious in humans.

Exactly which claims are extrapolated from studies on model organisms depends on the stage of research and whether the causal claims in question concern efficacy or harm. At the outset of a first-in-human trial, claims of *efficacy* in humans are not yet considered established, by way of extrapolation or otherwise. At this stage the model organism results are treated as exploratory; the established efficacy in non-human animal models suggests a hypothesis about efficacy in humans. While the model organism results support this hypothesis about a corresponding effect in humans, they fall short of establishing the hypothesis. Nevertheless, significant support is required here, in order to justify the cost and risks of human trials, and extrapolation from animal studies grounds this inference.

Once trials on humans have been performed, these trials will provide the primary evidence in favour of efficacy in humans. However, extrapolation then comes into play in another way: mechanistic evidence, obtained by studies on model organisms, is used to explain and support an observed correlation between the drug and the clinical benefit. As we shall argue, the evidence obtained in model organisms remains useful at this stage because it is typically the case that the tests that can be carried out on humans are limited in key ways.

With regard to *safety and harm*, extrapolation plays an even more substantial role in the process. At the outset of a first-in-human trial, one needs to be sufficiently confident that the drug being tested has no serious negative side-effects in humans. Thus, claims about the toxicity or other side-effects of a compound need to be extrapolated from model organism studies to justify human trials.

The plan of the chapter is as follows. In Sect. 3.2 we discuss some of the challenges that face extrapolation in pharmacology. Section 3.3 provides some examples of extrapolation in pharmacology that inform the rest of the chapter. In Sect. 3.4 we present four strategies for extrapolation and in Sect. 3.5 we see—by appealing to a thesis concerning the role of evidence of mechanisms in establishing causal claims in the biomedical sciences—why these strategies work when they do work. In Sect. 3.6 we argue that this analysis supports a recent movement to evaluate evidence of mechanisms in a more rigorous way in the biomedical sciences, and we discuss a recent objection to mechanism-based extrapolation.

3.2 Model Organisms in Pharmacology

Model organisms as diverse as yeast, rats, and non-human primates are used extensively in pharmacological research, both for identifying potential targets for drugs, and for safety and efficacy testing. In the first case, the purpose of studying a model is to identify a component of a pathophysiological mechanism as a potential target for intervention. In the latter case the task is to test whether intervening on a mechanism in a particular way produces desired outcomes. In both cases, one must deal with the uncertainty inherent in transferring the results from the model organisms to humans.

Evaluating the evidence from model organism studies relies on judgements about relevant similarities between the model and humans. These considerations differ slightly depending on the intended purpose of the model study. In target identification, one attempts to establish the biochemical properties of a component entity that would allow pharmacological interventions to be targeted on it. The required similarity here involves parts of a mechanism—model studies suffice to establish an entry point into a mechanism as long as the component(s) of interest are similar between humans and the model organisms in both structure and function, even if other parts of the mechanism are dissimilar. Component similarities are commonly established by appeal to shared evolutionary ancestry that can establish homology, a strategy we describe and scrutinize below in Sect. 3.4. Note that relevant similarity here involves a particularly robust sense of functional homology, i.e., not just similarity of selected function, nor similarity of structure (Love 2007). For target identification or the study of molecular level effects of a drug, it is often preferable to work with a model in which the mechanism of interest is well understood, even if there is reason to believe that the mechanism as a whole is in some ways dissimilar in humans. Thus, models as distant to humans as yeast are used when the primary focus is discovering novel targets for pharmacological interventions or detecting fine grained effects.

In safety and efficacy testing the problem of extrapolation is slightly different and arguably more difficult. Here the model-based inferences are not about properties of individual components of a mechanism, but about the effects that interventions on the mechanism cause to the well-being of the organism as a whole. To infer an organism-level outcome in humans based on results obtained in a model organism, one would need to establish that the output of the whole mechanism under interventions is similar in humans despite possible differences in some of the mechanism's parts or its causal environment. This cannot usually be assumed, even if the model and human mechanisms are known to be structurally similar in parts and the mechanism is well understood in the model. Biological mechanisms are typically complex in the sense that many component functions are coupled—connections between components are abundant and exhibit feedback—such that even a detailed understanding of the component properties will not allow deriving specific predictions about what would happen to the output of the mechanism under interventions if some individual components were changed. Nor does establishing structural similarities guarantee similarity of function. In addition, unexpected interactions between the mechanism and its environment might modulate or mask the typical output of the mechanism in humans, and these effects cannot be predicted just by knowing that humans exhibit a similar mechanism to one that is well understood in the model. A tragic example of translational failure despite established similarity of the mechanism of action is the TGN1412 trial, in which an administration of an immunomodulatory drug whose target receptor was well characterized resulted in catastrophic conditions in humans, despite being proven safe in trials on monkeys that share the same receptor with minimal structural differences (Kenter and Cohen 2006).

TGN1412 is a humanised antibody that is a strong agonist of the CD28 receptor on human T cells, a type of white blood cell that is a part of cell-mediated immune system. TGN1412 is capable of activating T cells irrespective of the presence of other regulatory signals typically required for T cell activation. This capacity promised great therapeutic potential, as T cell regulation can play a role in the treatment of many autoimmune diseases and cancer. Before the human trial, TGN1412 was tested on cynomolgus and rhesus monkeys. These were considered valid models of humans due to perfect sequence homology of the extracellular domain of CD28—the part of the receptor that lies outside the cell and binds signalling molecules (Attarwala 2010). Based on these studies, a no-observed adverse effect level (NOAEL) was determined. In the first-in-human trial, six human volunteers were administered a dose that was one five hundredth of the estimated NOAEL. The result of the trial was a tragedy: one of the participants died, and the rest suffered possibly irreversible adverse effects due to a rare immune system response called cytokine storm. Cytokines are signalling factors that normally play an adaptive role in the immune system, activating immune cells to attack invasive pathogens and to produce more cytokines. In a cytokine storm, this feedback loop runs out of control, causing local activation of too many immune cells and subsequent damage to any affected organs. Nothing like such a reaction had been seen in the experiments on monkeys. It has been suggested that the drastic difference between the monkey and human outcomes is explained by the fact that the monkey equivalent of the cell type that drove the cytokine storm in humans lacks CD28 receptors, and was thus not activated by TGN1412 (Eastwood et al. 2010). To summarize, even though humans share TGN1412's mechanism of action as it operates in monkeys, humans have in addition other mechanisms sensitive to TGN1412, which were responsible the effects not seen in monkeys.

Extrapolating causal effects of pharmacological interventions from model organisms to humans is thus risky even when one has knowledge of some relevant mechanistic similarities. What one can do to alleviate this risk is to search for outcomes that are robust against changes in background conditions or parts of the supporting mechanism. This involves testing an intervention in a range of models that differ from each other as well as from humans, and searching for convergent results across the various model studies. If it can be shown that an outcome is independent of physiological features idiosyncratic to any particular model, the strict assumption about similarity between models and humans can be relaxed. This strategy is described and analysed in more detail in Sect. 3.4.

3.3 Case Studies

This section describes examples of model organism research in the discovery and development of statins for treating and preventing heart disease. Statins reduce the level of cholesterol in blood, thus reducing the risk of atherosclerosis and subsequent heart disease. Statins produce their effect by inhibiting the activity of HMG-CoA reductase (HMGCR), a rate-limiting enzyme in cholesterol biosynthesis. HMG-CoA reductase catalyses a reaction in which HMG-CoA is converted into mevalonic acid, a precursor of cholesterol. Statins mimic the structure of HMGCR and compete with it in binding HMG-CoA, but do not have the same enzymatic function. This reduces the rate at which mevalonic acid is produced and thus controls the rate of cholesterol synthesis in the liver. Statins are now widely used as preventive treatment for heart disease, and their efficacy has been demonstrated in large randomized clinical trials.

The focus on cholesterol in the treatment of heart disease is based on decades of experimental and epidemiological research on the connection between elevated blood cholesterol and cardiovascular events (Steinberg 2007). The earliest evidence suggesting that this connection is causal came from experiments on the effects of cholesterol feeding in rabbits, which indicated that high blood cholesterol level is linked to atherosclerosis—the thickening and hardening of the artery wall in a manner that occludes blood flow, causing the cardiac events that characterise coronary heart disease (Anitschkow 1913). These results were subsequently replicated in other species with certain exceptions: some model species such as rats and dogs failed to show similar susceptibility to cholesterol induced atherosclerosis (Bruger and Oppenheim 1951). At the time of these early experiments, the researchers lacked detailed knowledge of the pathophysiology of atherosclerosis, and therefore could not validate animal models by directly comparing the relevant mechanisms to humans. But once the number of successful experiments in many different model species grew, one could argue that cholesterol's atherogenic potential had been shown to be independent of the specific physiology of any particular model to such a degree that it warrants inferring the existence of a mechanism extrapolatable to humans (Parkkinen 2016). Such reasoning appeals to the robustness of evidence, a strategy we describe in the next section.

It is only after the experimental animal results had established a manipulable link between cholesterol and atherosclerosis that large-scale epidemiological studies on the link between cholesterol and heart disease were conducted (Steinberg 2007, pp. 33–39). Given experimental evidence from model organism studies, and epidemiological evidence of correlation between blood cholesterol and heart disease in humans, the 'cholesterol conception of atherosclerosis' was established as an explanation of the prevalence of atherosclerotic heart disease in populations characterized by high lipid consumption and high average blood cholesterol levels. This sparked an interest in the details of cholesterol's role in the pathophysiology of atherosclerosis and a search for effective cholesterol lowering drugs.

The first statin, known today as compactin, was isolated by Akira Endo and his collaborators at Sankyo Research Laboratories in 1972 (Endo 2010, p. 487). Compactin was shown to be a highly efficient inhibitor of HMGCR in mammalian cell cultures (Endo 2010, pp. 487–488). The first studies testing the efficacy of compactin for cholesterol lowering were carried out in rats. These studies, somewhat surprisingly, showed virtually no effect on blood cholesterol levels (Endo 2010, p. 488). The failure of the first animal tests led Sankyo to effectively drop compactin from the drug development pipeline, but Endo was allowed to carry on studying its mechanism of action. This research led to a hypothesis that could explain the results seen in rats, while suggesting that compactin would be efficacious in many other species including humans. The working hypothesis had been that inhibiting cholesterol metabolism in the liver would lead to a reduction in serum cholesterol due to an increase in the extraction of cholesterol from plasma lipoproteins to support normal cellular functions. What Endo and his collaborators discovered is that this is not what happens in rats: the rat liver is incapable of catabolizing lipoproteins. Instead, the competition between compactin and HMGCR was compensated by upregulating HMGCR in the liver, directly cancelling the effect of compactin (Endo et al. 1979). This suggested that compactin would be efficacious in species that do not exhibit similar regulation of HMGCR. Importantly, a sub-population of humans suffering from hypercholesterolemia completely lack this regulatory mechanism due to a genetic dysfunction. Subsequent experiments showed that this explanation is likely to be true: compactin was shown to have a significant cholesterol lowering effect in other animal models such as dogs and monkeys. The important point to note for our purposes is that the rat model that was initially used was validated by neither direct comparison of the relevant mechanisms to humans, nor auxiliary evidence to justify an assumption about similarity of mechanisms. Once Endo and collaborators came up with a testable mechanistic explanation for the results, it became apparent that the rat is not a valid model of humans for testing the efficacy of statin treatment, due to differences in feedback regulation of cholesterol biosynthesis.

The examples discussed above consider a use of model organisms where some outcome variable of clinical interest, e.g., blood cholesterol level, is measurable both in the model and in humans, so that the model results can be taken to represent the behaviour of a corresponding variable in humans. But model organisms need not express an outcome variable similar to humans to be useful, if they nonetheless host similar mechanisms. As an example, consider a study by Maciejak et al. (2013), investigating the molecular and cellular level effects of statins using the yeast Saccharomyces cerevisiae as a model. S. cerevisiae is one of the most extensively studied eukaroytes due to its short generation time, low-maintenance culturability, and well-understood genetic architecture that amends itself to manipulation. What makes it a feasible model for pharmacological research is the fact that despite remarkable evolutionary distance, many biochemical pathways are at least partly conserved between yeast and humans. This is the case with the sterol pathway responsible for cholesterol synthesis in humans: S. cerevisiae hosts two homologs of the human HMGCR coding gene, and the pathway is biochemically similar to the human version with the exception that the end product synthesized in yeast is ergosterol instead of cholesterol. Maciejak et al. studied the effects of four different statins on the behaviour of the sterol pathway and cell growth in three yeast models: two models that carried either one of the native yeast HMGCR coding genes, and one engineered to express the human version of the gene (Maciejak et al. 2013). This allowed them to draw a number of conclusions about the potency of each statin, and the number and severity of side effects on other cellular processes. Understanding these cellular level effects is important from a clinical point of view, as many other cellular functions are dependent on the sterol pathway. The most general conclusion was that statin treatment will trigger compensatory upregulation of many sterol and non-sterol pathway genes, a result that could be validated by comparison to human cell cultures. These results are informative about the clinical efficacy of statins even though the model system does not express anything like the clinical variable of interest, as they provide clues to how the underlying mechanism can be manipulated without producing unwanted side-effects. In this case, extrapolating the model results to humans rests on knowledge of evolutionary conservation of the relevant mechanisms. Note in addition that, as explained in Sect. 3.1, precautionary reasoning suggests that the threshold of evidence required for extrapolation should be considered lower in this case, as the inferences drawn based on the model study consider possible harmful side-effects rather than efficacy.

3.4 Strategies for Extrapolation

In this section we present four general strategies for extrapolating from animals to humans: enumerative induction, comparative process tracing, phylogenetic reasoning, and robustness analysis.

Enumerative Induction. The most straightforward way to extrapolate from animal models to humans is to collect evidence from many models and generalize the results to humans by simple enumerative induction. This strategy makes no appeal to evidence of details of mechanisms. It is simply the similarity of the observed phenomenon in many non-human instances that supports the inference to the human case. This strategy is risky: the fact that many non-human animals or in vitro models share some feature or respond similarly to an intervention is no guarantee that humans will respond the same way. To justify extrapolation, it is typically better to have some further evidence of similarity of the relevant mechanisms. In an ideal case, one knows the details of a relevant mechanism both in the model and in humans, and can establish the similarity of the mechanisms by direct comparison.

Comparative Process Tracing. In reality, one is rarely in a position where a model is verifiably identical to humans with respect to the relevant mechanisms. Instead, biomedical scientists typically argue for the validity of their models based on partial mechanistic similarity between the models and humans. For partial mechanistic similarities to support extrapolation, one needs a method for investigating how differences in parts of a mechanism might result in differences in the mechanism's output. Daniel Steel calls such a method 'comparative process tracing', and argues that in many cases this method will be able to justify extrapolation (Steel 2008). Comparative process tracing starts with a search for component activities that act as bottlenecks for causal influence within the mechanism, such that, once the behaviour of the bottleneck is held fixed, the behaviour of any components causally downstream from it is not influenced by components causally upstream to it. If one is able to identify such critical components, one only needs to establish that the components between the critical bottleneck and the endpoint of interest are similar in the model and humans; establishing full mechanistic similarity is not necessary for extrapolation to be reliable (Steel 2008, p. 89).

But direct evidence of even partial similarity of mechanisms is not always necessary; there are other strategies to establish similarity of mechanisms. Below we describe two such strategies: phylogenetic reasoning and robustness analysis. Despite differences in how these strategies are implemented in research, there are similarities in the rationale that underlies their use. Both strategies work by making an explanation of the behaviour of the model in terms of mechanisms idiosyncratic to the model less likely than an explanation that posits similarities between the model and humans in the relevant mechanisms.

Phylogenetic Reasoning. Choice of a model organism is often guided by evolutionary considerations: extrapolation from a model organism to humans is considered more secure if the relevant mechanism is conserved between the model and humans, than if the mechanisms have evolved independently. Directly establishing that a mechanism is conserved, however, would amount to directly observing that the mechanisms are similar—no specifically evolutionary reasoning would be needed to justify the extrapolation. But phylogenetic information can and is being employed in model organism research in a different way, to establish assumptions about similarity of mechanisms in the absence of detailed knowledge of the mechanism in the target of extrapolation. This involves a two-step inference, where each step relies on supplemental empirical evidence other than direct comparisons between the model and the target (Levy and Currie 2015). In the first step, one observes the trait of interest in the model and other closely related species to establish that the trait is widely shared among members of a clade. This evidence is used to project the trait to an ancestor species based on the assumption that a trait shared by many members of a clade has likely evolved from a similar trait in their common ancestor. In the second step, the trait is projected to the target species based on empirical evidence that the target species, too, shares a common ancestor with the model.

The inference from the model to the target relies on knowledge of the nature of the evolutionary process. The fact—established by supplementing phylogenetic evidence—that the trait of interest has evolved from the same ancestral trait constrains the amount of variation between the model and the target, as any evolutionary novelty in the trait must be achievable by piecemeal modification of the ancestral form and must result in a viable phenotype. Given this fact about evolution, the reasoning from the model to the target can be reconstructed as an argument for the truth of a common descent explanation that, if true, would imply trait similarities between them.

Common descent explains the similarity of traits between species by constraining the possible variation that the species can exhibit in the trait. The explanation has a contrastive structure: the explanans (common descent) favours the explanandum (trait similarity) at the expense of an exclusive alternative that would be the case if the explanans were false, i.e., had the trait of interest evolved independently in the two species, it would exhibit more (or be more likely to exhibit more) variation between the species than is actually the case. Thus, if a common descent explanation applies to the trait of interest in the model and target species, they are likely to be similar. A phylogenetic argument aims to show that the common descent explanation is more likely to be true of the trait of interest than an explanation that posits independent evolution of the trait in the two species. The first step is an overtly abductive inference: common descent would be the best explanation of trait similarities observed between the model and other species in the clade. In the second step this explanation is applied to the target species based on supplemental phylogenetic evidence that the target species shares a common ancestor with the model.

Some elaboration is in order concerning what can be learned about causality by means of this strategy. Phylogenetic arguments work best for establishing similarities in components of a mechanism. However, more than mechanistic similarity between a model organism and humans is needed to establish causal claims about the mechanism's output in humans, even if the mechanism's behaviour under experimental interventions is well characterized in the model. This is because of the problem of masking. The problem of masking refers to a situation in which there are two or more separate mechanisms linking a cause variable to some effect variable of interest, such that the influence of these mechanisms on the effect differs. The mechanisms might for instance partly or completely cancel out, or modulate each other's output in a non-additive manner. In such a case, establishing just one of the mechanisms, or even establishing the behaviour of all of them independent of each other, will not support reliable inferences about the behaviour of the effect under interventions on the cause. Thus, establishing similarity of a mechanism between an animal model and humans will not support the extrapolation of an organism-level causal effect from the model to humans, unless one can in addition rule out that humans exhibit masking mechanisms that are not present in the animal model. The TGN1412 catastrophe mentioned earlier is an example of how difficult this can be. Even though humans share the mechanism responsible for the results that were seen in the models, this alone was not enough to predict the outcome of the experiment in humans, as humans exhibit in addition other mechanisms sensitive to TGN1412, and the operation of these mechanisms causes the human immune system to react drastically differently. For establishing similarities in the output of a mechanism under interventions on its components, one would need to establish that the operation of the mechanism is insensitive to differences not only in some of its parts, but also the causal context in which it is embedded, which may include other mechanisms relevant to the outcome of interest. Phylogenetic reasoning can rarely establish this.

Robustness Analysis. Extrapolating organism-level outcomes of pharmacological interventions, such as the results of toxicological tests, rests on the claim that the mechanism responsible for the effect is not too dissimilar between the model and humans, and that there are no interfering mechanisms in humans that could mask or modulate the effect. These assumptions can sometimes be partly justified by phylogenetic arguments or by deliberately engineering the model to make it similar to humans, but typically there is considerable uncertainty about the representativeness of model results with respect to humans. Moreover, it is often unclear how much detail about the mechanism and its environment one would need to know in order to know that the extrapolation is reliable. What one can do is to try to weaken or discharge the underlying assumptions by demonstrating that similar results can be produced in many different model organisms that vary in relevant parts of their physiology, thus demonstrating that the outcome is independent of the idiosyncratic features of any particular model. When each model organism is treated as a source of evidence—fallible due to causal dissimilarity that introduces error in the extrapolation to humans—the reasoning can be seen as a kind of robustness analysis.

Wimsatt (2007) describes robustness analysis as consisting of four procedures:

> 1. To analyse a variety of independent derivation, identification, or measurement processes. 2. To look for and analyse things that are invariant over or identical in the conclusions or results of these processes. 3. To determine the scope of the processes across which they are invariant and the conditions on which their invariance depends. 4. To analyse and explain any relevant failures of invariance. I call things that are invariant under this analysis robust (Wimsatt 2007, p. 44).

Wimsatt's procedures are meant to capture a highly general reasoning strategy that is used to study the reliability of evidence and scientific inference; the processes to which robustness analysis is applied can be anything from sensory modalities to derivations of results from mathematical models (Wimsatt 2007, pp. 45–46). Here we consider *empirical robustness analysis*, where the processes studied are experimental procedures and the results obtained through them. Empirical robustness analysis can be reconstructed as explanatory reasoning that works by ruling out explanations of experimental results in terms of idiosyncrasies of the causal set-up of any particular experiment (cf. Schupbach 2016). Every experiment or detection method has its own sources of error that make the evidence obtained from it fallible: instead of reliably tracking a phenomenon existing independently in nature, our experimental results might be artefacts created by the methods themselves. However, it would be unlikely that many methods based on different causal principles would produce similar artefacts. As an illustration, consider two experimental set-ups, one that exhibits error source E, and one that exhibits error source E* that is independent of E. Often, if both of these experiments produce a concordant result, one can rule out that the result is an effect of error source E, on the grounds that it is unlikely that both experiments would be independently erroneous but nonetheless yield the same result. By collecting evidence from many different experiments that exhibit independent sources of error, one can then infer that the best explanation for any concordant result is that the result is caused by a real underlying mechanism, rather than being a side product of the causal set-up of the experimental procedures themselves.

Similar reasoning can be applied to the evaluation of model organism studies as evidence. Extrapolation from a particular model organism rests on the assumption that the model and humans are similar with respect to relevant mechanisms. But all model organisms are dissimilar to humans in some ways, and these dissimilarities constitute a potential source of error when the model result is taken as evidence for inferences about humans. Therefore, one should initially have low credence in conclusions one draws based on model studies, unless one has direct evidence of relevant mechanisms to rule out the possibility of error. However, even if all model organisms are causally dissimilar to humans, the same evolutionary processes responsible for their dissimilarity to humans make different model organisms dissimilar to each other as well. Model organisms that are not close phylogenetic relatives are thus likely to exhibit independent errors with respect to extrapolation of the results to humans. This fact makes it possible to apply robustness reasoning to the evaluation of model organism evidence: a result that can be reproduced across heterogeneous pool of model species is independent of the idiosyncratic causal make-up of any particular model.

Under favourable conditions, robustness analysis can complement more directly attainable mechanistic evidence in extrapolation of causal claims from animal models to humans. Transferring a causal claim based on mechanistic evidence alone is susceptible to masking; not only differences within the underlying mechanism, but also differences in its causal context can be error sources that may defeat extrapolation. But searching for robust results in a pool of models where the mechanism itself as well as its physiological context varies sufficiently can help with the masking problem. Here the important aspect of robustness analysis is Wimsatt's procedure 4., the analysis and explanation of discordant results. Any failures of robustness are suggestive of either species differences in the hypothesized mechanism of action, or the presence of masking mechanisms in some species. Analysing and explaining failures of robustness by searching for masking mechanisms in the models can help evaluate the reliability of extrapolation by guiding the search of relevant masking factors in humans. It should be noted that this requires some explicit comparisons of mechanistic detail between the models and humans: simply seeing robust results in heterogeneous models, without knowing anything about the relevant mechanisms, does not rule out the possibility of masking in humans.

3.5 The Logic of Extrapolation

In this section we shall show how one recent line of work on the epistemology of causality can shed some light on the logic of the sort of extrapolations exemplified by the pharmacological case studies of Sect. 3.3.

Russo and Williamson (2007) argued that, in order to establish a causal claim, one normally needs to establish both that the putative cause and effect are appropriately correlated and that there is some underlying mechanism by which one can explain instances of the putative effect in terms of the putative cause and which can account for the observed correlation. It is not sufficient just to establish the existence of an appropriate correlation, because some correlations are not causal—they may be attributable to confounding, bias, chance or some non-causal connection between the variables of interest. Establishing the existence of an appropriate mechanism rules out these other explanations of the correlation. On the other hand, it is not sufficient just to establish the existence of an appropriate mechanism, because there may be counteracting mechanisms which cancel out or reverse the influence from the putative cause to the putative effect—this is the problem of masking. Establishing an appropriate correlation rules out such cases.

Note that the thesis is that one *normally* needs to establish an appropriate correlation and an appropriate mechanism. Only 'normally' because there are some awkward cases. For example, there are cases of causation without a linking mechanism. If the putative cause and/or the effect is an absence—something *not* happening, or a quantity that is absent—then there can be no physical mechanism linking the two. The speaker failing to catch a flight to a conference causes the absence of her talk, even though there is no physical connection between the two. In such cases we attribute causation when we can expect a mechanism linking counterfactual presences: if she were to have caught the flight, that flight and connecting transport would have provided the mechanism that gets her to the auditorium to give the talk. Moreover, there are cases of causation without a correlation: if the ball can descend to the bottom of the pinball machine whichever route it takes, then taking one particular route causes it to get to the bottom even though it does not increase the chances of it getting to the bottom. In such cases we attribute causation when we can expect a correlation were we to counterfactually block one or more of the alternative pathways. There are even cases of causation without either a correlation or a mechanism, as Longworth (2006, Sect. 4.1) points out. Such cases show that the epistemological thesis stated above is a first approximation to a more nuanced thesis. Fortunately, these nuances will not be important in what follows and we may stick with the above formulation.

The epistemological thesis has generated some controversy (see, e.g., Weber 2007, 2009; Campaner 2011; Clarke 2011; Darby and Williamson 2011; Gillies 2011; Illari 2011; Howick 2011a,b; Russo and Williamson 2011a,b; Campaner and Galavotti 2012; Claveau 2012; Dragulinescu 2012; Clarke et al. 2013, 2014; Fiorentino and Dammann 2015). However, we shall neither offer a detailed justi-

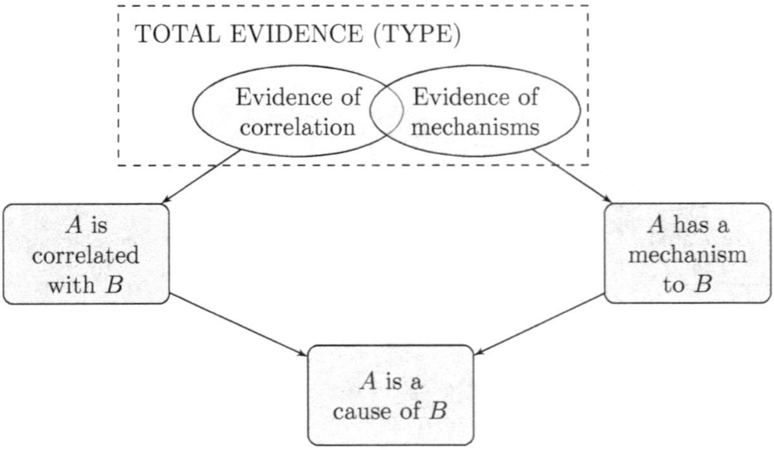

Fig. 3.1 The epistemology of causality motivated by Russo and Williamson (2007)

fication nor a defence of the thesis here. Instead we shall assume that the thesis is correct and show that it can shed any light on the process of extrapolation.

The epistemological thesis leads to the picture of Fig. 3.1: some of the total available evidence is evidence for or against the existence of a correlation; some of it is evidence for or against the existence of a suitable mechanism; these both provide evidence for or against causation. Note that some items of evidence can be both relevant to correlation and mechanism. Arguably, in the right conditions a large, well-conducted randomised controlled trial (RCT) can provide evidence both of correlation and that this correlation is not spurious, i.e., that there must be some suitable underlying mechanism that accounts for the correlation. It is very rare, however, for a single study on its own to *establish* both correlation and mechanism. More often, while a large enough correlation in the sample might be enough to establish a corresponding correlation in the population, significant doubts about mechanism will remain and other evidence of mechanisms needs to be obtained or invoked in order to establish mechanism and thereby establish causality. Figure 3.2 provides an alternative view of the epistemology of causality, splitting the relevant evidence into statistical trials—such as RCTs and observational studies—and other relevant evidence—such as evidence obtained from biomedical imaging, in vitro experimentation, and simulation.

The epistemological thesis also motivates a particular view as to the logic of extrapolation. This is depicted in Fig. 3.3. In this diagram, dashed arrows represent weak evidential relationships and full arrows represent strong evidential relationships. Normally, one can conduct more conclusive trials on a model organism population than on a human population—this is why model organisms are so important. For example, one can often perform trials on a model organism that, for ethical reasons, cannot be undertaken in humans. These trials can be better randomised in model organisms than in humans, and can involve more invasive

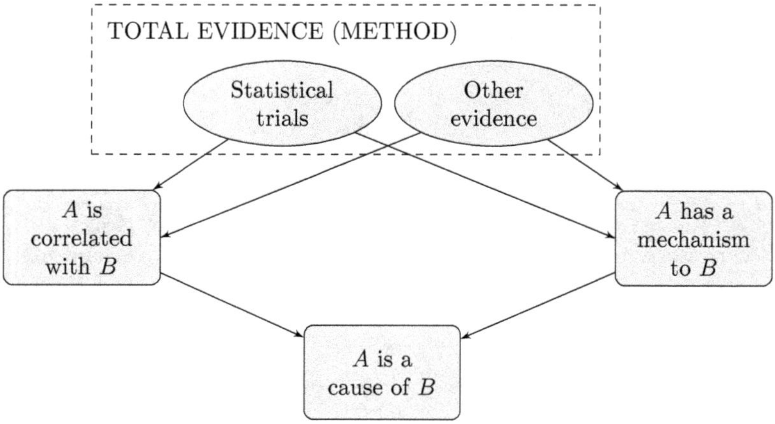

Fig. 3.2 An alternative view of the epistemology of causality

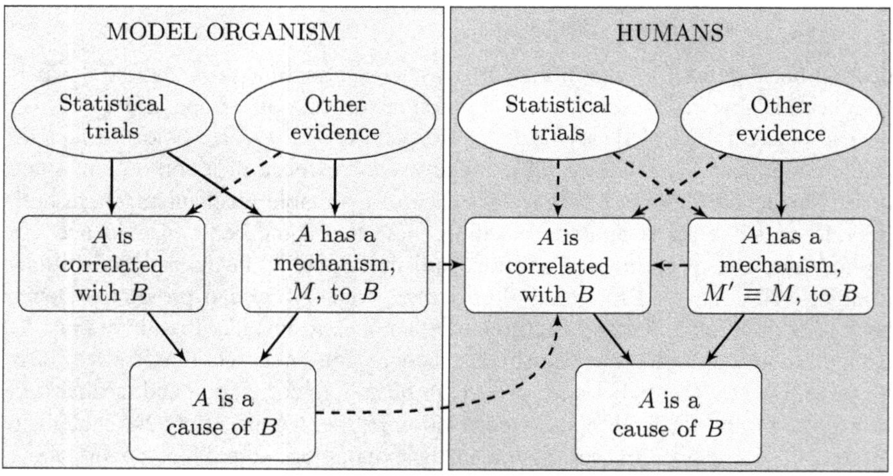

Fig. 3.3 The logic of extrapolation as motivated by the epistemological thesis

measurements, creating more conclusive evidence about causality in the model organism than what corresponding trials in humans could establish about humans. The obvious downside is that the evidential relevance of the model organism results to claims about humans may be uncertain. Thus, Fig. 3.3 depicts strong evidential relationships from statistical trials in animals, but weak connections in humans. Given these weak connections, it can be hard to establish a suitable correlation in humans, and thereby hard to establish causality in humans. Although studies in humans may suggest a correlation, this observed correlation may be spurious—e.g., due to confounding, bias, or chance. However, if one can establish causality in the model organism and one can establish that the mechanisms which underpin this causal relationship are sufficiently similar to those in humans, then this lends further credence to the claim that the correlation observed in humans is not spurious, i.e., that there is a genuine correlation in the underlying population. Although each consideration on its own provides rather weak evidence of correlation in humans, the combination of all these factors can, in the right circumstances, establish correlation, and thereby help to establish causality in humans.

Let us revisit the statin case studies of Sect. 3.3. Recall that compactin was first tested on rats. Here, the trial failed to establish a correlation. Thus, the model organism clearly could not be used as the basis for an inference to the effectiveness of compactin in humans, even in the weak sense of motivating a hypothesis that compactin would be effective in humans. Moreover, other evidence was inadequate to establish that the underlying mechanisms in rats and humans were sufficiently similar. Hence, the model organism could also not be used as the basis for an inference to the lack of effectiveness of compactin in humans, although it did provide some evidence against a correlation in humans. Further research on the mechanisms involved led to trials on dogs and monkeys. These trials did establish a correlation in the model organisms, and, at this stage, there was other evidence of both an appropriate mechanism of action in the model organisms and of a similarity in mechanisms between the model organisms and humans. Hence the model organisms could now be used as a basis for extrapolating effectiveness to humans, at least in the weak sense of raising one's confidence in a hypothesis about effectiveness in humans to a degree that warrants conducting a trial in humans.

Let us turn to the case study involving the yeast Saccharomyces cerevisiae. Yeast is not an animal and many of its mechanisms differ from those in humans; nevertheless, extrapolation from yeast to humans follows the pattern depicted in Fig. 3.3. In this case the causal claims in question are rather lower-level, involving the effect of statins on the upregulation of particular genes. There is sufficient similarity between yeast and humans to extrapolate certain of these low level causal claims from the yeast to humans, however, it should be noted that no claim about efficacy with respect to clinical endpoints of interest in humans is extrapolated here. Furthermore, the low-level claims themselves provide evidence for higher level claims of the efficacy of statins for lowering blood cholesterol in humans: they help to establish the existence of appropriate mechanisms of action of statins for lowering cholesterol, and they help to show that these mechanisms of action are shared between humans and model organisms such as monkeys. The reasoning behind this last inference to similarity of mechanism of action between humans and model organisms is typically phylogenetic (Sect. 3.4). Certain mechanisms are shared between yeast, monkeys, dogs and humans, despite evolutionary distance, because of common ancestry. Although experiments in rats undermine robustness of efficacy of statins across species, these results can be explained away, again by invoking evidence of mechanisms.

3.6 Conclusion

According to the above analysis, problems of extrapolation from model organisms to humans hinge upon evidence of mechanisms. Extrapolation works by establishing causation in the model organism and establishing similarity of the model organism to humans. Establishing causation in the model organism requires establishing the existence, if not the nature, of some suitable mechanism of action. The similarity that needs to be established is similarity of the mechanisms of action in the model organism to those in humans.

The strategies for extrapolation that lend most surety to the causal conclusion in humans are those that most directly establish similarity of mechanisms: comparative process tracing, phylogenetic reasoning, and robustness analysis. The first strategy employs evidence of partial similarities in mechanisms, and a search for crucial causal bottlenecks to limit the number of comparisons between a model and humans required to establish the validity of the model. The next two strategies involve an abductive inference as a key ingredient; both strategies work by ruling out explanations of model results in terms of mechanisms idiosyncratic to the model, and in favour of explanations in terms of similar mechanisms. Enumerative induction proceeds rather differently from the above three strategies: reasoning from the causal claim having been found to hold in previously observed species to the claim holding in humans. As a subsequent inference, one might also infer similarity of mechanisms as the best explanation of the causal claim holding across all the species under considerations. Similarity of mechanisms is thus inferred only indirectly, by chaining a simple induction and an inference to the best explanation. This form of inference is error-prone, if not entirely tenuous.

Given the importance of evidence of mechanisms for successful extrapolation, it becomes equally important to ascertain the quality of such evidence. Grading quality of mechanistic evidence is important for several reasons: in order to determine how credible are the mechanistic claims that the evidence supports (in particular, whether they can be considered *established* by the evidence); in order to avoid erroneous and fallacious mechanistic inferences, where possible; and in order to decide when more evidence is needed (when to commission further research). These considerations motivate the EBM+ approach to evaluating evidence in medicine (ebmplus.org), which seeks to evaluate the full range of mechanistic evidence alongside evidence of correlation, instead of focusing exclusively or almost exclusively on statistical studies, as is common in current EBM practice (Clarke et al. 2014; Parkkinen et al. 2018).

It is important to note that the evidential requirements of extrapolation vary depending on the level and specificity of the causal claim in question. Low level causal claims about interactions between individual components do not require establishing detailed similarities of the whole embedding mechanism. For example, if one is merely interested in the binding affinity of a compound to its intended molecular target, one clearly does not need a model that resembles humans with respect to the whole complex mechanism of which the target is a component of in humans. Thus, even model organisms that are phenotypically very dissimilar to humans may serve as fairly reliable sources of extrapolation, such as in the case of using yeast as a model for studying molecular level side-effects of statins. By contrast, extrapolating claims about phenotypically high level effects requires somewhat better understanding of the behaviour of the underlying mechanism as a whole, and more evidence of similarity of the mechanism as well as its causal context between the model and humans, as extrapolating such effects is susceptible to the problem of masking. Similarly, qualitative claims tolerate more differences between the model mechanisms and the human mechanisms than claims about specific quantitative effects of interventions. For example, inferences about causal relations where the exact timing of the effect in response to the intervention is of clinical interest require higher quality evidence of mechanistic similarities than extrapolation of claims about the mere qualitative effects of the intervention. In many cases, evidence of similarity of mechanisms can be rather indirect, involving functional similarity or confirmed theory, rather than detailed knowledge of bottlenecks and other components of mechanisms, as required by comparative process tracing (Guala 2010). Thus, the strategies of Sect. 3.4 should be thought of as examples of ways of generating evidence of similarity of mechanisms, but not exhaustive.

One prominent criticism of mechanism-based extrapolation overshoots its conclusions because it neglects the nuances mentioned above. Howick et al. (2013a,b) have argued that mechanistic evidence typically fails to support extrapolation in the intended way, and this is due to the inherent complexity of biological mechanisms, unexpected interactions between many mechanisms, and the uncertainty concerning how well the mechanism must be understood for one to know whether extrapolation is justified or not. They argue that predicting clinically relevant outcomes from mechanistic knowledge is often fallible, and then argue that this inherent uncertainty compromises most attempts to extrapolate from one context to another based on mechanistic knowledge (Howick et al. 2013b, pp. 281–285). We agree that one often cannot reliably predict outcomes of interventions from knowledge of mechanisms alone—evidence of correlation is also required (Sect. 3.5). The pessimistic conclusion of Howick et al. concerning extrapolation in general does not follow from this. In the pharmacological cases we have discussed, one has evidence of some causal effect of an intervention in a model already; as explained in Sect. 3.5, this is not inferred solely from knowledge of a mechanism, and indeed the details of the mechanism need not be known at all. The extrapolation task considers the use of mechanistic evidence to evaluate whether similar effects would be seen in humans. Howick et al.'s worries about complexity and contextual effects do apply to such inferences, but not with similar force in every case. As we have explained, it is the extrapolation of complex, whole organism level outcomes that is most susceptible to error due to contextual masking factors and insufficient mechanistic evidence, but even these problems can be mitigated by testing the outcome for robustness in a heterogeneous pool of models. When it comes to extrapolation of lower level effects, the evidential requirements of extrapolation are less demanding and more clearly delineated.

Acknowledgements This research was carried out as a part of the projects *Grading evidence of mechanisms in physics and biology*, supported by the Leverhulme Trust, and *Evaluating evidence in medicine*, supported by the UK Arts and Humanities Research Council. We are grateful to the referees and editors for helpful comments.

References

Anitschkow, N. (1913). Über die veränderungen der kanichnenaorta bei experimenteller Cholesterinsteatose. *Beitrage zur pathologischen anatomie und zur allgemeinen pathologie, 56*, 379–404.

Attarwala, H. (2010). TGN1412: From discovery to disaster. *Journal of Young Pharmacists, 2*(3), 332–336. ISSN: 0975–1483. https://doi.org/10.4103/0975-1483.66810. http://www.sciencedirect.com/science/article/pii/S0975148310230248

Bruger, M., & Oppenheim, E. (1951). Experimental and human atherosclerosis: Possible relationship and present status. *Bulletin of the New York Academy of Medicine, 27*(9), 536.

Campaner, R. (2011). Understanding mechanisms in the health sciences. *Theoretical Medicine and Bioethics, 32*, 5–17.

Campaner, R., & Galavotti, M. C. (2012). Evidence and the assessment of causal relations in the health sciences. *International Studies in the Philosophy of Science, 26*(1), 27–45.

Clarke, B. (2011). Causality in medicine with particular reference to the viral causation of cancers. Ph.D. diss., Department of Science and Technology Studies, University College London.

Clarke, B., Gillies, D., Illari, P., Russo, F., & Williamson, J. (2013). The evidence that evidence-based medicine omits. *Preventative Medicine, 57*(6), 745–747. https://doi.org/10.1016/j.ypmed.2012.10.020

Clarke, B., Gillies, D., Illari, P., Russo, F., & Williamson, J. (2014). Mechanisms and the evidence hierarchy. *Topoi, 33*(2), 339–360. https://doi.org/10.1007/s11245-013-9220-9. http://link.springer.com/article/10.1007%2Fs11245-013-9220-9

Claveau, F. (2012). The Russo-Williamson theses in the social sciences: Causal inference drawing on two types of evidence. *Studies in History and Philosophy of Biological and Biomedical Sciences, 43*(4), 806–813. https://doi.org/0.1016/j.shpsc.2012.05.004. http://www.sciencedirect.com/science/article/pii/S136984861200060X.

Darby, G., & Williamson, J. (2011). Imaging technology and the philosophy of causality. *Philosophy & Technology, 24*(2), 115–136. https://doi.org/10.1007/s13347-010-0010-7.

Dragulinescu, S. (2012). On 'Stabilising' medical mechanisms, truth-makers and epistemic causality: A critique to Williamson and Russo's approach. *Synthese, 187*(2), 785–800. https://doi.org/10.1007/s11229-011-0011-9. http://link.springer.com/article/10.1007%2Fs11229-011-0011-9

Eastwood, D., Findlay, L., Poole, S., Bird, C., Wadhwa, M., Moore, M., Burns, C., Thorpe, R., & Stebbings, R. (2010). Monoclonal antibody TGN1412 trial failure explained by species differences in CD28 expression on CD4+ effector memory T-cells. *British Journal of Pharmacology, 161*(3), 512–526. ISSN: 1476–5381. https://doi.org/10.1111/j.1476-5381.2010.00922.x.

Endo, A., Yoshio, T., Masao, K., & Kazuhiko, T. (1979). Effects of ML-236B on cholesterol metabolism in mice and rats: Lack of hypocholesterolemic activity in normal animals. *Biochimica et Biophysica Acta (BBA)—Lipids and Lipid Metabolism, 575*(2), 266–276. ISSN: 0005-2760. https://doi.org/10.1016/0005-2760(79)90028-6. http://www.sciencedirect.com/science/article/pii/0005276079900286

Endo, A. (2010). A historical perspective on the discovery of statins. *Proceedings of the Japan Academy, Series, B, 86*, 484–493.

Fiorentino, A. R., & Dammann, O. (2015). Evidence, illness, and causation: An epidemiological perspective on the Russo-Williamson thesis. *Studies in History and Philosophy of Biological and Biomedical Sciences, 54*, 1–9.

Gillies, D. A. (2011). The Russo-Williamson thesis and the question of whether smoking causes heart disease. In P. M. Illari, F. Russo, & J. Williamson (Eds.), *Causality in the sciences* (pp. 110–125). Oxford: Oxford University Press.

Guala, F. (2010). Extrapolation, analogy, and comparative process tracing. *Philosophy of Science, 77*, 1070–1082.

Howick, J., Glasziou, P., & Aronson, J. K. (2013). Can understanding mechanisms solve the problem of extrapolating from study to target populations (the problem of 'external validity')? *Journal of the Royal Society of Medicine, 106*(3), 81–86. https://doi.org/10.1177/0141076813476498. eprint: http://jrs.sagepub.com/content/106/3/81.full.pdf+html. http://jrs.sagepub.com/content/106/3/81.short

Howick, J. (2011a). Exposing the vanities—and a qualified defence—of mechanistic evidence in clinical decision-making. Proceedings of the Biennial PSA 2010. *Philosophy of Science*, *78*(5), 926–940.

Howick, J. (2011b). *The philosophy of evidence-based medicine*. Chichester: Wiley-Blackwell.

Howick, J., Glasziou, P., & Aronson, J. K. (2013). Problems with using mechanisms to solve the problem of extrapolation. *Theoretical Medicine and Bioethics*, *34*(4), 275–291. ISSN: 1573-1200. https://doi.org/10.1007/s11017-013-9266-0

Illari, P. M. (2011). Disambiguating the Russo-Williamson thesis. *International Studies in the Philosophy of Science*, *25*(2), 139–157.

Kenter, M., & Cohen, A. F. (2006). Establishing risk of human experimentation with drugs: Lessons from TGN1412 [in en]. *The Lancet, 368*(9544), 1387–1391. ISSN: 01406736. Accessed 30 Aug 2016. https://doi.org/10.1016/S0140-6736(06)69562-7. http://linkinghub.elsevier.com/retrieve/pii/S0140673606695627

Levy, A., & Currie, A. (2015). Model organisms are not (theoretical) models. *The British Journal for the Philosophy of Science, 66*(2), 327–348. https://doi.org/10.1093/bjps/axt055.

Longworth, F. (2006). Causation, pluralism and responsibility. *Philosophica, 77*, 45–68.

Love, A. C. (2007). Functional homology and homology of function: Biological concepts and philosophical consequences. *Biology & Philosophy, 22*(5), 691–708.

Maciejak, A., Leszczynska, A., Warchol, I., Gora, M., Kaminska, J., Plochocka, D., Wysocka-Kapcinska, M., et al. (2013). The effects of statins on the mevalonic acid pathway in recombinant yeast strains expressing human HMG-CoA reductase. *BMC Biotechnology, 13*(1), 68. https://doi.org/10.1186/1472-6750-13-68.

Parkkinen, V.-P. (2016). Robustness and evidence of mechanisms in early experimental atherosclerosis research. *Studies in History and Philosophy of Science Part C: Studies in History and Philosophy of Biological and Biomedical Sciences, 60*, 44–55.

Parkkinen, V.-P., Wallmann, C., Wilde, M., Clarke, B., Illari, P., Kelly, M. P., Norell, C., Russo, F., Shaw, B., Williamson, J. (2018). *Evaluating evidence of mechanisms in medicine: Principles and procedures*. Springer Briefs. Cham: Springer.

Russo, F., & Williamson, J. (2007). Interpreting causality in the health sciences. *International Studies in the Philosophy of Science, 21*(2), 157–170.

Russo, F., & Williamson, J. (2011a). Epistemic causality and evidence-based medicine. *History and Philosophy of the Life Sciences, 33*(4), 563–582.

Russo, F., & Williamson, J. (2011b). Generic versus single-case causality: The case of autopsy. *European Journal for Philosophy of Science, 1*(1), 47–69. https://doi.org/10.1007/s13194-010-0012-4.

Schupbach, J. N. (2016). Robustness analysis as explanatory reasoning. *The British Journal for the Philosophy of Science*. https://doi.org/10.1093/bjps/axw008. http://bjps.oxfordjournals.org/content/early/2016/05/04/bjps.axw008.abstract

Steel, D. (2008). *Across the boundaries: Extrapolation in biology and social science*. New York: Oxford University Press.

Steinberg, D. (2007). *The cholesterol wars: The skeptics vs. the preponderance of evidence* (1st ed.). San Diego: Academic. ISBN: 97801237397970123739799.

Weber, E. (2007). Social mechanisms, causal inference, and the policy relevance of social science. *Philosophy of the Social Sciences, 30*(3), 348–359.

Weber, E. (2009). How probabilistic causation can account for the use of mechanistic evidence. *International Studies in the Philosophy of Science, 23*(3), 277–295.
Wimsatt, W. (2007). *Re-engineering philosophy for limited beings. Piecewise approximations to reality*. Cambridge: Harvard University Press.
World Health Organization. (2014). *Ethical considerations for use of unregistered interventions for Ebola viral disease. Report of an advisory panel to WHO*. Technical report. World Health Organization.

Chapter 4
Mechanistic vs Statistical Extrapolation in Preclinical Research in Psychiatry: Challenging the Received View

Catherine Belzung, Etienne Billette de Villemeur, and Maël Lemoine

Abstract This chapter questions the received view that in medical research extrapolation from animal models mainly consists in establishing mechanisms of human pathological states in organisms, thanks to a step by step comparison of causal pathways. *Mechanistic extrapolation* takes the form: *(1)* cause C brings out effect E in animal through causal pathway M, *(2)* M is similar in animals and humans, *(3)* therefore C will likely bring out E in humans. As the example of psychiatric research shows, such mechanistic extrapolation may be replaced by *statistical extrapolation*, an inference of the form: *(1)* An animal model A has been successful in predicting the effects E of drugs $D_1 \ldots D_n$ of a certain class; *(2)* A will be successful again in predicting the effects of a new drug D_{n+1} of the same class. Statistical extrapolation relies on the predictive validity of a given animal model, without any knowledge of the mechanisms involved, on the sole ground of past successes of the model in predicting the effects of a class of drugs on their human target.

4.1 Introduction

In spite of growing opposition to their use in the general public and notwithstanding the emergence of alternatives such as *in silico* simulation, animal models still constitute the standard experimental system in preclinical studies. Evidence drawn from animal models is deemed necessary before it is acceptable to test a drug on

C. Belzung (✉)
University of Tours, INSERM U930, Tours, France
e-mail: catherine.belzung@univ-tours.fr

E. Billette de Villemeur
University of Lille, LEM (UMR CNRS 9221), Lille, France
e-mail: etienne.de-villemeur@univ-lille.fr

M. Lemoine
University of Bordeaux, Immunoconcept, Bordeaux, France
e-mail: mael.lemoine@u-bordeaux.fr

A. LaCaze, B. Osimani (eds.), *Uncertainty in Pharmacology*, Boston Studies in the Philosophy and History of Science 338,
https://doi.org/10.1007/978-3-030-29179-2_4

humans for the first time. This chapter focuses on the way standard preclinical studies provide such evidence: how do they establish that a drug is, or should be, effective for humans afflicted with a disease? This has often been summarized by philosophers as an extrapolative inference of the sort we may call *mechanistic*. It may take the following form:

(1) 'fluoxetine relieves depressive-like symptoms in observed rodents'
(2) 'the mechanisms of fluoxetine action in rodents are similar in humans'

---

(3) 'fluoxetine relieves depressive symptoms in humans'

Conclusion (3) is generally held to be a prediction. Assumption (2) is considered as a comparison based on the investigation of mechanisms, that is, mechanistic reasoning. A mechanism is an integral causal process or pathway leading from an initial state to a final state. It is constituted of entities. These entities interact with each other in a determined way (Machamer et al. 2000). They may be biological entities, such as proteins, receptors, enzymes, genes or small molecules, which engage in biological interactions such as 'binds with', 'upregulates', 'methylates', etc. They may also be cognitive constructs, such as 'memory', 'short term memory', 'long term memory', 'attention', and cognitive interactions such as 'inhibits', 'retrieves', 'activates', etc. They may at last be social factors, such as 'single parent', 'abused child', and social interactions, such as 'parental love' and 'parental abuse'. Mechanisms may also be biological, psychological and social at the same time. For example, major depression is characterized by alterations of some affective/cognitive processes including anhedonia, stress coping and over-generalization (Belzung et al. 2015). This can be counteracted by chronic treatment with fluoxetine. The mechanisms underlying this action are complex: fluoxetine inhibits the serotonin transporter, which leads to an increase of serotonin in the synaptic cleft, inducing stimulation of serotoninergic receptors of various types and causing the activation of a second messenger cascade in the cytoplasm of the post-synaptic neuron. This ends by the activation of transcription factors within the nucleus of the cell, leading to the synthesis of neurotrophic factors such as BDNF (Brain Derived Neurotrophic Factor), which might stimulate adult neurogenesis in brain areas in which newborn neurons can be observed during adulthood, such as the dentate gyrus of the hippocampus. The dentate gyrus displays specific computational properties such as sparse coding: a given stimulus induces the activation of very few cells, so that the probability that two different stimuli activate the same neuronal population is very low. This property in turn justifies a distinction between two closely resembling stimuli, improving pattern separation. Therefore, the fluoxetine-induced increase in adult newborn neurons may decrease the over-generalization displayed by depressive patients. In parallel, depressive patients display a disruption of the regulation of stress hormones. Again, the hippocampus and more precisely the dentate gyrus participates in the regulation of stress hormones: therefore, the fluoxetine-induced increase in adult neurogenesis might counteract the stress hormone dysregulation seen in depressed patients (for more details, see Willner et al. 2013).

We call the view that focuses on such extrapolation of mechanisms in preclinical studies the 'received view'. In contrast, our contention in this chapter will be that, at least in the case of treatments, inference of the following sort is sometimes considered sufficient to establish a claim of interest:

(1) 'fluoxetine relieves depressive-like symptoms in observed rodent models'
(2) 'the reaction in humans to similar drugs is frequently the same as that observed in rodent models'

---

(3) 'fluoxetine relieves depressive symptoms in humans'

In this alternative version, step (2) is not based on mechanistic reasoning, but on the observation of correlations between the effects observed in the model organism's, and the ones seen in patients. We call it *statistical* extrapolation. A correlation is an association found to be frequent enough between two events: if it is close to 0, it is likely that the two events are independent; if it is close to 1, it is likely that they are concomitant, and if it is close to -1, that they are exclusive (i.e. one cannot be observed when the other is). In the case at hand, the events are the exposure to a drug and the general state of the patient. The form of extrapolation under scrutiny involves a correlation with a high value, i.e. 'close to 1'. In other terms, it is based on statistical reasoning instead of mechanistic reasoning.

At first sight, all preclinical studies rely on statistical and mechanistic information. Yet the difference between statistical and mechanistic extrapolation depends on the justification of the inference. In mechanistic extrapolation, the observation that an effect is caused by a mechanism in the model justifies the prediction that a similar mechanism in the target will cause a similar effect. In statistical extrapolation, the observation of certain effects in a model which has consistently predicted effects of drugs of a certain class in humans, justifies the prediction that the effects will be observed in humans. Although knowledge of mechanisms and statistical reasoning may be present in both cases, they do not play the same role. The very existence of statistical extrapolation as we describe it implies a strong claim that it does not presuppose mechanistic extrapolation in the first place as a justification. However, this does not mean that mechanistic information has not been used in the first place to select the initial hypothesis. Finally, note that we explicitly focus on the case of animal models in psychiatry, where the lack of well-defined mechanisms is notable. This constitutes a good reason why it is more often resorted to in this domain. However, it remains an open question whether statistical extrapolation is also resorted to beyond this field.

This chapter thus suggests an important correction to the philosophical received view on animal studies, which misleadingly focuses on the role that the investigation of mechanisms plays in extrapolation through what Steel calls "comparative process tracing" (Steel 2008). Section 1 presents the received view. Section 2 presents a case study where statistical extrapolation is used in place of mechanistic extrapolation.

Section 3 raises objections to the existence of such statistical extrapolation and answers them.

4.2 The Received View on Evidence Based Upon Preclinical Studies

The received view claims that preclinical extrapolation consists in the attribution to humans, by analogy, of an extra entity or activity, unknown (or incompletely understood) in humans but observed in animals. An implicit claim follows that evidence of a mechanism is necessary to the extrapolation of any causal claim to humans. We will explicate the two claims in turn.

4.2.1 Inferences from Preclinical Studies Consist in Comparative Tracing

The received view consists in a certain picture of how extrapolation in preclinical studies works, which we dub 'mechanistic extrapolation'. It is necessary to describe it in order to understand when evidence of a mechanism is required in preclinical studies, and to contrast mechanistic extrapolation with statistical extrapolation.

Mechanistic extrapolation is analogical reasoning, whereby the presence of a property in the target is inferred from its presence in the model. Some knowledge of the mechanisms the model and target share is necessary for it to be plausible. Yet we do not by definition know all of it, and we have to assess the extent to which relevant differences impinge on extrapolation. More specifically, it consists in the following line of reasoning about the mechanism, that is, the causal chain linking cause (C) to effect (E):

1) C* causes E* (in model)
2) C (in target) is analogous to C* (in model)
3) E (in target) is analogous to E* (in model).

---

4) C will cause E (in target).

In her classic book about models in science, Mary Hesse emphasized the role that unknown factors may play in causal analogies. She suggested that based on both "*positive*" and "*negative*" analogies, respectively, what is known to be and not to be common in model and target, analogical reasoning on a model consists in investigating "neutral" analogies, i.e., what remains unknown (Hesse 1970). In other terms, the problem is:

> An analogue x containing the characters ABD is a model for an *explicandum* y containing BC but not A, where B represents the total known similarity between x and y. If A and B together are known to be causally connected to D in the model, are there any rational grounds for expecting or inferring the occurrence of D with B and C in the absence of A? (Hesse 1970)

Two other philosophical positions frequently cited when it comes to animal extrapolation in particular, LaFollette and Shanks (1995) and Steel (2008), phrase the problem in essentially the same terms as Hesse. Yet they solve it in opposing ways. First, LaFollette and Shanks state that for the inference to be valid,

> *there must be no causally relevant disanalogies between the model and the thing modeled* (LaFollette and Shanks 1995).

On the contrary, Steel concludes that

> a total absence of causally relevant disanalogies is not required for extrapolating (Steel 2008)

Whatever the answer, analogy of mechanisms plays a central role in that sort of extrapolation, according to the received view.

A further claim is that mechanistic extrapolation consists in an incremental comparison of mechanisms. The potential difference under investigation between mechanisms in model and target should be limited to one entity (or process) only. This might be called the 'increment condition'. Indeed, mechanisms are manipulated in experiments. An ideal intervention is supposed to act upon *one* variable of interest and isolate it from the rest of the mechanism, which in turn is supposed to go on working as it did before. To put all this briefly, we extrapolate causal claims of the sort 'X causes Y', X and Y being variables. The causal claim itself is justified insofar as it conforms to the conditions of an "ideal intervention" I, that is,

> I eliminates all other influences on X;
> I does not change anything to related variables except X;
> I does not depend on any variable related to X. (Steel 2008)

Then the mechanisms in the model, i.e. the manipulated mechanisms, are *compared* to the mechanisms in the target, i.e. the mechanisms of interest. More precisely, enough must be explicitly similar in both mechanisms for a further part of it to be extrapolated. Steel has proposed a view that rapidly became accepted, that this is achieved through a mode of inference he calls "comparative process tracing" (Steel 2008). Process tracing consists in trying to fill in a schema of a causal pathway in the causal structure under study. This in turn is achieved through tracing causal pathways, backward or forward, from a known starting point. Now *comparative* process tracing consists in establishing parallel processes in both the model and the target, then drawing the conclusion that the next causal link observed in the model, but not yet in the target, is likely to be found there. Steel also relaxes the condition from identity to relative similarity between known causal pathways in the target and in the model:

> If significant differences between the model and the target are likely to be restricted to a relatively small number of stages of the mechanism, then comparisons at those stages may provide good grounds for the suitability of the model. (Steel 2008)

That said, evidence of a mechanism in humans does not always rely on an incremental comparison. It sometimes relies on mere generalization of observations made on a single species to all living organisms, without any step-by-step tracing. However, in the health science in particular, such generalization is generally used as background biological knowledge – for instance about the mechanisms of gene expression, methylation, etc. –, but they are generally not intended to provide evidence about diseases in humans.

The investigation of mechanisms justifies the prediction about humans. It does not involve that statistics are absent from preclinical studies. Most philosophers agree with the difference, emphasized by (Illari 2011), between "mechanistic evidence" and "evidence of mechanisms" on the one hand, and "statistical evidence" and "evidence of difference making" on the other hand. It is a difference between object and method. Mechanistic methods involve manipulation, tracing steps in processes, and they certainly are dominant at the preclinical level. On the other hand, statistical methods are prominent at the clinical level. But it is easy to show that in the experimental method, evidence of correlation is critical to evidence that a mechanism exists (Campaner 2011).

4.2.2 Evidence of a Mechanism Is the Main Point of Preclinical Studies

The focus on mechanistic extrapolation has consequences as to the role played by evidence of mechanisms in preclinical studies. A widely accepted view in the philosophy of medicine is that evidence for causal claims in the health science is of two sorts, evidence of a correlation and evidence of a mechanism. Evidence of a correlation is evidence that a cause C makes a difference to an effect E (generally: the presence of C increases the probability of E: see Salmon 2006). Evidence of a mechanism is evidence that there is a possible continuous chain of events, generally called a 'pathway', linking the occurrence of C to the occurrence of E.

Among philosophers, the main debate about evidence of causal claims in the health science is about what is called the Russo-Williamson Thesis (RWT), which states that evidence that C causes E requires *(1)* statistical evidence that C makes a difference to E, and *(2)* evidence of at least one mechanism or pathway from C to E (Russo and Williamson 2007, criticized in Howick 2011; Broadbent 2011; see also LaCaze 2011; Worrall 2010). Are these two conditions necessary, or can either one suffice to establish a causal claim? This debate parallels the debate in the health science itself, about the status of evidence of mechanisms generally obtained in the so-called 'basic science' as compared to evidence of correlation, generally obtained in either epidemiological studies or clinical trials.

Let us ask the RWT one question: where does evidence of mechanisms come from in the health science? Obviously, if implicitly, it comes from basic science, in which animal models play a major role. The RWT does not explicitly refer to animal experiments in the original paper, but further developments of this view do (Illari 2011; Clarke et al. 2014). Now let us ask the reverse question: what sort of evidence does basic science in general, and studies on animal models in particular, provide? Whereas clinical studies are "aimed at establishing difference making", preclinical studies are "clearly aimed at establishing knowledge of mechanisms" (Clarke et al. 2014). Strictly speaking, this does not discard the possibility that there is statistical extrapolation in preclinical studies, but it leaves it in the dark.

Now evidence of any sort can be put to particular uses, namely, explanation on the one hand, to which evidence of mechanisms is critical; prediction and control on the other hand, to which evidence of difference-making is critical:

> Evidence of difference making is required because causal claims are used for prediction and control, and one can only predict an effect on the basis of the cause, or control the effect by manipulating the cause, if the cause makes a difference to the effect. Evidence that there is an underlying mechanism is required because causal claims are used to explain, but in order to explain some phenomenon one needs to point to the (functioning of the) mechanism responsible for it; so invoking a cause as an explanation for an effect is only successful to the extent that the cause is a part of the mechanism responsible for the effect. (Russo and Williamson 2012)

Evidence of any mechanism between C and E is therefore necessary to the causal claim, *at the level of demand required by explanation.* Along with evidence of difference making, it is also sufficient. Some, as Broadbent, have agreed that it is necessary to *explain*, but not to *warrant* the causal claim (Broadbent 2011) – to which evidence of difference making can suffice.

Our claim is that preclinical extrapolation of a certain kind, i.e. statistical extrapolation, does not aim to investigate mechanisms, but only to establish evidence of difference-making. Its goal is not to explain, but to predict and control. Our correction to the RWT is therefore more a development than a criticism.

That said, the complete received view on animal extrapolation may be phrased as follows:

1. Most preclinical studies consist in the extrapolation of the existence of an incremental part of a given mechanism as observed in the model, to the mechanism in the target, thanks to ideal intervention.
2. Preclinical studies provide evidence that a mechanism M explains why C causes E in animals, thereby indirectly predicts, in a weaker way than evidence of difference-making in humans would, that C will cause E in humans as well.

We will now challenge this view by examining how preclinical studies may also establish evidence of difference-making and thus predict a little more reliably than by establishing mechanisms.

4.3 Challenging the Received View on Preclinical Evidence

A challenge to this received view is introduced here through a case study of animal models of a major class of antidepressant drugs, selective serotonin reuptake inhibitors (2.1). Even if the logic of discovery was mainly based on mechanistic reasoning indeed in this case, the logic of justification is rather based on statistical reasoning (2.2).

4.3.1 The Effects of Selective Serotonin Reuptake Inhibitors Have Not Been Established by Mechanistic Extrapolation: A Case Study

Serotonin reuptake inhibitors are considered the standard treatment of major depression. Specific animal models have become standard in preclinical testing of this class of molecules. In this section, we first sum up the history of these models and molecules, then highlight how extrapolation has been used in this literature.

A first part of the context and background knowledge is constituted by animal models of depression. Models of learned helplessness, which consist in conditioning helplessness behavior in animals and had been developed in the mid sixties (Seligman and Maier 1967; Overmier and Seligman 1967), are generally considered to have started the practice of testing drugs for mood disorders on animal models. However, these models were soon criticized because they relied on the imposition of major stressors, like electric foot shock, a situation which was not deemed relevant for depression. The first study on one of the currently best validated models for depression, the chronic mild stress model (CMS), which consists in submitting animals to mild stressors for a long duration, was published in 1981 (Katz et al. 1981) while the first test, the forced swimming test (FST), which consists in measuring how long a rodent tries to escape water, was published in 1977 (Porsolt et al. 1977). Another animal model of depression, the olfactory bulbectomy in rats, which consists in removing a part of their brains important in the perception of smell, was first described in 1977 (van Riezen et al. 1977; Cairncross et al. 1977). CMS and FST are currently the most frequently used models in this area.

A second part of the context and background knowledge is constituted by available drugs. During the same years, a new class of antidepressants was developed: selective serotonin reuptake inhibitors (SSRIs). SSRIs have been the most popular treatments of major depression ever since. They stimulate neurotransmission in serotonergic circuits of the brain by increasing the availability of synaptic serotonin. This is achieved by downregulating the reuptake of serotonin in the post-synaptic cleft. When the first reports on these drugs were published in the mid-seventies, major depression was treated with inhibitors of the monoamine oxidase enzyme, which degrades monoamines such as serotonin or noradrenaline, or with molecules that blocked their uptake in the presynaptic neuron (tricyclics).

The first publication regarding the effects of SSRIs was a report by Wong and colleagues in 1974 (Wong et al. 1974, 1975a, b) showing that fluoxetine, a molecule of this class, had a highly potent effect on serotonin reuptake in nerve endings of drug-treated rats (*ex vivo*). This effect was confirmed using *in vivo* preparations (Fuller et al. 1974). In 1976, clinical studies started (for a review see Wong et al. 2005) and in 1978 Lemberger and colleagues published a paper in *Science* showing that fluoxetine is able to act as a SSRI in healthy human subjects as well, and they concluded that the drug could be effective in treating "mental diseases," but without providing any evidence of such effects, in animals or in humans (Lemberger et al. 1978). One year later, in 1979, Meltzer and colleagues administered the drug to 3 patients with severe depression, reported the effects on 28 others, and tried to explain the paradoxical effects on dopaminergic activity in only one of them (Meltzer et al. 1979). Finally, the drug appeared on the European market in 1986 and received Food and Drug Administration (FDA) approval one year later.

In the same period, other SSRIs have been tested, such as citalopram. The effects of this molecule were first described in Hyttel (1977). Three years later, the results of the first Phase II study were published (Gottlieb et al. 1980), showing an antidepressant efficacy of citalopram. A similar story repeats with paroxetine as, in a first paper, authors insist on the serotonergic effects of paroxetine (Lassen 1978); one year later a tolerance study is published in human subjects and in 1982 the results of the Phase II Open trial are published, again with no reference to preliminary data using animal models. In 1983, Koe and colleagues showed that sertraline is a very potent SSRI modifying brain serotonin content, and that this is sometimes associated to antidepressant effects in standard animal tests such as the FST (Koe et al. 1983). Moreover, they showed that it acts also on noradrenergic variables. They concluded that the drugs could have "potential antidepressant activity". This was indeed shown in a first clinical trial published in 1988 (Reimherr et al. 1988; Burrows et al. 1988).

A third and last part of the context and background knowledge is constituted by more recent treatments. Few new antidepressants have been introduced in clinical practice in the last years. Vilazodone is one of them. It was approved by the FDA in 2011. The first report is an unavailable communication at a meeting, suggesting antidepressant-like effects. One year later, authors report that the compound acts as a SSRI with selective presynaptic 5-HT1A receptor agonistic properties (Bartoszyk et al. 1997). As 5-HT1A ligands are known to reduce anxiety behavior, such effect is investigated in this article and in another one (Treit et al. 2001). Both show that the drug reduces anxiety behavior in some tests but not in others. One year later, a study reported antidepressant-like effects of this compound on animals (Page et al. 2002). A first randomized, double-blind, placebo-controlled trial was published in 2009 (Rickels et al. 2009), indicating that the treatment induces relief after 1 week. The drug received FDA approval in January 2011.

A brief chronology of the essential facts is presented in Fig. 4.1. Let us focus first on Hyttel's study on citalopram. The main result is to establish that this then new compound has a stronger and more selective effect on the level of serotonin available in the synaptic cleft in animal models, through inhibition of the reuptake. It draws

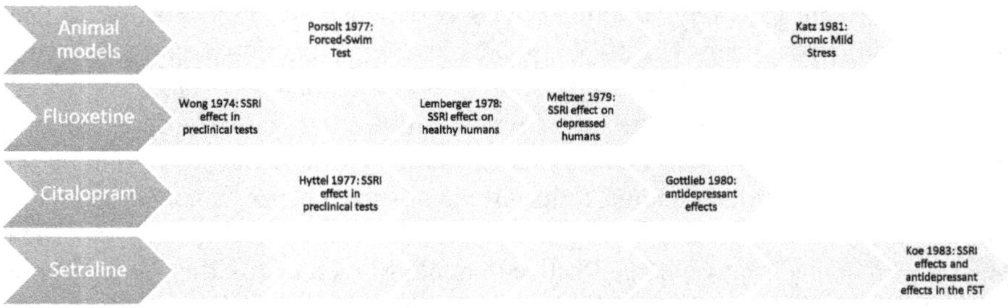

Fig. 4.1 A chronology of essential facts in preclinical studies of SSRIs

from pre-existing knowledge of general mechanisms of monoamine, serotonin, uptake, etc., but predates standard animal models and tests of depression. It is a typical investigation of one part of the mechanism of action of a potential new drug, in comparison with the mechanism of action of other, known drugs (namely, tricyclics). However, note that there is not direct extrapolation to humans here. Hyttel's conclusion is that if the effect of citalopram was confirmed in humans with depression, it might help investigate the etiology of the disease. For this reason, it is not a typical example of "comparative process tracing", but a simpler investigation of possible mechanisms with a remote analogy with humans in mind.

Let us now turn to a later study. Hyttel's paper could not rely on any standard model for depression, there being none at the time. It could only rely on the consistency of the result in various experiments. The argument is very different in Koe's paper on sertraline in 1983:

1. Depressed patients have low levels of serotonin and compounds with a strong serotonin component are effective in treating depressed patients;
2. Sertraline selectively inhibits serotonin reuptake;
3. Sertraline has a significant effect on depressive behavior in the standard FST;

----

4. Sertraline will have a significant effect on depressive behavior in humans.

Koe's paper does not meet the conditions of the received view. First, its point is not to establish a causal pathway between selective inhibition of serotonin reuptake and relief of depressive symptoms. As we have evidence that there probably is one, it is worth investigating the effects of a new molecule that we class in the same category. The goal is not to explain *how* sertraline would produce such an effect. Nor is such an explanation required for prediction. That we know that similar drugs have an antidepressant effect on standard models, is sufficient to predict that this drug will have the same effect. This is mere correlation. As in Hyttel's paper, this seems to presuppose that the mechanism is similar. However, the extrapolation itself is statistical: because the FST should be trusted in general, and not because the effect of sertraline in the mechanism involved in this test is well known or comparatively traced in both humans and animals, sertraline should be effective on human depression. For this reason, it is difficult to apply the standard view here.

Indeed, this sort of animal extrapolation is of a different nature than that emphasized by the standard view. We call it 'statistical', and claim that it is actually often resorted to, not only in the case of 'me-too' drugs, as one could expect, but also when innovative drugs are under investigation. Let us take a third example. Around 2000, basic research suggested that methylation of genes providing a protective effect against stress could play a role in vulnerability to depression. It became an interesting idea to investigate whether molecules with a demethylation effect could have a therapeutic effect. A standard strain of a rodent known to be responsive to standard childhood stress protocol was chosen to test this hypothesis, because if there were any behavioral alteration, it would be both obvious and predictive of what would happen in human subjects. Some mechanisms are indeed known in this standard model, but precisely not those hypothesized on the basis of basic research: behavioral effects would be strong piece of evidence that they exist. The basis of evidence about humans in this case is the well-established correlation between animal and human behaviors, in other terms, the existence of a standard model.

Note however that not all innovative drugs can be studied in standard models. When basic research had suggested for instance that a family of molecules called neuropeptides intervened in serotonergic neurotransmission as a catalyst in case of very intense stimulation, these effects could not be investigated on standard animal models. Indeed, some of the pharmacology was different and the distribution of the receptors of these neuropeptides was not homologous between humans and the standard rodent species used in this kind of research. Researchers had to find different, non-standard species, with entirely different behavioral patterns, responsive to different pathogenic protocols (Salomé et al. 2006; Griebel and Beeské 2012).

This brief overview of the history of antidepressant since the 1970s emphasizes how problematic the application of the received view to typical cases of preclinical extrapolation may be. However, we also found that statistical extrapolation, although different from mechanistic extrapolation according to the received view, seems to also rely on the presupposition that there is a shared mechanism. It seems impossible to conclude that drug D will have an effect on humans H on the basis that it has on animal A, if the mechanism linking D to H is not supposed to be the same as the one linking D to A, even if we do not know this mechanism. In other terms, it seems that statistical extrapolation in preclinical studies is not independent from mechanistic extrapolation. The rest of the paper investigates this question and clarifies the nature of statistical extrapolation.

4.3.2 The Effects of Selective Serotonin Reuptake Inhibitors Were Established by Statistical Extrapolation

These papers, and specially Koe's, show instances of the sort of animal extrapolation we dub 'statistical'. In several ways, they look like they involve mechanistic reasoning, at least partially. We do not deny it. But it is important to specify what role mechanistic reasoning plays in such studies. First, for any study of animal models to be potentially published in a scientific journal, either a mechanistic or a statistical rationale is needed. By 'rationale', what is meant is a piece of evidence of either a working mechanism, about which we want to know whether it underlies a set of cases, or of a correlation we know exists and want to use as a signal of the effect of an intervention on a given mechanism. In Koe's study, there is evidence that sertraline binds with serotonin transporters and prevents them from reuptake by the presynaptic neuron, leading to an increase of the quantity of serotonin available in the synapse. But such evidence establishes the relevance of the hypothesis of a study, not its result. To put it in a nutshell, mechanistic knowledge here does not play a direct role in the logic of *justification* of a hypothesis, but only in the logic of *discovery* of this hypothesis. What justifies the result, is the trust researchers place in the model based on the basis of its passed successes (statistical evidence). Yet, the accumulation of successes, i.e. the very fact that a model becomes standard, means that it supposedly involves the same mechanism. It is indeed correct that the model may not be relevant anymore when a significant difference in the mechanism is involved, as was the case for neuropeptides. But results from this model are deemed conclusive even when a different mechanism is known to be involved, as was the case for demethylation drugs referred to above. Moreover, researchers sometimes incorrectly focus on one mechanistic effect of a drug and misinterpret it as the explanation for its efficacy, possibly without any effect on the reliability of the model.

Second, it seems that no direct statistical extrapolation is possible from a given signal in an animal model to an effect on a human disease, if there is no assumption that the same mechanism links treatment to signal in the animal and treatment to effect on the human disease. We found that such assumptions were likely to be held in the studies we analyzed in the previous section. Thus, it seems that statistical extrapolation is always mixed with at least a bit of mechanistic extrapolation, be it a hypothetical mechanistic extrapolation. We suggest however that a distinction has to be made between the assumption that a specific mechanism links cause to effect in both model and target, and the assumption that cause and effect are mechanistically linked in both model and target. Suppose two scientists agree that a study reliably predicts the effects *E* of drug *D* in humans, but hypothesize different mechanisms *M* and *M'* to explain why. Or suppose one scientist who does not know whether *D* elicits *E* through *M*, *M'*, or *M"*, or sometimes *M*, sometimes *M'*, sometimes *M"*, or always a mix of *M, M'*, *M"*, etc. It would matter if the question was to produce an explanation. It does not if the question is to produce a prediction. A prediction assumes that there are mechanisms in the world that allow it to be reliable, but only an explanation needs that one is picked up, even if it still is unspecified. The only mechanistic assumption in a statistical extrapolation from animal models, is therefore that interventions and effects in both models and targets are mechanistically linked, even though a specific mechanism is very often provided. This hardly makes it more than a general *credo* about causality in the world.

The rationale of statistical extrapolation would therefore not be such a general assumption. It is rather that the model, in Koe's paper, the FST, is considered a reliable, predictive test of such drugs. In other terms, there is a correlation between the effects of such drugs in such animal models, and the effects of the same drugs on their human targets.

We propose to define such *statistical extrapolation* as a prediction based on the past successes of the model to predict what will happen in the target. Koe's paper, we claim, is an example of an extrapolation of this latter sort. Indeed, sertraline should have a significant effect on depressive behaviors in humans because it has this effect in the FST and because the FST is a reliably predictive test of the effects of monoaminergic compounds on human behavior.

As compared to mechanistic extrapolation, statistical extrapolation is an alternative way to escape what Steel (2008) calls the 'extrapolator's circle': as we do not know everything about model and target systems, there may still be significant differences that impair the former's predictive validity as regards effects in the latter. On the other hand, in order to know that the model system is identical to the target system in every relevant aspect, we would need to know everything about them, and experiments would be unnecessary. One way out of the circle, he justly says, is "comparative process tracing". Another way out of the circle, we claim, relies in establishing the predictive validity of a model system, the precise mechanisms of which may remain unknown. Indeed, a model is predictive insofar as a mechanism is shared, but one does not need to know what the mechanism is to know whether the model is predictive. In other terms, statistical extrapolation is incremental in a different sense that mechanistic extrapolation is: it consists in induction to the next case, based on former cases, and escape the extrapolator's circle because it is not causal knowledge that is searched, but only prediction.

The rule of statistical extrapolation may be summed up like this: the more a model of a disease has successfully predicted the effects of treatments of a certain class on human subjects, the more it can be considered predictive of the action of new molecules of the same class. For lack of space, a formalized expression of this sort of inductive inference is not provided here. Suffice it to say that the model is primarily proposed on the basis of its resemblance to known phenomena involved in the human disease, from etiology to symptoms through pathophysiology. It is then put to test. If a drug does produce on humans the effects predicted by the animal model, then the latter is corroborated and finally becomes standard. Statistical extrapolation may be summarized thus:

1. Animal model of a disease M has successfully predicted the effects of drugs D_1, ... D_n of a similar class C on human targets T
2. Drug D_{n+1} belongs to C

3. Drug D_{n+1} has effect E on M
4. Drug D_{n+1} will have effect E on T

Note here that 'class' does not necessarily mean a mechanistically homogeneous family of molecules (e.g. SSRIs). It may be more or less extensive – for instance, it could be the class of 'antidepressant medication' as well as the class of 'antidepressant medication except SSRIs', etc.

4.4 Objections to the Existence of Statistical Extrapolation

In this last section, we raise, and respond to, 4 objections to the view that statistical extrapolation exists and is distinct from mechanistic extrapolation. The first objection is that standardization, i.e. the frequent success of animal models in predicting effects in humans, is not evidence in and by itself. To this objection we reply that at least in the sense of 'used to represent the proper construct', standardization provides evidence in and by itself. The second objection is that statistical extrapolation is rare: on the contrary, we argue that it is the most frequently resorted to, at least, in preclinical studies of mental disorders. The third objection is that statistical extrapolation must in principle rely on mechanistic knowledge, an argument that ignores the possibility of statistical clustering. To the fourth objection, namely, that statistical extrapolation is typical of applied research and eventually relies on basic research, we respond that it does rely on basic research in the logic of discovery but not in the logic of justification, that they are strategically interdependent, but that predictive validity and mechanistic validity are theoretically independent: it is one thing to assess whether a drug is effective and another to discover why it is.

4.4.1 Standardization Is Not Evidence

'Standard model' is a phrased often resorted to in the scientific literature. Philosophers have also discussed the implications of standardized models (Meunier 2012). Note that what is at stake, at least in psychiatry, is not so much a model organism (Ankeny and Leonelli 2011) than a standard situation for a given organism, say, mice under chronic mild stress. At least five different meanings of this term must be mentioned:

1) 'arabidopsis thaliana is a standard model' may mean that it is 'frequently resorted to', independently from what is to be established;
2) 'Skinner box #1 at Columbus is a standard model' means that the experimental apparatus is 'set up with a normalized protocol', i.e., one can trust the reproducibility of the procedure;

3) 'the standard protocol resorted to in learned helplessness' means that it is an 'agreed on protocol', i.e. everybody agrees on what the procedure *should* consist in, although confounding factors in the *actual* setup in one investigation may have influenced the result (e.g. the perfume worn by an investigator may influence the outcome of the experiment);
4) 'UCMS is a standard model' means that it is 'robust', i.e., the model always gives the same results under the same circumstances (which is not strictly implied by the reproducibility of the procedure per se);
5) 'latent inhibition is a standard model of attention deficit in schizophrenia' means that this experimental process, consisting in teaching animals not to take a stimulus into account, has been 'used to represent the proper construct', i.e., the model allows to establish something about the construct this model is supposed to represent.

We are especially focused on the fifth meaning of the term standard here, notwithstanding the importance of the first four sense to the success of statistical extrapolation. Indeed, that a model is 'used to represent the proper construct' cannot mean anything else than it 'has frequently been successful in predicting causal claims about a certain construct'. It could not mean anything along the lines of 'it involves the right mechanism', because the fact that a mechanism is the right one, that is, instantiates all instances of a disease and only them, cannot be proved mechanistically and has to be established statistically.

Many scientific papers are refused because they justify their use of a given model by its being standard in any of the other four meanings, but fail to convince that it actually represents the proper construct. Yet the results of some articles are accepted as valid although they do not use standard models in this sense. A striking example is a study by Cryan and colleagues (Cryan et al. 2005). Since its introduction in 1977, the FST did not always detect the effects of SSRIs on rats. Cryan introduced two changes in the test, a slightly deeper tank so that rats cannot reach the ground, and a camera instead of the observer's presence, so that rats cannot see humans. This non-standard test detected effects of SSRIs, and has become the standard FST ever since, precisely because it then became successful in predicting causal claims about effects of SSRIs.

Now, the objection goes, the fact that a model has been standardized in the sense of 'validated as the proper model of a given construct', is not in itself evidence of any other claim that those already proven.

We claim that it is. First, statistical extrapolation is not just an instance of induction in the classical sense, in the sense that it does not simply rely on the reproduction of the same results in the same circumstances. It is not an instance of 'expect more of the same': remember that different drugs ($D_1 \ldots D_n$) of the same class C must be tested for the model to become 'standard' in that sense. Second, what matters to a standard model of a construct, is not the number of articles having established the same claim about the same dimension of the construct, it is on the contrary the number of different dimensions of the construct the model has been shown to instantiate. Indeed, scientists make a difference between a test or a bioassay and a model. Whereas tests and bioassays are valid representation of one dimension of a construct, models are valid representations of different dimensions of a construct. In this field, ‘model’ is frequently used in cases where animals have been subjected to manipulations inducing a pathological state (although, strictly speaking, many symptoms, such as deficit attention in schizophrenia or anhedonia in depression, are also constructs rather than dimensions). The tail suspension test is not a standard model of depression properly speaking, but rather a standard *test* for (one dimension of) depression. On the other hand, it could be considered a standard model of resignation, as a dimension of depression, although the term is generally not used because resignation is not a disease. Hence, the more dimensions of a construct a given model instantiates, the more predictive of the target’s behavior it can be considered. This includes both known and unknown aspects of the target’s behavior. Statistical extrapolation relies on the multidimensional resemblance of a model to its target to make new predictive claims that have not formally been proven. This does not mean that the mechanistic investigation of the model as well as that of the disease cannot play a key role here in guiding researchers in the selection of the relevant dimensions of the model.

4.4.2 Statistical Extrapolation Is Rare

A second objection is that if statistical extrapolation exists, it is rarely resorted to, and that standard reasoning in preclinical studies is mainly mechanistic. Here we exclusively focus on the literature in psychiatry. Although we have reasons to think that statistical extrapolation is also frequent elsewhere, some may challenge our views by claiming that research in neuroscience is a particular case.

The pharmacologist Guy Griebel described the major trends over the past 50 years in the search for new anxiolytic drugs (Griebel and Holmes 2013), carefully analyzing and summarizing 10 000 preclinical experiments, on approximately 1500 novel drugs.

It can be clearly seen that preclinical research in this field has developed progressively during the 1980s, peaking in the 1990s, with a number of published experiments/year ranging between 400 and 800. These papers always mention the neurotransmission system that is targeted: half of them focused on the serotoninergic system, a third of them focused on different kind of neuropeptides, and, more recently, a trend toward an increase in studies investigating the effects of ligands of the glutamatergic or of the endocannabinoid system appeared. These studies thus described quite precisely the molecular target that was investigated (a given neurotransmission system, or more precisely a subtype of a receptor) as data exist on the molecular substrate to which these compounds are binding. In some cases, these studies eventually used genetically-engineered mice to show that in the absence of the molecular target, the pharmacological effects were absent: in this case, they really provide data on the primary target of the putative drug. However, they less frequently describe the neural circuit that was targeted. For example, on the 10

000 experiments that are listed in this paper, only 2740 consisted in the local administration of the drug in a given brain area belonging to the fear circuit: the other studies did not target a specific sub-system. Thus, these investigations focus on the primary molecular target, not on more sophisticated underlying mechanisms.

Indeed, a difference is to be made between the definition of a class of molecules, and the mechanism involved in a disease. The former can, and actually is, specified without the latter in most studies. This is strong evidence that in fact, evidence in preclinical studies is, in psychiatry at least, mostly based on statistical extrapolation.

Note that the main shortcoming of preclinical studies may be blamed on statistical extrapolation based on poor knowledge of the mechanisms involved. As a result, the clinical outcome of most studies cited above was really poor: for example, among the serotoninergic compounds tested, only one made it to go to clinical use (the 5-HT1A agonist tandospirone, and only in Japan and China). In fact, in their conclusions, the authors recommend to shift from a target validation strategy, to an advanced strategy, focusing on the underlying mechanisms: "*Anxiolytic drug discovery (...) will be greatly facilitated by concerted efforts to elucidate the underlying neurobiology of anxiety*". They precisely make this recommendation because this strategy was poorly used up to recently, which indicates that most of the studies only rely on target-related drug discovery strategy.

4.4.3 *Statistical Extrapolation Must Rely on Mechanistic Thinking*

Models turn into standard models when their validity becomes public knowledge. Thus, one may argue that although their validity might appear to some as statistically proven, their widespread use actually relies upon prior work, based upon mechanistic thinking.

The function of animal models in scientific reasoning, and in particular when assessing the effect of a new molecule, can be summarized as follows. A drug D_n is known to have an effect upon both animal model A and human target H. Mechanistic reasoning allows to establish that this can be explained by the same mechanisms being at work on both sides. A is thus said to be an animal model of the action of D_n upon H. When a new molecule D_{n+1} is under scrutiny, mechanistic reasoning again suggests that its possible effects upon A are to be explained by the same mechanisms that explain the impact of D_n on M. When D_{n+1} indeed displays an impact upon A, the very fact that A is an animal model of the action of D_n upon H allows to predict, by transitivity, that D_{n+1} should also have an impact upon H.

The very existence of the animal model allows to dispense with that part of the evidence which consists in tracing the (mechanistic) link between the action of D_{n+1} upon A and the action of D_{n+1} upon H. Yet, whatever the number of molecules whose effects have been assessed, everything started with, and still relies on, mechanistic thinking.

We do not deny the accuracy of this description of animal model use in scientific reasoning, but we claim that the adjective 'mechanistic' is dispensable at every stage. It is possible to establish statistically that whenever a molecule D_n triggers an effect upon A, it does trigger that same effect upon H, without having the slightest idea of what could explain that effect, on either side (A or H). It is also possible to group molecules according to their effects, without having any clue of what can explain them. And the use of transitivity is not restricted to chain mechanistically justified logical propositions, but is legitimate piece of any rational reasoning.

"Statistical clustering" by which objects are grouped by properties that make them similar is nowadays a common technique of statistical ("model free") data analysis. Correlations are just an instance of statistical measures, which do have a recognized scientific status. Moreover, relying upon the results of others to extend the existing knowledge is at the basis of modern science. Although mechanics certainly played an important role in the development of science, the latter need not be based upon mechanistic reasoning, not even partially nor initially.

4.4.4 Statistical Extrapolation Is Applied Research in a Specific Case But Eventually Relies on Basic, Mechanistic Research on General Processes

It is one thing to show that in particular cases, no mechanistic knowledge is used, but it is quite another to claim that no mechanistic knowledge in general is involved in the justification of a causal claim. One may argue that even if no mechanistic extrapolation has been established in the first place for a particular study, general mechanistic knowledge, i.e. background knowledge gathered from basic research, is necessary to statistical extrapolation as an instance of applied research.

Applied research in preclinical studies focuses on *predicting* what the effects of a drug on a human disease will be, whereas basic research focuses on *describing* what the mechanisms of a disease are. The prediction of how effective a drug will be on humans is as reliable as the effects of similar drugs tested on animal models are correlated to the effects of these latter drugs on humans. This line of reasoning is generally adopted by pharmaceutical companies. Prediction is blind to *how* the disease is caused, maintained, cured, and pays attention only to *whether* effects of clinical interest will happen. In other words, it is a direct extrapolation from a sample of animals (humans not included) to a broader set of animals (including humans). It is atheoretical – it does not presuppose any theoretical model of how this all works.

Fundamental research on the other hand is more interested in the investigation of mechanisms. It seeks after an exact theoretical model of what happens. The theoretical model represents a part of what happens in the human disease, supposedly, the essential part of it. For that reason, it cannot be said that what happens in one model organism (animal) represents what happens in the human disease, but it only assesses one causal relation of possible interest to the understanding of the human disease.

We agree with this general claim that basic research is necessary to applied research, with qualifications. First, it is difficult indeed to make a conceptually clear distinction between basic research and applied research in the field of preclinical studies. The clear line between academic research and R&D in pharmaceutical companies is relevant, but insufficient. From a very general perspective, there is no essential difference between basic and applied research studies, for they both come down to exactly the same problem of assessing a causal claim: does *C* (etiological factor, treatment) have an effect *E* (elicits, relieves) on disease *D*?

Second, if a distinction is to be made, it must rely precisely on the theoretical independence of explanation and prediction. Applied research resorts to models with high *predictive validity*, that is, "the resemblance of the apparent impact of the etiological factors and of the treatment on the observable effects", two of us wrote in a previous review, whereas basic research resorts to models with high *mechanistic validity*, that is, "the similarity of the mechanism we suppose or know is working in the animal disease to the mechanism that is or is presumed to be working in the human disease." (Belzung and Lemoine 2011). We emphasized the difference thus:

> The fact that a therapeutic agent has a dramatic impact on a biological system does not imply that it dramatically reduces the symptoms. The reverse is also true of the dramatic action of an agent or a factor on the symptoms (or on the biological markers). The predictive validity of a model is assessed without looking into the mechanism which is really at work in the animal: generally it is assessed from a macro-observational point of view or through peripheral biological measurements (biomarkers). (Belzung and Lemoine 2011)

Again, there has been many mistaken claims about *why* many treatments were effective on given diseases, assorted with justified claims *that* these treatments were effective on these diseases. As *(1)* a treatment was effective on a disease and as *(2)* a mechanistic effect of the treatment was observed (say the increase of available serotonin in the synapse), the conclusion was mistakenly drawn that *(3)* the treatment was effective on the disease because of this mechanistic effect. In spite of that error, the model was predictively valid.

Third, what does rely on general mechanistic knowledge, is the formulation of a likely hypothesis (logic of discovery), not its predictive validity (logic of justification), which can only be assessed as a measure of predictive success in past attempts. It is a good strategy to rely on mechanistic knowledge to choose a model to put to test. Note that the reverse is also true: as a predictively valid model is likely to be mechanistically valid as well, predictive validity is a pragmatic measure of mechanistic validity. In other terms, it is a good strategy, and one frequently resorted to as well, to investigate the mechanisms of a given behavior in models with high predictive value of what happens in humans.

4.5 Conclusion

The upshot of this chapter is to question, at least in psychiatry, the prominence of the received view that extrapolation from animal models in medical research aims to establish mechanisms of human diseases and does so by investigating them in a step by step comparison with animal models of the same disease. We called this form of extrapolation 'mechanistic extrapolation'. We challenged its alleged prominence by describing another form of extrapolation from animal models, we called 'statistical extrapolation'. Statistical extrapolation relies on the predictive validity of animal models, a form of validity that can be established without any knowledge of mechanisms, nor of any resemblance of mechanisms. It relies on past successes of the model to predict the effect of a class of drugs on their human target.

Two caveats are important here. First, we think that our views should not be opposed to the RWT, but that the latter should be revised only on a minor point, namely, its implicit acceptance of the idea that preclinical studies are dedicated to establishing mechanisms. We do not assume that predictive success of preclinical studies, i.e. statistical extrapolation, is a sufficient condition to establish a causal claim, and thereby do not question the core of the RWT.

The second caveat regards mechanistic hypotheses. A rapid reading of the chapter may convey the impression that we think preclinical research may dispense with hypotheses on what the mechanisms of a disease may be. The fact that mechanistic extrapolation is not ubiquitous, and the fact that mechanisms mostly play a heuristic role, should not lead to the conclusion that preclinical studies, even in applied research, is not essential. The fact that mechanistic evidence can be dispensed with does not mean that it is a good thing to do so when it is possible to do otherwise. Whereas being *blind* to mechanisms would involve purely random testing of hypotheses, being *blinded* only involves selective ignorance of vast areas of possibly relevant mechanisms. Even drug screening is selective. Statistical extrapolation is therefore blinded, but not blind to mechanisms.

Alongside mechanistic extrapolation, which consists in the incremental establishment of a mechanism by comparing its instances in a model and in a target, comes statistical extrapolation, which consists in the incremental establishment that a model supports prediction of the target's behavior, whatever the underlying mechanisms maybe. Statistical extrapolation is frequent in the field of animal models, at least in psychiatry, and possibly elsewhere. It does not essentially depend on mechanistic evidence, investigation or even hypothesis. For this reason, it deserves more philosophical attention.

References

Ankeny, R. A., & Leonelli, S. (2011). What's so special about model organisms? *Studies in History and Philosophy of Science Part A, 42*(2), 313–323.

Bartoszyk, G. D., Hegenbart, R., & Ziegler, H. (1997). EMD 68843, a serotonin reuptake inhibitor with selective presynaptic 5-HT1A receptor agonistic properties. *European Journal of Pharmacology, 322*(2–3), 147–153.

Belzung, C., & Lemoine, M. (2011). Criteria of validity for animal models of psychiatric disorders: focus on anxiety disorders and depression. *Biology of Mood & Anxiety Disorders, 1*(1), 9.

Belzung, C., Willner, P., & Philippot, P. (2015). Depression: From psychopathology to pathophysiology. *Current Opinion in Neurobiology, 30*, 24–30.

Broadbent, A. 2011. Inferring causation in epidemiology: Mechanisms, black boxes, and contrasts. In P. M. Illari, F. Russo, & J. Williamson, éd. Causality in the sciences. Oxford Oxford University Press, p. 45–69.

Burrows, G. D., et al. (1988). Clinical effects of serotonin reuptake inhibitors in the treatment of depressive illness. *The Journal of Clinical Psychiatry, 49*(Suppl), 18–22.

Cairncross, K. D., et al. (1977). The olfactory bulbectomized rat: A simple model for detecting drugs with antidepressant potential [proceedings]. *British Journal of Pharmacology, 61*(3), 497.

Campaner, R. (2011). Understanding Mechanisms in the Health Sciences. *Theoretical Medicine and Bioethics, 32*(1), 5–17.

Clarke, B., et al. (2014). Mechanisms and the evidence hierarchy. *Topoi, 33*(2), 339–360.

Cryan, J. F., Valentino, R. J., & Lucki, I. (2005). Assessing substrates underlying the behavioral effects of antidepressants using the modified rat forced swimming test. *Neuroscience and Biobehavioral Reviews, 29*(4-5), 547–569.

Fuller, R. W., Perry, K. W., & Molloy, B. B. (1974). Effect of an uptake inhibitor on serotonin metabolism in rat brain: Studies with 3-(p-trifluoromethylphenoxy)-N-methyl-3-phenylpropylamine (Lilly 110140). *Life Sciences, 15*(6), 1161–1171.

Gottlieb, P., Wandall, T., & Overø, K. F. (1980). Initial, clinical trial of a new, specific 5-HT reuptake inhibitor, citalopram (Lu 10-171). *Acta Psychiatrica Scandinavica, 62*(3), 236–244.

Griebel, G., & Beeské, S. (2012). Is there still a future for neurokinin 3 receptor antagonists as potential drugs for the treatment of psychiatric diseases? *Pharmacology & Therapeutics, 133*(1), 116–123.

Griebel, G., & Holmes, A. (2013). 50 years of hurdles and hope in anxiolytic drug discovery. *Nature Reviews Drug Discovery, 12*(9), 667–687.

Hesse, M. B. (1970). *Models and analogies in science* (New edition). University of Notre Dame Press.

Howick, J. (2011). *The philosophy of evidence-based medicine*. Wiley-Blackwell, Bmj Books.

Hyttel, J. (1977). Neurochemical characterization of a new potent and selective serotonin uptake inhibitor: Lu 10-171. *Psychopharmacology, 51*(3), 225–233.

Illari, P. M. (2011). Mechanistic evidence: Disambiguating the Russo-Williamson thesis. *International Studies in the Philosophy of Science, 25*(2), 139–157.

Katz, R. J., Roth, K. A., & Carroll, B. J. (1981). Acute and chronic stress effects on open field activity in the rat: Implications for a model of depression. *Neuroscience and Biobehavioral Reviews, 5*(2), 247–251.

Koe, B. K., et al. (1983). Sertraline, 1S,4S-N-methyl-4-(3,4-dichlorophenyl)-1,2,3,4-tetrahydro-1-naphthylamine, a new uptake inhibitor with selectivity for serotonin. *The Journal of Pharmacology and Experimental Therapeutics, 226*(3), 686–700.

LaCaze, A. (2011). The Role of Basic Science in Evidence-Based Medicine. *Biology and Philosophy, 26*(1), 81–98.

LaFollette, H., & Shanks, N. (1995). Two models of models in biomedical research. *Philosophical Quarterly, 45*(179), 141–160.

Lassen, J. B. (1978). Influence of the new 5-HT-uptake inhibitor paroxetine on hypermotility in rats produced by p-chloroamphetamine (PCA) and 4,alpha-dimethyl-7-tyramine (H 77/77). *Psychopharmacology, 57*(2), 151–153.

Lemberger, L., et al. (1978). Pharmacologic effects in man of a specific serotonin-reuptake inhibitor. *Science (New York, N.Y.), 199*(4327), 436–437.

Machamer, P. K., Darden, L., & Craver, C. F. (2000). Thinking about mechanisms. *Philosophy of Science, 67*(1), 1–25.

Meltzer, H. Y., et al. (1979). Extrapyramidal side effects and increased serum prolactin following fluoxetine, a new antidepressant. *Journal of Neural Transmission, 45*(2), 165–175.

Meunier, R. (2012). Stages in the development of a model organism as a platform for mechanistic models in developmental biology: Zebrafish, 1970–2000. *Studies in History and Philosophy of Science Part C, 43*(2), 522–531.

Overmier, J. B., & Seligman, M. E. (1967). Effects of inescapable shock upon subsequent escape and avoidance responding. *Journal of Comparative and Physiological Psychology, 63*(1), 28–33.

Page, M. E., et al. (2002). Behavioral and neurochemical effects of 5-(4-[4-(5-Cyano-3-indolyl)-butyl-butyl]-1-piperazinyl)-benzofuran-2-carboxamide (EMD 68843): A combined selective inhibitor of serotonin reuptake and 5-hydroxytryptamine(1A) receptor partial agonist. *The Journal of Pharmacology and Experimental Therapeutics, 302*(3), 1220–1227.

Porsolt, R. D., Le Pichon, M., & Jalfre, M. (1977). Depression: A new animal model sensitive to antidepressant treatments. *Nature, 266*(5604), 730–732.

Reimherr, F. W., et al. (1988). Sertraline, a selective inhibitor of serotonin uptake, for the treatment of outpatients with major depressive disorder. *Psychopharmacology Bulletin, 24*(1), 200–205.

Rickels, K., et al. (2009). Evidence for efficacy and tolerability of vilazodone in the treatment of major depressive disorder: A randomized, double-blind, placebo-controlled trial. *The Journal of Clinical Psychiatry, 70*(3), 326–333.

Russo, F., & Williamson, J. (2007). Interpreting causality in the health sciences. *International Studies in the Philosophy of Science, 21*(2), 157–170.

Russo, F., & Williamson, J. (2012). EnviroGenomarkers: The interplay between mechanisms and difference making in establishing causal claims. *Medicine Studies, 3*(4), 249–262.

Salmon, W. C. (2006). *Four decades of scientific explanation* (Édition: 1). University of Pittsburgh Press.

Salomé, N., et al. (2006). Selective blockade of NK2 or NK3 receptors produces anxiolytic- and antidepressant-like effects in gerbils. *Pharmacology, Biochemistry, and Behavior, 83*(4), 533–539.

Seligman, M. E., & Maier, S. F. (1967). Failure to escape traumatic shock. *Journal of Experimental Psychology, 74*(1), 1–9.

Steel, D. (2008). *Across the boundaries: Extrapolation in biology and social science*. Oxford: Oxford University Press.

Treit, D., et al. (2001). Systemic EMD 68843 injections reduce anxiety in the shock-probe, but not the plus-maze test. *European Journal of Pharmacology, 414*(2–3), 245–248.

van Riezen, H., Schnieden, H., & Wren, A. F. (1977). Olfactory bulb ablation in the rat: Behavioural changes and their reversal by antidepressant drugs. *British Journal of Pharmacology, 60*(4), 521–528.

Willner, P., Scheel-Krüger, J., & Belzung, C. (2013). The neurobiology of depression and antidepressant action. *Neuroscience and Biobehavioral Reviews, 37*(10 Pt 1), 2331–2371.

Wong, D. T., et al. (1974). A selective inhibitor of serotonin uptake: Lilly 110140, 3-(p-trifluoromethylphenoxy)-N-methyl-3-phenylpropylamine. *Life Sciences, 15*(3), 471–479.

Wong, D. T., et al. (1975a). A new selective inhibitor for uptake of serotonin into synaptosomes of rat brain: 3-(p-trifluoromethylphenoxy). N-methyl-3-phenylpropylamine. *The Journal of Pharmacology and Experimental Therapeutics, 193*(3), 804–811.

Wong, D. T., et al. (1975b). A new selective inhibitor for uptake of serotonin into synaptosomes of rat brain: 3-(p-trifluoromethylphenoxy). N-methyl-3-phenylpropylamine. *The Journal of Pharmacology and Experimental Therapeutics, 193*(3), 804–811.

Wong, D. T., Perry, K. W., & Bymaster, F. P. (2005). Case history: The discovery of fluoxetine hydrochloride (Prozac). *Nature Reviews Drug Discovery, 4*(9), 764–774.

Worrall, J. (2010). Evidence: Philosophy of science meets medicine. *Journal of Evaluation in Clinical Practice, 16*(2), 356–362.

Chapter 5
Analogy-Based Inference Patterns in Pharmacological Research

Roland Poellinger

Abstract Analogical arguments are ubiquitous vehicles of knowledge transfer in science and medicine. This paper outlines a Bayesian evidence-amalgamation framework for the purpose of formally exploring different analogy-based inference patterns with respect to their justification in pharmacological risk assessment. By relating formal explications of similarity, analogy, and analog simulation, three sources of confirmatory support for a causal hypothesis are distinguished in reconstruction: relevant studies, established causal knowledge, and computational models.

Keywords Scientific inference · Pharmacology · Epistemology · Bayesian confirmation · Evidence · Relevance · Similarity · Analogy · Computer simulation

5.1 Introduction: Scientific Inference in Pharmacology

Pharmacological research is often driven by many forces at once: Cost effectiveness must be balanced against extensive data-collecting, potential risk against probable benefit, and breadth of applicability against well-documented higher confidence for smaller target groups. Many such decisions must be taken during the development stages of a certain drug before the desired effectiveness and safety level is reached and the drug is allowed to be marketed. A language for expressing both benefit and safety is found in the probabilistic language of expected utilities and dis-utilities.

This work is supported by the European Research Council (grant 639276) and the Munich Center for Mathematical Philosophy (MCMP).

R. Poellinger (✉)
Department of Arts and Culture of the City of Munich, Munich, Germany
e-mail: roland.poellinger@muenchen.de

A. LaCaze, B. Osimani (eds.), *Uncertainty in Pharmacology*, Boston Studies in the Philosophy and History of Science 338,
https://doi.org/10.1007/978-3-030-29179-2_5

Nevertheless, the formalization of a given decision problem in such vocabulary can only be as informative about future drug users as the evidence it is rooted in. Yet, whether it is all the evidence about the drug's effects that is to be taken into consideration, or only the best evidence available, is the subject of an ongoing discussion in the philosophy of medicine. A recent paper by Landes, Osimani, and Poellinger (Landes et al. 2017) explores the possibility of amalgamating all available evidence in a Bayesian reconstruction of scientific inference for the integrated probabilistic assessment of a drug's causal (side-)effects. Key to this endeavor is the distinction of the conceptual levels involved: (i) the causal hypothesis, (ii) testable indicators of the causal claim, i.e., theoretical consequences of the causal hypothesis, and (iii) concrete evidence (actual data) speaking for or against the respective indicators (thereby indirectly supporting the hypothesis, or not). Close to pharmacological practice, it is furthermore useful to introduce an additional *meta-evidential* level for the purpose of encoding the qualified assessment of the data at hand. This allows to conceptually distinguish the *content* of an evidential report from its *weight* in the evaluation of the hypothesis: Different pieces of evidence may possess different *levels of significance* (iv).

One important justification of the confirmatory support a piece of evidence lends to a given hypothesis (by virtue of it being evidence for an indicator of the very hypothesis) is the postulate (or implicit assumption) of analogy between the circumstances generating the evidence and the hypothesis' intended (future) scope of application. Sir Austin Bradford Hill lists analogy as one of his famous guidelines towards an informed assessment of potential causes in epidemiology:

> In some circumstances it would be fair to judge by analogy. With the effects of thalidomide and rubella before us we would surely be ready to accept slighter but similar evidence with another drug or another viral disease in pregnancy.[1] (Hill 1965, p. 11)

The aim of this paper is to explore different analogy-based inference patterns in the above-sketched layered reconstruction of scientific reasoning, with respect to their justification in pharmacological risk assessment. In particular, different aspects of similarity shall be made transparent in order to compare conceptually different types of evidence in analogical reasoning. Three interrelated questions shall be addressed in the following sections:

1. How can analogy considerations be used to explicate the relevance of evidence for a hypothesis under consideration? (Sect. 5.2)
2. How can an already well-established hypothesis be used as a supporting analog in confirming a hypothesis about a similar drug? (Sect. 5.3)

[1] In this passage, Hill refers to (i) severe disabilities (even death) among babies linked to the over-the-counter drug thalidomide, prescribed in the 1960s in Germany as Contergan to alleviate morning sickness in pregnant women, and to (ii) miscarriage or children born with the congenital rubella syndrome (CRS) due to infection by the rubella virus during pregnancy.

3. By what standards can a mechanistic computational model of a substance's causal effects as a theoretical analog lend evidential support to a causal hypothesis about the actual substance? (Sect. 5.4)

Before we can trace the role of analogy in pharmacological research, though, a formal reconstruction of the dynamics underlying scientific hypothesis testing shall be outlined in the following.

5.1.1 *Evidence Amalgamation and Hypothesis Confirmation*

Causal inference in pharmacology is a difficult task due to sometimes sparse, oftentimes very heterogeneous evidence, different standards for evidence evaluation and integration, and many sources of random and systematic error. In their pluralist, conciliatory approach, Landes, Osimani, and Poellinger (Landes et al. 2017) propose a blueprint for tracing the epistemological dynamics of evidence amalgamation, viewing risk assessment in pharmacology from a meta-perspective. In their framework, causal hypotheses (about adverse drug reactions) and evidential reports (about concrete studies) are related in such a way, that successively cumulating evidence allows for probabilistic causal inference. One important point of departure for this approach is Bovens' and Hartmann's general reconstruction of scientific inference (Bovens and Hartmann 2003) in which the epistemic dynamics of all the dependencies between a hypothesis, testable indicators implied by the hypothesis, and evidence for/against such indicators can be visually traced in the graph of a Bayesian network: The conceptual categories are depicted as layers of nodes, with directed edges between the layers marking those paths along which confidence in the hypothesis is boosted (or lowered). The probabilistic model underneath the graphical representation supplements the Bayesian network with quantitative information by encoding (i) conditional degrees of (un)certainty about all variables, and (ii) how these degrees will change under local updates.

The example in Fig. 5.1 illustrates the epistemological structure: The causal hypothesis (Hyp) entails n theoretical indicators (i.e., testable consequences Ind_1, $\ldots$, Ind_n), which are to be supported by concrete evidence reports $Rep_1, \ldots, Rep_n$ on the lowest level. Since evidence reports are based on concrete data (e.g., clinical trials, historical studies, or lab experiments), it will be desirable in many cases to modulate their significance for the assessment of a particular causal hypothesis for various reasons: The source of information seems undependable, the quality of a study might be doubtful, the reliability of method or measurement device is not guaranteed, test group and target group deviate in relevant details, and so on. To be able to express such levels of significance, the report nodes in the graph come with an additional weight node α (for the moment simply a blackbox placeholder for the aforementioned significance dimensions, used below to encode *evidential relevance* in particular, see Sect. 5.2). The degree to which a given evidence report supports the hypothesis depends precisely on its "weight", i.e., its relevance to the hypothesis,

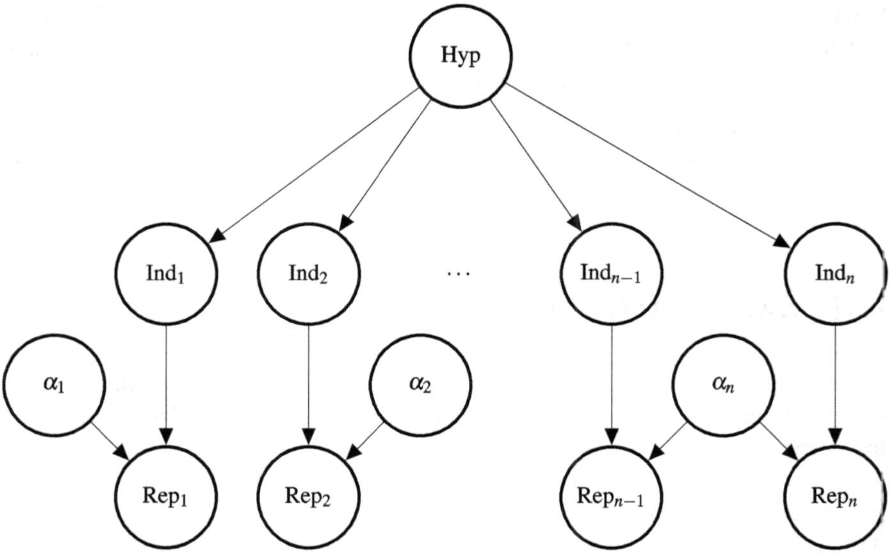

Fig. 5.1 Example of how a causal hypothesis, multiple testable consequences, and weighted evidence reports might be related in a Bayesian network

the reliability of the source, the error-proneness of the methods used to generate the data, and so on. Such weight nodes might in general be shared between reports – for example in cases where more than one report is based on data generated by the same measurement device. While Bovens and Hartmann utilize this weighting parameter to explicate the reliability of a report, Landes, Osimani, and Poellinger – aiming at hypothesis confirmation in pharmacology – split this weighting parameter into two variables (*reliability* and *relevance*) in order to distinguish the quality of method and information source from questions of external validity (by encoding the degree to which study results can be *extrapolated* to the target).[2]

What it means to be an observable consequence of the hypothesis is implicitly stated by the following inequality (see also Bovens and Hartmann 2003, p. 90):

$$P(Ind_i \mid Hyp) = p_i > q_i = P(Ind_i \mid \neg Hyp), \tag{5.1}$$

where 'Hyp' is to be understood as shorthand for 'Hyp = true', and analogously for the other variables (in the following, context disambiguates if a variable or its instantiation is referred to).

[2]As implied by this way of distinguishing reliability and relevance, the relevance weight (attached to a given evidential report) is really meant here to capture the degree of external validity. Of course, there are other ways in which a study can be relevant to the hypothesis – for example, if it is conducted by an acknowledged authority. In the framework of Landes, Osimani, and Poellinger (Landes et al. 2017), this way of being relevant to the hypothesis would be encoded in the reliability weight, which collects all sources of bias.

The conditional in/dependencies between the epistemic categories can be read off the structure in Fig. 5.1 by utilizing the graphical *d*-separation criterion[3]: All directed edges mark positive influence between adjacent nodes (variables, respectively), which makes all Hyp, Ind, and *Rep* variables positively correlated (conditional on the empty set). Following Bovens and Hartmann (2003), the weight variables are *independent* of the theoretical categories (i.e., hypothesis and indicators) and may be used to account for exceptions.[4] I will treat the α-weights as *binary modulators* throughout the paper (with $\alpha = 1$ indicating *significance* and $\alpha = 0$ indicating *insignificance*, respectively); nevertheless, the formal framework can easily be extended to allow for a more fine-grained (in many cases more appropriate) representation.

The in/dependencies visualized in Fig. 5.1 can be summed up probabilistically as follows, for all i $(1 \leq i \leq n)$:

$$Hyp, Ind_i \perp\!\!\!\perp \alpha_i \tag{5.2}$$

$$P(Rep_i \mid Ind_i, \alpha_i = 1) = 1 \tag{5.3}$$

$$P(Rep_i \mid \neg Ind_i, \alpha_i = 1) = 0 \tag{5.4}$$

$$\begin{aligned} P(Rep_i \mid\ & Ind_i, \alpha_i = 0) = \\ P(Rep_i \mid\ & \neg Ind_i, \alpha_i = 0) = \\ P(Rep_i \mid\ & \alpha_i = 0) = q_i \end{aligned} \tag{5.5}$$

Equation 5.2 precisely encodes the independent assignment of the weighting parameter. Equations 5.3 and 5.4 show that an evidence report marked as "significant" to the hypothesis is aligned with the respective indicator. The loose term "significance" is meant to be understood in this context as a quite general concept up for interpretation – it does not refer to statistical significance of study results, but rather to the quality of evidence (as meta-information), e.g., in terms of *reliability* or *relevance*, as discussed below. Equation 5.5 explicates what happens when an evidence report is qualified as "utterly insignificant" to the hypothesis: In that case, whether the report speaks for or against the respective indicator becomes probabilistically independent of the truth value of this very indicator variable (with $0 \leq q_i \leq 1$). In particular, an evidence report considered *irrelevant* for the

[3]The graphical *d*-separation criterion (with *d* for *directional*) distinguishes conditionally dependent (sets of) variables from conditionally independent ones by drawing on structural information, i.e., on how arrows are directed along the paths between the (sets of) variables under consideration; see, e.g., Geiger et al. (1990).

[4]In this case, independence between the weight variables and the hypothesis "may or may not be a realistic assumption", as Bovens and Hartmann concede, and they extend their discussion to cases where such weight nodes (reliability of evidence reports) and the hypothesis are made dependent through auxiliary theories (see Bovens and Hartmann 2003, pp. 107ff.). For the purpose of this paper, though, the standard for assigning values to weight variables is assumed to be fixed prior to hypothesis testing.

investigated hypothesis should not influence one's confidence in the truth of the hypothesis' indicators.[5]

With this Bayesian framework at hand, explicating what it means for a given piece of "significant" evidence to confirm a causal hypothesis is straightforward: An evidence report Rep_k is said to confirm the hypothesis Hyp iff it raises Hyp's probability, i.e., iff

$$P(Hyp \mid Rep_k) > P(Hyp). \tag{5.6}$$

Obviously, this inequality does not hold for $\alpha_k = 0$, i.e., for "insignificant" evidence reports (unreliable sources of evidence or evidence irrelevant for the hypothesis etc.). Consequently, what it means for a given piece of evidence to be "significant" to a hypothesis must be defined before we can start confirming this very hypothesis. Analogical reasoning shall be exploited towards that goal in the following.

5.1.2 Analogy as Inferential Pattern

At the heart of analogical arguments lies the assumption of (sufficient) similarity between the two relata (or aspects thereof, respectively). The concept of similarity comes with a mixed bag of notorious epistemological issues of its own, though. For example, if two drugs are to be related in an analogical argument, the following questions come to mind: What does it mean to be sufficiently similar in the case under consideration? In what way does the difference between the first and the second drug influence changes in expected outcome values? How specific are a drug's properties? If they are highly specific – to what extent can this drug be used in an analogical argument, if at all? Despite these conceptual difficulties, science and history are full of successful examples of analogical reasoning, even in the case of scientific discovery: In the nineteenth century, secured knowledge about acoustics was employed in the discovery of spectral lines. Guided by the image of a harmonic oscillator, physicists were able to focus their attention to groups of spectral lines with specific frequency patterns from the beginning (see Bartha's in-depth overview of analogical arguments in Bartha (2013), as well as Unruh (2008), Dardashti et al. (2017), and Hesse (1952) for discussions of analog arguments in physics).

In pharmacology and epidemiology, reasoning by analogy is a key mode of knowledge transfer from study to target population. Indeed, because of the context sensitivity of many causal associations in the biological realm, these associations can hold only in specific populations, and therefore evidence about causal effects related to one population may not license similar conclusions about another population, unless the two populations have been established as *analogous* (in

[5]The Bayes net structure in Fig. 5.1 for example illustrates that Ind_1 is influenced by (since d-connected to) α_1 once we know the value of the "collider variable" Rep_1.

relevant respects). Knowledge about an agent's mechanisms and about its impact on the biological environment might be sparse and come from quite heterogeneous sources, however. Yet, already if only little information about the agent's class of molecules is available, for example, this can justifiably be exploited for causal assessment *via analogy*.

As Bartha (2010) points out, analogical arguments fall into a different category than regular evidence as such. Consider the general form of an analogical argument (see Bartha 2013):

1. y is similar to x (in certain known respects), **(A)**
2. x has additional property A,
3. therefore: y has property A' (similar to A).

This argument obviously encodes basic evidence about x which is then used to make inferences about y. It is unclear, as Bartha points out, how such a structured argument could be encapsulated in a single *evidential proposition* to figure as a condition in the scheme of Bayesian confirmation, stated in Eq. 5.6 above. Moreover, according to Bartha, even if we were to condense a full analogical argument into one such evidential proposition A, the postulated similarity between source and target domain must have been established *before* A can be used in confirmation, which makes A "old evidence" and useless for confirmation (since it does not boost the degree of belief in the hypothesis). This becomes obvious once Eq. 5.6 is relativized to fixed background knowledge K. With the analogical argument A entailed by K, we get $P(Hyp \mid A, K) = P(Hyp \mid K)$.

With the above-sketched Bayesian reconstruction of scientific hypothesis testing at hand, we already have a structure-rich framework utilizable for a different formalization of analogical reasoning: The key idea is to understand analogy not as single nodes, but to trace confirmation by analogy along the edges in the graph. In other words, analogical arguments shall be understood and expressed rather as inferential patterns than as "evidence nodes".

The evidence-amalgamation framework can be operationalized for this purpose in the following ways:

1. The question of applicability of a study's findings is best phrased in terms of correspondence between study and target populations, where correspondence shall be understood as *similarity* (above a certain threshold): If study and target populations are *sufficiently similar*, researchers are licensed to reason about causal links in the target population by analogy with their test cases. The extent to which this kind of transfer is licensed is encoded in the framework as *relevance* of available reports (such that independent reports are each assigned a different degree of relevance with respect to the investigated hypothesis). The more characteristics are shared between study and target, the higher the relevance of evidence obtained in this study for the hypothesis under investigation. This case is treated in Sect. 5.2.
2. The investigated causal hypothesis Hyp may be related to a second causal hypothesis Hyp^* which has already been established in previous studies. Now,

if scientists have sufficient reason to postulate analogy between Hyp and Hyp^* (e.g., a high degree of chemical or functional similarity), knowledge about this second causal hypothesis supports the first hypothesis Hyp *via analogy*. This case requires an extension of the layered network introduced above and will be treated in Sect. 5.3.

3. As a special sub-case of 2., I will discuss how Hyp can possibly be confirmed by way of computational modeling (instead of by relating it to a second causal hypothesis Hyp^* about a biological system): In cases where scientists want to confirm mechanistic assumptions, computer simulation has become a viable alternative to costly, unethical, or otherwise inaccessible experimental studies. Assumptions about the efficacious mechanistic relationships may be modeled in a *virtual analog* to find out more about the protein a drug binds to or to simulate an agent's interaction with the biological system, for instance, w.r.t. dosage (see, e.g., Britton et al. 2013 and Carusi et al. 2012). In order to treat this special case I will tweak the network structure even further in Sect. 5.4.

My aim in this paper is to exploit the concept of analogy in explicating these different knowledge transfer strategies. To this end, I will investigate the confirmatory dynamics of analogical reasoning in a formal way: All three above mentioned cases shall successively be located in the evidence-amalgamating framework for the purpose of unifying the inferential patterns and emphasizing their structural differences at the same time. I will start by discussing ways of designing possible (comparative or numerical) distance measures for expressing similarity (between substances, populations, etc.).

5.2 Learning from Relevant Evidence

5.2.1 Heterogeneous Evidence

The evidence-amalgamating framework, as introduced above, presents itself as a tool for describing specific cases in that it can be adapted to accommodate specific pieces of evidence for (or against) specific theoretical consequences of the causal hypothesis under consideration. At the same time, it is meant to be understood as a normative statement about the quality of the causal assessment: The more information about the indicator variables is available, the more reliable the assessment of the hypothesis. This is especially true in the case of *causal* hypotheses, given the variety of methodological approaches towards discovering causal associations in the sciences.[6]

[6]Landes et al. (2017) contains a non-exclusive list of six causal indicators derived from Hill's guidelines in Hill (1965). See also Poellinger (forthcoming) for a discussion of the conceptual relationships of these causal indicators and the ramifications of theory choice in causal assessment.

The multi-layered framework introduces report nodes as placeholders for heterogeneous pieces of evidence. The reports can come from all levels of the "evidence hierarchy" – the framework treats all these levels as relevant in causal assessment: Systematic reviews, meta-analyses of randomized clinical trials or observational studies, comparative non-randomized studies (cohort or case-control studies), evidence of (sub-)mechanisms from basic science (lab experiments etc.), as well as expert judgment. Relevant information may come also from single case reports, case series, and animal studies. Amalgamating these different types of evidence has not only epistemological value (since it traces the epistemological dynamics from observed signals of Nature to the establishment of hypotheses) but also methodological value (since it readily accommodates different approaches towards evidence assessment and related disputes like the seeming tension between paradigms such as best-evidence vs. pluralistic approaches).

5.2.2 *Relevance*

However heterogeneous the evidence, different evidence reports (understood as data interpreted relative to the hypothesis under consideration) will be attributed different levels of significance for the hypothesis: Virtually no pharmacological study allows for straightforwardly transferring shown results from studied population to target population of possible future drug users – population size, inherent structure, specific circumstances, possible interactions with uncontrolled substances, etc. might be similar, but will virtually never be the exact same (see Chan and Altman 2005, p. 1180; Doll and Peto 1980; Worrall 2007, p. 992). Furthermore, patient inclusion criteria for participation in RCTs will skew the inference transfer from study to target population even more (see, e.g., Revicki and Frank 1999 and Upshur 1995, p. 483). When the results of animal studies are to be applied to a population of future human drug users it becomes quite clear that questions of applicability must be settled first before such transfer is licensed.[7] Moreover, in their recent analysis of the role of evidence about a substance's mechanisms in risk assessment, Luján, Todt, and Bengoetxea (Luján et al. 2016) point towards the lack of guarantee that similarity of modes of action may warrant extrapolation of phenotypic effects from one chemical to another – chemicals considered similar in important respects might not necessarily produce similar effects (for a given population). And finally, when laboratory experiments yield information about some component of the causal mechanism (e.g., at the molecular level or in terms of cell behavior), the significance of this (partial) result for the hypothesis about the entire mechanism must be determined first.[8]

[7] LaFollette and Shanks (1995) argue, e.g., that animal studies are limited to hypothesis *generation*.

[8] In particular, the question must be answered whether the partial result can be combined in an additive fashion with information about further sub-mechanisms, or whether complex interdependencies forbid partitioning the full mechanism into stand-alone modules.

In all these cases scheme (A) is applied implicitly or explicitly: The hypothesis is confirmed by analogy, i.e., its degree of confidence is raised by a given evidence report, once similarity between studied subjects (or objects) and intended application is established. In the framework of Landes, Osimani, and Poellinger, this type of significance of a given evidence report for a hypothesis to be tested is expressed as evidential relevance factor (whose value is to be determined outside the model), cf. Landes et al. (2017), Sect. 3.3[9]:

> Ideally, pharmacological studies would license the same inferences for the studied population and the target population of future drug users. In reality, studies are not conducted on the *entire population of future drug users* but on a much smaller number of patients, see Chan and Altman (2005, p. 1180), Button et al. (2013) for this problem in neuro science, Doll and Peto (1980) in cancer research and a philosophical discussion of this problem in Worrall (2007, p. 992). Additionally, studied populations, in particular in RCTs, often fail to be representative for the target population due to strict patient inclusion criteria, see Revicki and Frank (1999) and Upshur (1995, p. 483). Therefore, there is a need to reason by analogy from the studied population to the population of interest [...] The relevance pertaining to an item of evidence measures how well the observed results in a study population can be transferred to the target population of future drug users.

Figure 5.2 illustrates this relationship between an item of evidence and the target: When hypothesis Hyp is on the test bench, testable consequences $Ind_1, \ldots, Ind_n$ (indicators of causation) are supported (or not) by evidence reports $Rep_1, \ldots, Rep_n$. Since the framework is set up in Bayesian fashion, once the value of Rep_k is fixed (e.g., to $Rep_k = true$, indicating that there is evidence in support of $Ind_k = true$), the respective significance variable α_k will determine the relative strength of Rep_k's contribution to Hyp (since – in Bayesian terms – knowing a collider's value opens the flow of information along the path the collider lies

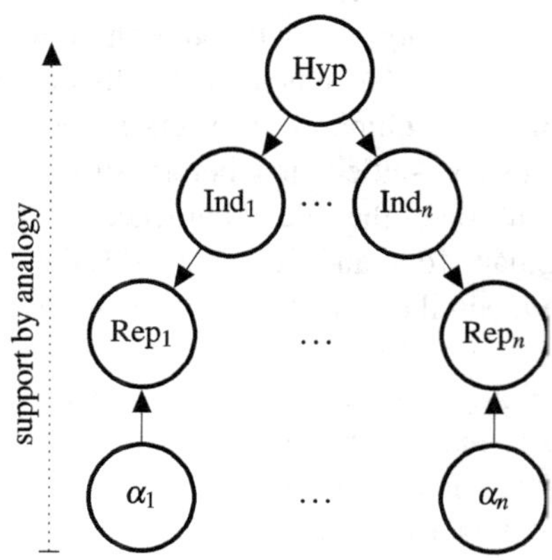

Fig. 5.2 Support by analogy from the level of evidence reports $(Rep_1, \ldots, Rep_n)$ upwards to the hypothesis under consideration

[9]Note that I am distinguishing *evidential* relevance (as a property of evidence relevant in causal assessment) from *causal* relevance (as a property of a variable causally relevant to a second variable in a causal model), cf. Footnote 17. Also see Footnote 2 above for a remark on other interpretations of relevance.

on). Note, that – in principle – two (or more) report nodes might share the same significance, as exemplified in Fig. 5.1, when inter-dependencies are to be indicated (in terms of either *reliability* or *extrapolability*, i.e., *relevance*).

If we now interpret the α-layer as encoding information about the relevance of our evidence reports, we can try to make the relationship between α_k and Hyp transparent in terms of similarity.[10]

5.2.3 Measuring Distance

In order to be able to compare the hypothesis and incoming evidence reports, the structural ingredients of these categories shall be made explicit in more detail. In risk assessment, Hyp is meant to be a causal hypothesis about a specific (harmful) side-effect E of a certain drug D – to be abbreviated as '$D \copyright E$' in the following. Since the evidence-amalgamating framework is intended to yield quantitative (probabilistic) information for decision-making (e.g., in terms of expected utilities), $D \copyright E$ must be evaluated in a quantitative framework that allows for the specification of all relevant causal relations in a suitable formal model M. One important modeling decision will be the separation of parameters influenced *within* the model (i.e., 'observed' or 'endogenous' variables – in particular, the modeled effects and side-effects of interest, along with further variables on the causal path from drug to potentially harmful outcome) from *ceteris paribus conditions* ('context' or 'world' or 'circumstances', encoded as 'exogenous' variables in a causal model). A fourth parameter shall be added to our language to express precisely these contextual conditions, i.e., the ceteris-paribus aspects of the population for which the causal mechanism is to be tested. The parameter U shall sum up those characteristics of population plus environment that will not be changed by, e.g., a clinical trial (as, e.g., average age, nutrition habits, aspects of the medical history, socio-economic status, etc.).[11] We can accordingly expand Hyp to encode the aforementioned ingredients:

$$Hyp: \; D \copyright E \text{ in model } M \text{ with context } U \tag{5.7}$$

[10]Note that for such an interpretation the prior of the network is required to be set up in such a way that the αs render the $Reps$ independent of (i.e., irrelevant to) the hypothesis Hyp in the extreme case. Formally: $P(Hyp \mid Rep_k = true, \alpha_k = irrelevant) = P(Hyp \mid \alpha_k = irrelevant) = P(Hyp)$. See also my discussion of the prior in Sect. 5.1.1 above.

[11]This suggests that the distinction between changeable and unchanged aspects of the study population will be a static one prior to modeling. Nevertheless, Paul and Healy discuss cases in which the modeler is forced to revisit her model because a clinical trial impacts on relevant characteristics of the population in a sort of feedback loop (see Paul and Healy 2016 on *Transformative Treatments*). For the present purpose it is uncritical to assume that the initial model can be refined at later stages to accommodate previously exogenous assumptions into the model as endogenous parameters/relations.

Explicating the causal hypothesis in this way is not meant here to express a specific commitment as to how I' is to be interpreted. In particular, it is not meant to confine the investigation to an epistemic, model-relative theory of causality. This specific way of encoding the causal hypothesis is rather meant to be as informative as possible about the intended scope of the causal claim and to mark deviations between studies and target with respect to population characteristics (information about age, multimorbidity, and so on, encoded in $U_{(k)}$) and structural assumptions (confounders, causal history, possible disablers, etc., encoded in $M_{(k)}$).[12]

Now, using a piece of evidence for the confirmation of a hypothetical causal effect *E by analogy* may be licensed for different reasons:

1. Similar substance D: The substance used on the study population is similar to the substance to be applied to the target population. Of course, if an experimental setup is designed with $D © E$ in mind, the experiment will generally test the causal effects of D itself. Nevertheless, what may vary is the administered dosage or indeed the final formula with respect to additional components.
2. Similar causal model M: Causal efficacy is judged with respect to similar modeling assumptions (variables and their order, causal in/dependencies, and so on) in the study and for the intended application. For example, when cohort studies point to the existence of a *causal path* between paracetamol and asthma, knowledge about these studies can only be transferred to a population of future drug users if the core causal assumptions are kept fixed. Postulating a deviating causal structure for the target, e.g., a *common cause* of paracetamol use and asthma instead of a causal path, would prevent using evidence from such cohort studies for predicting the effects of certain policy interventions towards decreasing asthma morbidity.
3. Similar context U: Contexts, i.e., study and target population plus respective environment, are similar in relevant respects. For example, when drugs are tested on animals first, pigs are one of the preferred species, because they share much of the human genetic make-up and consequently many of the complex genetic diseases, such that a drug's causal effects in pigs can to some extent be used in inference about human biology.

Scheme (A) above can now be adjusted for *learning from relevant evidence* in the evidence-amalgamating layered reconstruction of scientific inference:

1. $\langle D, M, U\rangle$ is similar to $\langle D_k, M_k, U_k\rangle$, **(A*)**
   therefore, by stipulation: Rep_k is relevant for Hyp, i.e., α_k is high,
2. Rep_k: $\Phi[D_k, E_k, M_k, U_k]$
   with E_k similar to the hypothesized E,

[12]If the causal hypothesis is thought of as a causal graph, D, E, and U are meant to represent designated (sets of) variables with token values in a causally interpreted structure M (possibly encoding the specifics of direct causal relations and assumptions about causal in/dependencies on type level). Note that, more generally, M and M_k can be thought of as *sets of structural constraints*, i.e., as *classes of causal graphs*.

3. therefore: $\Phi'[D, E, M, U]$
   with Φ' similar to Φ,

where Φ is a quantitative or qualitative statement about (the components of) the study (and Φ' a statement about the target, respectively). For example, Φ might encode the conditional probability $P(E_k \mid D_k) = q$; therefore, *by analogy*, Φ': $P(E \mid D) \approx q$ (for high α_k). Note that, by design, Rep_k will influence Hyp on a path through a suitable (set of) mediating indicator variable(s). If now Rep_k makes a positive contribution to the indicator level, it will boost confidence in Hyp, thereby confirming $D \copyright E$:

$$P(D \copyright E \mid Rep_k, \alpha_k) \geq P(D \copyright E).$$

The inference pattern (A*) crucially relies on a high relevance parameter α_k which is determined in assessing the *similarity* of the triples $\langle D, M, U \rangle$ and $\langle D_k, M_k, U_k \rangle$. The literature on similarity conceptualizations and measures is vast, spanning from Lewis' lexicographic *comparative similarity* of possible worlds (Lewis 1973a) to Shepard's *geometric account* in terms of vector distances (Shepard 1980) and to *contrast approaches* evaluating similarity by weighting and comparing properties (see, e.g., Tversky 1977; Weisberg 2012, 2013).[13] Some are formulated in a symmetric way, others encode asymmetric relations. All approaches seem tied to specific practices and purposes with seemingly none universally applicable. Nevertheless, with analogical arguments employed successfully in pharmacological research, general features of the similarity concept needed for (A*) shall be outlined in the following.

Claiming that $\langle D, M, U \rangle$ is similar to $\langle D_k, M_k, U_k \rangle$ (as in line 1 of the inference scheme for relevant evidence) quite generally means that the individual components are in the same equivalence class (by pairs), formally: $D_k \in [D]_\sim$, $M_k \in [M]_\sim$,

[13]These accounts of similarity share the intuition that comparing two things means (i) comparing certain aspects of those things and (ii) aggregating one's evaluation of those aspects in a certain manner.

Lewis' idea of comparative similarity is tightly connected to his concept of causation, where a cause–effect relation is evaluated in terms of the corresponding counterfactuals. The cause reveals its power in the effect event where the rest of the world remains unperturbed, i.e., as *similar as possible* to the state of world prior to the cause event. Lewis suggests a priority ordering for the assessment of similarity, where local changes in physical facts are understood as a lesser deviation from actuality than far-reaching global changes in natural laws, see Lewis (1973b).

The geometric account locates an object's properties (deemed relevant for comparison) in a multi-dimensional space by assigning a specific value to each of those properties. Similarity is then spelled out in terms of vector distance from a reference object.

The question of how to assign such values is circumvented in the contrast approach which deals well with similarity as *partial identity*, since in this approach degrees of similarity are assessed by assigning weights to *co-instantiated identical properties* (which might make the approach suitable rather for comparing different states of one and the same object than for comparing different objects).

and $U_k \in [U]_\sim$.[14] These equivalence classes are induced by (i) the components of the hypothesis, D, M, U, and (ii) highly purpose-oriented considerations as to balancing the properties to be compared (i.e., as to how properties are to be ordered in Lewis-style approaches, as to how which properties are metrized in geometric approaches, and as to how properties are to be weighed in contrast comparisons). Presenting similarity this way allows a research community to fill in a meaningful, informative understanding of similarity, and it separates purpose-driven considerations concerning how to suitably define an equivalence relation from the decision about if and when two components actually are equivalent (relative to a specific definition). Note that however dissimilar study and target, if some Hyp's components $\langle D, M, U\rangle$ and some Rep_k's components $\langle D_k, M_k, U_k\rangle$ were not understood to be comparable whatsoever, Rep_k would not be considered evidence for Hyp in the first place.

The threshold between sufficient similarity and unacceptable deviation is straightforwardly encoded as a component's membership of the respective equivalence class. Line 1 in (A*) can thus be understood as follows:

$$\langle D, M, U\rangle \text{ is similar to } \langle D_k, M_k, U_k\rangle :\Leftrightarrow \tag{5.8}$$
$$D_k \in [D]_\sim \text{ and } M_k \in [M]_\sim \text{ and } U_k \in [U]_\sim .$$

Meaningfully evaluating the different components' membership of a given equivalence class calls for the distinction between similarity of *numeric* properties and similarity of *structural* properties. In the following I will outline possible paths one might take towards assessing similarity both of the numeric and the structural type.

5.2.3.1 Similarity of Numeric Properties

I will first consider the question of how to possibly compare arrays of numeric properties (or properties represented numerically, respectively). What it means to be a member of an equivalence class in that case can be expressed in a weighted geometric approach as the following criterion (with X schematically representing D, M, U):

[14] In the formal notation used here, $\sim$ denotes some (reflexive, transitive, and symmetric) equivalence relation (equivalence w.r.t. a given property) such that for a domain A, some object $a \in A$, and an equivalence relation $\sim$ on A: $[a]_\sim := \{x \in A \mid x \sim a\}$. The expressions $[D]_\sim$, $[M]_\sim$, and $[U]_\sim$ are to be understood as encoding each a specific equivalence relation, since – to be precise – each category comes with its own standards for how equivalence classes are to be generated. If standards are set high, e.g., D_k might be in the class $[D]_\sim$ only if it is identical with D, while comparing U and U_k will naturally demand flexibility for possibly very different populations. (I will not add a further index, though, to avoid notational clutter.)

$$X_k \in [X]_\sim :\Leftrightarrow \mathrm{d}(\overrightarrow{\beta} \circ \overrightarrow{x_k}, \overrightarrow{\beta} \circ \overrightarrow{x}) \leq \sigma_X. \tag{5.9}$$

Equation 5.9 states that component X_k (e.g., some study population U_k) is similar to X (e.g., the target population U) for a purpose-oriented similarity evaluation $\sim$, explicated as distance between vectors of m properties $x_k^1, \ldots, x_k^m$ (of X_k) and $x^1, \ldots, x^m$ (of X) above a certain distance threshold σ (relative to component X). To keep the criterion as generally applicable as possible, the elements of $\overrightarrow{x_k}$ and $\overrightarrow{x}$ should represent normalized values of the properties measured for comparison. When comparing two populations, for example, their relevant characteristics, such as, e.g., average age and population size, might be mapped onto scales of equal magnitude, to arrive at a balanced aggregate assessment. Additionally, the vectors to be compared are weighted by componentwise multiplication (denoted by '$\circ$') with relative weights $\beta^1, \ldots, \beta^m$, indicating the *relative importance* of each property for comparison, i.e., its contribution to the similarity measure.[15]

Of course, the distance measure d must be specified (Euclidean distance, Manhattan distance, etc.) and the set of properties to be compared chosen carefully – in particular, a simple distance measure might not be straightforwardly applicable if the selected properties are dependent. What counts as independent and what not is, again, ultimately to be decided in a purpose-oriented, context-sensitive way. In particular, different similarity standards could lead to different in/dependence requirements. A toy example may serve as an illustration of such dependencies: Consider the task of comparing different apples. If shape and color are to be compared, then a big red apple is more similar to a big green apple than to a small green apple. If, however, "naturalness" is used as similarity standard, then a big red apple might be considered more similar to a small green apple than to a big green apple, since the natural apple naturally ripens from small and green to big and red, and not to "unnatural" big and green. The latter case precisely illustrates the dependency between properties, and how such dependency is relative to the similarity standard to be employed.

Example

The following simplified case equally shows advantages and shortcomings of such a geometric approach. Consider the assessment of a potential side-effect of a drug targeted at elderly people, who are likely to be using multiple medicines and likely to be affected by multiple diseases (note that those properties are dependent to a certain degree). Similarity between this specific target population and a test group (associated with report Rep_k) might be determined in comparison of property triples $\overrightarrow{u} = \langle u^1, u^2, u^3 \rangle$ and $\overrightarrow{u_k} = \langle u_k^1, u_k^2, u_k^3 \rangle$ where the properties to be compared are all

[15]Componentwise multiplication of two vectors (also referred to as "Hadamard Product") multiplies vectors A and B (both of length n) element by element and returns a vector C (also of length n). Example: $\langle a, a, a \rangle \circ \langle 0, a, b, \rangle = \langle 0, 2a, ab \rangle$.

numeric and partitioned into meaningful classes w.r.t. the current investigation. For example, it might be meaningful to distinguish the class of people who take one drug from the class of people who take two drugs, but it might not be particularly useful to draw the line between 11 and 12 multiple drugs. In our example, the selected characteristics $u_{(k)}$ of $U_{(k)}$ are partitioned as follows[16]:

$u^1_{(k)}$: (Average) age with integer values between 0 and 100 years of age;
$u^2_{(k)}$: (number of) multiple medicines in five classes:
0, 1, 2, 3–5, > 5 medicines;
$u^3_{(k)}$: (number of) multiple diseases in four classes:
0, 1, 2, > 2 diseases.

To be able to attach numbers to the vectors, we have to translate the description of U and U_k into concrete values step by step:

1. Suppose that the target population is described as 70 years old (on average), as taking 1 more medicine besides the tested drug (i.e., 2 in total), and as having more than 2 diseases considered relevant for the investigation (one targeted, the other possibly interacting).
2. The study is conducted on a population of 55 years on average with the same amount of medicines taken but with only one disease (the targeted one).
3. To be able to aggregate those properties on the same level, those characteristics are mapped onto the unit interval with respect to the partitions stated above. Moreover, for the current purpose, each class is simply represented by its central value (e.g., if the unit interval is partitioned into four intervals of equal length, the second interval, ranging from 0.25 to 0.5 is represented by its central value 0.375.).
4. Suppose further that – for the current investigation – average age is considered somewhat important (50%), whether multiple drugs are taken or not is considered highly important (100%), and the number of diseases is negligible (with importance of 0%).

The following tables show the step-wise calculation of the difference between weighted components needed for the determination of the distance between vectors $\vec{u}$ and $\vec{u_k}$:

[16]In this example, study and target are compared merely in terms of population characteristics $\overrightarrow{u_{(k)}}$. In general, though, differences between the substances and between the causal structures will also play a role in assessing the weight of a report, as explicated in Eq. 5.8. For sake of illustration it might be assumed here that substances and causal structures have been found equivalent w.r.t. the present purpose.

	Study population		
	Class	Normalized	Quantized
u_k^1:	55		0.55
u_k^2:	2		0.5
u_k^3:	1		0.375

	Target population		
	Class	Normalized	Quantized
u^1:	70		0.7
u^2:	2		0.5
u^3:	> 2		0.625

	β	Study	Target	Δ
$u_{(k)}^1$:	0.5	0.275	0.35	0.075
$u_{(k)}^2$:	1	0.5	0.5	0
$u_{(k)}^3$:	0	0	0	0

The upper tables show each property's class and its normalization in the unit interval (graphically with class partitions, and quantized using those classes' central values). The lower table shows the absolute differences between the individual components, weighted by corresponding β-values (relative importance). In our example, only *age* figures in the calculation of similarity between the two populations (since both groups take two medicines in total and the number of diseases is considered unimportant for the current comparison). If a Euclidean measure is to be used for calculating the distance, we arrive at the following value:

$$\mathrm{d}(\overrightarrow{\beta} \circ \overrightarrow{x_k}, \overrightarrow{\beta} \circ \overrightarrow{x}) = \sqrt{\left(\beta^1 \cdot (u^1 - u_k^1)\right)^2} = \sqrt{\left(0.5 \cdot (0.35 - 0.275)\right)^2} = 0.053 \tag{5.10}$$

If, for comparison, the number of diseases were indeed to be taken into consideration with a relative importance weight of 0.25, the dissimilarity increases:

$$\begin{aligned} \mathrm{d}(\overrightarrow{\beta'} \circ \overrightarrow{x_k}, \overrightarrow{\beta'} \circ \overrightarrow{x}) &= \\ \sqrt{\left(\beta^1 \cdot (u^1 - u_k^1)\right)^2 + \left(\beta^3 \cdot (u^3 - u_k^3)\right)^2} &= \\ \sqrt{\left(0.5 \cdot (0.35 - 0.275)\right)^2 + \left(0.25 \cdot (0.625 - 0.375)\right)^2} &= \\ \sqrt{0.0375^2 + 0.0625^2} &= 0.073 \end{aligned} \tag{5.11}$$

A σ_U benchmark value for testing *sufficient similarity* with the target population (as formulated in Eq. 5.9) will have to be defined with respect to the theoretical maximum distance and to allowable deviations from the target characteristics. Once σ_U is fixed and *sufficient similarity* between study and target properties is shown, the report Rep_k can be flagged as relevant (with a high degree of confidence in α_k, i.e., 1 in the binary formulation), and (A*) will allow inference by analogy from relevant evidence to the hypothesis under consideration.

There are various ways to refine the distance measure and adjust it to one's needs. Two remarks on the difficult task of choosing a suitable measure shall conclude this section:

1. The example above illustrates that the chosen Euclidean distance measure completely disregards identical properties, even such with high relative importance, since it is defined to simply collect *deviations*. In the example, the identical properties u^2 and u_k^2 could be missing from the vectors $\overrightarrow{u}$ and $\overrightarrow{u_k}$ and could receive high or low relative importance – the Euclidean distance would return the same value in all these cases. This goes against the intuition "the more *identical properties* two property vectors share, the lower their distance". If this intuition is to be encoded formally, though, an additional weighting factor must be added to the distance measure. Each component x^i $(0 < i < m)$ of $\overrightarrow{x}$ might additionally be multiplied by its *relative β-contribution*, e.g., in the following way:

$$\mathrm{d}'(\overrightarrow{\beta} \circ \overrightarrow{x_k}, \overrightarrow{\beta} \circ \overrightarrow{x}) = \sqrt{\sum_{i=1}^{m} \left(\left(\beta^i \cdot (x_k^i - x^i)\right)^2 \cdot \frac{\beta^i}{\sum_{j=1}^{m} \beta^j} \right)}. \tag{5.12}$$

Using d′ on our example case above returns a lower dissimilarity, thus reflecting the intuition that identical properties contribute to our similarity assessment (in decreasing dissimilarity). With $\sum_{j=1}^{m} \beta^j = 1.5$, we have:

$$\mathrm{d}'(\overrightarrow{\beta} \circ \overrightarrow{x_k}, \overrightarrow{\beta} \circ \overrightarrow{x}) = \tag{5.13}$$

$$\sqrt{\left(\beta^1 \cdot (u^1 - u_k^1)\right)^2 \cdot \frac{\beta^1}{1.5} + \left(\beta^2 \cdot (u^2 - u_k^2)\right)^2 \cdot \frac{\beta^2}{1.5}} =$$

$$\sqrt{\left(0.5 \cdot (0.35 - 0.275)\right)^2 \cdot \frac{0.5}{1.5} + 0} =$$

$$\sqrt{0.0375^2 \cdot \frac{1}{3}} = 0.022$$

And again, for comparison – for d′ with relative importance of 0.25 for u_k^3, we get:

$$\mathrm{d}'(\overrightarrow{\beta'} \circ \overrightarrow{x_k}, \overrightarrow{\beta'} \circ \overrightarrow{x}) = \tag{5.14}$$

$$\sqrt{\left(0.5 \cdot (0.35 - 0.275)\right)^2 \cdot \frac{0.5}{1.75} + \left(0.25 \cdot (0.625 - 0.375)\right)^2 \cdot \frac{0.25}{1.75}} = 0.031$$

2. If, e.g., a certain property is required to be *identical* in study and target population in order to grant transferability of study results, partial identity, being central to the contrast account of similarity, can be expressed in a straightforward generalization of the geometric approach exemplified above: Instead of determining *sufficient similarity* for full vectors $\overrightarrow{u}$ and $\overrightarrow{u_k}$, those vectors are partitioned into sub-vectors. $\overrightarrow{u}$ and $\overrightarrow{u_k}$ are then considered *sufficiently similar* iff all distances

of corresponding sub-vectors are lower than associated σ-values. For example, average age, number of medicines, and number of diseases (as in the case above) might be partitioned into sub-vector 1, average age plus number of diseases, and sub-vector 2 containing number of medicines only. If now the number of medicines taken is to be identical in target and study population (e.g., it might be required to be only the investigated drug), then the σ-value for sub-vector 2 is set to 0, while the σ-value for sub-vector 1 might be higher to allow for approximate correspondence.

5.2.3.2 Similarity of Structural Properties

When it comes to determining whether two structures (or, alternatively, sets of structural constraints) M_k and M are sufficiently similar, things are different: There is no straightforward way of encoding topology on a numeric scale. In particular, it depends very much on the given case what the comparison criteria are, and how they possibly interact to jointly suggest a similarity ranking over a set of causal structures.

Nevertheless, the relevance of a given study for the investigated hypothesis crucially depends also on structural knowledge about the (hypothesized) causal associations in both study and target. For example, when an RCT on animals supports a dose-response relationship between drug D and adverse drug reaction E, this report can only be considered relevant if assuming a causal path from D to E is *compatible* with knowledge about the target's causal structure. And conversely, as soon as further studies show the association to be spurious in human drug users, the relevance of the animal study for the current investigation will be downgraded.

The following compilation presents a (non-exhaustive) list of structural properties to consider when comparing causal topologies. The respective questions can be understood as a heuristic survey for assessing structural similarity:

1. Shared **variables**: How many and which variables are shared between study and target? How influential are the ones that are not shared?
2. Causal **in/dependencies** of shared variables: Are the causal (conditional) in/dependencies of the variables shared by study and target in agreement? In other words, are the *causal ir/relevance relations* in study and target the same?[17] If not, is the disagreement resolvable (e.g., by including suspected confounders)?
3. Presence of **co-factors**: If a causal structure marks contributing causes or necessary pre-conditions – can these be identified both in study in target? Are the co-factors shared between study and target? How weighty are those that are not shared?

[17] See Pearl (2000, Sect. 7.3.3), for a discussion of causal relevance. Note, that I am distinguishing *causal relevance* as part of the causal knowledge (encoded as report nodes) from *epistemic* or *inferential relevance* (encoded as attributional *Rlv* weight nodes).

4. **Distance** between D and E: How many mediating variables lie on the path between D and E? If one structure contains more such mediators, does this reflect temporally or spatially higher actual distance? And do such mediators mark possible points for disruption by disablers, such that the causal process must be considered "less stable"?
5. **Interacting influences**: If E is influenced through multiple paths which jointly (but non-additively) produce E, then identifying these influences (and their interaction) in study and target is crucial for transferring predictions about the effect of interventions on D – this is particularly important in the case of masked *back-up mechanisms* present in the target.
6. **Differences** between study and target: Going beyond the points above, one might also utilize knowledge about how and where two causal structures differ. Information about differences at *key stages* along the causal path from D to E will be most helpful if those stages screen off influences of minor stages "upstream" along the causal flow (i.e., above or before). This idea is exploited in *comparative process tracing*, where two processes are investigated in stepwise comparison of their entities. Steel relates this method to external validity in Steel (2008, p. 89):

    > [T]he greater the similarity of configuration and behavior of entities involved in the mechanism at [. . .] key stages, the stronger the basis for the extrapolation.
    >
    > The reliability of comparative process tracing depends on correctly identifying the points at which significant differences between the model and the target are likely to arise. Significant differences are those that would make a difference to whether the causal generalization to be extrapolated is true in the target.

    The strategy summarized here is especially suitable for causal chains with such key variables – identifying precisely those variables (and describing deviating behavior in study and target) will help in deciding whether given study results provide an insight into the target. One of the virtues of comparative process tracing is indeed that it points to the portions of a causal structure that allow for an effective assessment of the similarity between study and target.

This list of structural properties is meant to provide a guide in making *implicit* assumptions about study and target *explicit*. Just like in the case of comparing numeric properties above, all these considerations feed into the purpose-oriented explication of a similarity measure which ultimately determines what *sufficient similarity* means in light of the investigated hypothesis.[18] As above, once *sufficient similarity* between structural properties of study and target is shown, a particular

[18]The way it is presented here, one's assessment of such similarity between study and target is obviously relative to the set of aspects included in one's considerations. There is an argument to be made here that possible further differences not considered should lower one's confidence in the similarity assessment. In principle, in the Bayesian framework employed, it is possible to add an unspecified counter-weight (much like an error term) in order to encode one's uncertainty about potentially neglected, though relevant differences between study and target. Yet, assigning a number to this weight is a subjective task again. Indeed, I would like to argue that such analogy-based arguments are inherently perspectival: They rest on a specific choice of relevant aspects

report Rep_k about M_k can be flagged as relevant (with a high degree of confidence in α_k, i.e., 1 in the binary formulation), and (A*) will allow inference *by analogy* from relevant evidence, summarized in Rep_k, to the hypothesized causal association, $D \copyright E$ in the causal structure M, thereby facilitating Bayesian confirmation:

$$P(D \copyright E \mid Rep_k, \alpha_k) \geq P(D \copyright E).$$

5.2.4 *Extrapolating with Good Arguments and Breaking the Extrapolator's Circle*

I want to conclude Sect. 5.2 with a discussion of two points in the current debate on analogical reasoning in causal assessment. The first is Nancy Cartwright's worry about external validity, namely that even highly reliable RCT methodology only provides an insufficient warrant for the transferability of evidence from study to target (Cartwright 2011; Cartwright and Stegenga 2011). The second is a much-discussed riddle about the soundness of analogical arguments – the 'extrapolator's circle' (Steel 2008; Guala 2010).

5.2.4.1 What Can Make RCT Evidence Relevant?

In her discussion of predictions about the effectiveness of policy interventions in the context of *evidence-based policy* (EBP), Cartwright points out that "[t]he EBP literature rates positive outcomes in well-conducted RCTs as gold standard evidence for effectiveness predictions" (Cartwright 2011, p. 221). And she critically continues:

> Conventionally cited facts, like similarity between target and experimental situations, are then supposed to make this likely. But this is the wrong way to look at the relation between experimental results and the claims whose truth they bear on. Experimental results can help justify confidence that the same result – or that some different result – will hold elsewhere; i.e., they can be evidence for one of these claims. Whether they are evidence depends on whether they play the right kind of role in a good argument for that claim. Similarity, or just the right kind of dissimilarity, might play a role, but if so, only by fitting into a good argument.
>
> What then might a good argument look like that makes RCT results evidence for effectiveness predictions?

In this passage, Cartwright expresses precisely the worry that prompted Landes, Osimani, and Poellinger (Landes et al. 2017) to disentangle *reliability* and *relevance* as meta-evidential attributes. In order to address the question "What can make

(reasonably motivated) and a specific way of relating those aspects (non-arbitrarily). Thus, making the ingredients of such arguments explicit helps refining or potentially also refuting them.

RCT evidence relevant?", Cartwright presents the following argumentative template (Cartwright 2011, p. 222; formatting adjusted):

A1 *x* plays a causal role in the principle that governs *y*'s production there.
A2 *x* plays the same causal role here as there.
A3 The support factors necessary for *x* to operate are present for some individuals here.
Therefore: *x* plays a causal role here and the support factors necessary for it to operate are present for some individuals.

In this argument, *x* is to be understood as a cause of *y*, 'here' indicates the target, 'there' the study. The four lines evidently resemble scheme (A) above: A2 and A3 establish similarity in the relevant (numeric, structural) aspects, and A1 encodes the RCT evidence (e.g., some effect size) to be transferred to the target in the conclusion. In the Bayesian framework presented above, the existence of such a *good argument* would be encoded as high *relevance*, consequently making the RCT result evidence for effectiveness predictions by boosting the weight of the respective evidential report.

5.2.4.2 The Extrapolator's Circle

In his discussion of mechanistic reasoning for the purpose of extrapolation, Daniel Steel (see Steel 2008, p. 78) presents the following challenge any viable account of extrapolation ought to address (see also Guala's comments in Guala 2010):

> [A]dditional information about the similarity between the model and the target – for instance, that the relevant mechanisms are the same in both – is needed to justify the extrapolation. The extrapolator's circle is the challenge of explaining how we could acquire this additional information, given the limitations on what we can know about the target. In other words, it needs to be explained how we could know that the model and the target are similar in causally relevant respects without already knowing the causal relationship in the target.

In the account of Landes, Osimani, and Poellinger (Landes et al. 2017), this circle is broken since the proposed Bayesian framework can be used to probabilistically model *successive* evidence accumulation and amalgamation. At the beginning of the process, experimental results from basic science might contribute causal knowledge about a drug's metabolization – and even if this knowledge only illuminates part of the target mechanism, it can often be considered robust and highly relevant. Once animal studies come into play to answer more specific research questions, knowledge from previous unrelated studies (done on the same animals, maybe with similar substances) figures in forming a relevance estimate for the current investigation. Step by step, a picture of the target's causal structure emerges: The careful extrapolator can thus utilize the previous stage for his assessment of the next. While Steel proposes mechanism-based *comparative process tracing* (see also Sect. 5.2.3.2 above) as a solution to the extrapolator's circle, the Bayesian evidence-amalgamation approach presents a different type of 'process tracing' in providing a toolbox for tracing the dynamic process of evidence synthesis from a higher-level epistemological perspective.

The focus of this section lay on the confirmatory support of relevant pieces of evidence for a causal hypothesis ("upwards" in the evidence-amalgamating framework) – the next section will consider cases where an independently confirmed causal link serves for boosting one's confidence in a hypothesized relation.

5.3 Transferring Knowledge from Confirmed Causal Links

When Hill states that "with the effects of thalidomide and rubella before us we would surely be ready to accept slighter but similar evidence with another drug or another disease in pregnancy" (Hill 1965), he refers to the case of an independently confirmed causal link providing support to a hypothesis still unsettled. Unlike in the case of learning from relevant evidence along the vertical upward arrow in Fig. 5.2, it is not an evidence report about an indicator of Hyp itself, but instead a well-tested second hypothesis Hyp^* that boosts belief in Hyp. Even though little might be known about the mechanisms of substances D^* and/or D, Hyp is considered to share relevant *theoretical consequences* with Hyp^* (or variants thereof).

In order to formalize knowledge transfer across theoretical networks, our conceptual frame needs to be widened slightly. In a recent paper on analogical inference in physics, Dardashti et al. (2017) discuss analogies between experimentally accessible test setups and potentially less accessible target systems we want to gain insights about. The authors introduce the formal concept of *analog simulation* for this purpose which shall be introduced briefly in the following before it is applied in the context of pharmacology.

Analog simulation bridges two basic frames: The source system is prepared, manipulated, and observed to make inferences about the target system. Let us introduce some terminology first to relate all concepts in a formal way[19]:

1. The target system T (a class of situations of interest) is to be modeled as $\mathscr{M}_T$ in a suitably chosen modeling framework $\mathscr{L}_T$;
2. $\mathscr{M}_T$ is constrained by certain background assumptions $\mathscr{A}_T$, summarizing theoretical and empirical knowledge as well as the domain of conditions $\mathscr{D}_T$ to which the model is intended to apply;
3. $\mathscr{M}_T$ can be used to predict phenomena P_T and will in turn be validated by evidence in accordance with P_T;
4. The accessible source system S is to be modeled as $\mathscr{M}_S$ in a suitably chosen modeling framework $\mathscr{L}_S$;
5. $\mathscr{M}_S$ is constrained by background assumptions $\mathscr{A}_S$, containing the domain of conditions $\mathscr{D}_S$ to which the model is intended to apply;

[19]In the following, I deviate from Dardashti et al. (2017) in notational details.

6. Just as on the T side, $\mathscr{M}_S$ can be used to predict phenomena P_S and will in turn be validated by evidence in accordance with P_S.

The source system S will now allow *analog simulation* of target T's behavior if (i) there exist exploitable structural similarities between $\mathscr{M}_S$ and $\mathscr{M}_T$ sufficient to define a syntactic isomorphism robust within the domains $\mathscr{D}_S$ and $\mathscr{D}_T$, respectively, and if (ii) this isomorphism is prompted by and based on a set of *model-external empirically grounded arguments*, abbreviated as MEEGA.

Figure 5.3 relates these elements in a conceptual graph: The rounded box on the left side contains all elements of the target frame, while the right box contains all elements of the source system. MEEGA prompt the establishment of a bridge between theoretical networks in the form of a syntactic isomorphism as translation between the systems' components.

In Dardashti et al. (2017), the terminology is illustrated with an example from physics, where observations of phenomena P_S in table-top fluid systems boost confidence in theoretical assumptions $\mathscr{A}_T$ about gravitational phenomena described in framework $\mathscr{L}_T$. The syntactic isomorphism (motivated by additional knowledge about the underlying physics of both frames) allows for the transfer of knowledge about acoustic Hawking radiation in the fluid system to Hawking radiation in black holes.

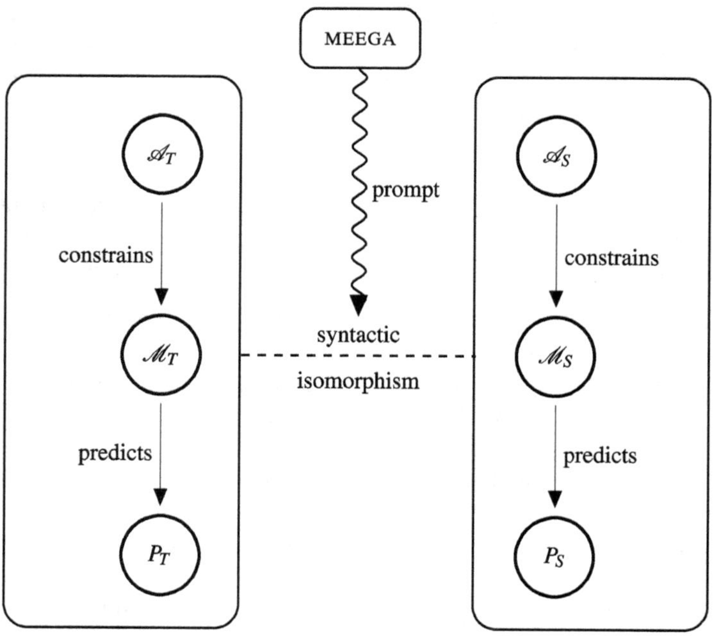

Fig. 5.3 The analog simulation scheme: Framework $\mathscr{L}_T$ (left box) is used to model target system T in model $\mathscr{M}_T$; source system S is accordingly treated in framework $\mathscr{L}_S$ (right box)

Now, the three-layered reconstruction of scientific domains on both sides of the syntactic bridge essentially represents the same conceptual categories as our layered reconstruction of inference in pharmacological research in Fig. 5.2: Scientific hypotheses entail system constraints which in turn predict (and are tested by) real-world phenomena. The syntactic isomorphism can be understood as a formal expression of similarity in terms of *partial identity under translation*. The set MEEGA can be understood to empirically, pragmatically, and semantically *prompt* the choice of relevant theoretical elements to be mapped onto each other. Syntactic isomorphism alone (unaware of semantic context) would be too weak a requirement for analogy since it can be used to translate far more models into each other than one would like to call *analog*. With the analog simulation scheme at hand, we can now trace the confirmatory boost of a well-tested causal link to an unsettled hypothesis in pharmacological research.

Figure 5.4 shows the independently confirmed hypothesis Hyp^* on the right side, e.g., as in Hill's example, the confirmed link between thalidomide (D^*), also known as Contergan, and phocomelia (E^*). Frame-external similarity assumptions (e.g., about the population of pregnant women, or even new *evidence* for relevant relations shared by both M and M^*) suggest an understanding of theoretical variables Ind_k in terms of Ind_k^*, e.g., a characteristic dose-response curve hinting at functional properties shared by D and D^*. Having established such a mapping between relevant theoretical elements, updating one's beliefs about the variables Ind_k^* amounts to updating one's beliefs about the variables Ind_k and, moreover,

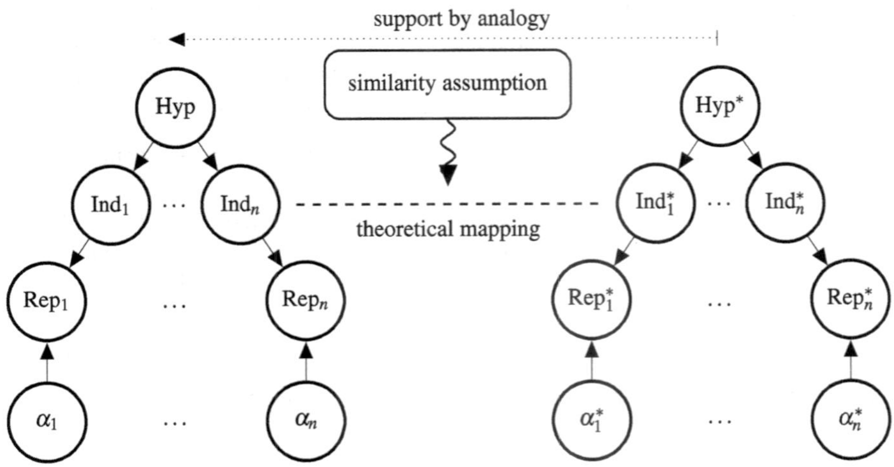

Fig. 5.4 Support by analogy from a second drug (i.e., from established knowledge about hypothesis Hyp^*) to the hypothesis under consideration

to updating one's belief about Hyp (quite in the Bayesian sense).[20] Consequently, *old evidence* $\overrightarrow{Rep^*}$ about Hyp^* can indeed provide novel support for Hyp across theoretical networks, with

$$P(Hyp \mid \overrightarrow{Rep^*}) > P(Hyp). \tag{5.15}$$

How exactly this theoretical bridge between two frames might look and how a relevance filter is to filter out precisely those properties and those indicator variables relevant for the posited analogy, is a subject of current discussion in the philosophy of science; see, e.g., Dardashti et al. (forthcoming) on analog simulation in Bayesian terms, or Beebe and Poellinger (forthcoming) on confirmation from analog models in formal extensions of Bayesian networks.

5.4 Confirmatory Support from In Silico Simulation

When insights are to be gained about mechanistic workings of biochemical phenomena that are not directly observable in vivo, one strategy of choice is computational modeling and simulation. As in the cases of learning from relevant evidence (Sect. 5.2) and transferring knowledge from confirmed causal links (Sect. 5.3), using computational models to learn about biological, medical, or pharmacological facts is based on analogy between simulation and target system.

The argumentative strength of analog simulation in physics is based on the fact that two *physical* systems are related on the grounds of model-external background assumptions about common underlying *physical* principles. In the case of computational simulation, we seemingly loose two important aspects:

(M) Computational simulation does not link two physical systems but rather a physical and a *virtual* system; and
(N) the set of *model-external, empirically grounded assumptions* (MEEGA) is replaced by *model-internal, theoretical principles* in the implementation phase: The symbolic system $\mathcal{M}_S$ is constructed directly from $\mathcal{M}_T$'s background assumptions.

These aspects raise questions about virtues of materiality (M) and novelty of virtual outcomes (N). Both shall be addressed in the following.

[20] As an illustration, consider the following: In many cases, *evidence for similarity* of the drug's causal effects comes from mechanistic knowledge, maybe in relating the molecular structure of the substances to known classes of biochemical processes. So, if D^* is known to be harmful because of its capacity to block some specific mechanism, and if this capacity is judged to be relevant in comparing D and D^*, then such blocking behavior should be part of Hyp's testable consequences Ind_k. Owing to differences in the investigated substances, the testable consequences of Hyp and Hyp^* are in general not *identical*, but can be related non-arbitrarily in motivating a specific *theoretical mapping*, i.e., some *isomorphism* at a suitably chosen level of description.

Computer simulations follow their source code and will behave like programmed as virtual, artificial environments, lacking the material link (M).[21] The position of the skeptic about computer simulation is pointedly summarized by Diez Roux in her discussion of the distinction between observation-based and simulation-based causal inference in epidemiology (Diez Roux 2015, p. 101):

> [...] there is a fundamental distinction between causal inference based on observations (as in traditional epidemiology) and causal inference based on simulation modeling. The traditional tools of epidemiology are used to extract (hopefully) reasonable conclusions from necessarily partial and incomplete (often messy) observations of the real world. [...] In contrast, when we use the tools of complex systems, we create a virtual world (based on prior knowledge or intuition) and then explore hypotheses about causes under the assumptions encoded in this virtual world. In the simulation model, we cannot directly determine whether X causes Y in the real world (because the world in which we are working is of our own creation); we can only explore the plausible implications of changing X on levels of Y under the conditions encoded in the model. In the real world, we have fact (what we observe) and we try to infer the counterfactual condition (what we would have observed if the treatment had been different). In the simulated world, everything is counterfactual in the sense that the world and all possible scenarios are artificially created by the scientist.

Nevertheless, even though computational models are virtual constructs, if they are to be employed for inference about our world they are required to be anchored in reality just as any classical scientific model. In a systematic review of successful agent-based computational models, Casini and Manzo (2016) aim to address Diez Roux's worries. In particular, they pin down various trends which have shown to increase the fruitfulness of such models. In the following I derive from their discussion three conditions that might even be understood as applicability criteria for causal inference from computational, agent-based models (ABMs) in general:

"Ideal" ABMs are to be

1. Based on problem-related theoretical knowledge (rather than merely common-sense, mathematical, topological etc. assumptions),
2. Shaped by data,
3. And iteratively calibrated by more data where the model shows weaknesses.

Figure 5.5 illustrates the relation between target system and computational model in the analog simulation scheme, which I modify here to accommodate the anchoring concept: The source system (our computational model) inherits all theoretical background assumptions from the target side. The arrow from $\mathscr{A}_T$ directly to $\mathscr{M}_S$ represents the preparation of the simulation system as logical cross-dependency, which can be called *model-internal* in a sense: Without *model-external* motivation provided by MEEGA, the rigid link from $\mathscr{A}_T$ to $\mathscr{M}_S$ expresses the

[21] This topic is a subject of current discussion in the philosophy of science with some authors regarding computational models as simply an implemented variant of scientific models as such (e.g., Frigg and Reiss 2009), while others emphasize as a special feature of computer simulations the possibility to experiment with such models as virtual test objects (e.g., Parker 2009 and Morrison 2015).

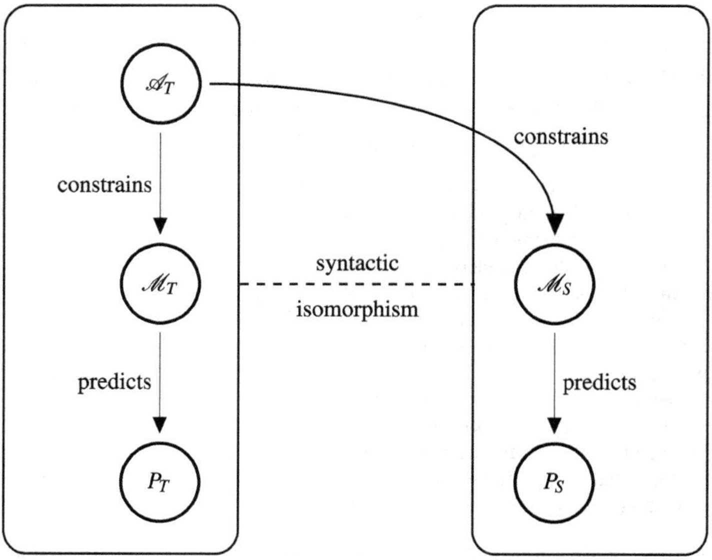

Fig. 5.5 The virtual source system inherits the target's theoretical background assumptions; the arrow from $\mathscr{A}_T$ to $\mathscr{M}_S$ represents the preparation of $\mathscr{M}_S$ as logical cross-dependency

instantiation of the theoretical assumptions $\mathscr{A}_T$ in a concrete (computational) model $\mathscr{M}_S$ *within* the constraints of $\mathscr{A}_T$. Whenever incoming data alters the background assumptions $\mathscr{A}_T$, $\mathscr{M}_S$ (the implementation) can (must) be revised accordingly.

What this picture reveals, though, is that – as formulated above in (N) – the *prediction* of P_S now turns into plain *inference* from $\mathscr{A}_T$. Consequently, it would not make sense to use P_S for the confirmation of hypothesis $\mathscr{A}_T$. Whatever theory of confirmation is to be employed, analog reasoning will only be justified if source and target systems are kept sufficiently independent: The computational model must not be determined by the target system in order to leave room for the possibility of *disconfirmation*.[22]

What are our options? – The direct, logical dependence between $\mathscr{A}_T$ and $\mathscr{M}_S$ may be disrupted by a reintroduced $\mathscr{A}_S$ in different ways: (i) $\mathscr{A}_S$ might be built upon a lower-level/finer-grained theory than $\mathscr{A}_T$, (ii) $\mathscr{A}_S$ might draw on a different set of parameters, (iii) $\mathscr{A}_S$ might incorporate theories from different domains, (iv) $\mathscr{A}_S$ might utilize databases generated from previously conducted material simulations. Of course, the list is certainly not exhaustive, and strategies might possibly be combined. Once the cross-link from $\mathscr{A}_T$ to $\mathscr{M}_S$ is disrupted by integrating external assumptions, P_S regains its confirmatory support towards $\mathscr{A}_T$.[23] In a case study

[22]In the context of modeling with Bayesian networks, this demand is captured in the requirement that all variables represent distinct events.

[23]This strategy introduces a secondary set of *model-external, empirically grounded arguments* in the picture which is first motivated by the logical cross-link between the two frames and later

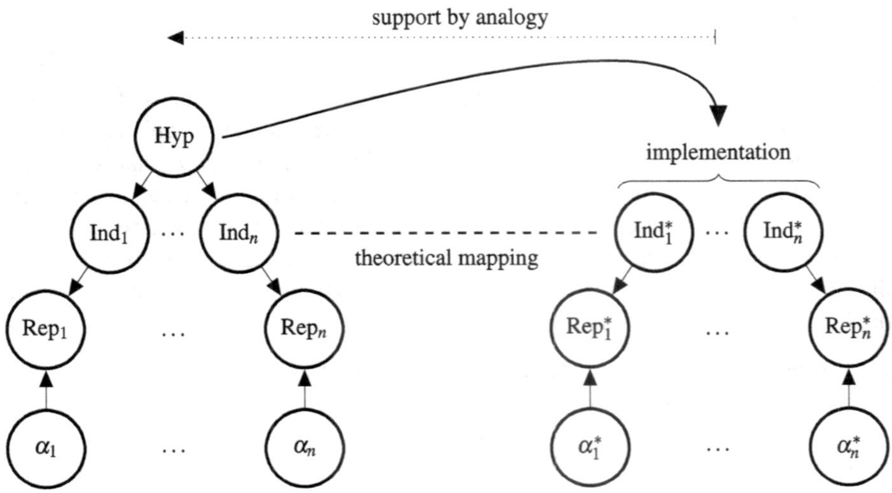

Fig. 5.6 Support by analogy from a computational model of the target system to the hypothesis about the target

on computational modeling of cell proliferation mechanisms in systems biology, Osimani and Poellinger (forthcoming) identify different ways in which a computer simulation can provide novel insights about the object of interest and confirmatory support to a scientific hypothesis:

1. The bottom-up combination of independently secured pieces of knowledge may produce unexpected results precisely because the components' interaction might influence the functioning of the whole mechanism.
2. Combining mechanistic knowledge from different sources might reveal surprising insights about hidden mechanisms in mismatches between virtual measurements and expectation.[24]
3. Moreover, novel insights might be generated when phenomena (potentially not predictable from a small set of basic rules) emerge in iterations of a complex computer simulation.

Figure 5.6 illustrates the conceptual dependencies in confirmation by analogy in our layered reconstruction of pharmacological research: Hyp is implemented by using the theoretical background assumptions about $D \circledcirc E$ in the first version of the computational model. By refining and enriching this model during verification and validation, the logical cross-dependency between Hyp and its implementation is disrupted to an extent that allows for novel 'observations' during runs of the simulation, such that analogy can be used for the confirmation of Hyp once more.

guided by anchoring considerations. See Osimani and Poellinger (forthcoming) for a detailed reconstruction of model creation, verification, and validation for computer simulation in systems biology.

[24]For a discussion of surprise in computer simulation see Parke (2014).

5.5 Conclusions

The evidence-amalgamation framework, introduced in Sect. 5.1, opens the possibility of tracing the dynamics of analogical reasoning across distinct epistemological categories: theoretical hypothesis (Hyp), testable indicators (Ind), and evidence reports (Rep). Within this layered reconstruction of scientific inference, the concept of Bayesian confirmation can be utilized to formulate precisely how a causal hypothesis about a drug's potentially harmful side-effects is confirmed or disconfirmed. This paper distinguishes three analogy-based inference patterns significant in pharmacological research:

1. **Inference from relevant reports:** When the conditions of a given study correspond to the intended application of the investigated hypothesis, reports about the study are marked as relevant for the hypothesis, thus facilitating knowledge transfer from evidence to hypothesis. In Sect. 5.2, study and target conditions are broken into three components: the drug itself (D), the causal model implicit in the hypothesis (M), and the respective population (U). Pair-wise comparison is based on a similarity measure to be chosen w.r.t. the nature of the investigation; e.g., the components might be compared using a geometric measure of similarity as given by the distance between property vectors like in the example case. Obviously, some important decisions are to be taken outside the framework presented, though: Questions as to how to arrive at selection criteria for the properties to be compared are left for another paper.
2. **Inference from established causal knowledge:** Section 5.3 discusses analogical inference from a second well-tested hypothesis Hyp^*. In this case, the connection between source and target is not established via similarity but across a syntactic isomorphism between the hypotheses' consequence sets, i.e., a theoretical mapping on the indicator level. Once this bridge is defined (motivated and justified by model-external empirically grounded arguments), evidence for Hyp^* (the established hypothesis) will also boost confidence in Hyp (the hypothesis under investigation).
3. **Inference from computational models:** If the analog system is a virtual, computational model of the investigated hypothesis, the bridge between source and target is not motivated by model-external considerations but much rather by model-internal constraints. Section 5.4 discusses how an *artificial* system can possibly provide support for a causal hypothesis about an *actual* drug with *real* risk.

While the paper focuses on hypothesis testing for the purpose of risk assessment in pharmacology, the second and third pattern in the list above make the role of analogical reasoning in the formulation of a hypothesis obvious. In particular, a computer simulation will support an investigated hypothesis if the codebase is not merely constructed from theoretical assumptions about the target hypothesis, but infused with further theoretical considerations or additional sources of knowledge, thereby breaking the logical dependency between source and target (see Sect. 5.4). Once a theoretical bridge is established on the indicator level, unexpected, surprising, possibly unpredictable evidence reports about an analog system will propel hypothesis discovery and theory revision. Indeed, this extends beyond computational modeling: When Otto Schaumann created meperidine, the first fully synthetic opioid pain medication, in 1937, he observed that meperidine and morphine produced similar physiological signs when administered to rats in lab experiments. In addition, meperidine was known to share chemical structural properties with morphine. Schaumann consequently (and rightly) hypothesized that meperidine also shares morphine's narcotic effects (see Bartha 2013). This further episode of successful inference by analogy illustrates the peculiar nature of pharmacology, integrating different levels of description from chemical structure to clinical observation. Justifying analogical arguments by reconstruction calls for a framework capable of accommodating heterogeneous sources of evidence that allows tracing confirmatory support across distinct epistemological categories – possibly also from analog systems. The collection of modules presented in this paper may serve as a toolbox for such justification in the scientific dialog between hypothesis testing and theory revision.

Acknowledgements This paper was presented at workshops and conferences in Munich, Sydney, Groningen, Bologna, Bochum, and Exeter. I greatly benefited from the comments and suggestions made by the audiences, and I am particularly thankful for personal discussions with Cameron Beebe, Lorenzo Casini, Radin Dardashti, Stephan Hartmann, Adam LaCaze, Jürgen Landes, Barbara Osimani, Jan-Willem Romeijn, Karim Thébault, Naftali Weinberger, and Michael Wilde, whose valuable comments helped me clarify my aims and shape the final version of this paper.

References

Bartha, P. (2010). *By parallel reasoning: The construction and evaluation of analogical arguments*. Oxford: Oxford University Press.

Bartha, P. (2013). Analogy and analogical reasoning. In E. N. Zalta (Ed.), *The Stanford encyclopedia of philosophy* (Fall 2013 ed.).

Beebe, C., & Poellinger, R. (forthcoming). Bayesian confirmation from analog models.

Bovens, L., & Hartmann, S. (2003). *Bayesian epistemology*. Oxford: Oxford University Press.

Britton, O. J., Bueno-Orovio, A., Van Ammel, K., Lu, H., Towart, R., Gallacher, D., & Rodriguez, B. (2013). Experimentally calibrated population of models predicts and explains intersubject variability in cardiac cellular electrophysiology. *Proceedings of the National Academy of Sciences of the United States of America*, (110), E2098–E2105.

Button, K. S., Ioannidis, J. P. A., Mokrysz, C., Nosek, B. A., Flint, J., Robinson, E. S. J., & Munafo, M. R. (2013). Power failure: Why small sample size undermines the reliability of neuroscience. *Nature Reviews Neuroscience, 14*, 365–376.

Cartwright, N. (2011). Predicting what will happen when we act. what counts for warrant? *Preventive Medicine, 53*(4), 221–224. Special Section: Epidemiology, Risk, and Causation.

Cartwright, N., & Stegenga, J. (2011). A theory of evidence for evidence-based policy. In P. Dawid, M. Twinning, & W. Vasilaki (Eds.), *Evidence, inference and enquiry* (Chapter 11, pp. 291–322). Oxford: Oxford University Press.

Carusi, A., Burrage, K., & Rodriguez, B. (2012). Bridging experiments, models and simulations: An integrative approach to validation in computational cardiac electrophysiology. *American Journal of Physiology – Heart and Circulatory Physiology, 303*(2), H144–H155.
Casini, L., & Manzo, G. (2016). Agent-based models and causality: A methodological appraisal. *The IAS Working Paper Series (Linköping University Electronic Press)*, (7), 1–80.
Chan, A.-W., & Altman, D. G. (2005). Epidemiology and reporting of randomised trials published in PubMed journals. *The Lancet, 365*(9465), 1159–1162.
Dardashti, R., Hartmann, S., Thébault, K., & Winsberg, E. (forthcoming). Confirmation via analogue simulation: A Bayesian analysis.
Dardashti, R., Thébault, K., & Winsberg, E. (2017). Confirmation via analogue simulation: What dumb holes could tell us about gravity. *The British Journal for the Philosophy of Science, 68*(1), 55.
Diez Roux, A. V. (2015). The virtual epidemiologist – Promise and peril. *American Journal of Epidemiology, 181*(2), 100–102.
Doll, R., & Peto, R. (1980). Randomised controlled trials and retrospective controls. *British Medical Journal, 280*, 44.
Frigg, R., & Reiss, J. (2009). The philosophy of simulation: Hot new issues or same old stew? *Synthese, 169*(3), 593–613.
Geiger, D., Verma, T., & Pearl, J. (1990). Identifying independence in Bayesian networks. *Networks, 20*(5), 507–534.
Guala, F. (2010). Extrapolation, analogy, and comparative process tracing. *Philosophy of Science, 77*(5), 1070–1082.
Hesse, M. B. (1952). Operational definition and analogy in physical theories. *British Journal for the Philosophy of Science, 2*(8), 281–294.
Hill, A. B. (1965). The environment and disease: Association or causation? *Proceedings of the Royal Society of Medicine, 58*(5), 295–300.
LaFollette, H., & Shanks, N. (1995). Two models of models in biomedical research. *Philosophical Quarterly, 45*(179), 141–160.
Landes, J., Osimani, B., & Poellinger, R. (2017). Epistemology of causal inference in pharmacology. Towards an epistemological framework for the assessment of harms. *European Journal for Philosophy of Science 8*(1), 3–49.
Lewis, D. (1973a). Causation. *The Journal of Philosophy, 70*(17), 556–567.
Lewis, D. (1973b). *Counterfactuals* (2nd ed.). Hoboken, New Jersey: Wiley-Blackwell.
Luján, J. L., Todt, O., & Bengoetxea, J. B. (2016). Mechanistic information as evidence in decision-oriented science. *Journal for General Philosophy of Science, 47*(2), 293–306.
Morrison, M. (2015). *Reconstructing reality: Models, mathematics, and simulations*. Oxford: Oxford University Press.
Osimani, B., & Poellinger, R. (forthcoming). A protocol for model validation and causal inference from computer simulation.
Parke, E. C. (2014). Experiments, simulations, and epistemic privilege. *Philosophy of Science, 81*(4), 516–536.
Parker, W. S. (2009). Does matter really matter? computer simulations, experiments, and materiality. *Synthese, 169*(3), 483–496.
Paul, L. A., & Healy, K. (2016). Transformative treatments. *Noûs, 52*(2), 320–335.
Pearl, J. (2000). *Causality: Models, reasoning, and inference* (1st ed.). Cambridge: Cambridge University Press.
Poellinger, R. (forthcoming). On the ramifications of theory choice in causal assessment: indicators of causation and their conceptual relationships.
Revicki, D. A., & Frank, L. (1999). Pharmacoeconomic evaluation in the real world. *PharmacoEconomics, 15*(5), 423–434.
Shepard, R. N. (1980). Multidimensional scaling, tree-fitting, and clustering. *Science, 210*(4468), 390–398.
Steel, D. (2008). *Across the boundaries. Extrapolation in biology and social sciences*. Oxford: Oxford University Press.

Tversky, A. (1977). Features of similarity. *Psychological Review, 84*(4), 327–352.
Unruh, W. G. (2008). Dumb holes: Analogues for black holes. *Philosophical Transactions of The Royal Society A, 366*, 2905–2913.
Upshur, R. (1995). Looking for rules in a world of exceptions: Reflections on evidence-based practice. *Perspectives in Biology and Medicine, 48*(4), 477–489.
Weisberg, M. (2012). Getting serious about similarity. *Philosophy of Science, 79*(5), 785–794.
Weisberg, M. (2013). *Simulation and similarity: Using models to understand the world*. Oxford: Oxford University Press.
Worrall, J. (2007). Evidence in medicine and evidence-based medicine. *Philosophy Compass, 2*(6), 981–1022.

Chapter 6
In Silico Clinical Trials: A Possible Response to Complexity in Pharmacology

Federico Boem, Ilaria Malagrinò, and Marta Bertolaso

Abstract The complex nature of human diseases, the significant differences between individuals, their own specificity and the inevitable variability in how a treatment is administered make pharmacological research extremely difficult. Due to the inability to completely predict how a product will affect individual patients, it's not unusual for a drug to show serious problems during clinical tests after having performed well in pre-clinical and laboratory tests. The strategy of *in silico* Clinical Trials (ISCT) comes as a possible bridge from the insights of molecular understanding to *ad personam* medical treatments. By refining and complementing traditional clinical trials, ISCTs should provide a proxy for biological conditions. The *in silico* strategy is typically justified by appealing to the power of big data, that computational strategies involve. Big data seem to imply a different methodology and thus a different way of doing research, thought to be more adequate to address biological complexity. In this chapter we analyze such a view by showing how these new computational strategies enhance and implement traditional approaches rather than changing the nature of clinical investigation. An epistemic look at ISCTs shows how such an issue is addressed in pharmacological studies. The moral of in *silico* approaches is that complexity cannot be fully addressed *per se*, only through heuristics. Contra enthusiastic claims arguing for a paradigm shift in scientific research, ISCTs do not change the theoretical account governing current scientific practice, nor the general framework according to which discovery strategies are adopted and justified. On the other hand, ISCTs do provide a change in the methodologies. However, such a novelty should be regarded as a complement and not as a replacement.

F. Boem (✉)
University of Milan, Milan, Italy

I. Malagrinò · M. Bertolaso
Università Campus Bio-Medico di Roma, Rome, Italy

A. LaCaze, B. Osimani (eds.), *Uncertainty in Pharmacology*, Boston Studies in the Philosophy and History of Science 338,
https://doi.org/10.1007/978-3-030-29179-2_6

6.1 Introduction: The Complexity of Pharmacology

Pharmacological research is facing new challenges that are changing its disciplinary and methodological setting.

Due to the risk of potential harm, any possible new compound must undergo an assessment process before being distributed into the market. This process is divided into a pre-clinical step and a clinical one. These two steps are both necessary and inevitably intertwined. Indeed "[t]oday, the only conclusive (and accepted) way to ensure the safety and efficacy of a biomedical product is to test it on living organisms, first on animals (preclinical evaluation) and then on humans (clinical evaluation)" (Viceconti et al. 2016a, p. 37). The pre-clinical process is very important because it allows to investigate how a biological target plays a relevant role in the diseases we want to treat. However, "due to the hugely complex nature of human diseases, the significant differences between individuals, and the inevitable variability in how a treatment is administered, it is not unusual for a product to perform exceptionally well in tightly controlled laboratory studies but show some serious problems during clinical trials" (*ibid.*)

Due to the inability to completely predict how a product affects individual patients, in clinical trials a drug might show effects that were latent in the pre-clinical stage. "Furthermore, at the present time, a clinical trial can usually evaluate only whether a product is safe/unsafe or effective/ineffective, but it often lacks indications on how to improve it" (*ibid.*). Therefore, if a product fails during a clinical assessment, it is completely abandoned, regardless the entity of changes that would be necessary in order to make it effective.

In face of this kind of problems, many attempts have been made to either reshape or improve pharmaceutical research. Such a change means also a new way of thinking in terms of methodologies and resources that can and should be adopted in order to assess new therapeutic drugs. Since the '90s, computer-based models started to get more popular within this kind of research. This was also due to the increase of computational power and to the precision [. . .] these approaches can actually provide. Over time, this type of implementation acquired more and more credit. Indeed, "[i]n the early 2000s, the computational ecology community started to debate the virtues of individual-based models for population ecology. Soon after, *in silico* medicine research also began to use the first patient-specific models and some analysts started to suggest that such approaches could be useful in the development of new medical products" (*ibid.*)

The advantage of adopting *in silico* approaches rests on supposed clearer methodological and epistemic gains. First, there is the possibility to deal with vast amounts of data, and to process it in a fast, controlled and standardized way. Second, standardization grants more solidity to scientific reproducibility of results. Last, but not least, computer modeling offers the chance to compare and analyze crucial aspects from a more global perspective, possibly mirroring more effectively some of the nuances and peculiarities of individual patients.

Pharma industry is facing this growing complexity, dealing with these new challenges and major changes. In this endeavor, new technologies such as *In Silico* Clinical Trials (ISCTs), defined as the use of patient-specific models to generate simulated populations on which new biomedical products can be safely tested, might play a central role. In this study we present an epistemic analysis of ISCTs by showing the reasons for their rising and applications, the potential they might have and their critical points. First we briefly examine what standard Clinical Trials (CTs) are. Next, we present the innovation portrayed by ISCTs, we discuss some case studies and a recent project (Avicenna Project) which is aimed at implementing these methodologies on a larger scale. Third, we critically evaluate ISCTs from a philosophical point of view, trying to highlight their interesting contributions but also their potential flaws. In the end we discuss some implications of ISCTs for biomedical research and we provide some hints for future debates.

6.2 RCTs: The State of the Art

In order to analyze some of the relevant aspects implied by the adoption of ISCTs, it would be convenient to briefly describe what is a standard Randomized Control Trial (RCT) and then trying to explain in what respect *in silico* strategies make the difference from an epistemic point of view. Roughly speaking a RCT is a study where a certain number of patients are randomly distributed to get one among different, possible interventions. "Typically, RCTs seek to measure and compare different events called '*outcomes*' that are present or absent after the participants receive the interventions. Because the outcomes are quantified (or measured), RCTs are regarded as '*quantitative*' studies" (Jadad 1998, p. 1). For instance, let us imagine a RCT comparing two putative drugs (let us call them A and B) which are believed to act as painkillers. In this trial the researchers can decide that the perception of pain should count as the main parameter for the outcome. Next, they will assess such a parameter evaluating the amount of people who were relieved from pain after a given period of treatment administration. RCTs rely for their methodological efficacy on the fact that they are a *comparative* kind of study. This means that, usually, the entire group of people involved will be randomly divided in two groups. One group will be administered the treatment that has to be tested, while the other will be the control group (this explains why random trials are also *controlled*). RCTs represent the other horn of experimental work (*i.e.* clinical and translational research) besides the lab research. RCTs present many features of experimental settings (such as empirical manipulation of both the material and the variable involved), and because of that they can be genuinely considered experimental procedures. This aspect is very important because it means that RCTs are not purely observational (*i.e.* those type of study in which there is no intervention) which is a clear indication of their experimental nature. From an epistemic point of view, Ian Hacking (1983) famously argued that the possibility of direct intervention (such as empirical manipulation) on the objects of scientific investigation represents the key feature of an experiment.

RCTs are said to represent a gold standard for pharmaceutical research. This might be true now but it was not always the case (for a history of RCTs and their impact see Bothwell et al. 2016). However, this does not mean that their adoption is immune to criticism nor that they cannot be implemented or made more precise by adopting complementary approaches. There are many academic and public discussions on the nature of RTCs and their role in current research[1]. Here we discuss the case of ISCTs which can be seen as a specific way to foster and corroborate standard RCTs.

In the next section we present a case study in which an application of *in silico* strategies is implemented to deal with glucose concentration in type I diabetic patients with an eye toward therapeutic interventions. Such an example will illustrate how these different methodologies can be applied to more traditional contexts. The description of how these approaches are pursued will be then analyzed against their traditional counterpart in the light of their epistemic contribution.

6.3 In Silico Trials for Disease Modeling

In recent years some researchers have tried to computationally address some scientific/practical issues in biological research that could not be easily solved by traditional experimental approaches. Some of these problems concern samples for clinical trials. In clinical trials, it is sometimes very difficult to obtain a cohort of patients which is both representative of the target population and methodologically well constructed. Even *in silico* models are often based on few cases and established only by few parameters (such as plasma concentration (see Magni et al. 2009). The rise of Big Data science offered the possibility to overcome such limitations by creating computational models (see for instance Dalla Man et al. 2007). The impact of Big Data in this field is constituted by the possibility to collect and compare data coming from different sources (from genomics to behavioural studies). Moreover, the computational power and the processing speed offered by Big Data science, can potentially reveal hidden patterns or connections that are invisible to other types of research (see for instance Kitchin 2014). Of course ICSTs are not the only way in which computational resources are implemented in current practice. Other kinds of computer-based approaches, such as *pharmacometrics* (see for instance Barrett et al. 2008) are currently adopted in pharmaceutical practice way before the creation of ISCTs. In this sense, the distinct prerogative and the ambition of ISCTs lies in the idea that not just sets of parameters can be modeled, but also the complexity of the patient itself can be somehow quantitatively addressed and computationally

---

[1] See for instance Bothwell et al. 2016 and Frieden 2017 at https://www.statnews.com/2017/08/02/randomized-controlled-trials-medical-research/

represented. The creation of an "*in silico* patient" is certainly an achievement but its nature is intrinsically problematic as there is no consensus to what extent a model as such (indeed a philosophical question) should or could represent a patient.

In a recent study, some researchers (Magni et al. 2009) took advantage of an *in silico* approach to model glucose-insulin system in patients affected by type I diabetes. One upshot is a detailed simulation of the many factors occurring in such condition, computationally related to the situation occurring in normal/physiological condition after a variegated meal composed by different nutritional elements. In their research, Magni et al. consider the feedback control of glucose concentration in type I diabetic patients. This kind of patients need to assume (via multiple injections) insulin since they lost their physiological capacity to regulate the glucose concentration in their blood. In the recent years, researchers have proposed implementations (towards forms of automatization) of current, manual, therapy. As also *Magni* et al. write "[t]he current open-loop therapy can be contrasted to glucose management by a closed-loop system controlling blood glucose in diabetic patients also known as an artificial pancreas"(Magni et al. 2009, p. 338). Computational resources play a key role in creating models which can establish and maintain controllable parameters for insulin automatic administration. Beyond traditional methods (such as glucose concentration) new data modeling whole meals made possible to obtain a virtual (a model algorithm) representation of insulin-glucose relationship during a single meal. In this study, Magni et al. took advantage of such a model to generate what they have called "virtual diabetic patients". A part of the research consisted in using these "patients" in a ISCT in order to test the algorithm. Despite the results, it seems that part of the problem, (*i.e.* what kind of patient is the virtual one and how it might differ from an actual patient) is often neglected. This could be due to the "philosophical nature" of the problem concerning the notion of representation. It is true that such kinds of trials are sometimes presented as a suitable, and more precise (with respect to the volume of data and to the number of variables and situations that can be considered together) tool to evaluate certain kind of therapeutic interventions (such as the compliance of a synthetic pancreas) but also to offer useful indications about subsequent clinical trials on actual patients. Nevertheless, due to the fact that all the relevant analyses have been conducted *in silico*, some concerns naturally rise in terms of reliability of those findings. As mentioned before, an aspect, in particular, is about the limitations that a model of that kind might have, compared to models coming from "real" patients.

Despite the hype on computational approaches in biomedicine, it is hard to consider a virtual patient a perfect image of a real one. However is this (i.e. "to be an image of") the aim of a model? Rather, the purpose of a model is precisely to select certain relevant properties and to emphasize them while discarding the others. A virtual patient then can be seen as an abstraction of some given parameters whose aim is precisely to mimic those features that play a role in determining a specific condition/situation. In the study under consideration (Magni et al. 2009, pp. 339–340) the virtual patient is represented by the global modelization (algorithmically intended) of different properties corresponding to distinct features of system mirrored by *subsections* of the model. Such subsections model, in sequence; glucose intestinal absorption, glucose kinetics, glucose elimination due to kidney activity, endogenous glucose production, the use of glucose by different areas of the organism between (*i.e.* brain activity) and its dependence/independence on insulin, insulin flow in both physiological condition (in which insulin is secreted by the pancreas directly into the portal vein) and diabetic patients (in which insulin is subcutaneously injected) and how also exogenous administration of glucose correlates with plasma glucose. Again, these properties do not fully resume a real patient (nor they are thought to do so) but they represent all the elements considered as relevant to represent a patient with type I diabetes at a given level of description. In order to realistically model diabetic condition, data obtained from healthy conditions were used as benchmark: the control algorithm must mimic the physiological situation in which pancreas beta cells contribute to maintain the glucose level within certain values. By modifying parameters it was then possible to represent a diabetic patient. Importantly, this approach preserves variability across different subjects. Once the model has been settled, a virtual protocol is established accordingly. Such a protocol should represent a simulation under the circumstances would hold in a real protocol. Next, testing the model means to run the algorithm in several *in silico* trials in order to simulate how it will respond to variations of parameters.

Studies like this certainly represent an innovation in drug testing and clinical trials. First, *in silico* simulation are presented as a proxy to consistently manage and display a large amount of pre-clinical information which, acting as a filter of plausible conditions, are presented as a more reliable source to set a therapeutic intervention. Moreover, the possibility to create a virtual patient, potentially representing (according to the specific needs of the study) all the desired relevant conditions, may be a way to overcome some of the limitations of animal models in traditional clinical trials. Even if *in silico* models cannot fully represent all the variables and possible outcomes deriving from animal models (an issue we will address in the next paragraphs), they can offer a tool to deal with a greater volume of data and thus they can help to refine and complement more traditional approaches.

In the light of these considerations, how should we evaluate the implementation offered by ISCTs?

6.4 ISCTs: What Kind of Issues Do They Address?

Computational tools should not be seen as a way to get rid of well established approaches. Sometimes they can be a useful instrument to *improve* the pre-existing ones. In this respect, another interesting case is constituted by another study (Bosmans et al. 2006) on lung cancer patients, where researchers had to search new ways to improve local treatments (which seem to be fundamental given the risk of metastases and recurrence), especially in those situations in which patients cannot undergo surgery (and thus must go for radiotherapy and/or chemotherapy).

The main sources of data come from a technology so called *Computer Tomography*[2]. As the researchers write, "[r]espiration correlated CT (RCCT) is a new development with which the systematic error (geographic miss) can be minimized and the random error (tumor motion during a radiotherapy fraction) can be estimated for an individual patient. With an RCCT, separate CTs can be reconstructed from multiple phases of the respiration, allowing estimation of tumor motion as well as the selection of a CT with the tumor in its central position" (Bosmans et al. 2006, p. 74). In order to manage all these conditions, the researchers settled an *in silico* strategy to analyze, in a global and consistent manner, several different therapeutic interventions on patients grounded on distinct datasets. Four different situations were considered: "free breathing CT (CTfb, with standard margins for motion), a slow CT (CTslow, no margins for motion), the half ventilation CT (CThv, with margins based on measured tumor motion), and the complete 10-phase RCCT (CT10ph, no margins for motion) (*ibid.*)". This study shows how the *in silico* simulation enabled, on the same cohort of lung cancer patients, a coherent comparison (in a single run) between 4 diverse therapeutic procedures. These datasets constitute the elements which have set the experimental space that could have been computationally explored. In other words, in this study researchers designed an ISCT to analyze many different conditions that were not fully tractable through traditional approaches (we will analyze what this kind of *virtual exploration* means in the further sections).

Despite these interesting results ISCTs did not receive a warm welcome in the practice of clinical research. Empirical concerns remain high and many physicians and researchers seemed to hesitate to rely on data coming from in silico modeling. As a matter of fact, both Bosmans et al. 2006 and Magni et al. 2009 constitute marginal applications of ISCTs. Their studies did not change the practice of clinical trials and their models seemed not fully accepted/assimilated (at least in the vast majority of cases) in the design of new trials.

Nevertheless, the interest towards this kind of research (maybe under its complementary vest) increased to the point that even the European Commission raised funds to support projects with the particular aim to develop and promote a research agenda concerning the adoption of ISCTs.

[2]"radiography in which a three-dimensional image of a body structure is constructed by computer from a series of plane cross-sectional images made along an axis — called also *computed axial tomography, computerized axial tomography, computerized tomography"* (definition retrieved from Merriam Webster Dictionary at https://www.merriam-webster.com/dictionary/computed%20tomography)

6.5 The Promise of *in silico* Strategies: The Avicenna Project and Its Relevance to Current Pharmacological Challenges

If we look back at the previous examples, we can see that they were developed by ingenious scientists who took advantage of new technological possibilities to overcome some limitations of their field study. These studies were not conducted under a common framework nor a common research agenda. They were rather exploratory studies. This means that many implications regarding their methodologies were not fully considered. Nor were their methods widely accepted or implemented in the standard research practice. This is not to say that scientists were not aware of those, but rather that this kind of scientific endeavour did not require a complex theoretical planning. Instead in the those cases, ISCTs were designed to surmount specific issues, often technical, which could not be easily addressed by conventional strategies. Therefore ISCTs did not constitute a deep change in the nature of research. Nor their use was consistently applied. However one should not see these attempts as a waste of time and resources. Rather, they represent a *problem-solving* tool, adequate to deal with those situations in which data-collection was not fully feasible in empirical terms or when specifically designed virtual patients may contribute to refine the research sample. Their pursuing should not be seen as theoretically justified but rather motivated by practical reasons. Indeed, ISCTs promised to reduce time and resources in clinical studies and their utilization was also grounded in the light of the growing cultural movement of "personalized medicine" which highly rely on Big-Data science and computer simulations. Such a situation, in which practice precedes theory, is typical of new circumstances in which researchers look like pioneers, who have to deal with puzzles which are not fully represented within the established science. Many solutions are *ad hoc*, not standardized, not regulated since the field is not theoretically structured and it is based more on local resolutions than on the attempt to provide a general framework.

However, when certain methodologies cross the boundaries of their standard applications and their adoption is considered within a more general scheme, things start to be different. Beyond being an exploration, if one wants to adopt computational resources to expand the horizon of clinical investigation he/she also needs to rethink the fundamental features of the research field itself. In this section we will present a European project aimed to create a physiological map of the patient that can virtually manage and address (in a comprehensive way) a larger number of elements that critically contribute to determine a specific condition. This effort should be seen more as a systematic approach to the issue of ISCTs rather than constituting a theoretical rethinking of the foundations of clinical trials as such. This is because the guiding, methodological principles of ISCTs are the same of standard RCTs. What is different here, lies more at the implementation level. Since ISCTs may fill certain gaps constitute by empirical difficulties, they could offer a way to overcome such obstacles, thus complementing the entire given research. Again, such a complement/implement is motivated by a more general aim to reduce research time and refine current methods due to the myriads of possible relevant details that *in silico* approaches might provide.

Since 2005, the "Virtual Physiological Human" (VPH) is defined as "a framework of methods and technologies that once established will enable the collaborative investigation of the human body as a single complex system" (Viceconti et al. 2016a, p. 38). This project was sponsored by the European Commission and, in a first step, was aimed at elaborating a research and technological development roadmap through a consensus process across a community: *Seeding the EuroPhysiome: A Roadmap to the Virtual Physiological Human.* Through quantitative information about the biology, physiology and pathology of a patient at different scales of space and time generated by using imaging and sensing technologies, this document, published in 2007, aimed to design multi-scale computer models containing all the knowledge available on a specific disease, in order to make patient-specific predictions for diagnosis, prognosis and treatment planning. Although the VPH project is not entirely realized yet, some VPH technologies already have practical and clinical applications in those cases where clinical data are missing.

To represent the emerging community of practice, in 2011 a not-for-profit organization, called the *VPH Institute*, was created. One of the first steps of the activity of this institute was the publication of a position paper showing three further lines of development of the Virtual Physiological Human project based on the concepts of "Digital Patient", a decision-support systems for personalized medicine to the medical professional, of "Personal Health Forecasting", patient-specific models constantly updated by personal health systems and providing decision-support systems for self-management to the patients/citizens, and of ISCTs.

Funded by the European Commission as part of the Seventh Framework Program for Research and Technological Development (FP7) under the Information Communication Technologies Programme, the 'Avicenna project: a strategy for *in silico* clinical trials' began in October 2013 and was closed in September 2015. As reported on the Avicenna project website, *in silico* clinical trials are aimed at "[t]he use of individualised computer simulation in the development or regulatory evaluation of a medicinal product or medical device/medical intervention" (see http://avicenna-isct.org). The project was conducted by a consortium co-ordinated by the University of Sheffield and included three partners: VPH Institute, Lynkeus srl and Obsidian Biomedical Consulting Ltd.

The consortium was tasked with three objectives: (1) creating a Roadmap addressing the steps needed for the introduction of *in silico* clinical trials, (2) establishing a partnership between the biomedical industry and European research organizations, with the aim of developing the technology, methods, protocols and standards required for *in silico* clinical trials to become a reality, and (3) identifying technologies and determining early examples of *in silico* clinical trials.

The Roadmap[3], published on ResearchGate in January 2016, shows that the use of ISCTs is already a reality in industrial practice, but only to a limited degree.

[3]Available here: http://avicenna-alliance.com/files/user_upload/PDF/Avicenna_Roadmap.pdf

The whole Roadmap document is constructed upon the aim to show that adopting the *in silico* approach in clinical trails is particularly advantageous. Given the hope to reduce time research and costs of specific materials, the Roadmap argued that the benefits of adopting ISCTs at a broader scale would have also benefits on potential ethical or social issues related with the research. Indeed, it promises improving clinical experimentations lowering costs and animal models, avoiding harmful effects and risks for patients, allowing to treat orphan diseases, refining prediction of long term or rare effects. According to their advocates, ISCTs seem an exciting perspective and opportunity to overcome our cognitive limitations to store, analyze and represent the complexity of biological and physiological processes happening in the human body, proposing at the same time a fruitful cooperation across disciplines. In this perspective, Roadmap stress the need to consider *in silico* technologies in clinical trials as the best way to reduce, refine, partially replace and complement them. In this aspect lies also the ambition of creating the so-called *virtual patient,* a computer model of a patient, that could undergo a *virtual treatment* which, in theory, could be perfectly matchable with standard procedures and real patients in order to integrate and refine actual studies. Thus, the VPH specified that ISCTs refer to "the use of individualised computer simulation in the development or regulatory evaluation of a medicinal product, medical device, or medical intervention. It is a subdomain of '*in silico* medicine', the discipline that encompasses the use of individualised computer simulations in all aspects of the prevention, diagnosis, prognostic assessment, and treatment of disease" (Viceconti et al. 2016a, p. 41).

ISCTs promise (1) to reduce the size and the time of clinical trials due to a more reasoned design and implementing a smaller clinical trial population by adding simulated patients that might serve as variability detectors missing in actual patients ; (2) to refine clinical trials by obtaining more precise data on potential outcomes, "understanding how a product interacts with the patient anatomy and physiology, predicting long-term and rare effects that clinical trials are unlikely to reveal" (Viceconti et al. 2016b, p. 7); (3) to complement traditional clinical trials by the ability to test the multiple variables (such as additional pathological conditions not taken into account by traditional approaches) concerning various clinical settings. The purpose of the project of performing "exploratory trials within the computer [...] would enable the simulation of a number of elements affected by the administration of the candidate biomedical product. In such a scenario, 'virtual' patients would be given a 'virtual' treatment, enabling us to observe through a computer simulation how the product performs and whether it produces the intended effect, without inducing adverse effects that might be potentially dangerous for the patient" (Viceconti et al. 2016b, p. 7). Proponents of the Roadmap argue that "ISCT will involve the generation of computer models that will be applied to each patient enrolled in a trial simulating his/her disease and the treatment being tested. These models will predict the outcome and will be used alongside, or as part of, an existing clinical trial. The predictive accuracy of the models can be tested against the observations produced by the parallel clinical trial. Once this process is repeated for a sufficiently large number of patients, these data can be used with other available information (for example, the distribution of genotypes that are known to be relevant to the course of the disease for product mode of action but which are not regularly recorded in clinical trials) to design 'virtual populations' (Viceconti et al. 2016b, p. 7).

One of the main innovations advocated by this type of project is that whereas most predictive models available today capture one generic behavior considered representative, the Avicenna Project aims to develop models parameterized for each patient, that is providing "when possible a fully mechanistic, quantitative model capable to predict accurately for each individual member of the population" (Viceconti et al. 2016b, p. 11).

ISCTs might offer, then, a possible way to dissect certain aspects of patient complexity. The real question is, as it often is in science: how do they do so and to what extent. From a technical point of view, individualized computer simulations specifically aim at taking single patients specificities into account. Such peculiarities cannot be properly detected by adopting current patient-stratification studies (which often are based on genetic information) as they also involve more layers of information such as "post-genomics, metabolomics, microbiomics, medical imaging, biomedical instrumentation (including implanted, wearable, or environmental sensors), self-reported, and clinical data" (Viceconti et al. 2016a, p. 41). Interestingly, the Roadmap sets (at least) three levels of individualization according to differences in the quantification of predictive accuracy of the model. The first level refers to a generic patient in a given population. In this setting, accuracy of prediction is obtained by comparing the model values to those gathered in the given population. The second level refers to all single patients in a given population. Here the model compares its values with certain properties (*e.g.* average, median etc.) of values detected in the given population. In the last level "[t]he model intends to represent each individual of the reference population. Its predictive accuracy is measured by comparing the predicted values to the observed values for each individual in the reference population" (Viceconti et al. 2016a, p. 41).

Yet, despite these efforts, this approach still does not solve two important, theoretical/empistemic issues related to modeling. The first one regards the the neutrality of computational models towards representations. In other words, even if we grant that certain computer simulations have the power (in terms of quantity) to fill certain gaps left out by empirical research, we still do not have clear indications that the way we computationally dissect individualization has no impact on our understanding of the phenomenon. The second problem might seem easier but still central: computational approaches definitely try to address completeness (in the sense of being more complete as possible) but there is not a clear justification on whether they can tackle complexity (which obviously differs from completeness).

The perspective offered by Viceconti and colleagues is surely intriguing but the complexity of such a multilayer approach also suggests that further work (also conceptual) should be done on what individualization is and how it could be properly addressed.

6.6 How Could *in silico* Approaches Extend and Integrate Traditional Clinical Trials

From a methodological point of view, ISCTs can be seen as a new case in which computational approaches, which often employ so called “Big Data”, are adopted to address biological complexity. In the case of biomedical research, as for the Avicenna Project, such approaches are often implemented and justified in order to unravel patient specificities, due to a more global understanding of underlying physiological conditions. Then, *in silico* methods are thought to be helpful in providing more precise possible therapeutic outcomes. A key aspect concerning new methodologies such as ISCTs involves the increasing importance of computational/*in silico* in the everyday-routine of both basic and clinical research. As a matter of fact, the rise of Big Data in biomedicine has generated an important discussion concerning the nature of scientific enterprise by itself. In recent years, for example, a heated debate has pervaded the scientific community on the epistemic plausibility of using these methodologies to make discoveries. In other words, some authors (such as Leonelli 2012) have claimed that such strategies can be genuinely seen as a “new mode of scientific research”. However, despite the increasing success of this new way of pursuing research agenda (both in terms of needs and methods), its impact on scientific practice has been rather controversial.

Many leading researchers (see among others Weinberg 2010; Brenner 1999, 2009) have raised doubts about the fact that computational approaches based on Big Data may constitute a legitimate, independent form of scientific investigation or a valid alternative to traditional ones. This claim should not be naively interpreted as a rejection of such strategies *in toto*, but rather as the establishment of the epistemic priority of experimental, empirically grounded, work over the *in silico* analysis. This is also because computational tools, often (but not always accurately) described either as ‘hypotheses-free’ or ‘data-driven’ science, are also considered unfit to provide a detailed understanding of the phenomena under investigation. Indeed, while there is no doubt on the efficacy in analyzing vast amounts of data, which eventually lead to a more global characterization of certain phenomena, nevertheless, according to its detractors, *in silico* approaches are accused to be lacking the possibility to fully describe in detail the underlying causal network[4].

Unfortunately, most of the debate, even within the scientific community, does not escape the level of scientific journalism and it is usually polarized between data-driven advocates and hypotheses-driven supporters as if these two positions would represent two alternative, not integrable visions of scientific work. Thus, such a situation demands a clarification of a certain type[5].

---

[4]On the other side, the proponents of Avicenna Project suggest that their strategy can also unravel mechanistic features of the phenomena under investigation. See above.

[5]Interestingly, unlike many other controversies over the nature of scientific work, which have been led mainly by philosophers, this time scientists themselves are embracing such a discussion. The unusual interest of many researchers for epistemological problems might be also explained by

Here we have no time to analyze and reconstruct the nature of this debate in depth. We limit ourselves to note the fact (somehow philosophically trivial) that nothing, in empirical sciences, can be really considered as based and generated only by data, neither hypotheses can rest on pure speculation. As recently proposed (see among the others Strasser 2011; Keating and Cambrosio 2012; Leonelli 2012; O'Malley and Soyer 2012), discovery strategies in contemporary biomedical research, due to presence of such an abundance/plurality of approaches, might be defined as *hybridized*. Such a perspective entails a more complex view by which scientific practice is not, too simplistically, crystallized in a specific set of rules or procedures but it is rather articulated in different phases or moments which enucleate different epistemic roles in its process.

It is important to recall that what is at stake here is not the theoretical stance of biomedical research as such (*i.e.* what makes biomedical research as such), but rather the investigation over the possibility that a different methodological approach could produce the same (or even better) result than more traditional ones (see Viceconti et al. 2016a for a proactive stance on this point. A more critical approach can be found in Malagrinò 2016). In other words, the innovation brought by *in silico* strategies should be evaluated, as argued for similar situations in which computational approaches changed/modified the practice of science and medicine, not in terms of imposing/promoting a different scientific paradigm (see among the others, for a general discussion about this point, Kitchin 2014; Ratti 2015) but, on the contrary, in terms of the reinforcement/complementing of the current paradigm by other means. Indeed, at the epistemic level, an investigation strategy like the one employed by *in silico* clinical trials is not a way to address biological complexity by adopting a distinct theoretical look on what complexity is, but a way to make some aspects of this complexity more tractable according to the current standards of biomedical research[6].

---

the fact that, precisely because of the economic impact that such new approaches have brought to research itself and the promise they present on the nature of diagnosis and treatment, in silico strategies may actually change not just what science and clinical research are in theory (thus affecting pure theoretical speculation) but also what they are in practice. Indeed, such a change could potentially reshape how scientific institutions, research centers, clinical hospitals and funding agencies (either public or private) are organized, structured and the way they act to foster their activities. Bruce Alberts for instance (2010), is worried about the possibility that so called "small-science" labs (those laboratories still not equipped with strong in silico technologies) might disappear thus creating a disparity among research projects and approaches in terms of funds and visibility. Indeed, if high impact-factor journals will privilege a certain kind of research, this will eventually lead to a sort of hierarchy of discoveries, according to which certain kinds of evidence (i.e. pure mechanistic description) might count less than global representations provided by computational approaches.

[6]Concerning this point, it is important to analyze what makes the difference here in terms of heuristic strategies and their justification. In their *Discovering Complexity: Decomposition and Localization as Strategies in Scientific Research* (1993), Bechtel and Richardson argue that the amount of information that is available to scientists when they start their investigation is, practically and theoretically, limited. This means that not all possible logical solutions could be pursued, nor even fully listed. Nevertheless, no research begins out of the blue. Scientists normally adopt a set of

From a methodological point of view, and beyond the technical aspects, the fact that ISCTs are *virtual trials* may constitute a legitimate source of concern about their belonging to any experimental procedure. If ISCTs lack empirical intervention then they are not experiment. If so, how should we evaluate the type of evidence they produce? Part of the issue can be solved by noticing that intervention strategies can be of many forms, of which empirical manipulation is just one. Boem and Ratti (2017) recently argued that, in the case of computational biology, ordering and reordering information within electronic databases can be counted as a genuine form of intervention (*a là* Hacking). This is because classification procedures (consider also, for instance, the case of specimen collections in natural history) are effectively a real form of scientific activity even if they are not intimately "experimental".

Nevertheless, in the case of ISCTs we face another different situation. The philosophical debate over computer simulation is vast (see among the others Guala 2002; Morgan 2003; Morrison 2009). Here we have no time to specifically enter in this controversy. To briefly sum up, philosophers struggle on whether simulations can legitimately count as real experiments or not. For our purposes it would be enough to note that, as a matter of fact, a simulation, in order to be run, needs an input source that has to come from an experimental setting (or at least an empirical one). This source creates the space of possibility that can successively be explored and analyzed by the simulation. A computer playing chess, for instance, does not invent anything (nor surprises us, as argued for example by Parke 2014), it rather explores thousands and thousands of feasible moves which cannot be easily examined otherwise. Indeed, a computational simulation is a tool of discovery not by adding new elements to the experimental setting but because it can explore many possible configurations of it in a reduced time.

---

variable rules and methods (i.e. heuristics) which help them in settling and understanding the type of questions they want to ask and how they want to answer them. However, heuristics do not have the power of logical principle. They are prone to err and they lack the power of generality since they are usually highly context dependent and bound to specific experimental conditions. Contrary to naive views, this fact is far from being a flaw of research strategy. As Bechtel and Richardson argue, "[l]imiting the relevant variables and imposing assumptions [...] are procedures for attacking the problem. They also describe partial solutions. Applied to scientific problem-solving, this would mean that the heuristic assumptions constitutive of explanatory models would be critical in a developing research program" (Bechtel and Richardson 1993, p 16). Heuristics then starts from the assumption that complexity cannot be fully addressed per se. 'Decomposition' here means that fundamental causal factors (mechanistically intended) and interactions are detected and considered more important than others. From a methodological point of view, decomposition can be structural, dissecting a system in physical parts, and functional, investigating the activity of a system in terms of its sub-activities. Localization then is the connection between a particular process and the entity that is responsible for that process. Only at a first glance computational approaches seem to proceed in a different way. The volume of data involved in such a perspective may let someone think that there is no need to decompose nor to localize systems, since the very system itself will be addressed as a whole. This is somehow true, but only in a specific sense. In fact, the global view provided by the capacity of computers in handling and displaying enormous amounts of data coming from different resources is not a way to overcome traditional discovery strategies but rather a way to broaden it and make it more efficient.

An *in silico* clinical trial follows an analogous scheme. By adding virtual patients to the cohort, these types of trials cannot provide indications on particular features or properties which were not already taken into account by investigators. However, due to its higher manipulability (i.e. the capacity of playing with its virtual parameter without other disturbing/not relevant constraints) such an approach can explore many different situations (as much as each variation of certain conditions) and provide related models (in the sense of modeling different outcomes) of the phenomenon under scrutiny. Nevertheless, while computationally pursuing thousands of different "paths" during a chess match concerns just the operational rules of the game, when scientists simulate different scenarios, these ones are shaped according to the specific available data and to the empirical constraints. Thus, computational explorations as such should be taken with caution, as they might not be fully reliable.

The Roadmap (Viceconti et al. 2016b) claims that ISCTs could refine standard clinical research by several reasons. First ISCTs, precisely because of their exploratory power, could offer important indications in determining earlier confirmation on how a drug follows expectations or not. Moreover, ISCTs could add *in silico* patients to traditional trials "that might fill gaps in the individual variability seen in 'real' patients" (Viceconti et al. 2016b, p. 7). Second, computational exploration settled by ISCTs might provide new hints on aspects which were not detectable nor manageable via traditional methods allowing also predictive analysis on possible multiple outcomes (especially in the long term). Third, ISCTs might partially replace CTs either when they mimic or resemble scientific evidence (as in other areas of research, where computational models are considered, under certain conditions, a good proxy for animal models) obtained via more traditional methods. In the end ISCTs, due to their virtual nature, can probably complement CTs allowing the inquiry of diverse and rare alternatives which can be hardly tested in real patients.

From these considerations, it should be clear that ISCTs can be seen as a way to foster, complete and enhance traditional experimental strategies of discovery. As recently proposed by Emanuele Ratti (2015), it could be argued that, actually, computational approaches can be definitely framed within more traditional heuristic strategies. In his work Ratti suggests that, from a methodological point of view, the main difference between new approaches and more traditional ones "is that the former provide quasi-algorithmic procedures to follow the strategies of decomposition and localization (Bechtel and Richardson 2010), that is, to identify components that are supposed to be involved in the phenomenon of interest" (Ratti 2015, p. 216). Another interesting aspect is that, following Ratti (2015), *in silico* methods can be seen as a filter to prioritize hypotheses. In other words, *in silico* methodologies may offer an additional help to biomedical research concerning the possibility of extending the traditional boundaries of experimental systems (in the sense of Rheinberger 1997) or, in this case, the boundary of actual samples in RCTs. Moreover, as Ratti argues "[v]ast data sets are useful mainly because, when analyzed by means of powerful computational resources, they are taken as more comprehensive starting points for typical decomposition and localization strategies" (Ratti 2015, p. 216). It is according to this perspective that *virtual patients* and relative *virtual treatments* employed by ISCTs can be seen as an attempt to computationally fill, through modeling, those gaps (made as such by unavoidable empirical constraints) traditional methods could not directly handle.

This means that, from an epistemic point of view, *in silico* studies do not constitute, *per se*, a way to overcome biological complexity as they address the same descriptive level of traditional, experimental approaches. The old philosophical question on the nature of the individual itself, so relevant for personalized medicine, cannot be answered just by looking at piles and piles of data. The promises of the Avicenna Project and its roadmap are still to be seen. Thus, there are not, at the moment, clear and visible benefits in the practice of clinical experimentation. However, there is definitely something new. What changes here, is the capacity to collect and manage an enormous amount of data (previously neglected or not obtainable via traditional methods) and to order them accordingly. Only when these kinds of practices will be consistently applied and compared with traditional ones on a larger scale we will be able to evaluate their impact (measuring real benefits and actual flaws).

In the end, an epistemic look to ISCTs shows how certain problems concerning biological complexity are addressed in pharmacological studies. However, *contra* certain overstatements arguing for a paradigm shift in scientific research, ISCTs do not change the theoretical stance governing current scientific practice on clinical trials, nor the general framework according to which discovery strategies are adopted and justified. Something relevant certainly changes in the methodologies, but it should be regarded as an improvement and not as a replacement of them. In other words, *in silico* practice represent a novelty in addressing elements of traditional understanding which cannot be treated through current methods. However such a practice *per se* does not guarantee fully epistemic requirements to be considered and implemented independently from traditional approaches.

6.7 Conclusion

To sum up, in this analysis, we have presented new technologies, such as ISCTs, that are designed help to manage some current issues of clinical studies. We have explained what ISCTs are compared to standard RCTs. First we have shown, with pioneering studies, how ISCTs were motivated and how they were used to overcome local problems. Successively, we have presented ISCTs within a more comprehensive and systematic strategy supported by epistemological considerations.

In this respect, we have tried to justify in what sense ISCTs can belong to experimental research and what kind of evidence they produce. We have argued that an *in silico* clinical trial, by adding virtual patients to the cohort, cannot provide full indications on particular features or properties which were not already taken into account by investigators. In this sense, *in silico* strategies are not replacing in toto standard trials but they present some features that can improve them. This is because, due to their higher manipulability (*i.e.* the capacity of playing with their virtual parameter without other disturbing/not relevant constraints), ISCTs can explore many different situations (as much as each variation of certain conditions) and provide related models (in the sense of modeling different outcomes) of the phenomenon under scrutiny. In this respect, we have tried to clarify that the purpose of virtually modeling patients is not to provide a realistic representation but rather to let some relevant (*i.e.* considered as such) features to emerge in order to be studied and operatively adjusted. As a matter of fact, the purpose of a model is precisely to select certain significant properties and to emphasize them discarding the others. The final aim of ISCTs application is to help investigators to simulate variability which definitely plays a pivotal role in this type of research. Analyzing the Avicenna Roadmap and its perspective to create patient-specific models to generate simulated populations on which new compounds can be safely and rapidly tested, we have furthermore stressed that this development should be taken as a refinement of the current paradigm, not as a real replacement of it. In fact, *in silico* methods represent, within a common stance, a new way to refine and better locate in respect to a more general picture the understanding of certain biological phenomena.

However, according to the current standards of biomedical research, it is very unlikely that ISCTs could be improvement of some current aspects of the pharmacological field and it certainly possible to think that they are likely to change how pharmacological research is organized.

References

Barrett, J. S., Fossler, M. J., Cadieu, K. D., & Gastonguay, M. R. (2008). Pharmacometrics: A multidisciplinary field to facilitate critical thinking in drug development and translational research settings. *Journal of Clinical Pharmacology, 48*(5), 632–649. https://doi.org/10.1177/0091270008315318.

Bechtel, W., & Richardson, R. C. (2010) (1993 1st ed.). *Discovering complexity: Decomposition and localization as strategies in scientific research. Discovering Complexity: Decomposition and Localization as Strategies in Scientific Research.* https://doi.org/10.1111/j.1467-9973.1995.tb00588.x

Boem, F., & Ratti, E. (2017). Toward a notion of intervention in big-data biology and molecular medicine. In G. Bonolio & M. J. Nathan (Eds.), *Philosophy of molecular medicine: Foundational issues in theory and practice* (pp. 147–164). New York: Routledge.

Bosmans, G., Buijsen, J., Dekker, A., Velders, M., Boersma, L., De Ruysscher, D., Minken, A. é, & Lambin, P. (2006). An “in silico” clinical trial comparing free breathing, slow and respiration correlated computed tomography in lung cancer patients. *Radiotherapy and Oncology, 81*(1), 73–80. https://doi.org/10.1016/j.radonc.2006.08.009.

Bothwell, L. E., Greene, J. A., Podolsky, S. H., & Jones, D. S. (2016). Assessing the gold standard – Lessons from the history of RCTs. Edited by Debra Malina. *New England Journal of Medicine, 374*(22), 2175–2181. https://doi.org/10.1056/nejmms1604593.

Brenner, S. (2009). Sequences and consequences. *Philosophical Transactions of the Royal Society B: Biological Sciences, 365*(1537), 207–212. https://www.ncbi.nlm.nih.gov/pubmed/20008397. https://doi.org/10.1098/rstb.2009.0221.

Brenner, S. (1999). Sillycon valley fever. *Current Biology, 9*(18), R671. https://doi.org/10.1016/s0960-9822(99)80432-1.

Guala, F. (2002). Models, simulations, and experiments. In *Model-based reasoning: Science, technology, values* (pp. 59–74). New York: Springer. https://doi.org/10.1007/978-1-4615-0605-8_4.
Hacking, I. (1983). *Representing and Intervening*. Cambridge: Cambridge University Press. https://doi.org/10.1017/cbo9780511814563.
Jadad, A. (1998). *Randomised controlled trials*. Oxford: Wiley-Blackwell.
Keating, P., & Cambrosio, A. (2012). Too many numbers: Microarrays in clinical cancer research. *Studies in History and Philosophy of Science Part C: Studies in History and Philosophy of Biological and Biomedical Sciences, 43*(1), 37–51. https://doi.org/10.1016/j.shpsc.2011.10.004.
Kitchin, R. (2014). Big data, new epistemologies and paradigm shifts. *Big Data & Society, 1*(1), 205395171452848. https://doi.org/10.1177/2053951714528481.
Leonelli, S. (2012). Classificatory theory in data-intensive science: The case of open biomedical ontologies. *International Studies in the Philosophy of Science, 26*(1), 47–65. https://doi.org/10.1080/02698595.2012.653119.
Malagrinò, I. (2016). In silico clinical trials: A new dawn in biomedical research? *Humana Mente Journal of Philosophical Studies, 30*, 87–104.
Magni, L., Raimondo, D. M., Dalla Man, C., De Nicolao, G., Kovatchev, B., & Cobelli, C. (2009). Model predictive control of glucose concentration in type I diabetic patients: An in silico trial. *Biomedical Signal Processing and Control, 4*(4), 338–346. https://doi.org/10.1016/j.bspc.2009.04.003.
Man, C. D., Rizza, R. A., & Cobelli, C. (2007). Meal simulation model of the glucose-Insulin system. *IEEE Transactions on Biomedical Engineering, 54*(10), 1740–1749. https://doi.org/10.1109/tbme.2007.893506.
Morgan, M. S. (2003). Experiments without material intervention. In H. Radder (Ed.), *The philosophy of scientific experimentation* (pp. 216–235). Pittsburgh: University of Pittsburgh Press. https://doi.org/10.2307/j.ctt5hjsnf.15.
Morrison, M. (2009). Models, measurement and computer simulation: The changing face of experimentation. *Philosophical Studies, 143*(1), 33–57. https://doi.org/10.1007/s11098-008-9317-y.
O'Malley, M. A., & Soyer, O. S. (2012). The roles of integration in molecular systems biology. *Studies in History and Philosophy of Science Part C: Studies in History and Philosophy of Biological and Biomedical Sciences, 43*(1), 58–68. https://doi.org/10.1016/j.shpsc.2011.10.006.
Parke, E. C. (2014). Experiments, simulations, and epistemic privilege. *Philosophy of Science, 81*(4), 516–536. https://doi.org/10.1086/677956.
Ratti, E. (2015). Big data biology: Between eliminative inferences and exploratory experiments. *Philosophy of Science, 82*(2), 198–218. https://doi.org/10.1086/680332.
Rheinberger, H.-J. (1997). *Toward a history of epistemic things: Synthesizing proteins in the test tube*. Stanford: Stanford University Press.
Strasser, B. J. (2011). The experimenter's museum: GenBank, natural history, and the moral economies of biomedicine. *Isis, 102*(1), 60–96. https://doi.org/10.1086/658657.
Viceconti, M., Henney A., & Morley-Fletcher E. (2016a). In silico clinical trails: How computer simulation will transform the biomedical industry (Technical report). Avicenna Consortium. http://avicenna-isct.org/wp-content/uploads/2016/01/AvicennaRoadmapPDF-27-01-16.pdf. Accessed 3 Mar 2018.
Viceconti, M., Henney, A., & Morley-Fletcher, E. (2016b). In silico clinical trials: How computer simulation will transform the biomedical industry. *International Journal of Clinical Trials, 3*(2), 37. https://doi.org/10.18203/2349-3259.ijct20161408.
Weinberg, R. (2010). Point: Hypotheses first. *Nature, 464*(7289), 678–678. https://doi.org/10.1038/464678a.

Chapter 7
Uncertainty in Drug Discovery: Strategies, Heuristics and Technologies

Chapter for *Uncertainty in Pharmacology*

Erman Sözüdoğru and Brendan Clarke

Abstract In this chapter we show how pharmacological researchers manage uncertainty when using high-throughput screening (HTS) techniques. HTS allows researchers to rapidly survey the activities of large numbers of chemical compounds against a biological target. However, not all molecules that show promising biological activity during screening will be suitable for clinical use. Researchers use the concept of druglikeness to manage this uncertainty about whether a biologically active molecule will be suitable for pharmacological use.

We introduce our approach in Sect. 7.1, during which we argue that the methodological choices we discuss are predicated on epistemic assumptions. We then give a brief overview of drug discovery in Sect. 7.2, and introduce HTS in more detail in Sect. 7.3. Section 7.4 then introduces druglikeness as an epistemic strategy for managing uncertainty, before discussing one of the heuristic tools, known as the rule of five, that is used by pharmacologists to reduce uncertainty about druglikeness. We conclude this chapter in more philosophical territory by considering druglikeness as a kind of extrapolation. Here, we argue that druglikeness does not seem to invoke biological mechanisms, but instead exploits the chemical capacities of molecules. The aim of this exploration is to provide a normative account of the role(s) played by values and philosophical assumptions in employing different epistemic and methodological strategies in scientific practice.

7.1 Introduction

This chapter explores one species of uncertainty in pharmacological research by engaging with the following question: is a particular molecule, identified by high-throughput screening, drug-like? High-throughput screening (HTS) is a technique

---

E. Sözüdoğru (✉) · B. Clarke
Department of Science and Technology Studies, UCL, London, UK
e-mail: erman.sozudogru.12@ucl.ac.uk; b.clarke@ucl.ac.uk

A. LaCaze, B. Osimani (eds.), *Uncertainty in Pharmacology*, Boston Studies in the Philosophy and History of Science 338,
https://doi.org/10.1007/978-3-030-29179-2_7

by which compound libraries are rapidly surveyed for their biological activity against a particular target. Molecules that show some promising biological activity during this screening, informally known as leads, progress to become the basis for more focused programmes of drug discovery in the laboratory and in the clinic.

However, this path from promising molecule to routine clinical practice is long, and uncertain. Most of the leads identified by HTS never become marketable drugs. One explanation for this is that biological activity against a target is (of course) not the only property that is required to make a drug useful in clinical medicine. For example, even if a lead shows promising biological activity *in vitro*, it will be discarded if it shows less desirable biological activities when tested *in vivo*. Dropping out on the basis of adverse effects happens relatively late in the process of travelling from molecule to market, generally during clinical testing. This chapter deals with a much earlier route by which a lead may fail to become a sellable drug, and the strategies used by researchers to manage uncertainty in the discovery and development phase of drug discovery. We introduce and explore an heuristic strategy that is used to manage uncertain inferences that take place comparatively early in the process of drug discovery. In particular, we unpack one set of rules, which are known as Lipinski's rule of five (Lipinski et al. 1997). These rules describe a set of simple molecular properties (like hydrophilicity, molecular weight, and so on) that jointly describe the *drugability* of a candidate molecule. By drugability, we refer to the presence of desirable biopharmaceutical properties of the molecule. Candidate molecules that fulfil the rule of five have a higher degree of drugability than those that do not.

We divide this chapter into six sections. In Sect. 7.2 we introduce the concept of the drug discovery pipeline as a way of describing the organisation of contemporary pharmacological research. Here, we also compare and contrast two competing approaches to drug discovery within the loose metaphor of the drug discovery pipeline, which we call the irrational and rational approaches. We then move on, in Sect. 7.3, to introduce a key technology in contemporary drug discovery: high throughput screening. This technology has greatly improved the certainty with which particular molecules are known to interact with a specified target system *in vitro*. However, as we note in Sect. 7.4, this greater certainty about *in vitro* behaviour has come at the cost of greater uncertainty about whether knowledge of the *in vitro* behaviour of a drug will be similar to its behaviour *in vivo*. We claim here that this question about druglikeness has arisen as a consequence of the widespread use of high-throughput screening. In Sect. 7.5, we then go on to describe a set of heuristic rules – Lipinski's rule of five – used to reduce uncertainty about this extrapolation. We approach the question using several philosophical approaches to explore the underlying philosophical assumptions in Lipinski's rule of five.

Our aim is to argue that methodological choices are predicated on epistemic assumptions. In the first half of this paper we examine the underlying philosophical assumptions and questions throughout the drug discovery process. Later in this paper we focus on how uncertainties arising from HTS are dealt with. We argue that Lipinski's rule of five is based on epistemic assumptions on the causal inference linking atomic properties to how molecules are expected to behave in *in vivo*. Note that these inferences do not depend on any kind of mechanistic understanding of the behaviour of molecules *in vivo*. We use Cartwright's account of capacities to show the how the underlying assumptions in the methodology can be exposed and scrutinised. Here we present how the heuristic rule of five is used to manage uncertainty about drugability based on the physico-chemical properties of compounds. This heuristic rule is an excellent example of a reductive and simple rule that is nevertheless successful in managing an inference of great complexity. In drawing our philosophical argument, our philosophical work depends on a descriptive account of the strategies used by Lipinski and colleagues. In other words, we aim here to unpick the way that an aspect of scientific practice actually comes about. This, we think, is an example of a species of naturalism, and in that we claim that philosophers of science should study science as it is actually practiced.

Our aim in doing this is we provide a normative account on the important role played by values and philosophical assumptions in scientific practice. We argue that uncertainty is managed in the following case by tapping into the capacities of molecules. For instance, given that there are no explicit statements in the Lipinski et al. (1997) paper on the rationale that this heuristic is based on, nor discussion of the underlying assumptions shaping the methodology of the authors, we can only assume that these rules ultimately depended on statistical data describing the physicochemical properties of licensed drugs, together with the assumption that the biopharmaceutic properties of these licenced drugs were likely to be desirable ones. This shows, we think, that non-verbalised assumptions held by scientists in making decisions such as these can be understood in ways that would make sense to the analytic philosopher of science (e.g. the way that these values may be connected to philosophical work on capacities). Our aim is to demonstrate the possible ways that philosophers interested in practices might themselves practise.

7.2 The Changing Face of the Drug Development Pipeline

In order to introduce our case-study, we will use the metaphor of the drug discovery and development pipeline. This represents drug discovery as linear process, sectioned into discrete phases, with a well-defined starting point and a similarly well-defined final outcome of a marketable drug. At the same time, we recognise that this pipeline is largely metaphorical, and take seriously calls that this simplification is excessive (e.g. Baxter et al. 2013).

This chapter largely concerns ways that drugs might drop out of this pipeline during stage 2 (pre-clinical development). Note that in this chapter, we are interested specifically in current methods of discovering and developing drugs. Of course, drug development is a constantly evolving process that has involved many distinct technologies (Drews 2000). Our discussion here therefore considers a snapshot of methods that are currently present – particularly high-throughput screening. We think that the reasons underlying drop-outs in Stage 2 are largely a consequence of

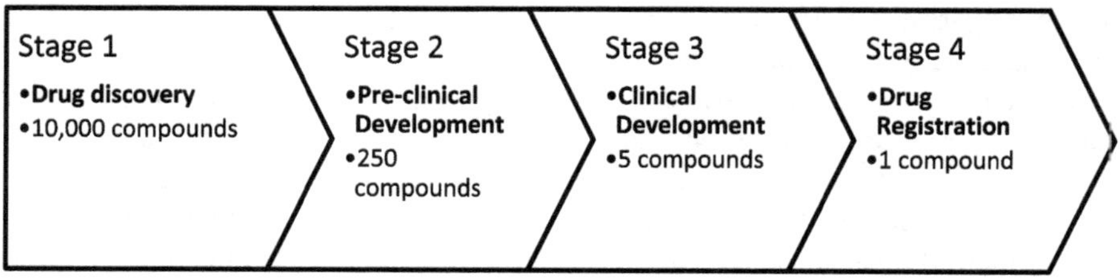

Fig. 7.1 A simplified overview of the drug discovery pipeline

Table 7.1 summary of the rule of five (Lipinski et al. 1997)

Lipinski's rule of five
"...Poor absorption or permeation is more likely when there are more than..."
no more than 5 H-bond donors no more than 10 H-bond acceptors MWT greater than 500 CLogP greater than 5

H-bond hydrogen bond, *MWT* molecular weight, *CLogP* calculated LogP (a measure of lipophilicity)

technologies used during stage 1. In order to make sense of how these consequences flow downstream, giving a brief overview of the pipeline as a whole will be helpful.

The pipeline is divided into different stages. A typical pipeline (see Fig. 7.1) begins with a screening process that is used to assess the potency of a very large number (tens, or hundreds, of thousands) of promising compounds. These are usually molecular variations on a theme, such as permutations of existing drugs. In the first stage, this multitude is winnowed by comparing the relative activities of each molecule against some selected target *in vitro*. The techniques used to measure the effects of compounds against targets are highly automated, as they have to be if this mountain of compounds is to be sifted. These techniques, which fall under the rubric of *high-throughput screening* are discussed in Sect. 7.3.

The best of these compounds – those leads that show highly promising patterns of activity on the isolated target – then pass into pre-clinical development. Here, laboratory techniques undergo a transition from the highly automated *in vitro* practices of HTS, to more hands-on work, typically in an animal model of the disease. In this second stage, the efficacy of the molecule as well as possible serious adverse effects are identified. The focus of this research concerns the biopharmacological properties of the lead molecules that were identified in the preceding discovery stage. Key properties, such as absorption, distribution, metabolism, and excretion are assessed. Additionally, maximum levels of drug safety (for instance, LD50) are determined along with the investigation of any side effects that would preclude the use of the drug candidate. The determination of these properties results in a further winnowing of candidate molecules, with only the very best (most effective, most safe) passing into the clinical phases of research. Here, the experimental substrate again changes as the drug is now tested in humans in a staged pattern of different test methods (clinical phases I – IV). All things being well, this clinical development end in registration and post registration trials of the new drug.

We now turn our attention to supply further details regarding the first part of the process – the discovery stage. Before drug discovery can begin, a necessary pre-requisite is *target identification*. This involves the selection of both a disease for investigation, and the subsequent identification of a drug target. This research is methodologically heterogeneous, but is largely non-automated, and performed in the wet laboratory. At this stage, the aim is to identify a target that can be modulated with a drug-like molecule to produce the desired therapeutic outcome. These drug targets are (typically) macromolecules that are necessary for disease proliferation in some way that can be modulated with a xenobiotic that results in the desired therapeutic effect. For instance, a suitable target in anti-parasitic drug research would typically be a protein expressed by the parasite that is in some way essential for its survival.

Following target identification, the next step in drug discovery – *screening* – involves a large-scale search for compounds that can interact specifically with the chosen target and modulate its function in some promising way. We will discuss the details of the methods by which this screening takes place in Sect. 7.3. This step entails the screening of compound libraries (Charles and O'Donnell 2006: 7), which are collection of molecules generated and maintained by pharmaceutical companies, in order to find suitable active compounds. An important point at this juncture is that compound libraries contain very many molecules that are structurally related to existing drugs. This is because they are (partly) derived from previous drug discovery processes carried out by the pharmaceutical companies that own and maintain these libraries.

We give these details about the discovery stage in order to introduce and defend a claim about the underlying epistemology of the drug discovery process. We think that this two-step strategy of target identification and screening is compatible with two very different epistemologies of pharmacological research. Because the remainder of what follows will only deal with one of these epistemologies, we would like to distinguish them clearly at this point so as to avoid possible later confusion. We call these two approaches to drug discovery *rational*, and *irrational* (we give a brief sketch of these terms below, also see Ng 2015: 54 for details). Note that we do not make any kind of claim that these two approaches to drug discovery are exhaustive of all possible approaches. Nor do we claim that the discovery of some particular drug will always involve the application of just one of these strategies. But, as our case-study shows, there is good evidence that these are at least plausible characterisations of the way that drug discovery is approached in some contexts.

Rational approaches are characterised by the synthesis (or rational identification) of a molecule specifically designed to achieve some specified biological operation on a target. A canonical example of this kind of rational strategy is the creation of targeted antibodies and other biological agents (Charles and O'Donnell 2006). This usually entails modelling of potential binding sites in a target, and the consequent modelling of a drug-like molecule with properties capable of suitably interacting with these target binding sites. In rational cases such as these, we think that most of the research effort is expended in creating a new molecule, and comparatively less in testing to see if that molecule has the desired biological effects.

By contrast, irrational approaches (which are the epistemology that our case-study is grounded in) emphasise efficacy testing over molecule synthesis. Here, potential drug candidates are identified through an automated search across many thousands of molecules in compound libraries. No, or very little, research effort is spent trying to create and identify molecules whose structures suggest that they will have some particular biological effect. Instead, the emphasis is firmly on selection (like natural selection) via the evaluation of the strength with which a molecule interacts in some desirable way with a target.

7.3 High-Throughput Screening

7.3.1 Introduction

In this section, we will introduce HTS methods, and note that this methodological change (using robotics to screen compounds quickly) was accompanied by a change to the method of finding candidate molecules for *in vitro* testing. This latter modifications, while arguably less visible than the methodological changes, have had profound implications for the drug discovery pipeline as a whole, particularly concerning the certainty of some inferences, as we discuss in the next section.

Measuring the activities of many thousands of substances from a compound library against a biological target is achieved using a technique called High-Throughput Screening (HTS). This is an automated way of screening thousands of molecules using robotics that is both cost-effective and fast. It is also extremely sensitive, in that it is very good at detecting specific interactions between a ligand and a target molecule. In the context of drug discovery, HTS has replaced slow, high-craft methods of testing where *in vivo* and *in vitro* assays were used to look for lead compounds. Traditionally, drug discovery began with the selection of small numbers of plausibly-active compounds that were drawn from natural products, clinical studies of the side effects of existing drugs, unexplored patents and so on (Lipinski et al. 1997).

These leads were then appraised using manual screening methods. To generalise, these were generally fairly large-scale (perhaps of the order of 1 ml of reagent per test), and were carried out in small numbers by trained laboratory workers – either one at a time in individual test tubes, or in small numbers using multiple-well plates. These methods were time consuming, laborious (for instance, in the manual weighing and aliquoting of reagents for test), and prone to human error. This limited the overall assay capacity to 20 to 50 compounds per week (Pereira and Williams 2007: 53).

Note though that the replacement of these earlier techniques, which began in the late 1980s (Pereira and Williams 2007: 56) was not a complete and direct one. As we discuss below, the molecule-target interactions detected by HTS are similar, but not exactly the same, as those that would have been detected using previous techniques. Note too that HTS is conceptually related to other biotechnologies – such as polymerase chain reaction and combinatorial chemistry (Hommon and Nelson 2006: 80) – through automation they have played such a pivotal role in enabling biochemical synthesis and evaluation to become both dramatically faster and cheaper during the last two decades.

7.3.2 HTS Methods

HTS was developed to overcome some of the frustrations encountered with manual methods of drug screening (Pereira and Williams 2007). HTS is a multiple step process for performing a very large number of assays very quickly, in a largely automated way. From the perspective of the user of HTS (say, a pharmacology researcher), the first step in the HTS process is assay development. As we discuss above, the identification of a suitable target is a non-trivial process. In assay development, a chosen target is developed into a suitable assay. HTS utilises plates containing reaction assays, which come in many different forms including binding assay, enzyme assays, cell-based assays, and so on (Macarron and Hertzberg 2009: 6). These assays are designed to be performed in very large numbers, and so they must use appropriate reagent volumes, require no manual tinkering, be reliable, and produce some kind of machine-readable output (colourimetric changes, for example).

Once a suitable assay has been developed, it is then implemented in the second step of HTS. Here, the developed assay is then implemented using well plates with very large numbers of wells (384, 1536, or 3456 are typical). Compounds under test (see below) and reagents are added to these wells in an automated process, and the resulting reaction rates between target and small compounds are detected with either luminescence or fluorescence depending on the details of the individual assay. The rate and order of change exhibited by a particular well corresponds to the potency of the interaction between the target and the particular small compound found in that well. This is then recorded by plate readers and the resulting data is passed for analysis.

The third and final step of HTS is data analysis. The method for detecting hits depends on the assay type (for instance colourimetric assays look for changes in the colour in each well). In general, though, the most important task at this stage is the distinguishing of true positive 'hits' from negative results. The most common statistical tool used for this role is statistical significance along with data visualization tools (Gunter et al. 2003). Data analysis is automated through the use of commercial and custom built software. This statistical analysis is tailored to the specific aspects of individual cases by including case-specific information in the statistical model, including considerations of potency of interaction between lead and target, or analysis of the chemical structures involved in HTS 'hits' (Gunter et al. 2003).

7.3.3 HTS and the Leaky Pipeline

In the context of the drug development pipeline, HTS allows the fast and cheap identification of compounds with promising biological activity against some biological target. However, for reasons that we discuss below, not all compounds that show promising activity against a target can be taken all the way to be marketable drugs. Besides having suitable biological activity against a target, molecules must also behave in an appropriate way in the human body if they are ever to aspire to be used in medicine. While the ability to screen vast quantities of molecules using HTS certainly increases the number of potential drug discovery projects, it has also had the paradoxical effect of increasing the attrition rate (that is, the number of molecules that drop out of the drug development pipeline during subsequent testing). This attrition is often (and pleasingly) described as *leakiness* of the drug development pipeline. However, the consequences of these leaks are certainly not pleasing. Why, then, is it the case that many compounds identified in HTS cannot be developed into marketable drugs? Why does high throughput not correspond with high output?

Evidence for this phenomenon can be seen in the trends in the output of the pharmaceutical companies. The anticipated increase in the discovery and development of new drugs was not realised despite the advent of HTS and aforementioned biotechnologies. Closer look at the trends in the drug discovery pipeline between 1993–2003 shows an increased number of drug candidates were entering the drug discovery pipeline, matching the anticipated increase in the productivity of pharmaceutical research (Parexel 2003). However, the vast majority of these candidates failed during the pre-clinical and clinical stages. Why did the pipeline leak?

We argue that the high degree of idealisation of biological systems that occurs as a necessary part of using HTS is largely responsible for this leakiness. HTS techniques are optimised for to take place in robust automated systems. This shift (from lab based assays conducted by hand) to HTS has led to many changes in the experimental environments used. In turn, these changed environments have tended to alter the physicochemical environment in which HTS assays are conducted in such a way that it becomes very much less similar to the physiochemical environment found *in vivo*. This growing dissimilarity has led to an apparent disjunction between the experimental questions asked of candidate molecules between the first stage (HTS) and later stages (*in vivo* and clinical) of the drug discovery pipeline. More specifically, although an HTS assay provides information on interactions between a target and a compound, it does not provide evidence as to the likely magnitude of that target-compound would be in a biological system (such as an experimental organism). It is this uncertainty, and the strategies used to manage it, that we explore in the following section.

7.4 Druglikeness, the Increasingly Leaky Pipeline

While HTS has solved many difficulties in drug discovery concerned with the effects of compounds on receptors, it has introduced new problems of its own. Before HTS, most assays of target potency were carried out in aqueous phase. Candidate molecules were dissolved in water, and this solution was then used in an assay. Producing these aqueous solutions was slow (often involving a 24 hour equilibration step), and hard to standardise. Yet the automated nature of HTS required highly standardised and convenient libraries of compounds that could be tested against the drug target in question. This pressure led to the development of compound libraries consisting of solutions of candidate molecules at a standard concentration (typically 20 or 30 mmol/L). This required the use of an appropriate solvent. Because of the widely differing solubilities of different candidate molecules, dimethyl sulfoxide (DMSO), an extremely potent solvent for both hydrophilic and lipophilic compounds, became the de facto standard for these compound libraries.

However, this standardisation has introduced a major source of uncertainty. Drug targets *in vivo* are typically localised in the body. This is a major topic, but to illustrate, SSRI drugs (such as paroxetine) must reach serotonergic neurones in the brain if they are to have an effect on depression, mefloquine must reach the bloodstream if it is to prevent malaria, and lansoprazole must reach the parietal cells in the lining of the stomach if it is to reduce acid secretion and thus heartburn. And – very obviously – drugs must reach their targets *in vivo* if they are to be effective. This places strictures on the kinds of structures that are capable of being used as effective drugs. Highly hydrophilic substances, for instance, will not readily pass through the lipid bilayers of membranes, meaning that they will not be effective on intracellular targets, or those behind the blood-brain barrier (like the neurones that paroxetine acts on). Yet investigating this requirement of appropriate solubility is not a feature of HTS techniques. Quite the opposite, in fact, because their employment of DMSO as a solvent is intended to remove considerations about solubility from the research process. How then can pharmaceutical researchers be sure that their compound, which works well on a target *in vitro*, will be able to reach that target in a clinical setting?

This difficulty is described as the question of desirable biopharmaceutical properties (the terms druglikeness and drugability are also used in the literature), which deal with the issue of whether a lead will be able to exert its effect on a target in the body. Cases where the lead shows excellent activity in the HTS system, but did not show such promising effects when tested *in vivo*, are common. Prior to the introduction of HTS, 39% of drugs failed during the development phase due to poor biopharmaceutical properties (Prentis et al. 1988). This figure is now very much higher, with un-druglikeness as a major contributor to biopharmaceutically-mediated drop-out. Worse, for pharmaceutical company shareholders, at least, this drop-out tends to happen relatively late in the process – during the expensive and intricate phase of preclinical testing.

We think that failures of drugability occur in cases where a) a specific molecule is a promising lead, and thus demonstrates strong biological activity against some intended target *in vitro* but b) that same biological activity is not seen when the molecule is administered in biological systems, and that c) this failure is not a result of competing mechanisms affecting the specific mechanism of action of the drug (i.e. that drugability describes cases where the drug does not work *in vivo* because it does not reach the salient site of action). We therefore regard this as being a species of failure of extrapolation (Steel 2008), in that failures of drugability describe a change in the relation between cause and effect when background conditions are changed. A (a lead) causes B (target modulation) in C (our HTS well-plate), but not in D (our experimental organism or human subject).

How can drugability be investigated without doing large numbers of expensive, slow, and cumbersome preclinical experiments? Our next section describes a strategy by which this uncertainty about extrapolation is managed in pharmaceutical practice.

7.5 Lipinski's Rule of Five

In this section, we introduce an heuristic rule that can reduce the chance that a lead might have undesirable biopharmaceutical properties. This rule, introduced by Lipinski et al. (1997) stipulates that a major reason that a molecule might lack these properties is as a consequence of the non-specific molecular interactions demonstrated by the molecule *in vivo*. Somewhat confusingly, the rule of five describes four parameters, which jointly predict whether a lead might have desirable biopharmaceutical properties (the 'five' refers to the happy coincidence that the value of each parameter is a multiple of five).

Note that these properties can be determined in a fully general way. That is, there is no need to perform a context-specific estimation of the rule of five. Instead, the rule of five can be accurately estimated once from a representation of the known molecular structure of a lead. This makes the rule of five inexpensive to use. It also means that a rule of five calculation is portable, in that it can be estimated once, but applied many times as a compound is tested by HTS against many different drug targets. Sources such as Drugbank (www.drugbank.ca), which is a publically accessible drug (and drug target) database, report an estimate of whether compounds satisfy this rule in just this way. It also reports estimates for many other similar rules (such as the Ghose filter, Veber rule, and the MDDR-like rule).

While the four points in the rule of five can be easily estimated from elementary chemical knowledge, note that collectively the rule of five is a stipulative definition of druglikeness. That is to say that there is no deductive entailment of these rules from some general chemical theory or anything resembling that kind of truth-preserving inference. Instead, the rule of five identifies some region of molecular space whose members have a much higher than usual probability of not being biopharmaceutically undruglike in the way specified above. For the Lipinski group, this region of molecular space was determined empirically using a selection from a database of 50,000 drugs in clinical use, drawn from the World Drug Index (WDI). The rule of five was calibrated in such a way that 90% of the most widely used drugs – their use presumably indicating that they have fairly good biopharmaceutical properties – were on the correct side of the druglike demarkation line. Their methodology for determining which drugs to include is worth repeating here:

> The (WDI) World Drug Index is a very large computerized database of about 50 000 drugs from the Derwent Co. The process used to select a subset of 2245 compounds from this database that are likely to have superior physico-chemical properties is as follows: From the 50 427 compounds in the WDI File. 7894 with a data field for a USAN name were selected as were 6320 with a data field for an INN. From the two lists, 8548 compounds had one or both USAN [United States Adopted Names] or INN [International Nonproprietary Names] names. These were searched for a data field 'indications and usage' suggesting clinical exposure, resulting in 3704 entries. From the 3704 using a substructure data field we eliminated 1176 compounds with the text string 'POLY', 87 with the text string 'PEPTIDE' and 101 with the text string 'QUAT' [excluding proteins and other large molecules]. Also eliminated were 53 compounds containing the fragment O=P-O [phosphate backbones]. We coined the term 'USAN' library for this collection of drugs. (Lipinski et al. 1997: 7)

This USAN set (of "superior" drugs) was then compared to the larger WDI set as a whole with respect to their properties. To illustrate for two of the four factors (molecular weight and H bond acceptor count):

> Molecular weight... is an obvious choice because of the literature relating poorer intestinal and blood brain barrier permeability to increasing molecular weight... and the more rapid decline in permeation time as a function of molecular weight in lipid bi-layers as opposed to aqueous media... In the USAN set 11% had MWTs > 500 compared to 22% in the entire data set. (Lipinski 1997: 7)

> The sum of Ns and Os is a rough measure of H bond accepting ability... When we examined the USAN library we found a fairly sharp cutoff in profiles with only about 12% of compounds having more than 10 Ns and Os. (Lipinski 1997: 8)

That's not to say that every drug that meets the rule of five will be a convenient drug to use. Several drugs – such as dantrolene sodium – are part of the pharmacopeia despite serious and well-known drawbacks in their use. Dantrolene is used for the treatment of malignant hyperthermia – a rare but serious complication of general anaesthesia (Kolb et al. 1982). The dose, which is given intravenously, looks standard enough: "Initially 2–3 mg/kg, then 1 mg/kg, repeated if necessary; maximum 10 mg/kg per course." (BNF n.d.). However, because of the extremely low solubility of dantrolene in water (0.0805 mg/mL), this dose required administration in very large quantities of diluent (at a minimum, several hundred mL). Yet – for emphasis – dantrolene meets the rule of five (and related schemes, including Veber, MDDR-like and Ghose filter). So fulfilling the rule of five does not mean that a molecule will have completely desirable biopharmacological properties.

Table 7.2 Data from Lipinski et al. 1997: 13 showing the increase in MLogP (measured LogP – a measure of lipophilicity) and MWT (molecular weight) of drugs

	MLogP		MWT	
Percentile	1986	1994	1986	1994
90th	4.30	4.76	514	726
75th	3.48	3.90	415	535
50th	2.60	2.86	352	412

Note that HTS was introduced in 1987, so this data shows the increase in both fat solubility and molecular weight that accompanied the introduction of HTS

So, the rule of five is used as a fallible heuristic device that reduces the chances that a drug will ultimately turn out to have undesirable biopharmaceutical properties in the way specified above. Lipinski et al. wrote that they (the authors) were motivated to produce the rule by the growing average MWT and LogP of drugs discovered by HTS during the early 1990s, which we have included below in Table 7.2.

Thus there's an historical aspect to the development of the rule of five that must not be overlooked in describing them as strategies for managing technologically-evoked uncertainty. As described above, HTS technologies – particularly the widespread use of DMSO in place of the prior aqueous equilibrium – provoked an alteration in the kinds of compounds that it was possible to evaluate for their pharmacological effects. Thus a trade-off between certainty about the questions "Do molecules of the following kinds act on such-and-such a receptor" and "Will such-and-such a lead be a useful drug" was introduced. In effect, improving certainty during the first stage of drug discovery came at the cost of decreasing certainty that these leads would ever make for clinically useful drugs.

This picture, we argue, is one of different certainties at different phases in the pipeline, accompanied by fallible methods for managing precisely defined kinds of uncertainty. Here, Lipinski's rule manages a species of uncertainty that – while not specific to HTS – became very much more significant than it had previously as a result of the widespread use of HTS.

The rule of five infers a link between the basic physico-chemical properties of compounds and their biopharmacological properties. As we note above, we think that this can be regarded as a technique to overcome the perennial insecurity of extrapolation. While, as might be expected, Lipinski et al. are not explicit about their philosophy of inference, we think that this case can be usefully connected to the philosophical literature on extrapolation in the sciences more general. Specifically, we think that an analysis of the rule of five in terms of comparative process tracing (Steel 2008: 88–92) does not adequately account for the way that Lipinski's rule of five assists working pharmacologists. This is a surprise, to us at least, because we (in general) strongly endorse Steel's development of comparative process tracing as a way of understanding extrapolation in many other areas of the biomedical sciences. In fact, we are minded to agree with Guala's claim that "Comparative process tracing is the best analysis of extrapolation inferences in the philosophical and scientific literature so far." (Guala 2010: 1070). While we think that this remains true, we wonder whether the extrapolation between molecular and functional properties that this case involves is better understood via an alternative technique. We have therefore developed a line of argument below that understands the rule of five as grounded in the study of capacities. In brief, the rule of five works by picking out a set of capacities of molecules. These capacities, in some general way, control how easily a drug can reach its active site, and it is therefore these capacities that provide the warrant for extrapolations that concern *in vivo* properties of those molecules. We will begin this line of argument below by showing how Steel's account of comparative process tracing does not adequately reflect the questions faced by pharmacologists in this case.

Steel defined extrapolation as an inference on the causal link between two factors, X and Y, in some context A based on the knowledge on the causal link between X and Y in some other context B. Note that, for Steele, this inference is not a question of induction, but depends on a mode of inference that he calls *comparative process tracing*. Comparative process tracing relies on the identification of mechanisms that are causally relevant in context A and comparing them to corresponding mechanisms in context B.

In this case, while we might be able to give an account of the way that a particular mechanism linking X and Y in our HTS system (Steel's context A), it is much less clear how we can generate knowledge about relevant mechanisms in some economic and reliable fashion about what will happen in context B: the human body. In fact, this is the ball-game: if we knew about what would happen between X and Y in B, we'd have no need to go drug discovering at all. Instead, we would already know just what the effects of the drug would be in a recipient.

Our argument in support of the claim that comparative process tracing does not explain the way that the rule of five is used in practice relies on two supporting claims. First, at the phases of drug discovery research that this paper discusses, researchers do not evaluate the mechanisms by which drugs get into the right place to have their therapeutic actions *in vitro*. Note that this biopharmaceutical question is distinct from questions about mechanisms of drug action, which are effectively modelled *in vitro*. Next, and in support of the first claim, we note that the rule of five is completely generic, such that it can be evaluated for any candidate molecule, and that this evaluation does not depend on claims about the mechanisms by which a particular molecule reaches its appropriate site of action, but instead depends on the properties of the molecule in question qua molecular chemistry.

We therefore conclude with a negative claim that follows from the discussion above: in the preclinical phase of research that we are discussing, evaluations of biopharmaceutical properties do not invoke knowledge of specific biopharmaceutical mechanisms that are thought likely to operate. Note that this negative claim is quite narrow: while researchers may discuss in detail mechanisms of drug action (in terms of drug-to-target interactions) they do not – in the preclinical phase – describe the specific mechanisms thought to govern biopharmaceutical properties of that drug. So we think that a view where comparative knowledge of mechanisms *in vitro* and *in vivo* just won't help us here, because knowledge of those mechanisms *in vivo* is just what we do not have at the start of the drug discovery pipeline. This means that we cannot follow the comparative process tracing procedure of "cross-checking the crucial nodes of the relevant causal mechanisms" (Guala 2010: 1071) here, just because we have no knowledge of such a thing as a unitary underlying causal mechanism for the biopharmaceutical properties of drugs in general. As Guala (2010: 1072–3) develops, comparative process tracing is useful where the similarities and differences of the two mechanisms under consideration can be compared:

> Comparative process tracing can be articulated in two preliminary steps and an inductive inference:
>
> (1) Learn the mechanism in the model organism [or experimental system], by means of process tracing or other experimental means.
> (2) Compare stages of the mechanism in which the two [the experimental and the target systems] are most likely to differ significantly.
> (3) In general, the greater the similarity of configuration and behaviour of entities involved in the mechanism at these key stages, the stronger the basis for the extrapolation. (Guala 2010: 1072–3)

By this method, researchers do not need to know everything about either the experimental or the target system in order to do good extrapolation by comparative process tracing (which is just as well). But they do need to know a bit about each, so that they can compare them. This is exactly what happens with respect to inferences about the specific effects of a drug candidate on its target. But the situation in inferences regarding biopharmaceutical properties is different: we have nothing at all to suggest the likely steps of a mechanism in the target system. For emphasis, that is not to say that we do not know about particular biopharmaceutical mechanisms, but that instead we do not have an overarching mechanism of biopharmaceutical properties that can be applied to all molecules in the generic fashion that the rule of five is. And note that this claim says say nothing mysterious about biopharmaceutical properties – just that we think that the properties picked out by the rule of five are highly disjunctive (in terms of their underlying biological structure). A simple piece of evidence for this claim is the portability of a rule of five estimate for a molecule. We can calculate it once, and expect it to provide useful guidance as to biopharmaceutical properties no matter what the intended site of action of a candidate molecule is. So, just for instance, an estimate of the rule of five for sildenafil (notoriously, Viagra) will guide biopharmaceutical practice no matter whether we are thinking about using that molecule to interact with a target in the heart (as was originally intended) or elsewhere in the body (as was the eventual application in practice).

Cases such as these seem to preclude an account of biopharmaceutical extrapolations that follow Guala's analogical lines, just because we do not have the same kind of knowledge about experimental and target system. Lemoine 2017 calls this the *transportation* view of extrapolation – we gain knowledge of some mechanism in an experimental system, and then infer that the experimental mechanism can be transported over to our target system. Yet we think that these biopharmaceutical extrapolations do not consist of sketching out some mechanism capable of explaining the biopharmaceutical properties of our candidate molecule. In fact, avoiding doing this mechanistic work is precisely what Lipinski's rule of five is useful for – we do not have to worry at all about the way that our particular molecule might wend its path to its receptor or whatever, just because the rule of five provide us with a totally generic measure of bioavailability. To return to Guala's work, Lipinski's rule of five puts the mechanistic details governing bioavailability into a black box (Guala 2010: 1080–1) – and yet permits us to make fairly safe extrapolations about the properties of a molecule.

This puzzle then moves us in a new direction, particularly if we are some kind of mechanism enthusiast. If the rule of five doesn't tell us about mechanisms, then why should we trust the extrapolations that are based on it? In her recent work on RCTs Nancy Cartwright addresses a similar question which revisits her earlier work on capacities (such as Cartwright and Pemberton 2013). Capacities, according to Cartwright, allow us to build links between two different contexts that permit comparative causal inference. Cartwright's examples of worries about extrapolation from the results of RCTs are typically of the following form:

Cartwright on Extrapolation If we take C to cause E, in context A (such as a RCT), then the question for those interested in external validity is *does C cause E in context B*? (after Cartwright 2009).

Cartwright argues that we cannot answer this question by fully inductive means, just because inductive inference does not permit us to take account of the differences between the different contexts in which C causes E. This is not to say induction does not play a role in extrapolation, but the induction (in its traditional sense) is not sufficient to explain the inference based on capacities given that it does not take differences in the context into account.

Instead, if we use methods of inference that are based on capacities then we can account for what C contributes in the causal picture, by solely focusing on C's ability to produce E. This property of production is C's capacity to produce E. Because strictly C does not cause E but instead C contributes E to some context, where the extent of that E might vary, we can now make inferences that are fitted to different contexts depending on the ways in which that context might modify the extent of E. This conception of capacities, where some C carries the capacity Q to produce E, is more clearly summarised by Steel as follows: "The central feature of capacities is their stability across changes in background conditions." (Steel 2008: 80). The implication of this stability is that is, capacities are not sensitive to changes in background conditions.

So far, so Cartwright. But, according to Steel, this stability precludes us using capacities to make extrapolations, because it is precisely in cases where background conditions might conceivably make a difference that we need to do extrapolations. Steel distinguishes capacities from other kinds of effects (see Steel 2008, chapter 8 in particular). Capacities, unlike other kinds of effects, are completely general – they are not, say, restricted to a group of cases which have similar causal structures. And, hopefully uncontroversial, we need to know whether we are dealing with capacities (or not) if we are to use capacities as a solution to problems of extrapolation.

How do we know whether we are dealing with a capacity? According to Steel at least, we can only decide whether something is a capacity via some kind of inductive inference. But (for critics of inductive solutions to extrapolation) this is no solution at all. For if we need capacities to avoid the problems inherent to inductive approaches to extrapolation, but the only way of having capacities is to use inductive means, then extrapolation based on capacities no stronger than an extrapolation that is based on nothing but simple induction. Here, we should note that we do not want to mount a general argument against the use of induction in inference. Instead, we want to examine a counterargument that is made by Steel about the capacities approach. Induction (Psillos 2002, chapter 1) requires some claim about the uniformity of nature if it is to avoid Goodman-ish traps. But extrapolations are cases where we know that some of our background conditions will differ between the systems involved in our extrapolation. This challenges the safety of the principle of the uniformity of nature (described in Psillos 2002). This might cause us to therefore distrust induction for epistemic reasons in extrapolation. However, if we distruct induction as a guide to extrapolation, we should also distrust capacities, just because we need to use some inductive methods in order to decide what counts as a capacity. Let's call this the argument from transitivity: capacities-based approaches to extrapolation suffer from the same problems as induction-based approaches, just because induction is necessary to decide whether some property really is a capacity.

That, at least, is the general shape of the argument in the *capacities or induction?* parts of the extrapolation literature. We don't want to rehearse these arguments in the abstract any further, because our solution (and resulting endorsement of capacities in this case) relies on showing that this argument from transitivity does not apply in this case. And this argument will not serve as a general refutation of the argument from transitivity.

Why does the argument from transitivity not apply in this case? Simply put, an extrapolation that is based on induction here picks out different properties than a capacities-based extrapolation. In constructing their heuristic rule, Lipinski and colleagues did use broadly inductive methods in order to decide what the relevant capacities capable of picking out molecules with desirable biopharmaceutical properties. On this, see the details given above regarding the use of the WDI and USAN databases in developing the rule of five. But this induction about capacities is much safer than a directly inductive extrapolation about biopharmaceutical properties would have been without this intervening capacities step just because it is less likely to be affected by changes in background conditions, than a direct induction-based extrapolation would be. Lipinski's rule of 5 is concerned with (broadly) absorption and permeation. This means that the physico-chemical effects that our causes have the capacity to produce various kinds of absorption and permeation. We know, all too well, that physico-chemical properties will vary depending on the chemical environments – as the example of using DMSO as a solute for drug development shows.

Contribution of hydrogen bond donors and acceptors, molecular weights and the distribution coefficient are known *in vitro* and expected to remain the same *in vivo*.

But this doesn't base the extrapolations that we might use the rule of five for squarely on the shoulders of this original induction. Swathes of background theory support the relation of these properties picked out by the rule of five, and support claims that these properties will modify absorption and permeation. If this is an inductive one, it is theory-laden in the best possible way, in that our background theories provide us with good reasons to think that these molecular properties will have the kinds of effects that we are interested in.

Here, we claim that inferences rely on properties that are intimately (read: physicoochemically) linked with causal capacities to behave in particular ways – desirable or not – in living systems. The way that the link that the rule of five makes between the druggability of compounds and their physiochemical properties (such as hydrophobicity, molecular weight, and lipophilicity) is made depends on these properties being capacities. This is because each of them are the result of a compound's molecular structure that is invariant both *in vitro* (as part of a DMSO-based compound library) and *in vivo* (animal models and in humans). For example, we take molecular weight as related to the capacity to be druglike in that it is a good predictor of the ability of a molecule to pass through a cell membrane. Hence, we propose that Lipinski's rule of 5 is a heuristic tool that relies on capacities of compounds in order to infer their drugability.

Herein, we argue that Lipinski's rule of five relies on the capacities, such as hydrophilicity and molecular mass, in order to extrapolate how these compounds are expected to behave *in vivo* during later stages of the pipeline. Lipinski's rule of five extrapolates pharmacological properties from the physico-chemical properties of compounds.

7.6 Conclusions

With the aim of furthering discussion about the possible and plausible ways of doing practically-engaged philosophy of science, we conclude this chapter with an outline of our own aims and values in engaging in this kind of philosophical work.

HTS has provided great methodological advantages: screening many compounds quickly, cheaply, and reliably. Yet, as we demonstrated above, this increase in screening efficiency has not been reflected in a matching increase in the overall productivity of the pipeline. Our value: modesty, or at least the avoiding of triumphalism about the likely impact of theoretical and technological changes in the sciences.

We think that part of the reason for this mis-match arises from the way that biological processes are abstracted such that they can be efficiently handled by HTS. In turn, this means that HTS does not give a reliable indication of all the biopharmaceutical properties that matter. This point is drawn into sharp focus by the difference between lead-target interactions and drug-receptor interactions *in vivo* (as we discuss above). Our value: contingency, and the valuing of the particular features of cases.

This fits with our broader claim about aims and values in the sense that *ad hoc* heuristics – i.e. those not based on well-developed and widely-used principles, such as parsimony – are used to shore up leaks in the drug development pipeline. This work is necessary as a direct result of the consequences of other short-term fixes (like the use of DMSO to standardise compound libraries). Our value: the valuing of complexity and its role in shaping scientific systems.

In conclusion, we have presented the heuristic rule of five as a means of managing uncertainty about drugability based on the physico-chemical properties of compounds. This heuristic rule is an excellent example of a highly reductive and simple rule that is nevertheless successful in managing an inference of great complexity. In sketching some philosophical points, our philosophical work depended on a descriptive account of the strategies used by Lipinski and colleagues. In other words, our aim here was to unpick the way that scientific practice actually comes about. This, we think, is an example of a species of naturalism, and in that we claim that philosophers of science should study science as it is actually practiced.

We say that uncertainty is managed in our case by tapping into the capacities of molecules. Since there are no explicit statements in the Lipinski et al. (1997) paper on the rationale that this heuristic is based on, nor discussion of the underlying assumptions shaping the methodology of the authors, we can only assume that these rules ultimately depended on statistical data describing the physicochemical properties of licensed drugs, together with the assumption that the biopharmaceutic properties of these licenced drugs were likely to be desirable ones.

Using the capacities literature, we argued that the rule of five relies on capacities to extrapolate how drug candidates are going to behave in different stages of the drug discovery pipeline. Following this argument, we think, that non-verbalised assumptions held by scientists in making decisions such as these can be understood in ways that would make sense to the analytic philosopher of science (e.g. the way that these values may be connected to philosophical work on capacities). Our value: in demonstrating the possible ways that philosophers interested in practices might themselves practise.

References

Baxter, K., Horn, E., Gal-Edd, N., Zonno, K., O'Leary, J., Terry, P. F., & Terry, S. F. (2013). An end to the myth: There is no drug development pipeline. *Science Translational Medicine, 5*(171), 171cm1–171cm1.

BNF. (n.d.). *British National Formulary entry: Dantrolene sodium*. Available online at https://doi.org/10.18578/BNF.207282907

Cartwright, N. (2009). What are randomised controlled trials good for? *Philosophical Studies, 147*(1), 59–70.

Cartwright, N., & Pemberton, J. (2013). Aristotelian powers: Without them, what would modern science do? In J. Greco & R. Gross (Eds.), *Powers and capacities in philosophy: The new Aristotelianism* (pp. 93–112). New York: Routledge.

Charles, G. S., & O'Donnell, J. T. (2006). Overview of the current processes of new drug discovery and development. In G. S. Charles & J. T. O'Donnell (Eds.), *The process of new drug discovery and development*. London: Taylor & Francis.

Drews, J. (2000). Drug discovery: A historical perspective. *Science, 287*(5460), 1960–1964.

Guala, F. (2010). Extrapolation, analogy, and comparative process tracing. *Philosophy of Science, 77*(5), 1070–1082.

Gunter, B., Brideau, C., Pikounis, B., & Liaw, A. (2003). Statistical and graphical methods for quality control determination of high-throughput screening data. *Journal of Biomolecular Screening, 8*(6), 624–633.

Hommon, C. A., & Nelson, R. M. (2006). High-throughput screening: Enabling and influencing the process of drug discovery. In G. S. Charles & J. T. O'Donnell (Eds.), *The process of new drug discovery and development*. London: Taylor & Francis.

Kolb, M. E., Horne, M. L., & Martz, R. (1982). Dantrolene in human malignant hyperthermia. *Anesthesiology, 56*(4), 254–262.

Lemoine, M. (2017). Animal extrapolation in preclinical studies: An analysis of the tragic case of TGN1412. *Studies in History and Philosophy of Biological and Biomedical Sciences, 61*, 35–45.

Lipinski, C., Lombardo, F., Dominy, B., & Feeney, P. (1997). Experimental and computational approaches to estimate solubility and permeability in drug discovery and development settings. *Advanced Drug Delivery Reviews., 23*, 3–25.

Macarron, R., & Hertzberg, R. P. (2009). Design and implementation of high-throughput screening assays. In W. P. Jansen & P. Bernasconi (Eds.), *High throughput screening methods and protocols*. Dordrecht: Springer.

Ng, R. (2015). *Drugs: From discovery to approval*. New Jersey: Wiley.

Parexel. (2003, August 4). Research and markets: PAREXEL's pharmaceutical R&D statistical sourcebook. *M2 Presswire*. Retrieved from http://search.proquest.com/docview/444265505?accountid=14511

Pereira, D. A., & Williams, J. A. (2007). Origin and evolution of high throughput screening. *British Journal of Pharmacology, 152*(1), 53–61.

Prentis, R. A., Lis, Y., & Walker, S. R. (1988). Pharmaceutical innovation by the seven UK-owned pharmaceutical companies (1964–1985). *British Journal of Clinical Pharmacology, 25*, 387–396.

Psillos, S. (2002). *Causation and explanation*. Chesham: Acumen.

Steel, D. P. (2008). *Across the boundaries: Extrapolation in biology and social science*. Oxford: Oxford University Press.

Chapter 8
"Caught in the Amber": A Sketch of Chemical Underdetermination

Klaus Ruthenberg

Abstract In the chemical sciences, theoretical uncertainties can have at least two different reasons. On the one hand, the theory in question might be just premature, on the other, the shift of the object range by introducing new substances can change the dispositional description of the phenomena one is aiming at. Mainstream philosophers of science sometimes discuss the former as what could be called "epistemological" underdetermination, whereas the latter variation is neglected to a large extend. Because this "ontological" underdetermination is central to (the history and philosophy of) chemistry it is worth a closer look. In the present chapter the attempt is made to clarify this aspect using a few pertinent examples, among them thalidomide.

Keywords Philosophy of chemistry · Ontological underdetermination · Pluralism · Nonknowledge · Substance dispositions

8.1 Introduction

Scientific theories are generally underdetermined by empirical evidence. In the philosophy of science, however, the notion of underdetermination is a terminus technicus that refers to the conflict of different theories for one stable empirical background. This technical term is strongly linked to big philosophical discourse topics, including realism, pluralism and holism. One of its very common expressions even carries the name of an influential physical chemist: the *Duhem*-Quine thesis (see below). In most of the discussions concerning underdetermination an epistemological perspective prevails, that is, the notion of underdetermination mainly refers

K. Ruthenberg (✉)
Coburg University of Applied Sciences and Arts, Coburg, Germany
e-mail: klaus.ruthenberg@hs-coburg.de

A. LaCaze, B. Osimani (eds.), *Uncertainty in Pharmacology*, Boston Studies in the Philosophy and History of Science 338,
https://doi.org/10.1007/978-3-030-29179-2_8

to scientific theories in an existing, unchanging ontological context.[1] According to such customary discussions, the world, that is the target of any experimental human activity, is considered fundamentally stable. The pieces of that scientific (and philosophical) game, to put it into an admittedly odd picture, are considered to be coming from one and the same toolbox. The chemical world, however, is different: Substances come and go, and some of them do change the lives of humans *and* their sciences dramatically. In other words: they change reality. This fact does not refer to theorizing alone, and it is not a merely metaphysical statement. Rather it is an ontological one. Hence if we call the customary description of theoretical uncertainty within one assumed stable ontological frame *epistemological underdetermination (EU)* on the one hand, it makes sense to call the theoretical uncertainty with respect to an ontological shift *ontological underdetermination (OU)* on the other. If a substance is new (at least new to the respective environment) its dispositions/properties cannot be explained or predicted conclusively by the received knowledge. Moreover, the newcomer can contribute to the increase of knowledge about so far already (but incompletely) known substances, because chemical properties are relational properties. In contrast to most parts of physics chemistry is generally changing the world (and its own working field) ontologically. In the first place, hence, the word underdetermination in OU addresses the material world, in the second place it might address changes of more general theoretical concepts as well.

The fundamental difference between chemistry and physics, namely that the former has a significant, designing influence on its own investigation subject, is almost completely neglected by mainstream, physicalist philosophy of science. By introducing the term OU I therefore propose to enrich that field of discourse to better accommodate the notion of underdetermination to the chemical sciences, including pharmacy and pharmacology. I take the pharmacological and biological behaviour of substances as important part of their chemical dispositions or properties. To illustrate the relevance for the discourse about uncertainty in pharmacology, I shall refer to the Contergan case, and, to obtain a broader picture and to emphasize the general aspect of my main point, to a few non-pharmaceutical chemical species as well. In Sect. 8.1, I give a short discussion of the definition of EU. Section 8.2 consists of an account of the importance of EU for the philosophy of chemistry, and in Sect. 8.3 I introduce and argue for what I call OU. Section 8.4 is an attempt to connect my proposal to the discussions in the philosophy of pharmacology, and the final Sect. 8.5 concludes this chapter.

[1] See Okasha 2002 for a textbook description and French 2009 for a recent discussion of the varieties of what I call *epistemological underdetermination* here, although the latter author addresses what he calls "metaphysical underdetermination".

8.2 Epistemological Underdetermination

A straightforward, merely trivial understanding of underdetermination refers to the plain lack of data supporting a certain theory, for instance a theory accounting for the nature of a particular substance. In professional philosophy of science, however, that notion is characterized in a more sophisticated manner. Pointing at theory dynamics, Bas van Fraassen, for example, says:

> Each scientific theory, caught in the amber at one definite historical stage of development and formalization, admits many different tenable interpretations. (van Fraassen 1991: 481)

One fact seems to be trivial: Because there always is some scientific progress, it is not likely to reach an ultimate scientific description of, for example, a chemical species. That must not mean that new theories are developed during any step. The progress of science, however, brings about empirical and theoretical results and claims which can easily come into conflict just because of the different interpretation of the same evidence. One of the best known theoretical conflicts in the chemistry of the nineteenth century, for example, referred to the correct compositional formula of water (HO vs. H_2O).[2] Today, most scientists agree that there is only one correct formula, although the empirical knowledge about this assumingly best known chemical species is still increasing (and the philosophical discussion brought up some intriguing results, too).[3] Modern philosophers of science do define the notion of underdetermination differently than as simple lack of data for one line of theoretical description. Larry Laudan, for example, differentiates two variants, namely, what he calls the *Humean* underdetermination and the *Quinean* underdetermination. The former he describes as follows:

> For any finite body of evidence, there are indefinitely many mutually contrary theories, each of which logically entails that evidence. (Laudan 2013: 290)

The latter version (closer to what is called the Duhem-Quine thesis) is:

> Any theory can be reconciled with any recalcitrant evidence by making suitable adjustments in our other assumptions of nature. (Laudan 2013: 295)[4]

8.3 Philosophy of Chemistry in Response to EU: An Example

How would philosophers of chemistry react to these versions of EU? Jaap van Brakel has presented the most elaborated investigation on that topic, though not

[2]See Chang 2012, chap.3, for a deep going study of that episode.

[3]See Ruthenberg 2012 for a brief survey of the modern "philosophy of water".

[4]Ladyman et al. 2007 give a similar account and discuss EU from a structural realist perspective. The Duhemian and Quinean interpretation can be differentiated, see Pietsch 2011. Except for Chang 2012, in the cited and otherwise studied literature on underdetermination neither chemistry related subjects or examples nor the possibility of what I call OU are discussed.

explicitly aiming at underdetermination or Laudan's formulations. Concluding his book *Philosophy of Chemistry*, he states:

> The fact that most of the posits of contemporary daily life in the western world have their origins in developments of science and its ocularcentric epistemology, doesn't change the fact that when their grounding status is disputed, adjudication will be governed by criteria that are not the product of science (. . .) it is to say that the judgement that one explanation is better than another is based on judgements grounded in the manifest image. (van Brakel 2000: 202)

While the received views on the underdetermination topic refer to the belief that only one of the concurring statements can be right or true, van Brakel advocates a pluralist view with a flat hierarchy: All descriptions, be it macroscopic, microscopic, or submicroscopic do in principle have the same right to characterize substances in the first place. According to van Brakel, there is no need to prioritize for example the elementary particle realm, and to colonize the other possible levels of description.[5] He also claims that pure chemical substances can be defined macroscopically as follows:

> (. . .) pure chemical substances are the relatively stable products of chemical analysis and synthesis: nodes in a network of chemical reactions. (van Brakel 1997: 253)

Let us consider an example: Nitric acid. Known since at least the early Middle Ages, this liquid substance was prepared from Saltpetre and Sulphuric acid by the alchemists by more or less "dry" distillation. In pure state, it has a quite low boiling point (84 °C), such at first glance it seems that it could be purified by distillation in workshops and laboratories, which was the customary way to obtain purified liquid substances.[6] However, it must be applied an elaborate manipulation to yield pure nitric acid: When the dilute acid is distilled, the boiling point is higher, and at the (azeotropic) boiling temperature of 121 °C the distillate contains 68% acid.[7] To remove the water, it is possible to perform the distillation with an addition of sulphuric acid. Although boiling points (which are substance constants) could not be measured in those days, and the here mentioned modern explanations were not accessible, the boiling behaviour of chemicals was part of the defining descriptive network of scientific statements. Other parts of that network of dispositional knowledge have been the colour, smell, taste, and affinity to other chemical species, for example its acidic behaviour, which consists of a whole group of dispositions, and its oxidation capacity, to use a modern notion. Obviously, the characterisation, naming and handling of that substance was possible long before the modern information of its compositional formula (HNO_3) and, for example, spectroscopic data, let alone the quantum mechanical mathematization of its molecules, were available. There are many ways to address the nature of the

---

[5] See Ruthenberg and Harré 2012 for a brief account of van Brakel's intercultural approach to the philosophy of chemistry.

[6] Cf. Moran 2007.

[7] The data are from Lowry and Cavell 1949.

same substance, although some of those theoretical and operational ways seem to be contradictory or not translatable into each other. The quantum mechanical approach, for example, does refer to the interaction of all the elementary particles (neutrons, protons, electrons) in the isolated molecule HNO_3 only and less so to the interaction of those molecules in condensed states and hence offers no way to explain what might be called substance properties or dispositions. The bridge between all the different perspectival accounts is the operational work in the laboratories. In other words, we have known the same substance (e.g., nitric acid) sufficiently well long before the advent of atomism and modern physicalist descriptions, and any "suitable adjustment" of the mentioned version of the Duhem-Quine thesis becomes nothing more than a pluralistic coexistence of various quite different – quotidian as well as sophisticated – approaches. According to van Brakel, there is no need to essentialize certain selected items of the network of substance characterization or to ask which part of the whole theoretical approach is "really" true. Particularly in an operational perspective, the chemical entity nitric acid has remained the same, and due to its historical stability in the laboratories, it certainly became a promising candidate for the list of natural kinds, although it started as a preparation or artefact.[8] When it comes to chemical entities, there is much more conceptual continuity than some naturalists and essentialists are willing to accept.[9] Opposing such essentialism, and similar to van Brakel, but explicitly addressing EU, Hasok Chang argues in favour of a pluralism and claims that "... the great achievements of science come from cultivating underdetermination, not by getting rid of it." (Chang 2012: 151) In fact, to cultivate the coexistence of different theoretical perspectives is exactly what has been the core activity of alchemy, chymia, iatrochemistry, and chemistry so far.

8.4 Ontological Underdetermination

After having presented one (important) line of argument from modern philosophy of chemistry which (indirectly) refers to the received theoretical or epistemological underdetermination theses I now shall discuss the importance of OU, which I characterize as the scientific uncertainty caused by the expansion of the explored part of reality. By expansion of reality I denote the objective enlargement of the possible field of enquiry, particularly the number of substances as epistemic objects of chemistry (followed by the vast increase of reaction possibilities). OU shall neither address the trivial cases of non-knowledge like the result of a football game

[8]It is possible, by the way, to find traces of nitric acid in nature, because discharges in the atmosphere may produce nitric oxides, which yield the acid when in contact with water.

[9]See Ruthenberg and Chang 2017, which comprises of a detailed case study about the significance of the quotidian knowledge in the chemical sciences. At the same time, though without an explicit purpose, it is an illustration of Chang's claim about the cultivation of underdetermination at the end of this section: Acidity seems to be a candidate to become a paradigm field of enquiry with respect to underdetermination.

nor the uncertainty of inductive knowledge like tomorrow's sunrise. OU rather refers to the nonknowledge of the rules of the "game". Moreover, the kind of scientific uncertainty regarding OU is different from the *productive* role of experimentation.[10] By ingenious experiments scientists and engineers can produce phenomena which are not occurring naturally, like the breaking of the sound barrier. Although it is related to that productive role in a certain sense, the production of a (novel) substance is different because of the material existence of that substance and the impact it has on the "nodes in a network of chemical reactions".

The object realm – or realm of epistemic interest – of chemistry is different to those of most parts of physics. One difference is the much lower grade of abstraction: Unobservable objects are rather rare: Although chemists talk about molecules, atoms, and electrons, and teach very abstract concepts like, for example, radicals, electronegativity, oxidation numbers, and hybride orbitals, in their experiments they apply at least tiny amounts of (observable) stuff samples and aim at the reactivity or affinity of substances.[11] Another, much more pertinent aspect is the following. In a famous statement, the eminent French chemist Pierre Eugéne Marcelin Berthelot (1827–1907) claimed that (synthetic) chemistry creates its own subject. That is, the preparation of substances literally forms what chemists are investigating, and thus synthesis becomes part of these investigations (and the interrelation of analysis and synthesis becomes much more complicated than described usually). Moreover, the nature of chemical substances can only be determined by empirical enquiry, which will include the reactions with other chemicals. Because of human activities, the number of chemical individuals on Earth is steadily growing, which means that the number of possible chemical reactions is growing as well. Hence, the chemical world is in steady flux. Nevertheless, the network of chemical reactions is active even without exact or complete knowledge about any of its nodes (see the example nitric acid).

As I pointed at above, the scientific efforts regarding the elucidation of the nature of chemical substances would be already infinite in a world without real material[12] changes. It seems that most participants in the EU discussion have one popular metaphysical presupposition in common: the One-World-One-Science argument (OWOS).[13] However, the chemical world is by no means stable. The flux I mentioned is a change in the real world and not only a flux of theoretical approaches only. For compounds that become part of our stuff inventory at a certain moment the net of knowledge about properties and dispositions is at least potentially enlarged.

---

[10]In his contribution to the theory of experiments, Michael Heidelberger 2003 differentiates three main functional roles for scientific experiments: they can be representative, productive, or constructive. Interestingly, the making of chemical substances seems to be generally neglected in the philosophy of experiments.

[11]Experimental physics also starts from material samples, but aims at the understanding of the ultimate parts, forces, and structures of the world.

[12]To avoid confusion: By "material" I here mean "referring to chemical substances", or, simply "substantial", in contrast to the description in the former footnote.

[13]Cf. the cited references from the philosophy of science and Kukla (1998, 2008).

This potential enlargement is typical for the chemical sciences. As to the field of human knowledge, some of these compounds, for example nitric acid, enjoy a very long, stable, and full life due to knowledge about their particular dispositions. The stability I refer to here is not the stability of every HNO_3 molecule that ever came to exist on earth. Rather, this stability refers to the referential and dispositional net of knowledge, as the disappearance in a natural or artificial sink just belongs to its dispositions, such that the actual amount of the substance is changing and irrelevant in this respect. Nitric acid, for example, soon was used in metallurgy to dissolve metals (although it is not able to dissolve all of them) and to serve testing (analytical) and preparatory (synthetical) means. All these aspects of reactivity did mainly contribute to the prescientific and scientific pictures about this substance. When aqua regia was invented, to refer to one prominent example, the dispositional net of knowledge about nitric acid was enlarged significantly. Aqua regia, in modern terms a mixture of hydrochloric and nitric acid found by lucky accident by alchemists, was able to dissolve gold, which no other known acid was capable of. Obviously, it is more than just a mere mixture. Hence, the impact of this new substance changed the dispositional description net of the involved acids – and that of gold. What has been a (unforeseen) potentiality in the former world became an actual property after the invention of this quite mysterious liquid.[14] As to this case there was no easily obtained new material product that changed the ontological frame, but a (until today) not fully understood reaction between two already baptized substances. Both gold and the two mixed acids were underdetermined ontologically – and they still are.

Another example with a more than local impact by so far unknown dispositions is the depletion of stratospheric ozone induced by chloro-fluorine-carbon compounds (CFC). The depletion of ozone in the upper atmosphere including the "ozone hole" by catalytic acceleration via chlorine atoms has been a purely natural (bio- and geogenic) phenomenon until the introduction and emission of these solvents into the environment by humans (Crutzen and Oppenheimer 2008).[15] As to this example, OU becomes a wider bearing:

> Beyond academic interest, the production of nonknowledge by the synthesis of new substances is of general concern if the new substances leave, through whatever door, the laboratories and become part of our material environment. (Schummer 2001: 111)

Hence, the chemical newcomer changes the ontological frame and may introduce unexpected capacities or opportunities, like the CFC did in the stratosphere. The event of the introduction of a new substance (which must not be a pure compound) may evoke some theoretical efforts, but the initial step is an ontological one, and therefore I call the resulting situation OU. Whereas what I called the *straightforward* EU is an unusual denomination and might be just called non-knowledge, the ontological variety goes significantly deeper and is peculiar for chemical entities, including, of course, pharmacologically active substances, because here the reactivity takes place in living beings.

---

[14] I thank one of the referees for helping me clarifying this point.

[15] The natural source of stratospheric chlorine atoms and hence the "ozone hole" is chloromethane.

8.5 A Notorious Pharmacological Example

The aqua regia example comprised of a mixture of two already well-known "pure" chemicals, at least in a modern understanding.[16] In the case of a newly synthesized pure substance, for example the CFC or an artificially made active pharmaceutical ingredient, the ontological change can go even deeper. To illustrate that case, I shall give a summarizing account and discussion of the *Contergan tragedy* in what follows.[17]

According to published results and official statements by researchers of the German company Chemie Grünenthal, the substance 2-(2,6-dioxo-3-piperidyl)isoindoline-1,3-dione, abbreviated *thalidomide*, was first synthetized in 1954. It is described as the 17th of a series of 18 similar compounds, and although it seems quite easy to prepare absolutely no information about the way of synthesis is given in the original scientific paper by the Grünenthal officials titled "N-Phthalyl-glutaminsäure-imid – Experimentelle Untersuchungen an einem neuen synthetischen Produkt mit sedativen Eigenschaften" (Kunz et al. 1956). In that article, "K17" is described as white solid with a melting point of 271 °C and severely restricted solubility in water, methanol, ethanol, acetone, diethyl ether, and benzene. Hence, the animal tests (applying the extremely unusual "jiggle-cage method", the results of which could never ever be reproduced) had to be performed by administration of suspensions. In those animal tests the state "sleep" had to be set arbitrarily, because due to the testing of groups of eight mice in the same cage at the same time the animals *could* not really fall asleep. This was only one of several major methodical drawbacks. The results of the comparison of seven sleep-inducing substances by that test were summarized in a table, in which unsurprisingly thalidomide was presented as almost harmless and at the same time very efficient, and which later was commented on quite tellingly as follows by the authors of the *Sunday Times Insight Team*:

> Without doubt this table is the origin of the illusion that thalidomide was something capable of overturning many decades of pharmaceutical experience: an effective drug and quite without side effects. (Knightley 1979: 18)

No other scientific paper on that remedy was published by the producers of the sleeping pills. Until today it is unclear when exactly the substance was synthesized and where and when the Grünenthal scientists got the information about the sleep-inducing effect in humans from in the first place. The prehistory of thalidomide remains somewhat mysterious. Official clinical trials began in 1955 (no standard test procedure was developed at that time), followed by the official approval by the Ministry of the Interior of Northrhine-Westfalia (no detailed formal

[16]The pre-modern interpretation, of course, afforded no reference to what we today mean by composition. Nevertheless, aqua regia was singled out and identified as a chemical individual in a similar way I described here for nitric acid.

[17]See Ruthenberg 2016 for a more detailed account of the early (German) part of that tragedy.

application was required) and in 1957 *Contergan,* which was the official product name in Germany, was launched. It was advertised aggressively, first and foremost addressing physicians and apothecaries. About 50,000 "therapeutic circulars" were distributed, which were composed as invitation to take part in a public project to achieve health improvement ("a good sleep without risk"). Particularly women were addressed as most suitable consumers of the drug, and sleep disorder, fear, contact weakness, nervousness, menopausal problems, and female sexual excitability were among the explicitly mentioned application cases. Grünenthal made a fortune with Contergan: In 1960, it amounted to one million Deutsche Mark per month, which was about 50% of the total turnover of the company. The economic attitude of the officials was particularly well developed, as can be inferred from the words of the "father of Contergan" Heinrich Mückter, who has been Head of the scientific development of the company at that time, in an internal document of the same year:

> From our side, everything must be done to avoid prescription enforcement, since already a substantial amount of our turnover comes from over-the counter sales. (cited after Sjöström and Nilsson 1972: 54)

The first hints to adverse effects (neuronal damage) during long-term administration of thalidomide came up in 1959, and in 1961 the evidence that this drug causes malformations of new-born babies were by no means contradictable anymore. Many thousands of babies suffered from phocomelia, and a similar if not bigger number of humans were stillborn. The chance to stop or to reduce the impact of this fatal disposition of that substance was given away because of thoughtlessness. What we today would call informed consent has been violated in many ways:

> Everyone who did not place compassion and an absolute commitment to the truth ahead of all other considerations in the use of this medicine loosed a destructive monster. *But the monster was never thalidomide itself.* (Stephens and Brynner 2001: 201)

Chemical species carry capacities, or dispositions, and might become therapeutic (or toxic) agents when brought into the places where these dispositions can be realized. This is a modern version of Paracelsus' well-known dictum about the possible negative causal powers (or dispositions) of any drug.[18] With Paracelsus in mind, I tend to claim that if an active substance has *any* power we should expect it to have adverse effects, too.

What we have in the case of Contergan is a kind of *blind* post-market drug monitoring, which yielded results not obtainable by a transparent clinical research.[19] Transparent clinical research would have to be omitted after the first appearance of adverse effects. The amendments to the drug legislations with respect to approval regulations since 1961 aimed exactly at this point in that in many countries a full

[18]In German: Alle Dinge sind Gift, und nichts ist ohne Gift; allein die Dosis machts, daß ein Ding kein Gift sei. Cf. the account of this topic in Deichmann et al. 1986.

[19]The Sunday Times Insight Team interprets the impact of thalidomide in the vein of Humean inductive uncertainty: "A working scientist does not need to have read the *Treatise of Human Nature* to perceive that our knowledge of the future is a series of more or less useful assumptions, any of which may be modified at any moment by experience." (Knightley 1979: 24).

preclinical toxicological testing (including teratogenicity assays) was required to be performed on two animal species including one rodent species equally sensitive to this particular drug as humans. Intriguingly, teratogenic effects are reduced to thalidomide-like causes in particular such that other imaginable effects might be excluded by choosing an inappropriate animal test species. In this way, the OU of chemical species is overlooked indeed. By all means, today there is no legal way of performing clinical tests on pregnant women particularly with drugs like thalidomide because of the available knowledge gained during the disaster. After the FDA approval in 1998, however, the application of thalidomide against leprosy, for example, is strictly regulated by the System for Thalidomide Education and Prescription Safety (STEPS). This program is described to be innovative, comprehensive, thoughtful, and precautious, but still cases are imaginable in which the active substance might reach a pregnant woman (through the semen of her sexual partner or that of a rapist, for example).[20]

Apart from ethical questions, which I do not address in detail here,[21] the ontological and epistemological perspectives considering drugs like thalidomide as epistemic objects are very interesting. Thalidomide was a newly made entity, that is a true artefact. Because it has not been there before, it brought new capacities (or dispositions) to the world. While in other cases new capacities and dispositions might be unknown, but not entirely unexpected (as for aqua regia), or they might be known in general but unexpected (as for the ozone depletion mechanisms caused by CFC), in the case of thalidomide they were neither known nor expected. Hence, there is an inclination of ontological novelty which is paralleled by an inclination of (non-)knowledge.

8.6 Conclusion

In contrast to other novel physical objects (cars, rockets, tooth brushes, electrical conductors, internet cables) chemical entities (as chemical entities, not just as constituents of physical objects) do have the potential not only to bring new forms and perhaps even new experiences and phenomena to the world. They change chemical knowledge. To discern this kind of scientific uncertainty referring to real ontological shifts from that of received versions of underdetermination, to enrich the discussion, and to bring the philosophy of chemistry closer to "science at it really is",[22] I propose to call it ontological underdetermination.

[20]See Stephens and Brynner 2001, ch.10, for an account of the more recent rebirth of thalidomide as remedy.

[21]See Ruthenberg 2016 for an attempt to evaluate the early part of that case ethically.

[22]I steal this slogan from Ladyman et al. 2007, although I do not subscribe to their ideas of a naturalized metaphysics.

As long as we are looking at epistemological underdetermination only the pluralist approach to the philosophy of chemistry, particularly the theoretical account of chemical substances, is most appropriate, and most promising, if future developments are accounted for.

Most of the novel chemical entities, like for example pharmaceuticals, are artificially brought into the world by humans, and some of these entities do enjoy an almost unlimited existence, at least with respect to the methodological knowledge of their preparation and characterization. If these new substances are brought into contact with humans or released into the environment, however, unknown or unexpected dispositions might be realized, as, for example, the teratogenic effects of thalidomide, which were unknown *and* unexpected. Whereas the common scientific endeavour aims at the investigation of the known and presumably stable empirical inventory, the empirical framework of chemical species is constantly changing, that is, it has an obscure evolutionary development of its own:

> (. . .) in synthetic chemistry, it is neither epistemological reflection nor knowledge about the world but the actual change of the material world, i.e. the object of knowledge, that results in nonknowledge. (Schummer 2001: 111)

Hence, van Fraassen's elegant expression "to be caught in the amber" becomes an ontological connotation to it when it comes to philosophizing about chemical entities.

Acknowledgements I am grateful to Hasok Chang for the discussions about underdetermination. I also thank two anonymous referees for their constructive criticism, as well as Barbara Osimani and Adam LaCaze for their patience.

References

Chang, H. (2012). *Is water H2O?: Evidence, realism and pluralism* (Vol. 293). Dordrecht: Springer.

Crutzen, P. J., & Oppenheimer, M. (2008). Learning about ozone depletion. *Climatic Change, 89*(1–2), 143–154. https://doi.org/10.1007/s10584-008-9400-6.

Deichmann, W. B., Henschler, D., Holmstedt, B., & Keil, G. (1986). What is there that is not poison? A study of the third defense by Paracelsus. *Archives of Toxicology, 58*(4), 207–213. https://doi.org/10.1007/bf00297107.

French, S. (2009). Metaphysical underdetermination: Why worry? *Synthese, 180*(2), 205–221. https://doi.org/10.1007/s11229-009-9598-5.

Heidelberger, M. (2003). Theory-ladenness and scientific instruments in experimentation. In H. Radder (Ed.), *The Philosophy of Scientific Experimentation* (pp. 138–151). Pittsburgh: University of Pittsburgh Press. https://doi.org/10.2307/j.ctt5hjsnf.11.

Knightley, P. (1979). *Suffer the children: The story of thalidomide*. London: Andre Deutsch.

Kukla, A. (1998). *Studies in scientific realism*. New York: Oxford University Press.

Kukla, A. (2008). The one world, one science argument. *The British Journal for the Philosophy of Science, 59*(1), 73–88. https://doi.org/10.1093/bjps/axn001.

Kunz, W., Keller, H., & Mückter, H. (1956). N-Phthalyl-glutaminsäure-imid: Experimentelle Untersuchungen an einem neuen synthetischen Produkt mit sedativen Eigenschaften. *Arzneimittel-Forschung, 6*, 426–430.

Ladyman, J., Ross, D., Spurrett, D., & Collier, J. (2007). *Every thing must go*. Oxford: Oxford University Press. https://doi.org/10.1093/acprof:oso/9780199276196.001.0001.

Laudan, L. (2013). Demystifying underdetermination. In M. Curd, J. A. Cover, & C. Pincock (Eds.), *Philosophy of science* (2nd ed., pp. 288–320). New York: W. W. Norton & Company.

Lowry, T. M., & Cavell, A. C. (1949). *Intermediate chemistry*. London: Macmillan.

Moran, B. T. (2007). *Distilling knowledge: Alchemy, chemistry, and the scientific revolution*. Cambridge: Harvard University Press.

Okasha, S. (2002). *Philosophy of science: A very short introduction*. Oxford: Oxford University Press. https://doi.org/10.1093/actrade/9780192802835.001.0001.

Pietsch, W. (2011). The Underdetermination debate: How lack of history leads to bad philosophy. In S. Mauskopf & T. Schmaltz (Eds.), *Integrating history and philosophy of science* (pp. 83–106). Dordrecht: Springer. https://doi.org/10.1007/978-94-007-1745-9_7.

Ruthenberg, K. (2012). What is water? Some philosophical considerations. In H. P. Hahn, K. Cless, & J. Soentgen (Eds.), *People at the well* (pp. 65–78). Frankfurt: Campus.

Ruthenberg, K. (2016). About the futile dream of an entirely riskless and fully effective remedy: Thalidomide. *HYLE–International Journal for Philosophy of Chemistry, 22*, 55–77.

Ruthenberg, K., & Chang, H. (2017). Acidity: Modes of characterization and quantification. *Studies in History and Philosophy of Science Part A, 65–66*(October), 121–131. https://doi.org/10.1016/j.shpsa.2017.04.003.

Ruthenberg, K., & Harré, R. (2012). Philosophy of chemistry as intercultural philosophy: Jaap van Brakel. *Foundations of Chemistry, 14*(3), 193–203. https://doi.org/10.1007/s10698-012-9161-1.

Schummer, J. (2001). Ethics of chemical synthesis. *HYLE–International Journal for Philosophy of Chemistry, 7*(2), 103–124.

Sjöström, H., & Nilsson, R. (1972). *Thalidomide and the power of the drug companies*. Harmmsworth: Penguin.

Stephens, T., & Brynner, R. (2001). *Dark remedy: The impact of thalidomide and its revival as a vital medicine*. Cambridge: Perseus Publishing.

Van Brakel, J. (1997). Chemistry as the science of the transformation of substances. *Synthese, 111*(3), 253–282.

Van Brakel, J. (2000). *Philosophy of chemistry*. Leuven: Leuven University Press.

van Fraassen, B. C. (1991). *Quantum mechanics: An empiricist view*. Oxford: Oxford University Press. https://doi.org/10.1093/0198239807.001.0001.

Part II
Methods

Chapter 9
A Millian Look at the Logic of Clinical Trials

Roberto Festa, Gustavo Cevolani, and Luca Tambolo

Abstract The use of a certain drug or treatment in the cure of a disease or condition should be based on appropriate tests of the hypothesis that said drug or treatment is efficacious in the cure of the disease or condition at hand. In this paper, we aim at elucidating how such "efficacy hypotheses" are tested and evaluated, especially in clinical trials. More precisely, we shall suggest that the principles governing the assessment of efficacy hypotheses, and, more generally, hypotheses of statistical causality, are provided by an appropriate statistical version of such a venerable procedure as the method of difference put forward by John Stuart Mill.

Keywords Mill's methods · Method of difference · RCT · Clinical trials · Placebo effect · Causal statistical hypothesis · Efficacy hypothesis · Cognitive structures · Scientific commonsense

9.1 Introduction

The use of a certain drug or treatment in the cure of a disease or condition should be based on appropriate tests of the hypothesis that said drug or treatment is efficacious in the cure of the disease or condition at hand. Let us call hypotheses of this kind "efficacy hypotheses". The problem of how to assess efficacy hypotheses is crucial not only in pharmacology and clinical epidemiology, but also in many other

R. Festa · L. Tambolo
Department of Humanistic Studies, University of Trieste, Trieste, Italy

G. Cevolani (✉)
IMT School for Advanced Studies Lucca, Lucca, Italy

Center for Logic, Language, and Cognition, University of Turin, Turin, Italy

A. LaCaze, B. Osimani (eds.), *Uncertainty in Pharmacology*, Boston Studies in the Philosophy and History of Science 338,
https://doi.org/10.1007/978-3-030-29179-2_9

fields, ranging from agriculture—where, to mention but one example, one needs to appraise the efficacy of new pesticides—to the implementation of any policy recommendation, whose efficacy needs to be properly evaluated. The principles governing the assessment of efficacy hypotheses are arguably the same in all these different areas. It then comes as no surprise that the problem of how to precisely characterize such principles has steadily attracted the attention of both practicing scientists and philosophers interested in the analysis of scientific reasoning and inference.

The great amount of work devoted to the above problem has so far produced some agreement on few crucial points. For instance, most scholars would agree that efficacy hypotheses are a special kind of causal statistical hypotheses, i.e., hypotheses aimed at identifying the cause of a statistical difference between two relevant groups (consider, as an example, the administering of a certain drug as the cause of a greater frequency of remissions among the members of the treatment group as compared to those of the control group in a properly conducted clinical trial). Granted this point of agreement, however, heated controversies concern, among other issues, the very meaning of efficacy hypotheses: philosophers and epidemiologists disagree, for instance, on the precise nature of the causal claims expressed by causal statistical hypotheses (see, e.g., Broadbent 2013). Still other debates concern the most adequate statistical methods to test such hypotheses, and many different proposals have been put forward within the two most important approaches to statistical inferences, the orthodox and the Bayesian one.

In this paper, we shall not delve into such debates. In particular, we shall not try to compare and evaluate the different methods of hypotheses assessment proposed in the literature. Instead, our aim will be to identify the common, qualitative "cognitive structures" (Kuipers 2001) underlying all of those methods, and hence to capture the "core" logic of efficacy hypotheses assessment. More specifically, we shall suggest that the basic principles governing the assessment of efficacy hypotheses, and of causal statistical hypotheses in general, are provided by an appropriate statistical version of such a time-honored procedure as the method of difference put forward by John Stuart Mill. To use Worrall's (2007a) apt phrase, our analysis will then provide a description of a key component of the "scientific commonsense", that is, a set of qualitative principles underlying such diverse theories of scientific method as Popperian falsificationism and Bayesianism, as well as the best scientific practice.

Roughly, Mill's method of difference (MD) aims at identifying the sufficient cause of a given phenomenon by eliminating all the other possible causes. The statistical version of Mill's method of difference (SMD), that we shall articulate here, aims at identifying the cause responsible for a significant statistical difference between two samples drawn from a given population by ruling out the alternative hypothesis that such difference is the result of chance. As we shall argue, SMD underlies various kinds of clinical trials and, more generally, methods of controlled inquiry widely deployed in the empirical sciences. Interestingly (as we shall see in Sect. 9.4), both philosophers and epidemiologists have often highlighted before, if only occasionally, the possible relevance of Mill's method for the assessment of causal statistical hypotheses. Still, quite surprisingly, to the best of our knowledge none of them has provided a detailed characterization of such method as applied in the statistical case.

We shall proceed as follows. In Sect. 9.2, we shall briefly survey the key features of randomized controlled trials (RCTs), which in spite of increasing criticism are still the method of choice for the assessment of efficacy hypotheses in clinical research. We shall focus on RCTs as the clearest instance of controlled inquiry, illustrating some crucial issues concerning efficacy assessment that are common to many other methods, like case-control and cohort studies. Our conclusions, in any case, apply more generally to all kinds of controlled trials, both inside and outside epidemiology. In Sect. 9.3, we shall put forward our Millian analysis of the assessment of efficacy hypotheses. Section 9.3.1 presents Mill's method of difference; we then introduce its statistical version in Sect. 9.3.2 and show how it applies to reconstructing the logic of clinical trials in Sect. 9.3.3. As we shall argue in Sect. 9.4, the Millian insight underlying our statistical method of difference uncovers a fundamental principle of scientific commonsense. In Sect. 9.5, we shall offer some concluding remarks and hints at future research.

9.2 Clinical Trials

In clinical research, one typically evaluates hypotheses concerning the efficacy of treatments on the basis of *controlled trials*. We can summarize the standard aspects of controlled clinical trials as follows. In order to assess the efficacy of some treatment T for a given disease or condition, researchers will go through the following fairly standard steps (see, e.g., Chow and Liu 2004; Day 2007):

(1) An experimental population P of patients suffering from the disease, and eligible for the trial, is identified. ("Eligible" means here, for instance, that members of P do not suffer from other pathologies, are not too old or too young, and so on.)
(2) A large sample S of patients, who agreed to participate in the trial, is drawn from population P.
(3) S is divided into two groups: the "treatment group", including the participants administered with T, and the "control group", including those receiving some control treatment, which may be a placebo, the standard treatment for the relevant disease, or even nothing at all.
(4) The outcome of the trial, i.e., the course of the disease in all participants, is ascertained by standardized procedures.
(5) The efficacy of T is assessed on the basis of an appropriate statistical analysis of the observed outcome of the trial.

Well-known problems arise with the procedure outlined above, especially in relation to steps 3–5. A crucial issue concerns step 3, i.e., how to split the sample in the treatment and control groups. In principle, one would require the two groups to be essentially similar with respect to *every relevant variable* which might affect the course of the disease, as, for instance, sex, age, and so on. The reason is clear: the possible differences in the outcomes within the two groups can be reasonably attributed to the action of *T* only if all the relevant variables are similarly distributed across the treatment and control groups.

However, a short reflection suggests that the above requirement is too strong in practice. Indeed, researchers can never guarantee that the treatment and control groups are essentially similar with respect to *every* relevant variable. At best, they will be able to control for those variables already known to be relevant for the course of the disease. This means that one can never exclude that the two groups will actually differ with respect to some variables, whose relevance is however unknown. In order to minimize the probability that this occurs, researchers commonly employ so called *randomized controlled trials* (RCTs),[1] that is, a kind of controlled trial where members of the relevant sample *S* are allocated to either the treatment or the control group by randomization, i.e., by using some random mechanism.

To put it differently, one crucial reason to employ RCTs is to reduce the so called selection bias, i.e., to minimize the probability that the treatment and control groups are unbalanced with respect to some relevant variable: randomization should maximize the probability that the two groups are essentially similar with respect to *any* relevant variable, including those which one does not know are relevant for the given disease. For instance, suppose that some food habit, mistakenly believed to be irrelevant for the evolution of the condition, is in fact relevant. Randomization, its advocates argue, makes it highly probable that the treatment and control groups will include approximately equal proportions of participants with the relevant habit.

The selection bias is not the only systematic error which may affect the outcome of clinical trials. Another source of biases is related to the participants' expectations concerning the effect of the treatment. In particular, knowing to which group one belongs may significantly affect the course of a disease. Moreover, also the expectations of the researchers may well distort the outcome of the trial. For these reasons, one should make sure that neither participants nor researchers will be able to learn to which group each participant has been allocated. To this aim, the standard technique is to design a *double-blind RCT*, i.e., an RCT where both participants and researchers are unaware of whether any given participant belongs to the treatment or the control group. This implies, in particular, that the patients should be unable to tell the studied treatment *T* from the treatment administered to the control group (see Jenkins 2004, Ch. 2, for a vivid illustration of the difficulties that this engenders).

In this connection, it seems beyond doubt that in the case of various diseases a treatment *T*, for the simple fact of being administered, will improve the condition of the patient; in other words, that *T* will have a more or less significant *placebo*

---

[1]Since the second half of twentieth century, RCTs have progressively gained the status of the method of choice first in clinical research and other fields of health care and later also in social and public policy. Their increasing popularity has recently attracted renewed critical attention on RCTs, and a growing body of literature has highlighted their limitations (see, e.g., Rothwell 2005, 2008; Bluhm 2007; Cartwright 2007, 2012; Howick 2011). Moreover, the alleged advantages deriving from randomization have been doubted (see, e.g., Urbach 1985; Howson and Urbach 2006. Ch. 6; Worrall 2007b; Howick 2011; Saint-Mont 2015), and a vocal minority has advocated instead Bayesian trials (see, e.g., Kadane 1996; Teira 2011).

effect (see, e.g., Benedetti 2008). For the moment, a placebo may be defined as any treatment that the practitioner believes to have no specific therapeutic efficacy for a certain disease.[2] A classic example of placebo is the *pillula panis* (bread pill), widely used in medicine since at least the Eighteenth century. Given that as hinted above the placebo effect manifests itself in connection with various diseases, assessing the efficacy of a treatment *T* for such conditions typically means to check whether the effect of *T* is merely a placebo effect. This is the reason why the members of the control group are usually administered an appropriate placebo, which is as similar as possible to *T*. This is crucial since the intensity of the placebo effect is affected by many variables, including the mode of administration and the context of medical intervention. As an example, generally a treatment administered at the patient's home does not have the same effect as when administered at the hospital (see Kaptchuk et al. 2000). For this reason, researchers testing a new treatment *T* look for a placebo such that its outlook, its mode of administration, and its collateral effects are identical to those of *T*. If this condition is satisfied, then the possible differences between the course of the disease in the treatment and control groups may be attributed with some confidence to the active component of the treatment. What this exactly means will be the topic of the next section.

9.3 Assessing Hypotheses with the Method of Difference

As recalled in the foregoing section, a clinical trial aims at assessing the efficacy of some treatment for a given condition. In this section, we shall argue that the logical structure of clinical trials is best reconstructed as a statistical version of the well-known "method of difference", an inference schema introduced by John Stuart Mill (1806–1873) in his *System of Logic* (1843). We shall start by briefly describing how the method of difference may be used in assessing hypotheses of sufficient causality (Sect. 9.3.1). We shall then introduce a statistical version of this method that, as we shall argue, may be applied to the assessment of hypotheses of statistical causality (Sect. 9.3.2) and, in particular, of hypotheses of efficacy as tested in clinical trials (Sect. 9.3.3).

[2] A more precise definition will follow in Sect. 9.3.3. There is a wide scientific and philosophical literature devoted to the attempt to provide an adequate definition of the concept of placebo: see, for instance, Grünbaum 1993, Kaptchuk 1998, Macedo et al. 2003, Howick 2011, Broadbent 2013, Miller et al. 2013, Holman 2015.

9.3.1 Hypotheses of Sufficient Causality

Imagine that a group of researchers is investigating the causal relations occurring among some features exhibited by the individuals[3] belonging to a given population P. In the course of their inquiry, researchers might come to consider what we call hypotheses of sufficient causality. A hypothesis of sufficient causality H has the form "C is a sufficient cause of E (in P)", where both C and E are features that may characterize the individuals of P. It is common to construe H as a universal hypothesis, restating it as follows:

> All the individuals (of P) exhibiting C also exhibit E,

or, equivalently, "If an individual (of P) exhibits C, it also exhibits E". In the following, we shall often omit the reference to the relevant population P, which we take as tacitly understood. Moreover, we shall call "phenomenon" any feature E which is investigated as the possible effect of some cause, and "conditions" the features $C, C_1, \ldots, C_n$ which are considered as possible sufficient causes of E. To illustrate, suppose that researchers are investigating a case of foodborne illness, so that the relevant phenomenon E is some kind of intoxication. The aim of the investigation is then to discover a sufficient cause of E among a number of possible conditions $C, C_1, \ldots, C_n$, like having consumed particular kinds of food (an example will follow in moment).

Dealing with the inference of causal hypotheses, Mill proposed his famous five "canons" of induction, which included the *method of difference* (Mill 1843, book III, ch. VIII, sec. 2, p. 256):

> if an instance in which the phenomenon under investigation occurs, and an instance in which it does not occur, have every circumstance in common save one, that one occurring in the former; the circumstance in which alone the two instances differ, is the effect, or the cause, or an indispensable part of the cause, of the phenomenon.

Since Mill's formulation may not be fully transparent, in the literature one finds various ways of restating the method of difference in more rigorous terms (see van Heuveln 2000 for a survey). Our proposed reconstruction of the method is the following:

(MD) Suppose you are investigating a phenomenon E (in population P) and accept the assumption A that "At least one of the conditions $C, C_1, \ldots, C_n$ is a sufficient cause of E". Moreover, suppose that, by considering two different individuals a and b in P, you make the observation O that "E is present in a and absent in b, C is present in a and absent in b, and $C_1, \ldots, C_n$ are all present both in a and in b". Then you can infer H, i.e., that C is a sufficient cause of E.

[3] We use here the term "individual" in a quite wide sense here, to denote not only persons or organisms, but also different kinds of objects or events. For instance, different portions of lands, or even different policies, may be individuals in our sense (cf., e.g., Holland 1986, p. 945).

Let us illustrate how MD works by means of a simple example. Suppose that Adam and Eve are two friends who had dinner together last night. In the morning, Adam feels sick, while Eve is fine: thus, there is a phenomenon E (the illness in question, or better its symptoms, like, say, vomit, aches, and fever) affecting Adam but not Eve. A reasonable assumption (A) is that Adam's illness may depend on some of the foods served at dinner being noxious for Adam. After a short reflection, the two friends realize that only Adam, but not Eve, had mushrooms as garnish for the steak, while all other courses — say soup, steak, and cake — were the same for both. In other words, there is a condition C (having eaten mushrooms) which affects Adam but not Eve, whereas all other conditions — i.e., C_1 (eating soup), C_2 (eating steak), and C_3 (eating cake) — have been common for both of them. Such observation (O) is graphically represented in Table 9.1. On the basis of it, one has reasons to suspect that eating mushrooms (C) was the cause of Adam's intoxication. Moreover, one can presume that also other guests of the same restaurant, who also had mushrooms for dinner last night, will be probably intoxicated as well.

Let us now consider the general case. Table 9.2 represents the general form of observation O appearing in our formulation of the method of difference. Here the presence of E in a is indicated by "+" and its absence in b by "−"; similarly, "+" ("−") is used to indicate the presence (absence) of each condition $C, C_1, \ldots, C_n$ in a and b.

It should be clear where the name "method of difference" comes from. It refers to the key intuition underlying Mill's method: observation O shows that, as far as features $E, C, C_1, \ldots, C_n$, are concerned, there are only two *differences* between the individuals a and b: first, C is present in a while absent in b; second, E is present in a while absent in b. It is then natural to conclude that the former difference is the cause of the latter, i.e., that a has E just *because* it has C. Moreover, under appropriate circumstances, one may generalize this inference and conclude that, presumably, C is a sufficient cause of E in the whole population P.

Table 9.1 A toy application of the method of difference MD: eating mushrooms is identified as the sufficient cause of Adam's intoxication

Individuals	Possible sufficient causes				Phenomenon
	Mushrooms	Soup	Steak	Cake	Illness
Adam	Yes	Yes	Yes	Yes	Yes
Eve	No	Yes	Yes	Yes	No

Table 9.2 General form of the method of difference MD: observation O shows that the presence of condition C in individual a, but not in b, is a cause of the presence of E in a, but not in b.

Individuals	Possible sufficient causes						Phenomenon
	C	C_1	…	C_i	…	C_n	E
a	+	+		+		+	+
b	−	+		+		+	−

The plausibility of the inference licensed by MD is relatively uncontroversial. Still, long debates have addressed the question of how to rigorously characterize and justify such kind of inference. In particular, one may ask under what conditions MD can be properly applied, so that H can be reliably inferred from observation O. For what concerns us here, it will be sufficient to recall the following points, which are generally agreed upon (cf. Mackie 1980; Skyrms 2000; Hurley 2012).

First, it should be clear that H is a deductive consequence of A & O. In fact, if assumption A concerning the possible sufficient causes of E is true, and if observation O is correct, then H must be true, i.e., C must be a sufficient cause of E. In our example, if one is sure that the symptoms affecting a may *only* depend on one of the foods served at dinner, then eating mushrooms has to be the cause of the illness, since all the other possible causes have been excluded. As already this simple example suggests, however, in typical situations one cannot be completely certain that A is true. Consequently, one cannot be sure that H is true, in spite of the fact that H deductively follows from A & O.

Second, H is *not* a deductive consequence of observation O alone. In fact, it is logically possible that H is false even though O is true: we cannot rule out that, sooner or later, one will stumble upon an individual of P exhibiting C but not E. This would mean that assumption A is false, i.e., that none of the conditions C, $C_1, \ldots, C_n$ is actually a sufficient cause of E. For instance, suppose it turns out that, on the morning Adam feels sick, other guests of the restaurant, who also had mushrooms for dinner, are however perfectly fine. In such scenario, eating mushrooms can be excluded as a sufficient cause of a's illness: some other cause, such as a previously acquired gastroenteritis, has to be identified.

Third, even if H cannot be deductively inferred from O alone, it can be however *inductively* inferred from O. Depending on the underlying epistemological framework, one can defend this claim in different ways. Within the Bayesian approach, for instance, one may argue that H inductively follows from O in the sense that O increases the initial probability of H or, in equivalent terms, that O confirms H (see, e.g., Crupi 2016). This can be proved as follows. Let $p(A)$ be the initial probability attributed to A in the light of the available background knowledge. Since H logically entails A, but not vice versa, inequality $p(A) > p(H)$ holds. However, O logically implies that H and A are, in fact, logically equivalent. In turn, this implies that $p(H|O) = p(A|O)$. Since there is no reason to think that O affects the initial probability of A, one can plausibly assume that $p(A|O) = p(A)$ holds. From the (in)equalities so obtained—i.e., $p(H|O) = p(A|O)$, $p(A|O) = p(A)$, and $p(A) > p(H)$—it follows that $p(H|O) > p(H)$. This means that O increases the initial probability of H or, which is the same, that O confirms H. Finally, one may also argue that H inductively follows from O in another sense: i.e., that the final probability $p(H|O)$ of H given O is significantly high. Indeed, given the above equalities $p(H|O) = p(A|O)$ and $p(A|O) = p(A)$, $p(H|O)$ will be high just in case $p(A)$ is. In other words, if the initial probability attributed to A on the ground of the available background knowledge is high, the final probability of H will be comparably high.

In sum, while the method of difference MD, in typical circumstances, licenses conclusions that are not certainly true, it can provide good reasons to accept them with reasonably high confidence.

9.3.2 *Hypotheses of Statistical Causality*

In most real-life circumstances, and especially in the medical domain, it is typically very difficult, if not plainly impossible, to identify a sufficient cause of a phenomenon *E*. For instance, in many cases, eating mushrooms of a certain species is not a sufficient cause of illness: it may be that most, but not all, individuals eating them will feel sick, perhaps depending on their personal allergies, intolerance, medical history, and on the complex combination of these and other, possibly unknown, factors. Even in such cases, however, the search for causal relations is not entirely hopeless. In fact, researchers will often rest content with investigating what one may call hypotheses of statistical causality. There are many ways of interpreting a hypothesis *H* like "*C* is a statistical cause of *E* in population *P*"; one, particularly useful for our purposes, is the "comparative model" proposed by Giere (1979, sec. 9.5).

According to Giere, we may construe the hypothesis "*C* is a statistical cause of *E*" as comparing two "hypothetical populations", to be called P_C and $P_{\neg C}$. These are obtained by altering the actual population *P* in two different ways: P_C by imaging "everyone in the population [*P*] as having [*C*], but otherwise the same in all relevant respects" (Giere 1979, p. 177), and $P_{\neg C}$ by imaging no one in *P* as having *C*, but otherwise the same in all relevant respects. Then, "*C* is a statistical cause of *E*" simply means that the number of individuals exhibiting *E* is greater in P_C than in $P_{\neg C}$. For instance, suppose that *P* is the group of people who had dinner in our restaurant last night; probably, some of them did eat mushrooms (*C*), and others did not ($\neg C$). Within this framework, P_C is construed as the same as population *P*, with the only difference that everybody had mushrooms for dinner; and $P_{\neg C}$ is the same as *P* except for the fact that nobody had mushrooms. The meaning of hypothesis *H*, in this case, is that there would be more sick people in the former population than in the latter.

In the following, we shall express a hypothesis *H* of statistical causality in this form:

> *C* is a *statistical cause* of *E* in *P* if and only if $\%E_C > \%E_{\neg C}$.

Here, $\%E_C$ and $\%E_{\neg C}$ are the percentages of the individuals exhibiting *E* in P_C and $P_{\neg C}$, respectively; in other words, *H* says that the frequency of *E* in P_C is greater than in $P_{\neg C}$, or that *C* is a "positively relevant factor" for *E*. This latter terminology is justified by noting that a statistical cause, in the sense just defined, is but a special case of a more general notion, that of a "causally relevant" factor. In words, *C* is causally relevant for *E* just in case the frequencies of *E* in P_C and $P_{\neg C}$ are not identical (i.e., $\%E_C \neq \%E_{\neg C}$). In the opposite case, when $\%E_C = \%E_{\neg C}$, *C* is causally *irrelevant* for E. Thus, a statistical cause is a particular kind of causally relevant factor, i.e., a *positively* relevant factor.

Assessing a hypothesis of statistical causality H, then, amounts to assessing whether C is a positively relevant factor for E. The most direct way to do this would be to check whether the number of individuals having E is greater in P_C than it is in $P_{\neg C}$, i.e., whether $\%E_C > \%E_{\neg C}$. This, however, is impossible, for the plain reason that both P_C and $P_{\neg C}$ are hypothetical, and not actual, populations. Since the only actual population is P, what one can do is to draw a large sample S from P, and then divide the sample into two parts, S_C and $S_{\neg C}$, such that all the individuals in S_C have C and none of the individuals in $S_{\neg C}$ has C (more below on the appropriate procedures to adequately split S in this way). Arguably, S_C and $S_{\neg C}$ "play the role of samples from the two hypothetical populations" P_C and $P_{\neg C}$, since these samples "are just *as if* they had been sampled from the two hypothetical populations" (Giere 1979, p. 249).[4] Finally, one can check whether the percentage $\%e_C$ of the individuals exhibiting E in S_C is higher than the percentage $\%e_{\neg C}$ of individuals having E in $S_{\neg C}$.

The intuition outlined above underlies what we call the statistical method of difference, SMD for short:

(SMD) Suppose you are investigating a phenomenon E (in population P) and you accept the assumption A that "Conditions $C, C_1, \ldots, C_n$ include all causally relevant factors for E". Moreover, suppose that, by considering two samples S_C and $S_{\neg C}$ from P, you make the observation O that "$\%e_C$ is much greater than $\%e_{\neg C}$, while the two samples are otherwise 'balanced' with respect to $C_1, \ldots, C_n$". Then you can infer H, i.e., that C is a statistical cause of E.

Here, that the two samples are balanced means that they include roughly the same percentage of individuals having C_i as population P. More precisely, note that, by their very definition, the two hypothetical populations P_C and $P_{\neg C}$ contain exactly the same percentage $\%C_i$ of individuals exhibiting C_i, for any condition C_i different from C; as far as the latter condition is concerned, P_C and $P_{\neg C}$ contain, respectively, 100% and 0% of individuals having C. This latter point holds also for the two

[4] Giere's passage is worth quoting at length:

> [...] we *create* samples that play the role of samples from the two hypothetical populations [P_C and $P_{\neg C}$]. We do this by taking a sample from the real population and then dividing it in two parts [i.e., S_C and $S_{\neg C}$]. All the individuals in one part [i.e., S_C] are experimentally manipulated so that they have the condition [C] that defines the hypothetical population [P_C]. The individuals in the other part [i.e., $S_{\neg C}$] are experimentally manipulated so that they have the condition [$\neg C$]. These two groups, then, are just *as if* they had been sampled from the two hypothetical populations [P_C and $P_{\neg C}$].

Here Giere refers to an experimental inquiry, where researchers can control the condition C by manipulating individuals taken from P in order to create, as it were, C-individuals. However, Giere's insight can be easily adapted to purely observational contexts, where researchers lack full control on C, but still can observe whether an individual drawn from P exhibits, or not, C. Also in this case, in fact, they can adequately form the two relevant samples S_C and $S_{\neg C}$.

samples S_C and $S_{\neg C}$: all individuals in the former, and none in the latter, exhibit C. What distinguishes the hypothetical populations from the actual samples is that the percentage of individuals having C_i (for $C_i \neq C$) may differ in the two samples. This means two things: first, that the frequency of individuals having C_i in S_C may differ from the corresponding frequency in $S_{\neg C}$; and, second, that both those figures may differ from the actual frequency $\%C_i$ of the conditions in P. In short, the frequencies of individuals exhibiting C_i in S_C, $S_{\neg C}$, and P will often be, at best, only approximately equal; this is normally expressed by saying that S_C and $S_{\neg C}$ are "balanced" with respect to any $C_i \neq C$. Table 9.3 displays the ideal special case where the two samples are "perfectly balanced".

The structural similarity between Table 9.3 and Table 9.2 makes it clear why SMD can be construed as a statistical version of MD. In fact, the key intuition underlying both methods is the same. In the present case, this amounts to the following: there are only two *substantial* differences between the two samples S_C and $S_{\neg C}$. The first, as just recalled, is that, by definition, all members of S_C and no member of $S_{\neg C}$ exhibit C. The second substantial difference, highlighted by observation O, is that the frequency of E in S_C is much higher than in $S_{\neg C}$. Hence, one can plausibly infer that the second difference depends on the first, in the sense that the higher frequency of E in S_C is caused by the higher frequency of C in S_C. Moreover, under appropriate circumstances, one may generalize this inference to the whole population P, and conclude that C is presumably a statistical cause of E in P.

For the sake of clarity, let us slightly modify our restaurant example. Suppose that the relevant population P includes some hundreds workers of a big factory, some of which suffered from a severe form of intoxication after consuming lunch at the factory's cafeteria. Suppose that the cafeteria's menu had soup, steak, mushrooms, and cake for lunch, and that these are all causal factors considered as possibly relevant for the intoxication. Now suppose that S_C and $S_{\neg C}$ are two groups of 100 workers each, the former including only workers who had mushrooms for lunch, the latter including only people who did not have mushrooms for lunch. Table 9.4 displays a distribution of all possible causal factors among the members of such groups. In such a case, SMD would allow one to infer that eating mushrooms was probably a statistical cause of foodborne illness.

Our statistical version of Mill's method provides, we submit, an appropriate characterization of the logical structure, i.e., of the qualitative conceptual aspects, of a wide variety of statistical procedures deployed in empirical sciences. As we shall

Table 9.3 General form of the statistical method of difference SMD

Samples	Possible causally relevant factors						Phenomenon
	C	C_1	...	C_i	...	C_n	E
S_C	100	$\%C_1$		$\%C_i$		$\%C_n$	$\%e_C$
$S_{\neg C}$	0	$\%C_1$		$\%C_i$		$\%C_n$	$\%e_{\neg C}$

Observation O shows that the fact that all individuals in one sample, and none in the other, share condition C, is a cause of the difference in the proportion of individuals exhibiting E in the two samples: it is assumed that $\%e_C$ is much greater than $\%e_{\neg C}$ and that the frequency of each C_i in both samples is (at least approximately) equal to $\%C_i$. See Table 9.4 for an illustration

Table 9.4 A toy application of the statistical method of difference SMD (see Table 9.3 for the general form): eating mushrooms is identified as a statistical cause of intoxication among the factory's workers

Samples	Possible causally relevant factors				Phenomenon
	Mushrooms	Soup	Steak	Cake	Intoxication
S_C	100/100	51/100	68/100	27/100	89/100
$S_{\neg C}$	0/100	55/100	71/100	25/100	22/100

better detail in Sect. 9.4, several authors have noted that some statistical version of Mill's method of difference should be applicable in assessing hypotheses of statistical causality. Still, to the best of our knowledge, no explicit formulation of principles such as SMD is to be found in the philosophical and statistical literature. This is rather surprising, among other things, since it seems quite clear that SMD is widely, if tacitly, accepted by statisticians, philosophers, and working scientists. One reason for neglecting SMD (or similar formulations of the same principle) may be that scholars have focused on the differences, more than on the fundamental similarities, among the various statistical procedures for the assessment of hypotheses of statistical causality, which have been the topic of heated debate.

In this connection, we believe that all these procedures may be construed as alternative ways to specify and implement the central intuitions expressed by SMD. To be sure, one can construe each step in SMD in different ways, and this leads to apparently wildly diverse conceptions of the assessment of hypotheses of statistical causality. Of particular importance is the disagreement on three related points: the selection of samples S_C and $S_{\neg C}$ from population P, the nature of the inference licensed by SMD, and the "strength" of this inference in relation to the size of the samples. Let us briefly comment on these points in turn.

Selection of Samples S_C and $S_{\neg C}$ As the reader will recall, observation O in SMD requires that, for any causally relevant factor C_i different from C, the frequencies of C_i in the two samples S_C and $S_{\neg C}$ are approximately equal to the frequency of C_i in P. Typically, observing that this is indeed the case is only possible because S_C and $S_{\neg C}$ are formed in advance exactly in order to meet such requirement. There are basically two possible ways of forming balanced samples, briefly described below.

Under the first, "restrictive" construal of S_C and $S_{\neg C}$, conditions $C_1, \ldots, C_n$ are just the conditions that, on the basis of the available background knowledge, one knows to be causally relevant factors for E in P (or, more inclusively: the conditions that, given one's background knowledge, one believes are certainly or likely causally relevant). Under this construal, one typically knows, at least approximately, the frequencies $\%C_1, \ldots, \%C_n$ of the corresponding conditions in P. Given this restriction, one will draw S_C and $S_{\neg C}$ from P in such a way that the two samples will contain roughly the same percentages of individuals exhibiting each of the conditions $C_1, \ldots, C_n$.

A second, "unrestrictive" construal of S_C and $S_{\neg C}$ is that conditions $C_1, \ldots, C_n$ are *all* the possible, known and unknown, causally relevant factors for E in

P. This means that the set $C_1, \ldots, C_n$ is actually split in two parts: the class of causally relevant factors $C_1, \ldots, C_k$ which are known to be causally relevant, and the set of causally relevant factors $C_{k+1}, \ldots, C_n$ which are not known to be causally relevant. Under this construal, one cannot know, even approximately, the actual frequencies of all the conditions $C_1, \ldots, C_n$ in P. Hence, in order to guarantee that (very probably) S_C, $S_{\neg C}$, and P contain approximately the same percentages of individuals exhibiting each condition, one needs to apply appropriate procedures to randomly draw the two samples from P. A huge literature discussing such randomizing procedures exists, and we shall not discuss this issue further.

Inferring H from O in SMD It is well-known that proponents of different approaches to statistical inference disagree quite strongly on the nature of the inference leading from observation O to the hypothesis of statistical causality H. Still, they agree at least on one point, i.e., that such inference is not deductive, since H follows neither from O alone nor from A & O (contrary to what happened with MD). In turn, this means that the inference of H from either O or A & O is an inductive inference, which, in any case, cannot afford full certainty to H.

Granted this, heated debates concerning the most appropriate way to characterize such inductive inference divide the supporters of different versions of both the orthodox and the Bayesian approach (e.g., Romeijn 2017). Even a quick look at the various positions defended within the debate is sufficient to recognize two "families" of approaches. The first construes the inference licensed by SMD as leading to some form of "inductive acceptance", meaning that H is fully, if only tentatively, believed as the conclusion of that inference.[5] The second view is that such inference assigns H a (high) degree of credibility, where "credibility" is here a catchall term used to denote epistemic probabilities, p-values, indexes of significance, and other quantitative notions of credence.

Sample Size and Inference Reliability In all accounts of statistical inference, the question arises as to what is the "sufficient size" that samples S_C and $S_{\neg C}$ are required to exhibit in order to guarantee that the inference from O to H is indeed reliable. The issue is intertwined with the specific interpretation of inductive inference adopted in each case, so opinions diverge also in this matter. At the very least, all accounts agree on one simple point: that *ceteris paribus*, the greater the sample, the more reliable the corresponding inference. A further point of agreement concerns another important feature of the sample, which affects the reliability of the corresponding inference. This is the following: the greater the extent to which $\%e_C$ exceeds $\%e_{\neg C}$, the stronger our confidence that $\%E_C$ is greater than $\%E_{\neg C}$. In other words: when $\%e_C$ is *much greater* than $\%e_{\neg C}$, this increases the probability that $\%E_C$ is *greater* than $\%E_{\neg C}$, i.e., that C is a statistical cause of E. Again, granted this, opinions differ on the proper way of measuring how much e_C exceeds $\%e_{\neg C}$,

[5]This notion of inductive acceptance, where acceptance may be construed in several, cognitive or pragmatic, senses, has been thoroughly analyzed by philosophers of science, starting already with Hempel (1960; see also Levi 1967 and Hilpinen 1968).

and hence how strong is the corresponding inductive inference (Broadbent 2013, Ch. 9).[6]

We conclude by emphasizing again that, however important these points of disagreement are, they should not obscure the fact that different views concerning sample size and selection, as well as the nature and strength of inductive inference, are but variations on a common logical structure, which is provided, as we argued, by SMD.

9.3.3 Hypotheses of Efficacy

An "efficacy hypothesis" H is the attempt of identifying the cause responsible for a statistical difference between two relevant groups. In a typical case, H says that some drug or treatment T is efficacious in the cure of some disease or condition (cf. Sect. 9.2). Performing an RCT amounts to testing the corresponding claim that the administration of T is the cause of a greater frequency of remissions among the members of the treatment group as compared to those of the control group. In what follows, we shall argue that the statistical method of difference SMD is an adequate tool for assessing efficacy hypotheses, which are but a special kind of causal statistical hypotheses as discussed in Sect. 9.3.2. To this purpose, we shall focus on an imaginary RCT in which a given treatment T is administered to the treatment group and a placebo to the control group. Let us start by clarifying what a placebo is in relation to some treatment T.

In general, one can view a treatment T as a combination of two components: a *specific* or *distinctive* component D, believed to favor the remission of the relevant disease, and a *generic* component G that is constituted, as it were, by the kind of treatment to which T belongs (see Grünbaum 1993, Ch. 3, who calls D and G, respectively, the characteristic and the incidental factors of T). For instance, suppose that treatment T consists in administering some substance, say vitamin C, which is believed to favor quick remission of some condition, like the common cold. Suppose that the substance is administered orally: for 7 days the patient takes, every 8 h, a small, circular, white, bitter-tasting pill. Of course, the pill will contain a given amount of vitamin C; but, besides this, it will also contain certain quantities of various excipients (e.g., lactose, flavors, etc.) that, as far as we know, can in no way

[6]Interestingly, this issue seems an instance of the so called problem of measure sensitivity. In the theory of Bayesian confirmation, such label refers to the existence of a plurality of non-equivalent ways of measuring how much some piece of evidence confirms a given hypothesis, and to the fact that the soundness of theoretical arguments surrounding confirmation crucially depends on the specific measure adopted (see Fitelson 1999; Brössel 2013; Crupi 2016, sec. 3.4; Festa and Cevolani 2017). As the present case suggests, the problem may arise in virtually any area where one explicates informal concepts (like confirmation, inference strength, and the like) by using formally defined models or measures. On probabilistic measures of causal strength see Fitelson and Hitchcook (2011) and Sprenger (2018).

influence the course of the disease. We shall say that vitamin C, taken in the dose mentioned above, is the specific component D of T, while the pills are its generic component G, i.e., the "vehicle" through which vitamin C is administered to the patient.

The distinction just introduced allows us to define the concept of a "placebo corresponding to T"—for short: a T-placebo. We shall say that a certain treatment is a T-placebo if it has the same generic component of T, so that it appears identical to T, but lacks the specific component of T. By designing an RCT, one aims at testing the hypothesis H that T is efficacious, i.e., that if patients were administered T, the frequency of positive outcomes would be higher than if they were administered a T-placebo. Based on our characterization of H as a causal statistical hypothesis, we can proceed to describe the situation as follows. The relevant population P is formed by all the patients suffering from a given condition, and the phenomenon E under investigation is the possible remission or positive outcome recorded for any patient. Then our efficacy hypothesis concerning treatment T has the following form:

D is a causally positive factor for E in P,

where, to recall, D is the specific component of T. As we know, SMD implies that testing such a hypothesis means comparing the relative frequencies of E in two balanced samples drawn from P. Indeed, this is exactly what an ideal RCT does. The first sample, S_D, corresponds to the experimental group, containing patients from P who are all administered both the specific component D and the generic component G of the treatment. The second sample, $S_{\neg D}$, is instead the control group, where all patients receive G (via the T-placebo), but none of them receives D.

Table 9.5, which is but a small variation on Table 9.3, shows how SMD is applied to the testing of the efficacy hypothesis in our imaginary RCT.

The causally relevant factors $C_1, \ldots, C_n$ appearing besides D and G in the table are the so called "prognostic factors", which can influence the probability of a positive outcome. Since the only substantial difference between the two samples concerns the frequency of patients treated with the specific component D, one can infer that this is the cause of the observed difference in the frequencies $\%e_D$ and $\%e_{\neg D}$ of remissions in the two cases, i.e., that treatment T is efficacious.

Table 9.5 General form of the statistical method of difference SMD as applied to clinical trials

Samples	Possibly relevant causal factors							Phenomenon
	D	G	C_1	...	C_i	...	C_n	E
S_D	100	100	$\%C_1$		$\%C_i$		$\%C_n$	$\%e_D$
$S_{\neg D}$	0	100	$\%C_1$		$\%C_i$		$\%C_n$	$\%e_{\neg D}$

Observation O shows that the fact that all individuals in one sample, and none in the other, are given the specific component D of the relevant treatment, is a cause of the difference in the proportion of individuals exhibiting E in the two samples: it is assumed that $\%e_D$ is much greater than $\%e_{\neg D}$ and that the frequency of each C_i in both samples is (at least approximately) equal to $\%C_i$. See Table 9.6 for an illustration

Table 9.6 A toy application of the statistical method of difference SMD to an imaginary RCT (see Table 9.5 for the general form): the assumption of vitamin C is identified as a statistical cause of early remission from the common cold

Samples	Possible causally relevant factors					Phenomenon
	Vitamin C	White pill	Sex	Sporting	Vegetables	Early remissions
S_D	100/100	100/100	48/100	29/100	27/100	89/100
$S_{\neg D}$	0/100	100/100	51/100	32/100	25/100	22/100

Table 9.6 (which is structurally identical to Table 9.4) illustrates an imaginary RCT designed to test the efficacy of vitamin C in favoring the quick remission from the common cold. Both the treatment and the control group include 100 participants; all the members of the former are administered vitamin C in the form of a white pill, while all the members of the latter receive the corresponding placebo. The only other prognostic factors are sex (expressed as the number of men in each sample), doing sport and eating vegetables. Given the displayed figures, SMD would allow one to conclude that vitamin C is probably a positively relevant factor for the remission from common cold.

Table 9.6 should make it clear that, as we argue, SMD describes the qualitative structure of all kinds of RCTs. Indeed, SMD arguably describes the logical structure of any kind of controlled trial, randomized or not. In fact, to briefly sum up the central point of our discussion so far, the aim of all controlled trials is to assess hypotheses of statistical causality, i.e., to identify the likely cause of an observed difference between two sufficiently large samples of the relevant population. As we argued here, the logic governing such assessments is essentially the same underlying Mill's method of difference.

9.4 Mill's Method and the Cognitive Structure of Scientific Commonsense

In the previous section, we argued that most standard methods for the assessment of hypotheses of statistical causality, including RCTs and other kinds of controlled experiments widely used in the medical and social sciences, essentially rely on a statistical version of Mill's method of difference, i.e., SMD. In this section, we shall briefly survey some Millian insights in the philosophical and scientific literature, and show how the logic of controlled inquiry, as summarized in SMD, meshes well with recent reflections on the fundamentals of the scientific method.

Millian Insights in Epidemiology and Statistics To the best of our knowledge, a study of the influence of Mill's methods on the development of modern statistical methods, and of their reception in the philosophical, epidemiological and, more generally, scientific literature of the last 150 years is still to be done. Here we shall restrict our attention to the works of a number of philosophers, epistemologists of medical sciences, and epidemiologists who noticed the relevance of Mill's methods for standard statistical procedures.

To begin with, let us point out that, as a matter of historical coincidence, Mill wrote the *System of Logic* during what Morabia (2004) calls the "pre-formal" phase of epidemiology. During this phase, lasting until the end of nineteenth century, "scientists used population thinking and group comparisons, spontaneously, without referring to some theory [. . .] The mathematical and philosophical bases existed but no formal theory [. . .] none of the concepts and methods had been *formally* defined" (Morabia 2004, p. 107). Some main figures of this era include such pioneers as British physicians William Farr (1807–1883) and John Snow (1813–1858), French physiologist Claude Bernard (1813–1878), and Hungarian physician Ignaz Philipp Semmelweis (1818–1865). Studying the possible relations and interactions between Mill, who employs a number of medical examples in his work (cf., e.g., Mill 1843, Book III, Ch. IX, §§ 1, 6, 7), and the epidemiologists of his time would be of great interest, especially in order to establish whether Mill was motivated (at least partly) by the epidemiological research of his time in developing his methods, and whether the latter influenced some working scientist.

In light of our discussion of SMD, one can even speculate that Mill's methods, and in particular his method of difference, might have provided the formal reconstruction of the basic "concepts and methods" of epidemiology that, according to Morabia, was missing. Indeed, Morabia acknowledges, without pursuing the issue, that "from Mill on, group comparisons combined with population thinking became a philosophically valid principle of knowledge acquisition" (2004, p. 102). Some encouraging suggestions in this direction are indeed to be found in the recent literature. For instance, Scholl (2013, p. 73) convincingly argues that Semmelweis, in his famous investigation of puerperal fever carried out in the years 1846–49, "made conscious use of something like the method of difference, and that he did so in full awareness of the method's logic and force".[7] Scholl provides a detailed reconstruction of Semmelweis' reasoning, based on the tables of mortality and morbidity appearing in his original publications, and concludes that "Semmelweis's data and arguments align neatly with [Mill's] methods" (2013, p. 67) and, in particular, that "the causal role of cadaverous matter was established according to the method of difference" (2013, p. 74). Similarly, Lee (2012) discusses Snow's study of the 1854 cholera outbreak in London (based on the so called Broad Street pump experiment), concluding that "Snow's experiment, as well as drug trials conducted today, appears to rely, in the main, on the method of difference" (2012, p. 153). A consonant assessment appears also in Tulodziecki's (2012) detailed discussion of the principles of reasoning underlying Snow's method.

[7] Semmelweis' discovery has been a favorite case study of philosophers of science, from Hempel (1966) to Gillies (2005) and beyond. Tulodziecki (2013) offers a very useful critical survey of all the most important philosophical reconstructions of this historical case. Both Salmon (1984, sec. 29) and Lipton (2004, pp. 74 ff.) discuss Semmelweis' method in relation with Mill's method of difference.

As the above cases show, Mill's method of difference is close to the kind of reasoning routinely implemented in science since the earliest days of epidemiology and statistics.[8] Indeed, it is not difficult to find references to Mill's methods also in later epidemiological writings (cf. Morabia 2004, p. 101, who refers in particular to the work of Mervyn Susser; see also Broadbent 2013, p. 67). To mention but one example, epidemiologist Raj Bhopal (2002, p. 117) recently claimed that "Mill's canons [...] are of paramount importance to epidemiology and are essentially incorporated into its own widely used criteria" and that, in particular, "the method of difference is at the core of epidemiological thinking (e.g. why do some people get heart disease and others of the same age and sex do not)". As Morabia (2004, p. 101) puts it, the method of difference is "typically the rationale for epidemiologic designs, either cohort or case-control studies, aiming to compare like with like. In epidemiology, however, the method has to be reformulated in probabilistic terms" (see also Morabia 2013). Our presentation of SMD provides precisely this kind of reformulation of Mill's MD.

Early Insights on a Statistical Version of Mill's Method of Difference As we already noted in Sect. 9.3, several logicians and philosophers have recognized that a probabilistic version of Mill's methods may illuminate some basic forms of inference used in scientific research. For instance, in a well-known textbook on critical thinking the author plainly claims, without further comments, that "the method of difference is virtually identical to the method of controlled experiment employed in such fields as biology, pharmacology, and psychology" (Hurley 2012, Ch. 10. pp. 540–41). Indeed, this seems to have become a commonplace in philosophical literature on evidence-based medicine, policy, and social science of the last few years. As an example, Cartwright and Hardie (2012, 33–34), introducing the reader of their *Evidence-based policy* to the very concept of an RCT, note that "an RCT is a study design based on John Stuart Mill's method of difference for making causal inferences", specifying that "an RCT is a Mill's method-of-difference group study in which individual units, all of which are supposed to be governed by the same causal principle, are randomly assigned to the treatment and control groups". Following them, Holman (2015, p. 1334), aiming at a "methodological definition of a placebo [...] determined by the underlying logic of ideal RCTs", plainly states that "the ideal RCT is a manifestation of Mill's method of difference". Similarly, in his introduction to the philosophy of economics, Reiss (2013, p. 201) explains that "RCTs aim to implement a probabilistic version of Mill's method of difference".

[8]Statisticians are familiar with Mill's method of difference at least since Sir Ronald Fisher (1890–1962) famously attacked it in his groundbreaking work on the *Design of experiments* (1935). In his analysis of statistical causality, Holland (1986) discusses both Fisher and Mill as two of the central contributors to the topic. Rosenbaum (2009, Ch. 15) provides an assessment of the debate. For a compact but highly informative survey on the origins of control groups and RCTs, see Dehue (2005).

In view of the fact that most commentators appear to agree that the logic of RCTs and controlled inquiry is essentially Millian, it is surprising that, to the best of our knowledge, no previous attempts to formulate detailed statistical versions of the method of difference, or of the other Millian methods of eliminative induction, have been made. To be sure, one can find in the philosophical literature a couple of interesting suggestions on the need to integrate those methods with corresponding statistical versions. In his *Teatrise on probability* (1921), for instance, John Maynard Keynes claimed that an adequate inductive logic should be based on the theory of probability, since most "laws" established in scientific research are not universal but only statistical, i.e., what Keynes called "inductive correlations". As Keynes notes, his inductive correlations are the same as the "approximate generalizations", propositions of the form "Most *A* are *B*" discussed by Mill in his book (ch. XXIII). While recognizing Mill as a leading figure in inductive logic, Keynes also suggests that his ignorance of probability theory may have prevented him to develop an adequate theory of the inductive inference of approximate generalizations (1921, pp. 267–8).[9]

Von Wright's *Treatise on Induction and Probability* (1951) is probably the most systematic attempt to provide a logical reconstruction of Mill's methods, although perhaps not the clearest one (cf. Mackie 1980, p. 298; van Heuveln 2000, p. 28). In passing, von Wright (1951, pp. 183 ff.) notes that it would be interesting to investigate whether the "logic of induction" underlying Mill's methods could be applied to the assessment of "statistical inductions" or "statistical generalizations", his labels for Mill's approximate generalizations and Keynes' inductive correlations. However, he drops this suggestion without developing it.

Later, Salmon (1984, pp. 103 ff.) argued that, in order to account for the fundamental role played by statistical considerations within investigations concerning causal relations, one needs to put forward appropriate refined versions of Mill's methods. According to Salmon, one such refinement is the method of controlled experiment, whose hardcore he characterized as follows: if the presence of a certain condition *C* makes a difference for the onset of *E*—for instance, if *C* increases the probability that *E* is present—then *C* is a cause of *E*. His illustrative example, involving vitamin C and the common cold, is essentially identical to the one used to illustrate SMD in Sect. 9.3.3. Following Salmon, Grünbaum discusses Mill's methods in his analysis of inference patterns in medical and psychological sciences, noting that "the establishment of a causal connection in psychoanalysis, no less than in 'academic psychology' or in medicine, has to rely on modes of inquiry that

[9]Keynes' harsh assessment of Mill's contribution is worth quoting at length. After having said that "Mill has not justly apprehended the relativity of all inductive arguments to the evidence, nor the element of uncertainty which is present, more or less, in all the generalisations which they support", Keynes adds in footnote: "This misapprehension may be connected with Mill's complete failure to grasp with any kind of thoroughness the nature and importance of the theory of probability. The treatment of this topic in the *System of Logic* is exceedingly bad. His understanding of the subject was, indeed, markedly inferior to the best thought of his own time" (Keynes 1921, Pt. III, Ch. 23, pp. 267–68).

were refined from time-honored canons of causal inference pioneered by Francis Bacon and John Stuart Mill" (1984, p. 47). His conclusions are as clear as telling: "modernized or refined statistical versions of the famous four methods of controlled inquiry articulated by Mill serve to test the causal relevance of an *X* to a *Y*: such controlled inquiry shows whether the presence of X *makes a difference* to the occurrence of Y" (Grünbaum 1993, p. 163).

The Cognitive Structure of Scientific Commonsense There is a simple way to explain the wide, if often implicit, consensus on the plausibility of Mill's method of difference amongst epidemiologists and philosophers of different schools and tendencies. The reason, we suggest, is that Mill's method illuminates some of the essential aspects of what, following Kuipers (2001, p. x), we may call the cognitive structures of scientific inquiry. In different terms, the method of difference is a particularly clear illustration of what Kuipers (2000), Worrall (2007a) and others call "scientific commonsense", that is, a set of qualitative principles underlying such diverse theories of scientific method as Popperian falsificationism and Bayesianism, as well as the best scientific practice. According to Worrall one such principle, "an obvious (but in practice immensely powerful)" one, is the "idea that really telling evidence for any claim is evidence that at the same time tells against plausible rival alternative hypotheses" (Worrall 2007a, p. 1015; cf. also Howick 2011, Ch. 4). We can reformulate in a more precise way this fundamental methodological principle, widely embraced by scientists and philosophers of science, as follows: a hypothesis is strongly confirmed by empirical evidence if, and only if, the evidence is explained very well by the hypothesis, and it is not explained well by any plausible alternative hypothesis. This idea might be labeled as the "principle of explanatory differential".

It should be clear that the principle of explanatory differential underlies the idea of any controlled (clinical) study, and RCTs in particular (cf. Table 9.5 in Sect. 9.3.3). In this regard, Worrall (2007a, pp. 1015–1016) claims:

> In the case of therapeutic claims in medicine, this means evidence that tells in favour of the claim that the treatment at issue (or rather the 'characteristic features' of that treatment over and above any placebo effect) is responsible for the observed outcome, but that at the same time tells against plausible rival explanations of that observed outcome. [...] If the control group received no treatment at all (making it a 'natural history' group) then – depending, I would suggest, on the nature and size of the observed difference in average outcome – the alternative hypothesis may be plausible that the difference is a placebo effect; that is why clinical trials invariably involve a control group that is given either placebo or conventional treatment.

As readers will have noticed, similar considerations apply, more generally, to all inductive strategies for the testing of hypotheses of statistical causality as discussed in this paper. In fact, the principle of explanatory differential underlies both Mill's original method of difference MD (cf. also Howick 2008, Ch. 3) and its statistical version SMD. In the former case, the evidence provided by the relevant observation (cf. Table 9.1 in Sect. 9.3.1) rules out all other possible sufficient causes besides *C*, which is then inferred as the probable sufficient cause of phenomenon *E* of interest. In the latter case, things are only slightly more complicated. Indeed, what is ruled out by applying SMD (cf. Table 9.3 in Sect. 9.3.2) is the hypothesis that the much greater frequency of E in one sample with respect to the other is due, not to the causal factor C under investigation, but to some other factors, for instance plain chance. In short, the conclusion that C is a statistical cause of E is made probable by excluding all other plausible alternative hypotheses concerning E as improbable in the light of the relevant observation. In this way, the (statistical) method of difference illuminates a crucial aspect of the cognitive structures of scientific commonsense.

9.5 Concluding Remarks

In this paper, we offered a rational reconstruction of the logic of clinical trials and, more generally, of the assessment of hypotheses of statistical causality, along distinctively Millian lines. More precisely, we argued that an appropriate, statistical version of Mill's method of difference underlies most current methods used to test and evaluate efficacy hypotheses in many different fields, from pharmacology to evidence-based policy and social sciences. Very roughly, Mill's method of difference (MD) aims at identifying the sufficient cause of a given phenomenon by eliminating all other possible causes. Similarly, the statistical version of the method of difference (SMD) aims at identifying the cause responsible for a significant statistical difference between two samples from a given population by ruling out the alternative hypothesis that such difference is the result of chance. As we argued, this is the crucial insight underlying RCTs and many other kinds of clinical trials and, in general, methods of controlled inquiry widely adopted in different fields; indeed, SMD highlights a basic aspect of scientific commonsense or, in other terms, of the qualitative cognitive structures of scientific inquiry.

The reader may well wonder whether and how, as a matter of fact, Mill's work exerted a more or less direct influence on the historical trajectory that led to the development of randomized clinical trials. The brief historical overview offered in Sect. 9.4 shows that SMD-like intuitions have been around for quite a while in both philosophical and scientific literature—and this, we submit, renders our rational reconstruction all the more plausible. However, here we choose to remain noncommittal about the specifics of Mill's historical role.

We wish to conclude by mentioning that the foregoing discussion of SMD is just a preliminary contribution to a more extensive exploration of the basic logic of controlled inquiry, and that much work lies ahead in this regard, as the following examples suggest. First, in its present formulation SMD is just a sketch of a proper inference method, a sort of "inductive inference schema" which one needs to reconstruct in detail to put it at work. A more detailed discussion than the one offered in Sect. 9.3.2 is needed in order to assess how SMD is applied within the different statistical methods on the market and in the various kinds of test of statistical causality. Secondly, it is a task for the future to explore and define statistical versions of the other methods put forward by Mill: the "method of agreement" and that "of concomitant variations" look particularly promising under this respect. Finally, in view of the consensus on the plausibility of Mill's insights in the field of hypothesis testing and causality assessment, it would be interesting to investigate particular examples where those insights are effectively applied by working scientists. For the moment, however, we remain content with having highlighted the Millian logic of clinical trials and controlled inquiry in general.

Acknowledgements We thank Theo Kuipers and four anonymous reviewers for very useful comments on a draft of this paper. Financial support from the Italian Ministry of Education, University and Research (R.F. and L.T.: PRIN grant "Models and Inferences in Science" no. 20122T3PTZ; G.C.: FIRB project "Structures and dynamics of knowledge and cognition" no. Turin unit: D11J12000470001; FFABR 2017 individual grant), and from the University of Turin and the Compagnia San Paolo (G.C.: project grant "Assessing information models: exploring theories and applications of optimal information search", no. D16D15000190005) is gratefully acknowledged.

References

Benedetti, F. (2008). *Placebo effects. Understanding the mechanisms in health and disease.* Oxford: Oxford University Press.

Bhopal, R. J. (2002). *Concepts of epidemiology*. Oxford: Oxford University Press.

Bluhm, R. (2007). Clinical trials as Nomological machines: Implications for evidence-based medicine. In H. Kincaid & J. McKitrick (Eds.), *Establishing medical reality. Essays in the metaphysics and epistemology of biomedical science* (pp. 149–166). Dordrecht: Springer.

Broadbent, A. (2013). *Philosophy of epidemiology*. London: Palgrave Macmillan.

Brössel, P. (2013). The problem of measure sensitivity redux. *Philosophy of Science, 80*, 378–397.

Cartwright, N. (2007). Are RCTs the gold standard? *BioSocieties, 2*(1), 11–20.

Cartwright, N. (2012). RCTs, evidence, and predicting policy effectiveness. In S. Chow & J. Liu (Eds.), *Design and analysis of clinical trials. Concepts and methodologies*. Hoboken: Wiley.

Cartwright, N., & Hardie, J. (2012). *Evidence-based policy. A practical guide to doing it better.* Oxford: Oxford University Press.

Chow, S., & Liu, J. (2004). Design and analysis of clinical trials. In *Concepts and methodologies*. Hoboken: Wiley.

Crupi, V. 2016. Confirmation, *The Stanford encyclopedia of philosophy* (Fall 2016 ed.), Edward N. Zalta (Ed.), URL = http://plato.stanford.edu/archives/fall2016/entries/confirmation/

Day, S. (2007). *Dictionary for clinical trials* (2nd ed.). Chichester: Wiley.

Dehue, T. (2005). History of the control group. In *Encyclopedia of statistics in behavioral science*. Chichester: Wiley.

Festa, R., & Cevolani, G. (2017). Unfolding the grammar of Bayesian confirmation: Likelihood and Antilikelihood principles. *Philosophy of Science, 84*(1), 56–81.

Fitelson, B. (1999). The plurality of Bayesian measures of confirmation and the problem of measure sensitivity. *Philosophy of Science, 66*, S362–S378.

Fitelson, B., & Hitchcock, C. (2011). Probabilistic measures of causal strength. In P. M. Illari, F. Russo, & J. Williamson (Eds.), *Causality in the sciences* (pp. 600–627). Oxford: Oxford University Press.

Giere, R. N. (1979). Understanding scientific reasoning. In *Holt*. New York: Rinehart and Wiston.

Gillies, D. (2005). Hempelian and Kuhnian approaches in the philosophy of medicine: The Semmelweis case. *Studies in History and Philosophy of Biological and Biomedical Sciences, 36*(1), 159–181.

Grünbaum, A. (1993). *Validation in the clinical theory of psychoanalysis: A study in the philosophy of psychoanalysis*. Madison: International Universities Press.

Hempel, C. G. 1960. Inductive inconsistencies. *Synthese*, *12*, 439–469. Reprinted in *Aspects of scientific explanation*, Carl G. Hempel (pp. 53–79). New York/London: The Free Press/Collier Macmillan, 1965.

Hempel, C. G. (1966). *Philosophy of natural science*. Englewood Cliffs: Prentice-Hall.

Hilpinen, R. (1968). *Rules of acceptance and inductive logic*. Amsterdam: North-Holland Pub..

Holland, P. W. (1986). Statistics and causal inference. *Journal of the American Statistical Association, 81*(396), 945–960.

Holman, B. (2015). Why most sugar pills are not placebos. *Philosophy of Science, 82*, 1330–1343.

Howick, J. (2008). *Philosophical issues in evidence-based medicine: Evaluating the epistemological role of double blinding and placebo controls*. PhD thesis. London: London School of Economics.

Howick, J. (2011). *The philosophy of evidence-based medicine*. Wiley-Blackwell.

Howson, C., & Urbach, P. (2006). *Scientific reasoning. The Bayesian approach* (3rd ed.). La Salle: Open Court.

Hurley, P. J. (2012). *A concise introduction to logic* (11th ed.). Boston: Wadsworth Pub.

Jenkins, S. (2004). *How science works. Evaluating evidence in biology and medicine*. Oxford: Oxford University Press.

Kadane, J. B. (1996). *Bayesian methods and ethics in a clinical trial design*. New York: Wiley.

Kaptchuk, T. (1998). Powerful placebo: The dark side of the randomised controlled trial. *Lancet, 351*, 1722–1725.

Kaptchuk, T., et al. (2000). Do medical devices have enhanced placebo effects? *Journal of Clinical Epidemiology, 53*, 786–792.

Keynes, J. M. (1921). *A treatise on probability*. London: Macmillan & Co..

Kuipers, T. A. F. (2000). *From instrumentalism to constructive realism*. Dordrecht: Springer.

Kuipers, T. A. F. (2001). *Structures in science*. Dordrecht: Springer.

Lee, K. (2012). *The philosophical foundations of modern medicine*. London: Palgrave Macmillan.

Levi, I. (1967). *Gambling with truth*. Cambridge: MIT Press.

Lipton, P. (2004). *Inference to the best explanation* (2nd ed.). London/New York: Routledge/Taylor and Francis Group.

Macedo, A., et al. (2003). Placebo effect and placebos: What are we talking about? *European Journal of Clinical Pharmacology, 59*, 337–342.

Mackie, J. L. (1980). *The cement of the universe*. Oxford: Clarendon Press.

Mill, J. S. (1843). *A system of logic, ratiocinative and inductive* (8th ed.). London: Longmans, Green, and Company. 1886.

Miller, F., et al. (Eds.). (2013). *The placebo: A reader*. Baltimore: The Johns Hopkins University Press.

Morabia, A. (2004). Epidemiology: An epistemological perspective. In A. Morabia (Ed.), *A history of epidemiologic methods and concepts* (pp. 1–125). Basel: Birkhäuser Basel.

Morabia, A. (2013). Hume, Mill, Hill, and the sui generis epidemiologic approach to causal inference. *American Journal of Epidemiology, 178*(10), 1526–1532.

Reiss, J. (2013). *Philosophy of economics: A contemporary introduction*. New York: Routledge.

Romeijn, J.-W. (2017) Philosophy of statistics, *The Stanford encyclopedia of philosophy* (Spring 2017 ed.), Edward N. Zalta (Ed.), URL = https://plato.stanford.edu/archives/spr2017/entries/statistics/. Accessed on 3 Mar 2018.

Rosenbaum, P. R. (2009). *Design of observational studies*. Springer.

Rothwell, P. M. (2005). External validity of randomised controlled trials: To whom do the results of this trial apply? *Lancet, 365*, 82–93.

Rothwell, P. M. (2008). Factors that can affect the external validity of randomised controlled trials. *PLoS Clinical Trials, 2006*.

Saint-Mont, U. (2015). Randomization does not help much, comparability does. *PLoS One, 10*(7), e0132102. https://doi.org/10.1371/journal.pone.0132102.

Salmon, W. C. (1984). *Logic* (3rd ed.). Englewood Cliffs: Prentice-Hall.

Scholl, R. (2013). Causal inference, mechanisms, and the semmelweis case. *Studies in History and Philosophy of Science, 44*, 66–76.
Skyrms, B. (2000). *Choice and chance: An introduction to inductive logic* (4th ed.). Scarborough: Wadsworth.
Sprenger, J. (2018). Foundations for a probabilistic theory of causal strength. *Philosophical Review, 127*(3), 371–398.
Teira, D. (2011). Frequentist versus Bayesian clinical trials. In F. Gifford (Ed.), *Handbook of the philosophy of science* (Philosophy of medicine) (Vol. 16, pp. 255–297). Amsterdam: Elsevier.
Tulodziecki, D. (2012). Principles of reasoning in historical epidemiology. *Journal of Evaluation in Clinical Practice, 18*(5), 968–973.
Tulodziecki, D. (2013). Shattering the myth of Semmelweis. *Philosophy of Science, 80*(5), 1065–1075.
Urbach, P. (1985). Randomization and the design of experiments. *Philosophy of Science, 52*, 256–273.
van Heuveln, B. (2000). A preferred treatment of Mill's methods: Some misinterpretations by modern textbooks. *Informal Logic, 20*, 19–42.
von Wright, G. H. (1951). *A treatise on induction and probability*. London: Routlegde & Kegan Paul.
Worrall, J. (2007a). Evidence in medicine and evidence-based medicine. *Philosophy Compass, 2*, 981–1022.
Worrall, J. (2007b). Why there's no cause to randomize. *The British Journal for the Philosophy of Science, 58*, 451–488.

Chapter 10
Learning by Difference: Placebo Effects and Specific Efficacy in Pharmacological RCTs

Marco Annoni and Giovanni Boniolo

Abstract The logic underpinning randomized controlled trials depends on the assumption that any non-specific effects of treatment are equal and additive across trial arms. We label this the "assumption of specific efficacy". There are a number of situations in which the assumption of specific efficacy does not hold. We describe the epistemic consequences that arise in these situations and discuss strategies for preserving the internal validity of randomized trials in such situations.

10.1 Introduction

The randomized double-blind placebo-controlled trial (RCT) is a fundamental pillar of the contemporary biomedical edifice. Ever since its first introduction, RCTs have been considered the gold standard to test the efficacy of medical treatments, and especially of new pharmacological interventions. For this reason, the approval of new drugs by regulatory agencies is conditional upon the proof that such interventions have been proven efficacious in at least one high quality RCT. More recently, the growing attention generated by the popularity of evidence-based medicine has provided the rationale to extend the logic of RCTs to other research fields such as sociology or economics. In this chapter we claim that the logic beneath

M. Annoni (✉)
National Research Council, Rome, Italy

National Institute of Biomedical Technologies, Milan, Italy

Fondazione Umberto Veronesi, Milan, Italy
e-mail: marco.annoni@itb.cnr.it

G. Boniolo
Dipartimento di Scienze Biomediche e Chirurgico Specialistiche, Università di Ferrara, Ferrara, Italy

Institute for Advanced Study, Technische Universität München, Munich, Germany

A. LaCaze, B. Osimani (eds.), *Uncertainty in Pharmacology*, Boston Studies in the Philosophy and History of Science 338,
https://doi.org/10.1007/978-3-030-29179-2_10

RCTs depends on the problematic assumption that non-specific effects are equal and additive across trial arms.

This chapter is divided in five sections. First, we explain what a RCT is and why it represents an essential epistemological device in contemporary clinical research. Second, we identify the "assumption of specific efficacy" (ASE) as a key-hypothesis for the logic of RCTs. Third, we describe various epistemic consequences that would ensue if the ASE does not hold. Fourth, we discuss a series of empirical studies supporting the hypothesis that the ASE cannot be taken for granted in RCTs whose primary clinical endpoints are prone to placebo modulation. Fifth, we present several epistemic strategies to preserve RCTs internal validity in these scenarios.

10.2 Why Are RCTs Used to Test Pharmacological Interventions?

RCTs are a particular kind of prospective scientific experiment used to assess the efficacy of medical treatments. Today there exist many different kinds of RCTs, but they can all be considered as particular tokens of the same general type.[1] As in this chapter we aim to analyze the underlying logic beneath RCTs, we shall discuss only the most representative case. This prototypical RCT features only two trial-arms and it is meant to answer the following superiority question: "Is treatment x more efficacious than placebo with regard to a specific clinical endpoint?"

Even in their simplest form, RCTs employ a series of sophisticated experimental techniques. However, for our limited purpose it is sufficient to describe the key-steps of a typical RCT. First, before the trial begins, the eligible research subjects are randomized either to the "experimental" or the "placebo" group. Second, during the trial, the participants in the experimental group receive the treatment "x", while those in the placebo group receive the "placebo of x", that is to say, something that is virtually indistinguishable from treatment "x" but lacks those features of "x" that are supposed to make "x" efficacious. Third, after the trial is completed, the results of the two groups are compared. If a statistically significant improvement is measured in the treatment group over the control group, then the tested treatment is said to be superior to placebo.

In difference to other comparative trial designs, RCTs control for a series of confounding factors and biases that might otherwise distort the comparison. In particular, RCTs are useful to control for the "selection bias" and "performance biases".

A "selection bias" might occur whenever the choice over who is assigned to the experimental or to the control group is left to the experimenters. For example, assigning all the most healthy subjects to the experimental group would likely

[1]In this article we discuss only the case of explanatory and superiority RCTs; for a detailed introduction to all the different kinds of RCTs see Schulz and Grimes (2006).

increase the benefits measured at the end of the trial–as healthy subjects would tend to score better results independently from their receiving or not a treatment. To prevent this possibility RCTs employ a series of randomization techniques to ensure that participants' allocation is not skewed by researchers' selection biases.

"Performance biases" occur instead whether during a trial the experimenters and/or the participants alter their behavior because of their awareness of someone's group allocation (Howick 2011). For example, if the experimenters know who belong to the experimental group they may selectively devote more attention to them, biasing the final outcome. Likewise, knowing that one is receiving a placebo may influence one's judgments and self-reports, similarly distorting the final result. To prevent these biases, RCTs employ a series of blinding technique for which participants' group allocation remains concealed to both the experimenters and the participants till the end of the experiment.

In sum, the combination of randomization and blinding techniques allow RCTs to control for a series of biases that would otherwise distort the assessment of the effects of tested treatments. As we argue in the next sections, however, there are a few critical cases in which the use of these techniques may not be enough to preserve the internal validity of RCTs.

10.3 The Logic Beneath RCTs and the Assumption of Specific Efficacy (ASE)

When treatments are tested in RCTs their overall effects are typically larger than their specific effects. The difference between overall and specific effects is attributed to "non-specific" effects. "Non-specific effect" is an umbrella term denoting all the improvements observed at the end of a trial that are not ascribable to the characteristic features of the tested treatments. For instance, some conditions like colds naturally heal over time, and thus one can expect to observe some improvements in related symptoms even without administering medicines. The natural course of diseases is a "non-specific" effect to be controlled for, as it might lead to observe an improvement at the end of a trial that is not caused by the specific action of the administered therapies. Other prominent non-specific effects include the statistical phenomenon of the "regression to the mean" and – as we shall see in the following sections – placebo responses.

Defining "specific" and "non-specific" effect, however, is problematic. As scholars have argued, despite the widespread use of these terms, they are seldom defined as precise technical terms (Grünbaum 1981, 1985, 1986; Howick 2016). Furthermore, previous attempts at defying these concepts have enhanced, rather than diminished, the confusion (e.g. the critique of Grünbaum 1986 of Shapiro and Morris 1978). Though the epistemic and philosophical debate is still ongoing on these issues, for the limited purpose of this article, we shall retain the terms "specific" and "non-specific" effects, adding the following qualifications.

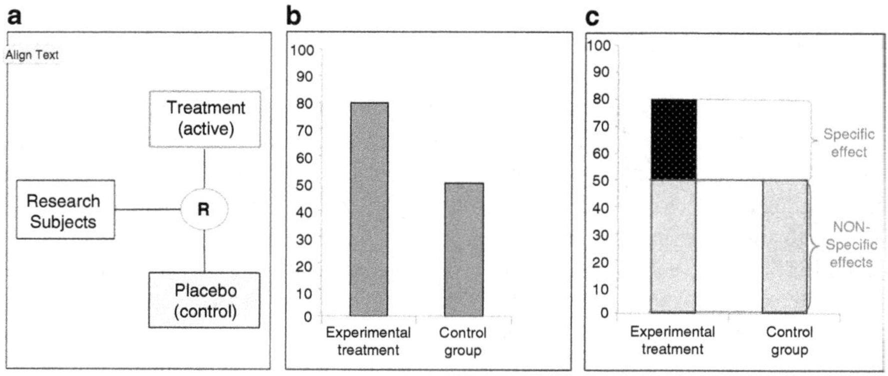

Fig. 10.1 In standard RCTs research participants are randomized between two trial arms: the "treatment" arm and the "control" or "placebo" arm (**a**). At the end of the trial the improvements in the two groups are measured and compared (**b**). Finally, if the improvement in the treatment arm is greater than the one in the control group by a statistically significant margin, then such difference is attributed to the specific effects of the tested treatment (**c**)

Following Grünbaum (1986), by "specific effects" we shall henceforth understand those effects that a treatment "t" has in a RCT on one or more clinical endpoints "c" in virtue of its characteristic features "f" and according to a therapeutic theory ψ. In turn, following Howick (2016), we define "f" a characteristic feature of "t" if and only if (i) "f" is not reducible to the expectancy that "t" is effective; and (ii) "f" has an incremental benefit on the target clinical endpoints "c" over a legitimate placebo control (for a definition of "legitimate placebo control", see Sect. 10.6.2). Likewise, by "non-specific effect", we shall understand all those benefits observed in a RCT on a target clinical endpoint "c" that are not attributable to the characteristic features "f" of "t".

Often, specific and non-specific effects may concur in determining a clinical outcome. However, RCTs are epistemically useful precisely because they permit to neatly distinguish the specific effects of treatments from the non-specific effects that might otherwise provide an alternative explanation for the differential improvements observed at the end of a trial in the experimental group.

To clarify this point, let us describe how the specific efficacy of a treatment is calculated at the end of a typical RCT testing for a new painkiller. Let us assume that at the end of the trial we observe an overall effect – i.e. a reduction in patients' reported pain – in the experimental group that it is clearly superior to the one in the placebo group (Fig. 10.1b). Now, to calculate the specific efficacy of the treatment, it is necessary to *subtract* the improvement observed in the placebo group from the overall improvement observed in the experimental group: the resulting difference is the *specific efficacy* attributable to the tested treatment – i.e. the amount of reduced pain that is attributable to the specific features of the painkiller (Fig. 10.1c).

The logic of RCTs is thus elegantly efficient, for it provides a precise estimation of the specific efficacy of tested treatments even if one is unable to explain why these effects obtain. In a RCT one does not require to understand why a given treatment is efficacious in order to prove its efficacy; similarly, one does not need to account for what non-specific effects are so long as they can be effectively evened out by the non-specific effects measured in the control group.[2]

The logic beneath RCTs, however, is also deceptively simple and depends on a series of epistemic assumptions that are often taken for granted. To unpack this claim it is useful to summarize the epistemological rationale beneath RCTs (and other comparative experiments) in the form of a hypothesis to be tested. To this end, we shall henceforth name as the *assumption of specific efficacy* (ASE) the hypothesis according to which in a RCT the specific efficacy (SE) of a treatment (x) equals the overall improvement measured in the experimental group (I_x) minus the improvement measured in the control group (I_c). So, if we are testing a treatment x:

$$\text{ASE} : \text{SE}_x = I_x\text{–}I_c$$

This simple equation depends on two further assumptions: that (i) the overall improvement in the treatment group (I_x) is the sum of the specific effects of the tested treatment (Se_x) plus its associated non-specific effects (Ne_x); and (ii) that the improvement in the control group is due to non-specific effects (Ne_c). Accordingly, the ASE can be rephrased as ASE_*:

$$\text{ASE}* : \text{SE}_x = [Se_x + Ne_x]\text{–}Ne_c$$

Now, in order to obtain the specific efficacy of x using this formula, it is also necessary to postulate that (iii) the non-specific effects measured in the experimental group are equal in magnitude to the non-specific effects measured in the control group ($Ne_x = Ne_c$). Let us now consider each of these assumptions in turn.

According to (ii) all improvements observed in the control group are caused by non-specific effects. This statement is tautologically true, for non-specific effects are precisely defined as those improvement observed in the control group. Hence, (ii) is uncontroversial.

According to (i) the overall improvement measured in the experimental group is the sum of the improvement due to the specific effects of the tested treatment *plus* the improvement due to non-specific effects in the experimental group.

---

[2]In fact, the evidence gathered through RCTs is considered of such high quality and reliability that it is possible to market drugs that have been demonstrated efficacious in RCTs even though their underlying biochemical mechanisms are still unknown (Howick 2011). For instance, diazepam (Valium) was first tested in a RCT in 1961 to treat "neurotic anxiety" (Randall et al. 1961). Since diazepam was found superior to placebo (and without significant side-effects), it was deemed efficacious and approved for the market. However, the underlying biochemical mechanisms of diazepam remained mostly unknown till the end of the '70. Similarly, today if a drug is found effective in RCTs it may be approved by regulatory agencies even though the precise mechanisms through which it brings about its specific effects are unknown.

However, it is important to specify that the term "sum" is here understood in the same way as it is understood in mathematics. With reference to RCTs, this is the so-called *additivity assumption* (AA) – i.e. the assumption that specific and non-specific effects "add" or "combine" like integers numbers or vectors (Kirsch 2000). As explained by Howick (2011, 108), the AA "is the assumption that the various treatment factors combine like vectors [. . .]. For example, we can dissect a force propelling a billiard ball in a northeasterly direction into its northward and eastward components. Since the component forces act independently we can deduce the resultant force if we know the magnitude and direction of its components". Likewise, the ASE_* assumes that we can calculate Se_x by *subtracting* the relative improvement measured in the control group from the total improvement measured in the experimental group.

In fact, if specific and non-specific effects do not add like vectors but *interact* like drugs then it would impossible to calculate Se_x with the ASE_*. Image two identical patients in pain: A and B. After taking painkiller α, patient A experiences a reduction in pain equals to x; after taking painkiller β, patient B experiences a reduction in pain equals to y. However, from this it does not follow that if patient A takes *both* painkiller α *and* painkiller β then she will necessarily experience a reduction of her pain of $x + y$. In fact, it could be that, after taking both painkillers, patients A report a pain reduction *greater* than x and *smaller* than $x + y$. One explanation could be that each person has a finite number of pain receptors. After taking α, there might not be a sufficient number of free pain receptors left for β to exercise its full effects. In this case, the effects of α and β would interact rather than "add" in the same way as vectors do. As a result, in such cases, it would be impossible to calculate Se_x knowing only the effect of α and the magnitude of the effect of α *plus* β. Therefore, for the ASE_* to hold, one must assume the AA, that is to say, one must assume that the specific and non-specific effects in the experimental group $Se_x + Ne_x$ add like vectors rather than interact like drugs.

Finally, according to (iii) the non-specific effects in the experimental group equal those in the control group. For simplicity, henceforth we shall label this as the *equality assumption*(EA). Like the AA, also the EA is a necessary requirement for the ASE to hold. As we shall see in the next section, if the non-specific effects in the two groups are different by a **Δ**, then the ASE would not provide the specific efficacy of x, but the specific efficacy of x *plus* or *minus* **Δ**. This is why researchers put a considerable effort in ensuring the non-specific effects are equal across trial arms.

It is important to realize that the EA and the AA are logically independent, for the issue of whether specific and non-specific effects interact or combine is conceptually distinct from the issue of whether they are equal across trial arms. However, it is possible that *because* of the nature of their relationship, the non-specific effects in the trial arms might anyway result unequal. For instance, it is possible that the non-specific effects are equal across trial arms – i.e. the EA holds true –, but then they become unequal because they interact rather than combine with the specific effects – i.e. the AA does not hold. Vice versa, it is also possible that specific and non-specific effects combine like vectors – i.e. the AA holds true– but are of unequal magnitude –i.e. the EA does not hold. Thus, regardless of which assumption does not hold in specific cases, the non-specific effects will eventually have unequal magnitudes across the trial arms.

In sum, the epistemic rationale justifying the use of RCTs depends on their ability to single out the specific efficacy of tested treatments. To measure treatments specific efficacy, RCTs rely on the *assumption of specific efficacy* (ASE). The ASE is an "assumption" because its truth is a function of the truth of two more specific assumptions: the *additivity assumption* (AA) and the *equality assumption* (EA). If one of these assumptions does not hold, then also the ASE does not hold. If specific and non-specific effects are not equal across trial arms, then it is no longer possible to infer with precision the specific efficacy of tested treatments by relying solely on the ASE.

10.4 Falsifying the ASE: Some Epistemic Consequences

In this section we identify various epistemic consequences that ensue if the ASE does not hold. The most important consequence is, of course, that if the ASE does not hold then RCTs cannot provide a precise estimation of the magnitude of the specific effects of tested treatments. From this general conclusion there follow a series of particular scenarios that are worth mentioning.

First, let us consider the implications that follow if non-specific effects are unequal across the trial arm. Here there are two possible scenarios (Howick 2011). The first scenario occurs when the non-specific effects in the experimental group are greater then the ones in the placebo group. As depicted in Fig. 10.2a, this would lead to overestimate the specific efficacy of the tested treatment. The second scenario, instead, is the one in which the non-specific effects in the experimental group are lower than the ones in the placebo group. As depicted in Fig. 10.2b, this would lead to underestimate the specific efficacy of the tested treatment. Thus, if the EA does not hold the effects of tested treatments would likely be either overestimated or underestimated.

A second series of implications regards instead the possibility of running into the *efficacy paradox*, of which there are two versions depending on whether we consider one or more treatments in the same trial. As for the single trial version, the paradox obtains when a given treatment is compared to a placebo and the non-specific effects are very significant in the placebo group but negligible in the experimental group (Fig. 10.3a). In this case, the non-specific effects may "hide" the treatment specific efficacy. Because of the unequal magnitude of the non-specific effects in the trails arms, the paradoxical result is that an effective treatment is nonetheless found ineffective in a RCT despite its being tested according to the standards currently adopted for evidence-based medicine.

The second version of the paradox has been firstly proposed by Walach (2001), and occurs when two treatments are compared within a single RCT (Fig. 10.3b). Suppose we have two treatments X and Y – e.g. painkiller and acupuncture – for the same condition – e.g. dental pain – and that each of them is tested against its

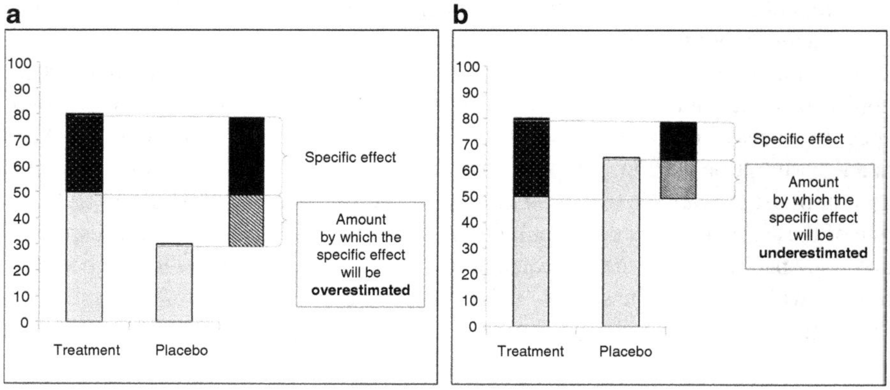

Fig. 10.2 If the ASE does not hold, then the specific effects of the tested treatment could be either overestimated (**a**) or underestimated (**b**) as the specific and non-specific effects may interact rather than add algebrically

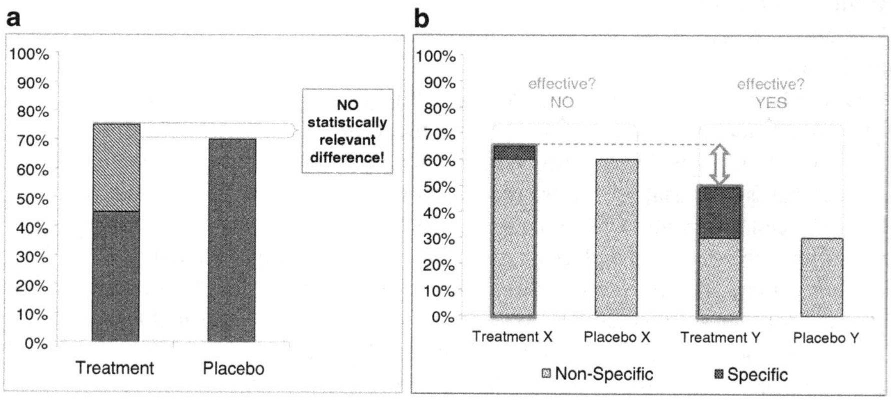

Fig. 10.3 Two instances of the efficacy paradox. The first case obtains if the non-specific effects of a treatment are very significant in the placebo group but negligible in the experimental group. In this case the comparison between the two groups could "mask" the real magnitude of the specific effects of the tested treatment (**a**). The second case obtains if two treatments are compared within a single trial: if the non-specific effects of the first trial are greater than the combination of the specific and non-specific effects of the second treatment, then the first treatment may be found efficacious in this trial, even thought it would be found inefficacious if compared just to a placebo (**b**)

own placebo. Treatment X is marginally more efficacious than placebo X, while treatment Y is largely more efficacious than placebo Y. Hence, in a separate and independent RCT, treatment Y would be found efficacious, while treatment X would not. However, suppose that we set up a comparative trial between these two treatments and that the non-specific effects of X are greater than the one for Y. In this case, the paradoxical situation is that treatment X would be found more effective than treatment Y. Because of the difference of the non-specific effects across the trial arms, an inefficacious treatment is found effective if compared with an efficacious treatment.

Interestingly, this latter scenario has been exemplified by the GERAC trials in which the conventional treatment for chronic low-back pain (anti-inflammatory drugs or NSAIDs, physiotherapy, and exercises) has been compared with both real and "sham" acupuncture (performed on pressure point identified as "non-specific" by acupuncture experts) (Haake 2007). This randomized, single-blinded, parallel, multi-centric trial with an impressive sample size of 1672 participants found that real acupuncture was almost as effective as the placebo acupuncture. However, sham and real acupuncture were almost twice more effective than conventional therapy. In other words, an inefficacious treatment – i.e. a treatment that do not show any statistically significant benefit over a placebo in a well-controlled RCT – was found more effective than an efficacious treatment (NSAIDs had been proven superior to placebo in other trials; see Roelofs et al. 2008). Thus, the GERAC trial proved that the *efficacy paradox* is not a mere theoretical possibility, but that such scenarios may actually occur because the non-specific effects of an inefficacious treatment may be greater than the combined specific and non-specific effects of a different efficacious treatment.[3]

It is important to stress that if also the other scenarios were true, there would follow significant implications not only for clinical research. Today many pharmacological interventions are marketed because they have been proved to be marginally more effective than placebos. This is, for instance, the case of antidepressants (Kirsch 2010). But if their specific efficacy was instead due to overestimation (as depicted in Fig. 10.2a), then some of these interventions could have a lower efficacy than the one for which they were approved, and even be inefficacious. Accordingly, there would be millions of patients taking drugs that are either less efficacious than expected or inefficacious but still loaded with serious side effects. Likewise, if the scenario depicted in Fig. 10.3a turns out to be true, then it entails that some potentially beneficial treatments may be withhold from the market and patients not because they are inefficacious, but because the used testing procedures were unable to detect their true specific efficacy.

In sum, if the ASE does not hold, then it is impossible to measure the specific efficacy of tested treatments. As a consequence the specific efficacy of tested treatments would be either overestimated or underestimated. Moreover, if the non-specific effects are not equal across trial arms, one might run into paradoxical scenarios so that efficacious treatments may fail to prove so in RCTs, or inefficacious treatments may instead be found more effective than efficacious ones. These epistemic consequences demonstrate that the epistemic rationale of RCTs stands or fall depending on whether the ASE holds true or not.

---

[3]One could also image a RCT in which the conventional therapy is tested against only sham acupuncture. In this case, one would obtain the first version of the efficacy paradox (Fig. 10.3a) in which the non-specific effects of an inefficacious treatment would hide the specific efficacy of the tested treatment.

10.5 Placebo Responses as Plausible Confounders in Pharmacological RCTs

So far we have considered the ASE as an epistemic assumption. Indeed, in this article our aim is not that of providing a general proof or refutation of the ASE – a task that cannot be met by looking only at a few exemplar RCTs. Rather, our more limited scope is just to argue that the ASE cannot always be taken for granted in all RCTs. To justify this claim we need to further unpack the concept of non-specific effects. In RCTs non-specific effects comprise multiple variables that include the natural course of diseases, statistical phenomena like the regression to the mean, subjective reporting bias, and subjects' neurocognitive response to salient features of the healing context (Benedetti 2009). This latter variable is the *placebo response*.

In the last decades a growing series of clinical trials, brain-imaging studies, and meta-reviews have identified several neurocognitive and biochemical key-mechanisms of placebo responses (Benedetti 2009, 2010; Miller et al. 2013; Colloca et al. 2013). Collectively, these studies demonstrate that expectation-induced placebo responses may lead to significant improvements in RCTs. For instance, in one experiment by Pollo et al. (2001) thoracotomized patients were treated with a potent analgesic (buprenorphine) on request for three consecutive days together with a basal infusion of saline solution. After being randomized, participants' expectations about the basal infusion were manipulated in the following way: group 1 was not informed about any analgesic effects (natural history); group 2 was told that it was "a potent painkiller" (deceptive administration); group 3 was told that it was "either a placebo or a potent painkiller" (double-blind administration). The placebo effect was measured by the amount of buprenorphine requested over the following 72 hours. With respect to the natural history group, patients in the double-blind and in the deceptive group requested 20,8% and 33.8% less brunemorphine respectively. As the time course of pain was the same in the three groups, the trial proved that expectation-induced placebo responses significantly modulate analgesia. Analogous results have been observed in conditions other than pain, such as anxiety and Parkinson's disease (Colloca et al. 2004).

Thus, compelling evidence indicates that expectation-induced placebo responses may significantly modulate a series of health outcomes in RCTs. It follows that, in RCTs where the clinical endpoints are prone to placebo modulation (e.g., "pain reduction"), placebo effects may act as plausible confounders. By contrast, this risk is virtually absent in RTCs in which the clinical endpoints are not susceptible to placebo modulation (e.g., death or tumor regression).

In standard RCTs the general approach to control for effects of expectancy is to rely on blinding to ensure that experimental and the control group are treated equivalently, and then to rely on random allocation to distribute variables that are not explicitly controlled. These latter variables include the different expectations of trial participants, which can be further analyzed in expectations about the effects of treatment and expectations about the experimental design itself (as participants in randomized trials known from the outset that they have a chance of receiving a placebo, whereas patients in clinical settings presume that they will receive always an effective treatment) (Colagiuri 2010; Teira 2013).

Against this backdrop, in the rest of this section we discuss a series of empirical studies indicating that the ASE cannot be taken for granted in RCTs in which the clinical endpoints are prone to placebo modulation. To articulate this claim, we analyze a series of empirical studies indicating that (i) placebo non-specific effects may sometimes interact with drugs specific effects, thus potentially falsifying the AA; (ii) unsuccessful blinding may lead to unequal placebo non-specific effects across trial arms, thus potentially falsifying the EA.

10.5.1 Placebo Non-specific Effects May Interact with Drug Specific Effects

Gaining a deeper understanding of how placebo and drug effects concur in shaping clinical outcomes is a precondition for harnessing more effectively placebo responses in clinical research and practice (Annoni 2013). However, at present, it is an open empirical question whether placebo effects interact, interfere or synergistically add to the specific effects of treatments. As there are multiple mechanisms of placebo responses realized in different systems and apparatuses of the body, it is possible that placebo and drug effects relate in more than one way, interacting, interfering or combining depending on the circumstances (Benedetti 2010).

Despite a lack of definitive evidence, a few studies suggest that sometimes placebo and drug effects interact rather than combine. For instance, Hughes et al. (1989) manipulated participants' expectations to test the efficacy of nicotine gum for smoking cessation. Participants were randomized to 2×3 balanced-placebo design. Half of the participants received the nicotine gum treatment, while half received the placebo gum. However, participants were also randomized to three different information conditions: "treatment" (positive information); "either treatment or placebo" (neutral information); "placebo" (negative information). Thus, one group received the treatment deceptively labeled as "placebo"; while one group received the placebo deceptively labeled "treatment". The trial found that the nicotine gum was slightly more effective than placebo if both were presented as the effective treatment, but it had a very significant effect when both were presented as the placebo. These results suggest that the specific effect of the nicotine gum decreased as the strength of the expectations increased. Other studies in pain (Levine and Gordon 1984), self-intoxication reports (Ross et al. 1962), and caffeine (Hammami et al. 2010) similarly indicate that placebo and drug effect may interact, so that the magnitude of their combined outcome is greater than each of their individual effects, and yet smaller than their sum.

Whenever this is the case, placebo non-specific and drug specific effects would not be additive, and thus the AA would not hold. Consequently, it would be impossible to assess the precise specific efficacy of tested treatments by relying only on the ASE.

10.5.2 Unequal Placebo Non-specific Effects Resulting from Unsuccessful Blinding

RCTs employ double blinding to rule out performance biases. These biases include the placebo effects induced by patient's expectations about the efficacy of what they assume during the trial. In ideal conditions, blinding procedures assure that the participants should not be able to guess someone' group allocation beyond chances (50–50% in a two-arms trial).

Unfortunately, the available evidence indicates that blinding procedures are seldom checked, and if checked rarely successful. For instance, Hróbjartsson et al. (2007) created a database of blinded randomized trials published in 2001 and indexed in the Cochrane Central Register of Controlled Trials. Of the 1599 identified trials, only 31 reported testing for blinding success (2%). The tests were successful in 14 of the 31 trials (45%), unsuccessful in seven (23%), and had unclear results in ten cases (32%). Of all the trials that did not mention testing for blinding success, the researchers randomly identified 200 studies, and submitted a questionnaire to the corresponding authors about the blinding procedures. Of the 130 questionnaire returned (65%), only 15 (12%) reported testing for blinding success, with nine of them admitting that the data had been assessed without a formal statistical test. These findings are consistent with other studies that have addressed similar research questions (Ney 1986; Fergusson 2004; Boutron et al. 2005). Thus, blinding is rarely tested, and when tested is often found unsuccessful.

The success of blinding is important in any RCT, but it is especially important in RCTs that assess health-outcomes prone to placebo modulation. In these trials, if blinding is unsuccessful then the non-specific effects may be unequal across the trial arms. Once aware of their allocation, participants in the placebo group will have lower expectations about their clinical improvements as compared to those in the experimental group. Such different expectations might induce placebo responses of different magnitude–as showed in the buprenorphine trial by the different effects of the saline basal infusion. In these cases, the EA would not hold and thus it would be impossible to calculate the specific efficacy of the treatment by relying on the ASE.

10.6 Epistemic Strategies for Saving the Assumption of Specific Efficacy

In the previous section we argued that the ASE cannot be taken for granted in RCTs where the primary endpoints are susceptible to placebo modulation. In such scenarios, the placebo non-specific effects may be unequal in the two arms of the trial either because they may interact with the drug specific effects (hence falsifying the AA), or because blinding is unsuccessful (hence falsifying the EA). Either way, it would be impossible to rely on the ASE to calculate the specific efficacy of tested treatments. In this section we discuss some epistemic strategies that may aid in preserving RCTs internal validity in these scenarios.

10.6.1 Different Trial Designs to Assess the Contribution of Specific Drug Effects and Participants' Expectations

The standard "additive" RCT design runs into two main critiques if used to test treatments whose specific efficacy is assessed in terms of clinical endpoints prone to placebo modulation.

First, if placebo and drug specific effects interact rather than add, then the AA would not hold, and it this might lead to underestimate the specific efficacy of the tested treatment (as in Fig. 10.2a). Thus, when testing treatments it is crucial to know whether placebo and drug effect combine or add in those specific experimental conditions. However, the standard two-arms superiority RCT design does not provide an estimation neither of the placebo effect without the drug, nor of the drug specific effect without the placebo component. Consequently, from the results of standard RCTs it is impossible to infer whether the overall effect is due to the interaction or to the combination of specific and non-specific effects. Therefore, with the traditional additive RCT design researchers cannot be sure whether the AA holds or not.

Second, even if the two effects add synergistically, by using the standard RCT additive design researchers would miss a chance to know what is the overall effect of treatments in different informational contexts. Assuming that blinding is successful, the expectations of participants in RCTs would be stirred toward a condition of uncertainty for which they have a 50% of receiving the drug and a 50% of receiving the placebo. However, also this informational regime may induce expectancy-based placebo effects that are presumably different from those that would accompany the administration of the same treatment in clinical contexts, where the expectations of receiving something "effective" are typically higher. While to learn more about the magnitude of placebo responses in different informational contexts may not be a primary objective of typical superiority RCTs, if there were a way to test for treatment superiority while at the same time gaining more clues about their overall effectiveness, then, other things being equal, such model would be preferable to the still widely used additive RCT.

For these reasons, scholars have proposed to adopt different trial designs to test for the efficacy of treatments with placebo-susceptible, continuous, subjectively reported outcomes. Among others, Kirsch and Weixel (1988) have argued for a more systematic adoption of the *balanced-placebo design* (BPD). Similarly to standard RCTs, in a BPD trial half of the participants receive the tested treatment and half a placebo. However, similarly to the nicotine gum-trial, participants are further randomized to other two groups: one in which they are told that they will receive the treatment, and one in which they are told that they will receive the placebo. Hence, in a standard BPD, there are four experimental groups: group 1 receives the treatment labeled as such; group 2 received the treatment labeled as placebo; group 3 received the placebo labeled as the treatment; and group 4 receive the placebo labeled as such. The advantage of the BPD design is that it allows for the comparison of the relative contribution of participants' expectations (group 2 vs. group 4) and of the drug specific effects (group 1 vs. group 3), eventually assessing whether the two add or interact in those experimental conditions.

While BPD trials are epistemologically sounds, they are problematic under two respects. First, in order to manipulate participant's expectations according to the design, researchers need to inform the participants before the trial begins about what they will receive. Since this disclosure procedure is different from the one of standard RCTs, it might raise second thoughts that might alter participants' expectations in unintended ways. To avoid this drawback, Enck et al. (2011) have proposed to adopt the *balanced-cross-over design* BCD, in which all participants are informed that they participate in a RCT with a cross-over design in which they will receive both the drug and the placebo at two different times in a randomized and blind fashion. However, only two groups receive both regimes, while the other two receive either the drug or the placebo twice. The BCD design retains the ability to test whether the AA holds in those specific experimental conditions without instilling second thoughts in the trial participants, but it looses the ability to compare the drug effects under different informational regimes.

Second, the BPD and the BCD are ethically questionable because they involve deception. Deceptive trial designs violates participants' right to informed consent, and thus are generally prohibited unless their epistemic usefulness is deemed sufficiently great to justify an infringement of the respect of participants' autonomy. To circumvent this moral dilemma, Miller et al. (2005) have proposed to resort to "authorized deception", so that participants entering BPD or BCD trials may agree in advance to receive misleading information during the trial (all participants are then debriefed at the end of the experiment). Interestingly, Martin and Katz (2010) have shown in a trial in placebo analgesia that including authorized deception in informed consent does not affect the magnitude of the placebo effects for experimentally induced pain.

In sum, different trial designs such as the BPD or the BCD allow to test whether placebo and drug specific effects interact or add in conditions that are prone to placebo modulation. With respect to the standard RCT additive model, these alternative trials designs allow to disentangle the specific contribution of drugs from its possible interaction with expectancy-induced placebo effects. Using these trial designs may aid researchers not only in preserving the trial internal validity, but also in gaining a deeper understanding of how specific and non-specific placebo effects relate across different experimental and pathological conditions.

10.6.2 Active Placebos and Retrospective Tests for Securing and Testing Blinding Success

In RCTs with clinical endpoints that are prone to placebo modulation, if the blinding is unsuccessful the risk is that of overestimating the specific efficacy of treatments (Fig. 10.2b). In this section we discuss some epistemic strategies to prevent the premature breach of blinding and to assess whether blinding has been successful.

One of the primary causes for the premature breach of blinding is that placebos are often distinguishable from tested treatments. Accordingly, trial participants are able to guess their group allocation. Borrowing a concept from Howick (2011), in order for blinding to be successful it is necessary that placebo controls are "legitimate" and respect two conditions. First, the control must have all the non-specific characteristics of the tested treatment (form, color, smell, etc.) but it must also lack all of its specific characteristics (e.g., the active analgesic principle). Second, the control must not possess other characteristics than the ones of the tested treatment. If the placebo is "illegitimate" then the trial participants might be able to tell the difference with the tested treatment, hence breaking the blinding prematurely.

Incidental *nocebo responses* or the intentional use of *active placebos* may aid in preserving blinding during a trial. As for nocebo effects, placebo controls may sometimes induce "negative" nocebo effects (Colloca et al. 2008, 2013; Colloca and Miller 2011). A trial by Kaptchuk et al. (2006) compared the placebo effects induced by sham acupuncture versus the placebo effects induced by an inert pill in patients with persistent arm-pain. One of the surprising finding of this trial was that patients in the two groups reported different side-effects, and those reported side-effects entirely mirrored the information provided to participants. These results suggest that placebo controls can "mimic" some of the side effects of tested treatments that are prone to nocebo modulation–provided that trial participants are informed about what side effects they have to expect. Aside from the ethical hurdles associated with these "nocebic" information disclosure (Colloca and Nestoriuc 2016), even if nocebo effects can mimic some of the non-specific characteristics of tested treatments, others effects like colored urine would still not be replicable, as there is no evidence that nocebo responses can induce such effects. Thus, despite nocebo effects, there would still be cases in which it would be easy to identify the tested treatment breaching blinding.

Researchers may use "active placebos" to mimic the side effects of tested medications. In an experiment meant to test whether psilocybin could reliable induce mystical experiences in religiously predisposed subjects, participants in the experimental group received psilocybin, while those in the control group received as the "active placebo" a large dose of niacin, an organic compound that may induce clear psychological effects (Pahnke 1966). Active placebos may aid in preserving blinding, but they are difficult to construct as legitimate controls because treatments may have side effects that may be hard to replicate. While researchers can easily replicate the appearance of a pill, it is much harder to accurately replicate all the side effects of a treatment like skin-rushes, possible sexual dysfunctions, or colored urine without adding also *other* (side) effects that the tested drug does *not* possess (Howick 2011). Furthermore, it is also unclear to what extent the use of active placebos is ethical, as the participants in a trial with active placebos assigned to the control group will experience only negative effects without any benefit. Thus, other things being equal, active placebos are epistemically better than inert placebos, but in many cases it is still difficult if not impossible to construct perfectly legitimate active placebo controls.

Another technique that can be deployed to control for biases in RCTs is testing for blinding success at the end of the trial. However, it has been argued that participants and clinicians' "hunches of efficacy" may confound these end-of-trial tests (Sackett 2007). Observers (i.e., doctors and participants) often have preconceived expectancies about the efficacy of treatments. Accordingly, they tend to assume that patients who had done well in the trial were the ones on the treatment, and that those who did poorly were on the placebo. However, this might not be the case. For instance, if a treatment contributes in worsening patients' condition, then those who do well in the trial would be those on the placebo rather than those getting the treatment. In any case, asking at the end of the trial for doctors' or participants' guess about someone's group allocation would not be a test for blinding, but rather a survey of the observers' "hunches of efficacy". Sackett (2007) has thus proposed to substitute end-of-trial tests with a series of tests checking for the consequences of unblinding. For instance, if many participants in the placebo group assume medicines outside the trial, this might be a sign of their having correctly guessed their allocation. This presence of this confounding factor may be empirically ascertained by running periodic tests on participants' blood-samples.

As noted by Howick (2011), the best way to test for blinding success is probably to revise and combine the two strategies. As for end-of-trial tests, it is possible to ascertain whether observers' hunches of efficacy have acted as confounders by simply incorporating questions about observers' previous expectations. If participants and doctors' hunches of efficacy correlate with their guess about participants' allocation, then probably the test would be uninformative about blinding success. As for the tests of other possible confounders due to unblinding, they also have problems on their own. For instance, if those in the placebo group righty guess their allocation, they may start adopting other habits (work less, exercise more, etc.) to cope better with their ailments. These possible behavioral modifications may too bias the results of the study, but they are difficult to ascertain through objective procedure like blood tests. Again, a possible solution may be that of expanding the end-of-trial tests a to include targeted questions as to identify other kinds of plausible confounders. As the two strategies are not incompatible, the most effective way for testing blinding success in a RCT is likely that of combining refined end-of-trials qualitative tests with other quantitative tests that may reveal the presence of other confounders caused by unblinding.

Finally, published articles reporting the results of RCTs should include clear descriptions of both the placebo control and of the procedures implemented to test for blinding. At present, most of the published studies still do not report in full details how the placebo was selected and manufactured, thus making it difficult to evaluate to what extent the control used in the study was legitimate. Furthermore, omitting a description of the physical placebo that has been used may hinders other attempts at replicating the results of the study, or at reconstructing the same experimental conditions. Similarly, researchers should provide either a description of the methodologies that they have used to test for the efficacy of blinding, or the rationale for not undergoing such tests. Both aspects are important in any RCTs, but they are especially so for those RCTs assessing clinical endpoints that are prone to placebo modulation.

In sum, there are different ways in which researchers may mitigate the risk that unequal placebo non-specific effects confound RCTs outcomes. First, researchers should strive to use only legitimate placebo controls. For this purpose, active placebos are epistemically preferable to inert placebos. Second, researchers may check whether blinding has been successful, either by testing for the presence of possible confounders that would result from a breach of blinding, and/or by testing for its success at the end of the trial, while being aware of the possible confounders due to the observers' "hunches of efficacy". Thirdly, the results of RCTs should be published with a transparent and accurate report informing about both the epistemic rationale for selecting the placebo–together with a description of its manufacturing–, and on the various procedures that have been deployed to maintain and check for blinding success.

10.7 Conclusion

Historically, most of the pharmacological treatments that have been administered and quaffed by patients in the past millennia are now known to be either inefficacious or detrimental for their health. It is only during the last century that pharmacology and clinical medicine have begun to systematically rely on experimental methods to test for the efficacy of old and new treatments. One of the most important upshots of this process has been the invention and institutionalization of a shared experimental methodology able to control for the most malicious biases: the RCT. Today RCTs serve the double role of functional epistemical devices and indispensable regulatory yardsticks for distinguishing those treatments that have specific therapeutic utility from those treatment that instead are not better than a placebo.

Despite their unquestionable merits, in this chapter our intent has been that of suggesting that under a deceptively simple surface the epistemological rationale justifying the use of RCTs hides a series of unquestioned assumptions. In particular, in their prototypical form, RCTs are epistemically justified because they allow singling out the specific effects of the tested treatments. In order to do so, we argued, RCTs rely on what we have defined as the *assumption of specific efficacy* (ASE), according to which the specific efficacy of a treatment equals the difference between the improvement measured in the experimental group minus the one measured in the control group. We have thus explained that this assumption depends on two further assumptions: the *additivity assumption* (AA), for which specific and non-specific effects in RCTs add rather than interact; and the *equality assumption* (EA), according to which the non-specific effects must be equal across trial arms. If one or both of this assumption does not hold, nor does the ASE, and so RCTs cannot provide a precise estimation of treatments specific efficacy.

We have then argued that the ASE cannot be taken for granted in RCTs assessing clinical endpoints that are prone to placebo modulation. Placebo non-specific effects can confound RCTs in different ways. In particular, placebo effects may invalidate the ASE by interacting with the drug specific effects (hence falsifying the AA), and by resulting unequal as a consequence premature unblinding (hence falsifying the EA). When the ASE does not hold, however, the specific efficacy of tested treatments is likely to be either underestimated or overestimated, sometimes leading to paradoxes.

Therefore, while conducting RCTs with primary clinical endpoints that are prone to placebo modulation, research should resort to a series of epistemic techniques as to preserve the trial internal validity. These techniques include the adoption of other trial designs such as the balanced placebo design, the use of legitimate placebo controls, and the combination of diverse testing procedures for assessing blinding success.

References

Annoni, M. (2013). Highlights from the 2013 science of placebo thematic workshop. *Ecancermedicalscience, 4*(7), 346.

Benedetti, F. (2009). *Placebo effects: Understanding the mechanisms in health and disease*. New York: Oxford University Press.

Benedetti, F. (2010). *The Patient's brain*. Oxford: Oxford University Press. https://doi.org/10.1093/med/9780199579518.001.0001.

Benedetti, F., Maggi, G., Lopiano, L., Lanotte, M., Rainero, I., Vighetti, S., & Pollo, A. (2003). Open versus hidden medical treatments: The patients knowledge about a therapy affects the therapy outcome. *Prevention & Treatment, 6*(1), 1–19. https://doi.org/10.1037/1522-3736.6.0001a.

Boutron, I., Estellat, C., & Ravaud, P. (2005). A review of blinding in randomized controlled trials found results inconsistent and questionable. *Journal of Clinical Epidemiology, 58*(12), 1220–1226. https://doi.org/10.1016/j.jclinepi.2005.04.006.

Colagiuri, B. (2010). Participant expectancies in double-blind randomized placebo-controlled trials: Potential limitations to trial validity. *Clinical Trials: Journal of the Society for Clinical Trials, 7*(3), 246–255. https://doi.org/10.1177/1740774510367916.

Colloca, L., & Miller, F. G. (2011). The nocebo effect and its relevance for clinical practice. *Psychosomatic Medicine, 73*(7), 598–603. https://doi.org/10.1097/psy.0b013e3182294a50.

Colloca, L., & Nestoriuc, Y. (2016). Nocebo and the PatientPhysician communication. In *Ethical counselling and medical decision-making in the era of personalised medicine* (SpringerBriefs in applied sciences and technology) (pp. 29–37). Cham: Springer. https://doi.org/10.1007/978-3-319-27690-8_4.

Colloca, L., Lopiano, L., Lanotte, M., & Benedetti, F. (2004). Overt versus covert treatment for pain, anxiety, and Parkinsons disease. *The Lancet Neurology, 3*(11), 679–684. https://doi.org/10.1016/s1474-4422(04)00908-1.

Colloca, L., Sigaudo, M., & Benedetti, F. (2008). The role of learning in Nocebo and placebo effects. *Pain, 136*(1), 211–218. https://doi.org/10.1016/j.pain.2008.02.006.

Colloca, L., Flaten, M. A., & Meissner, K. (2013). *Placebo and pain: From bench to bedside*. Amsterdam: Academic.

Enck, P., Klosterhalfen, S., Weimer, K., Horing, B., & Zipfel, S. (2011). The placebo response in clinical trials: More questions than answers. *Philosophical Transactions of the Royal Society B: Biological Sciences, 366*(1572), 1889–1895. https://doi.org/10.1098/rstb.2010.0384.

Fergusson, D. (2004). Turning a blind eye: The success of blinding reported in a random sample of randomised, placebo controlled trials. *BMJ, 328*(7437), 432–430. https://doi.org/10.1136/bmj.37952.631667.ee.

Grünbaum, A. (1981). The placebo concept. *Behaviour Research and Therapy, 19*(2), 157–167. https://doi.org/10.1016/0005-7967(81)90040-1.

Grünbaum, A. (1985). Explication and implications of the placebo concept. In L. White, B. Tursky, & G. E. Schwartz (Eds.), *Placebo: Theory, research and mechanisms* (pp. 9–36). New York: Guilford Press.

Grünbaum, A. (1986). The placebo concept in medicine and psychiatry. *Psychological Medicine, 16*(01), 19–38. https://doi.org/10.1017/s0033291700002506.

Haake, M. (2007). German acupuncture trials (Gerac) for chronic low back pain. *Archives of Internal Medicine, 167*(17), 1892–1898. https://doi.org/10.1001/archinte.167.17.1892.

Hammami, M. M., Al-Gaai, E. A., Alvi, S., & Hammami, M. B. (2010). Interaction between drug and placebo effects: A cross-over balanced placebo design trial. *Trials, 11*(1). https://doi.org/10.1186/1745-6215-11-110.

Holman, B. (2015). Why most sugar pills are not placebos. *Philosophy of Science, 82*(5), 1330–1343. https://doi.org/10.1086/683817.

Howick, J. (2011). *The philosophy of evidence-based medicine*. Wiley-Blackwell. https://doi.org/10.1002/9781444342673.

Howick, J. (2016). The relativity of "placebos": Defending a modified version of Grünbaum's definition. *Synthese, 194*(4), 1363–1396. https://doi.org/10.1007/s11229-015-1001-0.

Hróbjartsson, A., Forfang, E., Haahr, M., Als-Nielsen, B., & Brorson, S. (2007). Blinded trials taken to the test: An analysis of randomized clinical trials that report tests for the success of blinding. *International Journal of Epidemiology, 36*(3), 654–663. https://doi.org/10.1093/ije/dym020.

Hughes, J. R., Gulliver, S. B., Amori, G., Mireanlt, G. C., & Fenwick, J. F. (1989). Effect of instructions and nicotine on smoking cessation, withdrawal symptoms and self-Administration of nicotine gum. *Psychopharmacology, 99*(4), 486–491. https://doi.org/10.1007/bf00589896.

Kaptchuk, T. J., Stason, W. B., Davis, R. B., Legedza, A. R. T., Schnyer, R. N., Kerr, C. E., Stone, D. A., Nam, B. H., Kirsch, I., & Goldman, R. H. (2006). Sham device V inert pill: Randomised controlled trial of two placebo treatments. *BMJ, 332*(7538), 391–397. https://doi.org/10.1136/bmj.38726.603310.55.

Kirsch, I. (1997). Specifying non-specifics: Psychological mechanisms of placebo effects. In A. Harrington (Ed.), *The placebo effect: An interdisciplinary exploration*. Cambridge, MA: Harvard University Press.

Kirsch, I. (2000). Are drug and placebo effects in depression additive? *Biological Psychiatry, 47*(8), 733–735. https://doi.org/10.1016/s0006-3223(00)00832-5.
Kirsch, I. (2010). *The Emperor's new drugs: Exploding the antidepressant myth*. New York: Basic Books.
Kirsch, I., & Weixel, L. J. (1988). Double-blind versus deceptive administration of a placebo. *Behavioral Neuroscience, 102*(2), 319–323. https://doi.org/10.1037//0735-7044.102.2.319.
Levine, J. D., & Gordon, N. C. (1984). Influence of the method of drug administration on analgesic response. *Nature, 312*(5996), 755–756. https://doi.org/10.1038/312755a0.
Martin, A. L., & Katz, J. (2010). Inclusion of authorized deception in the informed consent process does not affect the magnitude of the placebo effect for experimentally induced pain. *Pain, 149*(2), 208–215. https://doi.org/10.1016/j.pain.2009.12.004.
Miller, Franklin G, Luana Colloca, R. A. Crouch, and T. J. Kaptchuk, eds. 2013. The placebo: A reader. Baltimore: The Johns Hopkins University Press.
Miller, F. G., Wendler, D., & Swartzman, L. C. (2005). Deception in research on the placebo effect. *PLoS Medicine, 2*(9), e262. https://doi.org/10.1371/journal.pmed.0020262.
Ney, P. G., Collins, C., & Spensor, C. (1986). Double blind: Double talk or are there ways to do better research. *Medical Hypotheses, 21*(2), 119–126. https://doi.org/10.1016/0306-9877(86)90001-0.
Pahnke, W. N. (1966). Drugs and mysticism. *International Journal of Parapsychology, 8*(2), 295–313.
Pollo, A., Amanzio, M., Arslanian, A., Casadio, C., Maggi, G., & Benedetti, F. (2001). Response expectancies in placebo analgesia and their clinical relevance. *Pain, 93*(1), 77–84. https://doi.org/10.1016/s0304-3959(01)00296-2.
Randall, L. O., Heise, G. A., Schallek, W., Bagdon, R. E., Banziger, R., Boris, A., Moe, R. A., & Abrams, W. B. (1961). Pharmacological and clinical studies on valium (TM) a new psychotherapeutic agent of the benzodiazepine class. *Current Therapeutic Research, Clinical and Experimental, 3*, 405.
Roelofs, P. D., Deyo, R. A., Koes, B. W., Scholten, R. J. P. M., & van Tulder, M. W. (2008). Nonsteroidal anti-inflammatory drugs for low Back pain. *Spine, 33*(16), 1766–1774. https://doi.org/10.1097/brs.0b013e31817e69d3.
Ross, S., Krugman, A. D., Lyerly, S. B., & Clyde, A. D. J. (1962). Drugs and placebos: A model design. *Psychological Reports, 10*(2), 383–392. https://doi.org/10.2466/pr0.1962.10.2.383.
Sackett, D. L. (2007). Commentary: Measuring the success of blinding in RCTs: Dont, must, cant or Neednt? *International Journal of Epidemiology, 36*(3), 664–665. https://doi.org/10.1093/ije/dym088.
Schulz, K. F., & Grimes, D. A. (Eds.). (2006). *The Lancet handbook of essential concepts in clinical research*. Spain: Elsevier.
Shapiro, A. K., & Morris, L. A. (1978). The placebo effect in medical and psychological therapies. *Handbook of Psychotherapy and Behavior Change, 2*, 369–409.
Teira, D. (2013). Blinding and the non-interference assumption in medical and social trials. *Philosophy of the Social Sciences, 43*(3), 358–372. https://doi.org/10.1177/0048393113488871.
Walach, H. (2001). The efficacy paradox in randomized controlled trials of CAM and elsewhere: Beware of the placebo trap. *The Journal of Alternative and Complementary Medicine, 7*(3), 213–218. https://doi.org/10.1089/107555301300328070.
Walach, H. (2011). Placebo controls: Historical, methodological and general aspects. *Philosophical Transactions of the Royal Society B: Biological Sciences, 366*(1572), 1870–1878. https://doi.org/10.1098/rstb.2010.0401.

Chapter 11
An Evidence-Hierarchical Decision Aid for Ranking in Evidence-Based Medicine

Jürgen Landes

Abstract This chapter addresses the problem of ranking available drugs in guideline development to support clinicians in their work. Based on a pragmatic approach to the notion of evidence and a hierarchical view on different kinds of evidence this chapter introduces a decision aid, HiDAD, which draws on the multi criteria decision making literature. This decision aid implements the wide-spread intuition that there are different kinds of evidence with varying degrees of importance by relying on a strict ordinal ordering of kinds of evidence. In order to construct a ranking every pair of drugs is first compared separately on all kinds of evidence. Next, these quantitative comparisons are then aggregated into an overall comparison between drugs based on all the available evidence in a way which avoids that evidence of less importance is trumped by evidence of the higher levels. Finally, these overall comparisons are used to determine the final ranking of drugs which then informs the process of guideline writing. Properties, modifications and applicability of the decision aid HiDAD are discussed and assessed.

11.1 Introduction

Evidence in medicine has been a hot topic ever since the late 1980s. Special interest has been paid to the questions "What constitutes best or legitimate evidence in medicine?", "How does one amalgamate evidence for medical decision making?" and "How important is the most important kind of evidence?", see, e.g., Clarke et al. (2014), GRADE Working Group (2004), Howick and co workers (2011), Osimani (2014a,b), Russo and Williamson (2007), Sackett et al. (2000), Worrall (2007a). This chapter builds on previous work providing answers to these questions and introduces a ranking heuristic for comparing drugs termed *HIerarchical Decision AiD* (HiDAD) and addresses, to various degrees, all these three questions. As an

J. Landes (✉)
Munich Center for Mathematical Philosophy, LMU Munich, Munich, Germany
e-mail: Juergen.Landes@lrz.uni-muenchen.de

A. LaCaze, B. Osimani (eds.), *Uncertainty in Pharmacology*, Boston Studies in the Philosophy and History of Science 338,
https://doi.org/10.1007/978-3-030-29179-2_11

illustrating example I shall consider the following decision problem: a medical body is re-writing its guidelines to treat migraines. The problem arises to rank drugs for treating migraines to guide clinicians in their work.

By bringing formal machinery, which was developed in decision science, to bear on this ranking problem I aim to create a decision heuristic which is transparent to the stakeholders (patients, doctors, guideline developers, politicians, drug developing and manufacturing companies, journalists etc.) and to direct attention of the evidence-in-medicine movement to parts of the, so far, under-appreciated decision science literature. I aim to achieve these aims by delineating the importance of different kinds of evidence and by making all subjective judgements during the decision making process explicit and transparent.

Clearly, to successfully complete this ranking task all the *available evidence* ought to be taken into account and, ceteris paribus, the more evidence that is available the better the final decision. Philosophically, these principles have been expressed as the *Principle of Total Evidence* (Carnap 1947) and as the *Value of Knowledge Theorem*, see Savage (1954, Chap. 7) and Skyrms (1990, Chap. 4). Unsurprisingly, hiding of important information can have disastrous consequences, e.g., in the infamous Vioxx case (Horton 2004; Jüni et al. 2004; Krumholz et al. 2007; McGauran et al. 2010). The ranking problem is further complicated by reporting biases in medicine distorting the available evidence, see Bes-Rastrollo et al. (2013), Every-Palmer and Howick (2014), McGauran et al. (2010), which are not addressed here.

While the "available" part in "available evidence" is uncontentious, there is much philosophical debate about the notion of "evidence", e.g., Kelly (2015), Reiss (2015), Williamson (2015). Here, I am concerned with a concrete decision problem and take a pragmatic approach. I shall take all information to be evidence which *can on its own or jointly with other information conceivably influence the decision problem* of ranking drugs. This notion of evidence is hence depending on the decision problem at hand and the decision making entity's epistemic state at this time.[1] This approach allows, even compels, one to take into account and aggregate all possibly relevant information; in particular evidence comes in different kinds or sorts such as Randomised Controlled Trials (RCTs), cohort studies, expert testimony, case reports, etc.[2]

[1] Information known to be false or irrelevant is thus ignored. Which information is deemed relevant and which is deemed irrelevant is a complicated question outside the scope of this contribution. The answers will depend on the epistemic state, as well as cognitive limitations and the exact framing of the decision problem.

[2] In the more applied sciences, the term *information fusion* rather than *evidence amalgamation* or *evidence aggregation* is often used. Definitions of the term *information fusion* are surveyed in Boström et al. (2007). Further often-used terms are "research synthesis" and "evidence synthesis", see also Sect. 11.2.1.3.

There are further reasons arising from the concrete decision problem to take such a liberal view on what constitutes evidence: Ideally, studies would license the same inferences for the studied population and the target population. In reality, studies are not conducted on the *entire population of interest* but on a much smaller number (even too small a number) of patients, see Chan and Altman (2005, p. 1160), see Button et al. (2013) for this problem in neuro-science, Doll and Peto (1980) in cancer research, Etz and Vandekerckhove (2016, p. 10) in psychology and a philosophical discussion of this problem in Worrall (2007a, p. 992) and Bertamini and Munafó (2012) for a general discussion. Additionally, studied populations, in particular, RCTs often fail to be representative for the target population due to strict patient inclusion criteria, see Revicki and Frank (1999) and Upshur (1995, p. 483). Furthermore, observational and/or cohort studies are sometimes much larger than the largest conducted RCT.

Yet another concrete reason for taking such a liberal view is that smaller studies using different protocols rather than a large single study are a good way to counter biases due to subject selection, study design, and execution strategy in a single study, as argued by epidemiologists in Borm et al. (2009). Philosophically, this idea has been expressed as the *Variety-of-Evidence Thesis* which states that the more varied the body of evidence is, ceteris paribus, the more informative it is. Earman called this a "truism of methodology" (Earman 1992, p. 77).

With such a liberal view on evidence and a highly diverse body of evidence, there is a lot of evidence to be taken into account. One ubiquitous intuition regarding different kinds of evidence is that some kinds or sorts of evidence matter more than others. This intuition manifests itself in an ever-growing number of evidence hierarchies which rank kinds of evidence from most to least important, see Canadian Task Force on the Periodic Health Examination (1979), Evans (2003), GRADE Working Group (2004), Howick and co workers (2011) and see http://cjblunt.com/hierarchies-evidence/ for a much more complete list of close to 100 evidence hierarchies. A kind of evidence is termed "level" in Howick and co workers (2011).

I develop the decision aid based on this construal of evidence, a hierarchical (ranked) view on kinds of evidence and a qualitative approach to appreciate evidence I develop the decision aid HiDAD.

The rest of this chapter is organised as follows: Next, I spell out the decision problem I tackle in detail and argue that current approaches are inadequate for resolving it. Then, I show HiDAD can be used to help tackle this decision problem, discuss some of its properties, possible variations and its applicability. Finally, I conclude.

11.2 The Decision Problem

The formal decision problem I here consider is a regulatory body tasked with writing guideline recommendations.[3] This regulatory body aims to:

> Rank a number of available drugs to treat migraines in non-pregnant adults according to their respective outcome given all the available evidence.

The term *outcome* is to be widely understood, it includes treatment efficacy, side effects (negative as well as positive effects) and monetary as well as non-monetary costs.[4] I denote the set of available drugs to be ranked by $\mathscr{A} = \{A_1, \ldots, A_a\}$. The option of not treating migraines at all is, of course, also a possible course of action. For ease of exposition, I shall assume that this *null act* is included in the set of alternatives $\mathscr{A}$. The literature on decision making also knows the words "options", "acts" or "actions" instead of "alternative".

A ranking is simply some formal way of expressing that some alternatives are judged (based on the available evidence) to be more preferable than some other alternatives. A ranking in this sense is not necessarily transitive, acyclic nor complete, e.g., A_1 might be ranked higher than A_2 which is ranked higher than A_3 which is ranked higher than A_1 (cycle of length three) and A_1 may not be ranked relative to A_4 (ranking is incomplete).

As is tradition, the decision making entity (in this the case the regulatory body) is referred to as the *Decision Maker* (DM). In reality, the DM consists of a number of different individuals with possibly conflicting preferences. I here ignore this layer of complexity and point the reader to the group decision making literature in general and to Urfalino (2012) for an analysis of the actual procedures by which group decisions are made by regulatory bodies. How regulatory bodies do approach such a ranking problem is nicely described in Kelly and Moore (2012, p. 4–5).

---

[3]These recommendations are intended to *guide* doctors in their daily work. I emphatically do not want to suggest that a recommendation of a regulatory body ought to be followed at all times. There are good reasons to deviate from general medical guidelines when it comes to the treatment of individual patients. Patients have individual circumstances such as: co-morbidities, known or suspected (drug)-intolerances and treatment preferences as well as outcome preferences. For deciding on a treatment in an individual patient at a particular time, these patient-specific circumstances ought to matter, too.

[4]There is no principled reason for which I could not construe the decision problem as a multi-outcome problem. For migraines, these outcomes might be: hours with headache, headache severity, days of sick leave and adverse events. In order to keep the complexity of the problem and of the presentation manageable, I abstain from doing so.

11.2.1 Related Methodological Work

Closely related work to this approach is the literature on evidence hierarchies, GRADE in particular, and on medical decision support. This literature is discussed next.

11.2.1.1 Evidence Hierarchies

I now briefly explain why evidence hierarchies alone do not solve the drug ranking problem. Typically, the first level of evidence hierarchies is made up of systematic reviews (many of which, but not all, are meta-analyses). The lower levels are, typically, populated by the case reports and the expert opinions. Howick (2011, p. 5) describes a very simple hierarchy with three levels: the first level is made up of randomised trials, the second level is made up of observational studies while the third and lowest level is made up of expert judgement and mechanistic reasoning.

There are two ways of interpreting and subsequently applying current evidence hierarchies.[5] (1) A trumping interpretation of an evidence hierarchy entails that high level evidence *trumps* lower level evidence. That is, the right decision is determined *only* by high level evidence, if such evidence is available and favors one alternative over another. Lower level evidence hence plays no role in case high level evidence is available and favors one alternative over another. It matters not how strongly the high level evidence favors one alternative over the other, nor does it matter how strongly the reversal is at the lower levels.

(2) According to the less strict interpretation, these hierarchies only rank kinds of evidence. This interpretation of evidence hierarchies only entails a *ranking of kinds of evidence* without trumping. That is, a DM faced with a concrete decision problem is only provided with a ranking which assigns every item of evidence a level. In general, these rankings alone will be much too little to help a DM deciding between different alternatives A_k and A_l in $\mathscr{A}$. What the DM is lacking is a clear decision process (guideline) which allows the aggregation of items of evidence falling into different ranking levels.

Concerning (1) Recently, philosophers have argued for putting more weight on the lower levels of evidence hierarchies. Hence, higher level evidence should not automatically trump lower level evidence. In Clarke et al. (2013, 2014), Clarke et al. argue that the evidence hierarchies currently on the market under-value mechanistic evidence. Worrall questions the uniquely privileged epistemic role of RCTs in Worrall (2002, 2007a,b, 2010). Cartwright points to a low external validity of RCTs (Cartwright 2007; Cartwright and Munro 2010). Osimani and Vandenbroucke argue in Osimani (2014b) and Vandenbroucke (2008) that current evidence hierarchies are inadequate to capture potential risks of health interventions

[5] The GRADE approach is discussed in more detail in Sect. 11.2.1.2.

and advocate to put more weight on the lower levels of the hierarchies. Stegenga goes as far as calling for an end of all evidence hierarchies in medicine in Stegenga (2014). The legal scholar Twinning also expressed his dislikes for hierarchies at Twinning (2011, p. 76).

Solomon puts forward the idea that "ranking of evidence is done by reference to the actual, rather than the theoretically expected, reliability of results" (Solomon 2011, p. 463–464). For current purposes, this means that the hierarchy of evidence to be used depends on the decision problem and the available evidence. Solomon also questions the reliability of using actual RCTs in clinical decision making while acknowledging that this is controversial, Solomon (2011, Sect. 3.2).[6] LaCaze argues in LaCaze (2009) that an evidence hierarchy is best understood as a hierarchy of comparative internal validity.

It is not only philosophers who have criticised the trumping interpretation of evidence hierarchies. Epidemiologists and physicians also worry about hierarchies and evidence amalgamation. Borm et al. have claimed that the best way to evaluate the performance of a treatment is to use multiple, possibly smaller, trials (Borm et al. 2009, p. 711) which can be seen as further arguments against trumping. Upshur has expressed the worry that RCTs may be under-powered for secondary outcome measures and are hence severely limited in informing us about the harm/benefit ratio, see Upshur (1995).

Onakpoya et al. give a long table of approved drugs which were world-wide withdrawn from the market without(!) any high-level evidence (Onakpoya et al. 2016, Table 1). These withdrawals are difficult to square with the trumping interpretation in which low-level evidence is taken into account only if the higher-level evidence does not clearly indicate the drug causes an adverse drug reaction. Under the implicit assumption that it was the right decision to withdraw these drugs, their work speaks against the trumping interpretation.

In sum, I think that the cumulative force of these arguments is strong enough to demonstrate that the strict interpretation of evidence hierarchies is inadequate for this drug ranking problem.

11.2.1.2 GRADE

The most closely related decision method which grades evidence and subsequently makes recommendations in health care is the GRADE system (GRADE Working Group 2004; Guyatt et al. 2013, 2008, 2011). Like this approach, GRADE is a bottom-up approach which first draws comparisons between two health interventions based on studies or meta analysis and then aggregates these comparisons to

[6]Recently, it was alleged that it is impossible to amalgamate of evidence of different kinds (Stegenga 2013); an appropriate response was provided in Lehtinen (2013).

arrive at a recommendation of one health intervention over the other.[7] I agree with Guyatt et al. that the

> [...] merit of GRADE is not that it necessarily ensures reproducible judgments (observers will inevitably differ in close-call situations when rating up or down for individual domains or for the overall confidence per outcome) but that it achieves explicit and transparent judgment. (Guyatt et al. 2013, p. 155)

However, I think that GRADE suffers from a number of drawbacks which make it inadequate for solving the ranking problem. While I would argue that some of these drawbacks are almost insurmountable, I will only aim to establish the significantly weaker claim that taken together these drawbacks make GRADE inadequate for current ranking purposes.

(I) GRADE comes with a pre-described evidence hierarchy which only has three levels where evidence best suited to detect adverse drug reactions of newly released drugs (case reports, expert opinions), see Onakpoya et al. (2016), is entrenched as the least important kind of evidence. The DM hence cannot use her own favourite hierarchy. For example, DMs worried about safety of a newly released drug cannot move case reports and experts opinions up in the hierarchy. They will hence have a hard time to make safety signals weigh heavily.

(II) In GRADE, neither upgrading nor downgrading of evidence is clearly operationalised. For example, it is not clear in which cases inconsistency in the data is important, when data is sparse, when directness is major nor when the study quality is (seriously) limited.[8] There is no clear question formulated which upon answering would allow the DM to decide whether to upgrade or downgrade items of evidence. That is, no intuitive scales are given to which a DM may calibrate her judgements.

(III) GRADE does not make explicit – with or without further judgement – how the grading of evidence leads to a decision. For example, it is left open whether three studies downgraded for conflict of interest are as valuable as two studies with no conflict of interest, ceteris paribus. That is, it is not clear how the choice of up-grading or down-grading an item of evidence effects the decision making. In other words, GRADE does not offer a way to aggregate these (upgrade/downgrade) judgements to arrive at a resolution of the decision problem. This lack of clarity becomes the more severe the more the DM needs to aggregate heterogeneous judgements. The very recent 'Evidence to Decision Frameworks', see Alonso-Coello et al. (2016), offer great practical value in how to approach the decision problem in practise. However, they offer no way to resolve the problems discussed here.

---

[7] Without going into details here, GRADE and HiDAD use similar language to refer to different concepts and techniques.

[8] The ever-present difficulties from passing from a continuum to a discretisation (of judgements) are another layer of complexity (Guyatt et al. 2013, p. 154–155), which apply equally to GRADE and to HiDAD.

(IV) GRADE requires an explicit judgement for every study. For a comparison of two health interventions for which there exist a large number of studies, the number of choice points will be large. Given the large number of subjective choices feeding into the overall recommendation, the recommendation made appears to be based more on judgement than on evidence. This seems to be less than ideal in *evidence* based medicine which aims to reduce the number of subjective judgements.[9]

11.2.1.3 Medical Decision Support Literature for Evidence Amalgamation: A broader Perspective

There exist a number of decision support systems for medical decision making see, e.g., van Valkenhoef et al. (2013). A readable, though by no means complete overview, can be found in van Valkenhoef et al. (2013, pp. 463–464). These systems support the aggregation of medical data of one kind; normally RCTs only. The outputs of these systems often are (translatable into) preferences over health care interventions; based on evidence of a single kind. One particular strand of such support systems are rapid reviews which are tools for faster amalgamation of medical evidence, see Khangura et al. (2014). Practical issues arising from the need for amalgamation are discussed in Thomas et al. (2013).

A review of medical decision models developed in the UK during a seven years span (1997–2003) clearly expresses the need for the amalgamation of evidence of different kinds (Cooper et al. 2005, p. 249)

> Currently, the formal synthesis of evidence tends to be limited to RCT data and applied using standard meta-analysis techniques,[16] where appropriate. However, with a move towards identifying all relevant sources of evidence for model inputs, the application of generalized evidence synthesis methods that combine both randomized and non-randomized data are needed. Some methods for generalized evidence synthesis have been proposed,[17 – 19]...

All three models referred to in this quote are Bayesian models: they use precise numbers to represent the DM's epistemic uncertainty; see also the very recent (Landes et al. 2018). In the present setting, I want to avoid such precise quantification, as I shall explain in Sect. 11.2.2 below.

I take the apparent lack of proposals assigning cardinal weights to levels of evidence hierarchies to be evidence that decision makers are not able and comfortable with precise quantifications in this context.

Important episodes in the history of how the amalgamation of evidence (in medicine) became a field on its own is summarised in Chalmers et al. (2002, p. 25). Chalmers et al. identify DMs as a driving force behind this development

> Consumers of research have begun to point out more forcibly that "atomized", unsynthesized products of the research enterprise are of little help to people who wish to use research to inform their decisions.

---

[9]Under the construal of evidence offered here, expert clinical judgement is evidence, too.

Kelly & Moore provide philosophical underpinnings for the amalgamation of evidence

> The idea is that the greater the number of observations, the greater the degree of accuracy about that which is being observed and the greater the chance of the elimination of uncertainty. Further, if multiple studies with multiple results are pooled, then there is an even better chance of the results being averaged out in an optimally accurate way. This is a way of dealing with uncertainty. It recognises explicitly that single observations may be unreliable and that multiple observations offer protection against outliers. (Kelly and Moore 2012, p. 7)

11.2.2 Multi Criteria Decision Making for this Ranking Problem

I develop HiDAD with the goal in mind to help the DM decide on how to compare two alternatives A_k and A_i given the available evidence. Comparing the evidence for the overall outcome of A_k to that of A_i can be a daunting task, in particular, when, say, the RCTs speak in favor of A_k and the cohort studies and observational studies are more favourable towards A_i while case reports are only available for alternative A_k. In applications, such comparisons of alternatives with such heterogeneous alternatives are not straightforward.

Less difficult are comparisons between two alternatives when attention is *restricted* to one specified kind of evidence. I take it, that in the current ranking problem to be the case that the DM is *able and comfortable* to assess whether one kind of evidence supports the conclusion that (a) alternative A_k is much better than an alternative A_i, (b) an alternative A_k is better but not much better than A_i (c) alternative A_k is as good as A_i or (d) that the strength of the evidence of this kind which supports A_k over A_i is *incomparable* to the strength of the evidence of this kind which supports A_i over A_k. This last case, (d), might be deemed to be the case when there are a great number of contradictory cohort studies available investigating A_k but no cohort studies have been conducted for alternative A_i.

I call such a comparison restricted to evidence of one particular kind of evidence a *marginal* comparison.[10] The decision science literature refers to a decision problem which can be analysed in terms of different evaluation criteria as a multi-criteria decision problem. In this ranking problem, the different marginal evidential support relations play the role of evaluation criteria. The reader who is unacquainted with Multi Criteria Decision Making (MCDM) is referred to the excellent overviews

[10]There is no suggestion here that even such limited comparisons are always feasible. I would like to refer the reader to Footnote 14 for further discussion. To help determine marginal comparisons the DM may choose to avail herself to further (medical) decision aids. For example, (a) to assess (systematic reviews of) RCTs the DM may use decision support systems put forward in the medical decision literature which were discussed in Sect. 11.2.1.3, (b) means to make sense of multiple, possibly conflicting, expert opinions are put forward in the literature on judgement aggregation.

(Belton and Stewart 2002; Bouyssou et al. 2006; Figueira et al. 2005; Keeney and Raiffa 1993).[11]

Naturally, the interesting cases are the cases in which two alternatives A_i and A_k are judged differently in different criteria. Say, A_i is preferable to A_k in three criteria and A_k is preferred to A_i when evaluated on four other criteria. Not making any recommendations in the face of such heterogeneous evidence is often not an option for a regulatory body. So, a decision making process has to be followed which leads to recommendations for treating migraines.

11.2.2.1 Landscapes of Multi Criteria Decision Making

One way to divide the multi-criterial decision methodology landscape is along the following fault line. (A) Every alternative is first evaluated in every evaluation criterion, these evaluations are then aggregated into one final overall score for an alternative. The overall score of an alternative A_i is then compared to the overall score of other alternatives A_k in order to recommend (or come to) a decision [top-down]. (B) The second type of approach is to first draw all marginal comparisons between alternatives and then aggregate these marginal comparisons between alternatives to support the decision making process [bottom-up].

In general, it makes a difference whether a top-down or a bottom-up approach is pursued. For example, a top-down approach assigning every alternative a score which is a real number can easily determine a transitive, acyclic and complete ranking by ranking alternatives according to their score. A bottom-up approach on the other hand aggregates marginal comparisons to determine a ranking between pairs of alternatives. It is highly unclear how to aggregate the marginal comparisons into a ranking over alternatives which is transitive, acyclic and complete in general.

Another way to carve up the set of multi-criterial decision methods is by distinguishing (1) quantitative (cardinal) approaches from those (2) which refrain from using precise quantitative evaluations. Qualitative Decision Theory (QDT) is a different paradigm than the "usual" expected utility maximisation, the latter is due to Savage (Savage 1954). QDT supports decision making processes by non-cardinal approaches. The term "qualitative decision theory" dates back at least two decades to Boutilier (1994), Tan and Pearl (1994), refer to Doyle and Thomason (1999) for a very readable albeit slightly outdated overview and to Dubois et al. (2002) for a compact contrasting of QDT and expected utility maximisation. A recent overview of qualitative decision rules under uncertainty may be found in

[11] The term multi criteria decision *analysis* is also often found in the literature which is, at times, used interchangeably.

Dubois et al. (2009).[12] Normally, it is not advisable to use a qualitative approach, if quantification is sensible.

11.2.2.2 Purely Ordinal MCDM

A naïve qualitative decision theoretic approach to the decision problem represents the marginal comparisons of alternatives in terms of "better supported", "equally supported" or "less supported". That is, all marginal comparisons are purely ordinal, it only matters which alternative is better supported by a sub-body of evidence but how much greater the support is, is irrelevant. The arising decision problem is said to be a purely ordinal multi-criterial decision problem.

Dubois et al. showed in Dubois et al. (2002, Corollary 2) that, given very natural axioms formalising the essence of purely ordinal decision making, purely ordinal multi-criterial decision problems are *oligarchical*. That is, there have to exist oligarchies of criteria which decide the overall comparison of two alternatives. Only if all stronger oligarchies are indifferent between two alternatives, may criteria of lower importance influence the overall ranking of two alternatives.

As discussed in Sect. 11.2.1.1, such oligarchical decision making (trumping) is not appropriate for the decision problem under consideration. Rejecting one of the natural axioms of ordinal decision making in an ordinal decision making problem is also not an option I entertain here.

It seems that I am caught between a rock and a hard-place: On the one side I face a complex decision problem with heterogeneous evidence which only supports relatively weak comparative claims and on the other hand the results by Dubois et al. show that there are no decision rules appropriate for purely ordinal multi-criterial decision making.

---

[12] A reluctance to use precise numbers has not only manifested itself in the analysis of decision problems but also in the related, but by no means equivalent, epistemological problem of determining rational degrees of beliefs. This reluctance has given rise (among others) to the framework of imprecise probabilities, see Troffaes and de Cooman (2014) for a very recent treatment, Dempster-Shafer Theory, see Shafer (1976), and fuzzy logic as championed by Dubois and Prade, see Dubois et al. (1997). In Shafer and Srivastava (1990, p. 129), Shafer & Srivastava argued in favor of qualitative approaches [those with "fewer inputs" in their terminology] thusly:

> *When fewer inputs are required, we have a better chance of finding reasonably solid evidence on which to base these inputs, and thus, we have a better chance of producing an overall argument based on evidence rather than mere fancy.*

11.2.3 Methodology of this Approach

I take it as my starting point here, that given the available evidence for the ranking problem, the DM is not able and comfortable to give an overall assessment of an alternative nor is the DM able to articulate meaningful quantitative judgements.[13] Thus, HiDAD first draws comparisons and then aggregates marginal comparisons and it refrains from using sharp quantitative evaluations. It is the decision problem at hand which motivates HiDAD's location in the third quadrant (bottom-up and qualitative) of the coordinate system drawn in Fig. 11.1.

I escape the predicaments of purely ordinal multi-criterial decision making by allowing for a final ranking that may be cyclic, incomplete and intransitive and by

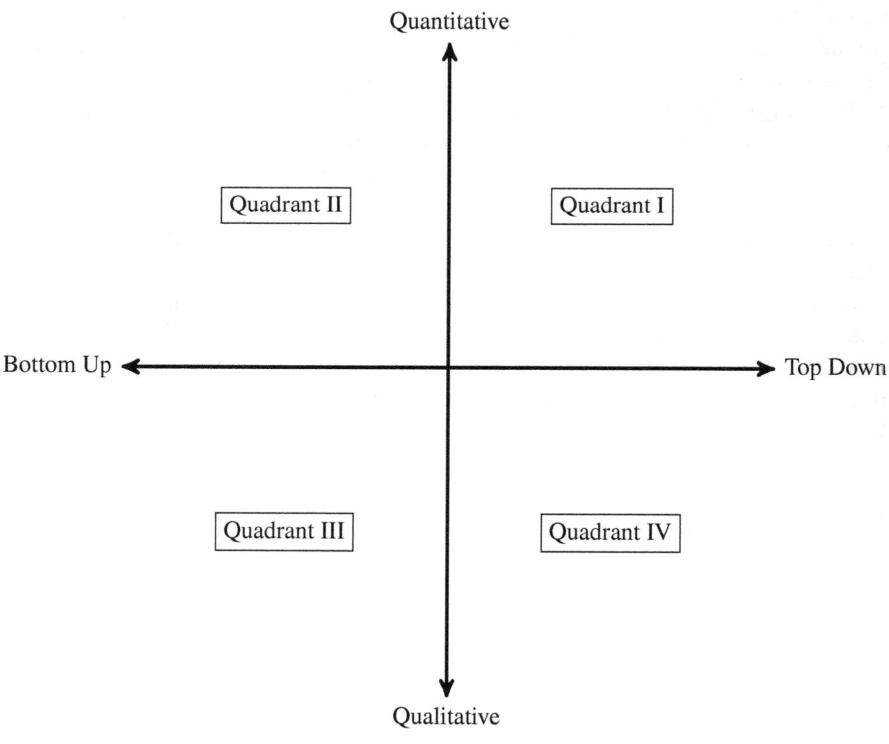

Fig. 11.1 Coordinate system for multi-criteria decision making

[13]An ideal rational agent, the protagonist of many a philosophical piece, may be in a position to give meaningful precise quantitative assessments. A (group of) human decision makers is in a significantly different epistemic situation. The applicability of HiDAD depending on the DM's situation is discussed in Sect. 11.6. Section 11.6.1 focuses on applications of HiDAD to other problems, while Sect. 11.6.2 provides conditions under which HiDAD should not be applied.

representing marginal comparisons in a more nuanced, i.e., not purely ordinal, way, see Sect. 11.3.1 for an operationalisation of the marginal comparisons.

11.3 HiDAD

As mentioned above, a number of evidence hierarchies have been put forward which disagree on the kinds of evidence to take into account and how to order kinds of evidence. HiDAD supposes that the DM has already established a hierarchical ordering of kinds of evidence, whatever this ordering may be as long as the ordering is strict and total, i.e., for every two different kinds of evidence one of them is deemed strictly more important than the other. The choice of a particular hierarchy is left to the DM. There is some genuine subjective choice on the DM's part here, e.g., whether case reports and expert clinical judgement combine for a level of the hierarchy, or if case reports and expert clinical judgement constitute levels on their own; in the latter case they also need to be ranked.[14] I opt for the second interpretation of the evidence hierarchy, see Sect. 11.2.1.1.

Having set-up the decision problem and spelled out the assumptions I now present HiDAD. To do so I shall suggest solutions to two questions: (1) How can one operationalise the marginal comparisons, i.e., how to elicit the marginal comparisons from the DM? (2) How can one aggregate marginal comparisons into an overall ranking?

The answer to the first question is found by appreciating the available evidence (Sect. 11.3.1), the second question is answered in Sects. 11.3.2 and 11.3.3. A schematic representation of HiDAD is given in Fig. 11.2.

The need for a systematic way of aggregating marginal comparisons arises from the motivation to design a decision aid which requires relatively few subjective judgements as inputs. Surely, there is no absolute need for an systematic aggregation in general and aggregation of marginal comparisons via case-by-case reasoning is possible. Such a method would lack the systematic approach presented here. Furthermore, if aggregation is performed via case-by-case inferences, then subjective

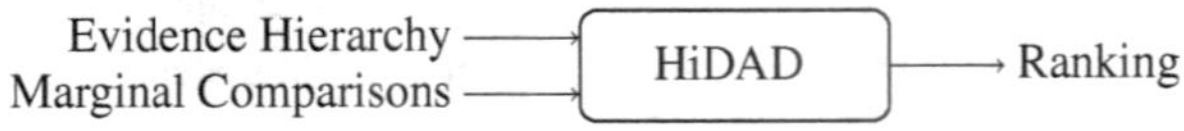

Fig. 11.2 Given an evidence hierarchy and all marginal comparisons (the input) HiDAD deterministically outputs the overall ranking

[14]Clearly, it may not always be the case that the DM is able and comfortable to do so. This does not mean that HiDAD is wrong, it simply means that it should not be applied in such a case. Mutatis mutandis, the same is true for further assumptions I make: If the assumptions I make do not hold in another concrete decision problem, then HiDAD should not be applied, see also Footnote 10.

judgements used to determine aggregates are neither explicit nor transparent to stakeholders.

11.3.1 Evidence Appreciation

The way evidence is appreciated is given in Tables 11.1 and 11.2 using the following bit of notation. Fixing the evidence hierarchy provided by the DM, I denote the sub-body of evidence consisting of all evidence of Level L by $\mathscr{E}_L$. Furthermore, define the sub-body of evidence of the most important L levels by $\mathbb{E}_L := \bigcup_{l=1}^{L} \mathscr{E}_l$. By definition, it holds that $\mathscr{E}_1 = \mathbb{E}_1$. The entire body of evidence is denoted by $\mathbb{E}$.

Intuitively, to obtain a marginal comparison on a level between two drugs one considers the available evidence of this kind of evidence and assesses only this relatively small body of evidence. This assessment results in a preference judgement which is elicited from the DM as indicated in Tables 11.1 and 11.2.

Every item of evidence is assigned a level L in the DM's evidence hierarchy. Ceteris paribus, the higher the level, the more influential the item of evidence will be for the decision. Furthermore, different studies on the same level may carry different evidential weights. For example, the external and internal validity as well as the effect size of the studies populating level L feed into the subjective judgement as to which of these six relations holds on Level L. So, while items of evidence only wield influence within their assigned level, they can do so to different degrees. For example, a set of RCTs may be all but ignored due to suspected biases, $A_1 \sim_2 A_2$, while a large group of observational studies on Level 3 may sway the DM to put $A_1 \gg_3 A_2$.

A DM worried about adverse drug reactions may set $A_i \gg_1 A_k$, only in case there are safety RCTs which clearly establish that A_i is safer than A_k. See Price et al. (2014) for a recent description of design and analysis of safety trials.

Table 11.1 Operationalisation of the marginal comparisons on the first, most important, level of evidence

Marginal Comparison	The sub-body of evidence of the first (most-important) level, $\mathscr{E}_1$, supports the conclusion that when A_i is compared to A_k that
$A_k \gg_1 A_i$	A_k is so much better than A_i that the DM prefers A_k over A_i no matter all the other evidence
$A_k >_1 A_i$	A_k is better than A_i but not that much better that $A_k \gg_1 A_i$ holds
$A_k \sim_1 A_i$	A_k and A_i are roughly equally good
$A_k \ll_1 A_i$	A_i is so much better than A_k that the DM prefers A_i over A_k no matter all other evidence
$A_k <_1 A_i$	A_i is better than A_k but not that much better that $A_k \ll_1 A_i$ holds
$A_k \bowtie_1 A_i$	A_k and are A_i incomparable

Table 11.2 Operationalisation of the marginal comparisons on all but the first level of evidence, $L > 1$

Marginal Comparison	The sub-body of evidence $\mathscr{E}_L$ $(L > 1)$ supports the conclusion that when A_i is compared to A_k that
$A_k \gg_L A_i$	$\mathbb{E}_{L-1}$ equally supports A_k and A_i, A_k is so much better than A_i that the DM prefers A_k over A_i no matter all other evidence in $\mathbb{E} \setminus \mathbb{E}_L$
$A_k >_L A_i$	A_k is better than A_i but not that much better that $A_k \gg_L A_i$ holds
$A_k \sim_L A_i$	A_k are and A_i are roughly equally good
$A_k \ll_L A_i$	$\mathbb{E}_{L-1}$ equally supports A_k and A_i, A_i is so much better than A_k that the DM prefers A_i over A_k no matter all other evidence in $\mathbb{E} \setminus \mathbb{E}_L$
$A_k <_L A_i$	A_i is better than A_k but not that much better that $A_i \gg_L A_k$ holds
$A_k \bowtie_L A_i$	A_k and are A_i incomparable

I take it that for every level L and every pair of alternatives A_i and A_k the DM is able and comfortable to judge that A_k and A_i stand in exactly one of the relations $\gg_L, >_L, \sim_L, \ll_L, <_L, \bowtie_L$, that is, I take it that these six relations are exhaustive and mutually exclusive.

Example 1 Consider a case with a large number of high-quality meta-analyses which all show a that A_1's treatment outcomes are significantly better than those of A_2. The DM may hence judge that $A_1 \gg_1 A_2$. HiDAD will hence recommend A_1 over A_2.

In a case in which the available expert opinions (Level 6) regarding A_3 vary from "widely effective and no side effects" over "as good as the standard treatment" to "ineffective with significant adverse drug reactions", while there are no expert opinions on the newly approved drug A_4 available, the DM may judge that $A_3 \bowtie_6 A_4$.

In case there is no evidence on Level L concerning A_k and also no evidence concerning A_i available, a reasonable DM will set $A_k \sim_L A_i$.

Regarding the first case in this example, HiDAD does not maintain that higher-level evidence always trumps lower-level evidence, but higher level evidence determines preferences in certain cases. For example, a large number of high-quality meta-analyses (internally and externally valid) which all show a clearly better outcome for A_1 than for A_2. In such a case, HiDAD maintains that A_1 is preferable to A_2.

There are cases in which there are no RCTs available where good observational studies alone are sufficient to swing the pendulum in one way; for a forceful argument of this point see the discussion of the ECMO case in Worrall (2007b, Sect. 2). Hence, it makes sense to formalise decisive evidence on the lower levels, say, $\gg_5$ in HiDAD.

Incomparability occurs when the evidence on Level L regarding two alternatives is strongly heterogeneous and the DM cannot say which alternatives is better supported or whether both alternatives are equally-well supported, see Aumann (1962) for an influential piece on the notion of incomparability. In such a case, it seems plausible to me, that higher level evidence is required to determine the DM's preferences. Hence, the lower-level evidence will be of no use, since the evidence which matters most is so heterogeneous and the lower-level evidence does not hold sufficient sway.

A DM has a hard time deciding between two alternatives, if she has equal preference for these two alternatives or if the alternatives are incomparable. In the first case, an assessment of the body of available evidence suggests that both alternatives are equally good; in the second case the available body of evidence does not allow such an assessment. So, while both cases lead to a difficult choice problem, they do so for different epistemic reasons. When further but less influential evidence becomes available, then the epistemic state in both cases may, in general, change in different ways; see further Table 11.5.

11.3.2 Aggregation of Comparisons

With the evidence appreciated I now turn to evidence amalgamation of different kinds by aggregating the marginal comparisons. I aggregate the marginal comparisons step-by-step and begin with the first, most important, level. Example 2 illustrates the notation for and the aggregation of comparisons.

I shall use relations $\succ_L, \succeq_L, \approx_L, \prec_L, \preceq_L, \perp_L$ to formalise aggregated comparisons on the first L levels combined. The intended meanings are given in Table 11.3.

With these meanings in place, it is straight-forward to define the first step of the aggregation. Given the marginal comparisons on Level 1 I put forward definitions of $\succ_1, \succeq_1, \approx_1, \prec_1, \preceq_1, \perp_1$ in Table 11.4.

Next, I inductively define the aggregated comparison relations up to Level L given two ingredients: (i) the marginal comparisons on Level L and (ii) the overall aggregated comparisons up to and including Level $L-1$ in Table 11.5. Broadly speaking: the more the marginal comparison on Level L is in favour of one alternative, the stronger the aggregated preference for this alternative on the first L levels combined; the stronger the aggregated preference on the first $L-1$ levels, the

Table 11.3 Meanings of the relations representing relative evidential support on the first L levels

Ranking up to and including level L	The sub-body of evidence $\mathbb{E}_L$ supports the conclusion that when A_i is compared to A_k that
$A_k \succ_L A_i$	A_k is so much better than A_i that the DM prefers A_k over A_i no matter all other evidence in $\mathbb{E} \setminus \mathbb{E}_L$
$A_k \succeq_L A_i$	A_k is better than A_i but not that much better that $A_k \succ_L A_i$ holds
$A_k \approx_L A_i$	A_k are and A_i are roughly equally good
$A_k \prec_L A_i$	A_i is so much better than A_k that the DM prefers A_i over A_k no matter all other evidence in $\mathbb{E} \setminus \mathbb{E}_L$
$A_k \preceq_L A_i$	A_i is better than A_k but not that much better that $A_k \prec_L A_k$ holds
$A_k \perp_L A_i$	A_k and are A_i incomparable

Table 11.4 Evidential support relations given marginal comparisons on the first and most important level

Marginal Comparison	Overall Ranking up to Level 1
$A_k \gg_1 A_i$	$A_k \succ_1 A_i$
$A_k >_1 A_i$	$A_k \succeq_1 A_i$
$A_k \sim_1 A_i$	$A_k \approx_1 A_i$
$A_k \ll_1 A_i$	$A_k \prec_1 A_i$
$A_k <_1 A_i$	$A_k \preceq_1 A_i$
$A_k \bowtie_1 A_i$	$A_k \perp_1 A_i$

Table 11.5 Overall evidential support relations given marginal comparisons on the Level L and aggregated marginal comparisons up to and including Level $L-1$. The fields marked by * are explained in detail in the text

	$A_k \succ_{L-1} A_i$	$A_k \succeq_{L-1} A_i$	$A_k \approx_{L-1} A_i$	$A_k \prec_{L-1} A_i$	$A_k \preceq_{L-1} A_i$	$A_k \perp_{L-1} A_i$
$A_k \gg_L A_i$	$A_k \succ_L A_i$	$A_k \succ_L A_i$	$A_k \succ_L A_i$	$A_k \prec_L A_i$	$A_k \approx_L A_i$	$A_k \perp_L A_i$
$A_k >_L A_i$	$A_k \succ_L A_i$	$A_k \succ_L A_i$	$A_k \succeq_L A_i$	$A_k \prec_L A_i$	*	$A_k \perp_L A_i$
$A_k \sim_L A_i$	$A_k \succ_L A_i$	$A_k \succeq_L A_i$	$A_k \approx_L A_i$	$A_k \prec_L A_i$	$A_k \preceq_L A_i$	$A_k \perp_L A_i$
$A_k \ll_L A_i$	$A_k \succ_L A_i$	$A_k \approx_L A_i$	$A_k \prec_L A_i$	$A_k \prec_L A_i$	$A_k \prec_L A_i$	$A_k \perp_L A_i$
$A_k <_L A_i$	$A_k \succ_L A_i$	*	$A_k \preceq_L A_i$	$A_k \prec_L A_i$	$A_k \prec_L A_i$	$A_k \perp_L A_i$
$A_k \bowtie_L A_i$	$A_k \succ_L A_i$	$A_k \succ_L A_i$	$A_k \perp_L A_i$	$A_k \prec_L A_i$	$A_k \prec_L A_i$	$A_k \perp_L A_i$

less influence the marginal comparison on Level L wields; if there is overwhelming evidence on the first $L-1$ levels, then the less important evidence on Level L does not allow for a change of overall preference; in case of conflicting evidence the evidence on the first $L-1$ levels matters more but does not trump lower level evidence.

I now briefly discuss the entries of Table 11.5.

By the definition of $A_k \succ_{L-1} A_i$ (conclusive evidence for preferring A_k over A_i on the most important $L-1$ levels, $\mathbb{E}_{L-1}$) and $A_k \prec_{L-1} A_i$ (conclusive evidence for preferring A_i over A_k), the marginal comparisons on Level L do not have any influence on the aggregates, that is, $A_k \succ_L A_i$ (respectively $A_k \prec_L A_i$) always holds – no matter the evidence on Level L [second and fifth column of Table 11.5]. This is in line with the definition of $\succ_{L-1}$ and $\prec_{L-1}$ in Table 11.3.

For alternatives which are equally supported by the sub-body of evidence $\mathbb{E}_{L-1}$ the marginal comparisons on Level L determine the aggregate preference [fourth column of Table 11.5].

If the sub-body of evidence $\mathbb{E}_{L-1}$ is such that the strength of the evidence which supports A_k over A_i cannot be compared to the strength of evidence which supports A_i over A_k, then no evidence of less importance can break this incomparability [last column of Table 11.5].

I now address the third column of Table 11.5; and by symmetry also the second to last column. If $\mathbb{E}_{L-1}$ supports A_k over A_i but not as much such that $A_k \succ_{L-1} A_i$ holds, then the further evidence on Level L which supports A_k over A_i is enough to tip the balance in favor of A_k over A_i for good. In case the evidence on Level L equally supports A_k and A_i, then the overall ranking of A_k and A_i will remain unchanged. On the other hand, strong evidence in favor of A_i over A_k ($A_k \ll_L A_i$)

will lead to a modification of the overall ranking of A_k and A_i; I here suggest to change it to $A_k \approx_L A_i$.

Bottom entry of the third column: A sub-body of evidence $\mathscr{E}_L$ which does not allow a comparison between the evidence in $\mathscr{E}_L$ which supports A_k over A_i and the evidence in $\mathscr{E}_L$ which supports A_i over A_k renders marginal comparisons on lower levels meaningless. It is hence at this Level L that I will have to set the overall ranking of A_k and A_i. Given that $\mathbb{E}_{L-1}$ supports A_k over A_i it appears sensible to set $A_k \succ_L A_i$.

The intuition behind this aggregation is the same as above, in order to break ties between incomparable alternatives higher level evidence is required. In this case, this means that the lower level evidence is – sadly – ignored and the higher level evidence carries the day.

*: In case there is weak evidence in favor of A_i over A_k ($A_k <_L A_i$), then this should not automatically result in a change of the overall ranking since the evidence on Level L is less important than the evidence on higher levels.

I suggest the following procedure to determine *: Let $1 \leq j \leq L-1$ be the level at which the overall ranking switches for the last time in favor of A_k, formally let $1 \leq j \leq L-1$ be maximal such that $A_k \succeq_j A_i$ holds but $A_k \succeq_{j-1} A_i$ fails to hold. If for all $j+1 \leq r \leq L-1$ $A_k \sim_r A_i$ holds, then there is not enough evidence to change the overall ranking and hence $A_k \succeq_L A_i$ holds. But if there exists one $j+1 \leq r \leq L-1$ such that $A_k <_r A_i$ holds and if for all other $j+1 \leq s \leq j-1$ either $A_k \sim_s A_i$ or $A_k <_s A_i$ hold, then $A_k \approx_L A_i$ holds as there are two levels of evidence which support A_i over A_k.[15] For an example see the example below for A_5 and A_6.

Example 2 Consider an evidence hierarchy with five levels. Comparing two alternative drugs $A_1, A_2 \in \mathscr{A}$ such that $A_1 >_1 A_2$ and $A_1 >_2 A_2$ one finds $A_1 \succeq_1 A_2$, $A_1 \succ_2 A_2$ and thus $A_1 \succ_5 A_2$. That is, based on the entire body of evidence, HiDAD ranks A_1 higher than A_2, no matter the evidence on the three lower levels. This may reflect the DM's subjective judgement that, say, RCTs (Level 1) and cohort studies (Level 2) combined, are deemed much more important than case reports (Level 3), expert testimony (Level 4) and animal data (Level 5) combined.

For alternatives A_3, A_4 and the same hierarchy with $A_3 >_1 A_4$, $A_3 \ll_2 A_4$, $A_3 <_3 A_4$, $A_3 <_4 A_4$ and $A_3 <_5 A_4$ one has $A_3 \succeq_1 A_4$, $A_3 \approx_2 A_4$, $A_3 \preceq_3 A_4$, $A_3 \prec_4 A_4$ and finally $A_3 \prec_5 A_4$. Hence, HiDAD ranks A_4 higher than A_3. This is an example of non-trumping: the evidence on the first level is not strong enough to suppress the evidence on the lower levels which suggests that A_4 is preferable to A_3. Lower level evidence has changed the DM's preferences.

Concerning *: Assume that $A_5 >_1 A_6$ and thus $A_5 \succeq_1 A_6$ and assume furthermore that $A_5 <_2 A_6$. Since the evidence on the second level matters less than the evidence on the first level, it seems right to set $A_5 \succeq_2 A_6$. In case there is further evidence supporting A_6 over A_5 in the form of $A_5 <_3 A_6$, then my

[15]Note that all other marginal comparisons lead to a change of the overall ranking of A_k and A_i and hence all possible cases have been considered here.

proposal is to set $A_5 \approx_3 A_6$. That is, the lower level evidence has changed the ranking from a moderately more supported, $\succeq_1$, to equally supported, $\approx_3$. HiDAD's recommendation will, in this example, depend on the lower level evidence.

If and only if either $A_k \preceq_{L-1} A_i$, $A_k \succeq_{L-1} A_i$ or $A_k \approx_{L-1} A_i$ hold, then $A_k \gg_L A_i$, $A_k \ll_L A_i$ and $A_k \bowtie_L A_i$ will lead to a change of the aggregated comparisons. So, strong evidence on lower levels plays an important role in HiDAD.

11.3.3 The Ranking

Having determined evidential support relations on the entire body of evidence, $\mathbb{E}$, I can now determine the overall ranking of alternatives. That is, I now say which recommendations the body of evidence $\mathbb{E}$ supports according to HiDAD. The final ranking relations are given in Table 11.6. I use $\prec, \approx, \succ, \perp$ to denote these mutually exclusive and exhaustive relations.

Based on the final ranking the DM then makes the guideline recommendation to doctors treating migraines in non-pregnant adults:

> Use the *maximal elements* of the final ranking.

A maximal element is an alternative A_k such that there is no alternative $A_i \in \mathscr{A}$ with $A_k \prec A_i$. The rationale for this recommendation is simple: The maximal elements appear to be best, given all the available evidence, see Fig. 11.3 for a visualisation.

In case there are no maximal elements, a number of plausibly sensible courses for actions suggest themselves. I shall here only list some these options without endorsing a particular one: (i) Exploit information encoded in $\succeq_L$, (ii) recommend all alternatives A_k for which the number of other alternatives A_i which are ranked lower ($A_k \succ A_i$) is maximal or (iii) recommend all but the minimal elements in the final ranking. An alternative A_k is not minimal in the final ranking, if and only if there exists an alternative A_i which is ranked lower than A_k, $A_k \succ A_i$.

Table 11.6 The overall ranking of pairs of alternatives based on the aggregation of all marginal comparisons

Overall support	The entire body of evidence $\mathbb{E}$ supports the conclusion that
$A_k \succ_L A_i$	A_k is better than A_i, $A_k \succ A_i$
$A_k \succeq_L A_i$	A_k is better than A_i, $A_k \succ A_i$
$A_k \approx_L A_i$	A_k are A_i equally good, $A_k \approx A_i$
$A_k \prec_L A_i$	A_i is better than A_k, $A_k \prec A_i$
$A_k \preceq_L A_i$	A_i is better than A_k, $A_k \prec A_i$
$A_k \perp_L A_i$	The strength of the evidence in $\mathbb{E}$ supporting A_k over A_i is incomparable to the strength of evidence in $\mathbb{E}$ which supports A_i over A_k. A_k and A_i are incomparable, $A_k \perp A_i$

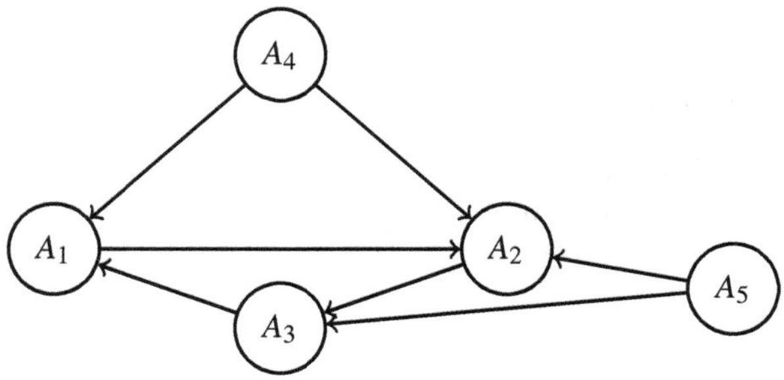

Fig. 11.3 Example of a final ranking, directed arrows are intended to mean that the target of the arrow is ranked lower than alternative from which the arrow originates. In the depicted configuration, A_4 and A_5 are the only maximal elements (no arrows end there). There are no minimal alternatives, an arrow originates from every alternative. This ranking is cyclic (A_1, A_2, A_3 form a cycle) and intransitive ($A_4 \succ A_2$ and $A_2 \succ A_3$ both hold, but $A_4 \succ A_3$ does not hold)

Example 3 Continuing Example 2: HiDAD recommends A_1 over A_2 and A_4 over A_3. Given the ranking depicted in Fig. 11.3 HiDAD recommends the two alternatives A_4 and A_5.

11.4 Properties of HiDAD

I now briefly discuss properties of the final ranking depending on the marginal comparisons.

11.4.1 Compatibility

Let $A_1, A_2, A_3 \in \mathscr{A}$ be pairwise different alternatives and consider the set of marginal comparisons of A_1 to A_2 and the marginal comparisons of A_1 to A_3. If for all these marginal comparisons A_1 does better when compared to A_3 than when it is compared to A_2, then overall ranking of A_1 relative to A_3 ought to be better or equal than the ranking of A_1 relative to A_2. I now show that this does indeed hold.

To simplify notation I introduce the concept of a better or equal marginal comparison. For a fixed level L, I say that the set of better or equal comparison(s) of

- $\gg_L$ is $\{\gg_L\}$,
- $>_L$ is $\{\gg_L, >_L\}$,
- $\sim_L$ is $\{\gg_L, >_L, \sim_L\}$,
- $<_L$ is $\{\gg_L, >_L, \sim_L, <_L\}$,
- $\ll_L$ is $\{\gg_L, >_L, \sim_L, <_L, \ll_L\}$.

If for all levels L the marginal comparison of A_1 relative to A_3 is better or equal than the comparison of A_1 relative to A_2, I then say that the ordered pair (A_1, A_3) is *marginally better or equal* than the ordered pair (A_1, A_2).

Proposition 1 (Compatibility) *If for all levels L neither $A_1 \bowtie_L A_2$ nor $A_1 \bowtie_L A_3$ hold and (A_1, A_3) is marginally better or equal than (A_1, A_2), then*

- $A_1 \succ A_2$ *entails* $A_1 \succ A_3$,
- $A_1 \approx A_2$ *entails* $A_1 \succ A_3$ *or* $A_1 \approx A_3$ *and*
- $A_1 \prec A_2$ *entails* $A_1 \succ A_3$, $A_1 \approx A_3$ *or* $A_1 \prec A_3$.

In plain English, if no marginal is worse nor incomparable, then the overall ranking is no worse.

11.4.2 Further Properties

Properties often discussed in ranking tasks are *transitivity* and *acyclicity*. In general, the body of evidence is such that the marginal comparisons are not transitive – here understood in the sense that if $A_1 >_L A_2$ and $A_2 >_L A_3$ hold, then it can well be the case that one of $A_1 \approx A_3$, $A_1 <_L A_3$, $A_1 \bowtie_L$ or, in exceptional cases, $A_1 \ll_L A_3$ holds.[16] It should hence come as no surprise that the final ranking is, in general, not transitive. Furthermore, it may be the case that the final ranking is cyclic. Intransitive and/or cyclic final rankings can arise even if all marginal comparisons are transitive and acyclic.

Indifference over alternatives is often taken to be an *equivalence relation*. Here, $\perp$ and $\approx$ are by definition symmetric neither is, in general, transitive. Again, this is to be expected in this decision problem with heterogeneous evidence.

Finally, consider a hierarchy and two alternatives A_1, A_2 where there is no evidence for A_1 nor evidence for A_2 on Level $L \geq 2$. We have $A_1 \sim_L A_2$. Now imagine a hypothetical scenario in which the same information is a available but a different hierarchy is used. This new hierarchy is obtained from the old one by ignoring Level L. That is, all information previously categorised as evidence of Level L is now not considered to be evidence at all. In this hypothetical scenario, HiDAD will rank A_1 and A_2 the same way as in the first scenario. This due to the fact, that $\sim_L$ in does not influence aggregated comparisons (Table 11.5).

In a case where there is a level of evidence in the hierarchy for which there is no evidence available at all one might as well remove this level from the hierarchy.

[16] I think that such cases are *very* rare. However, should the DM assess the evidence thusly, then there have to be good reasons for doing so.

11.5 Modifications of HiDAD

There are a number of ways in which HiDAD can be modified to better suit the needs of the DM in a particular decision problem. Firstly, there are different ways in which marginal comparisons could be represented and/or operationalised (modification of Tables 11.1 and 11.2). More nuanced (finer) representations of marginal comparisons are one option, *if* the DM is able and comfortable with making such more nuanced assessments. Invariably, this will lead to a more involved aggregation of marginal comparisons. In case the evidence does not support more nuanced marginal comparisons but only allows for weaker comparisons I would like to recall the problems purely ordinal multi-criterial decision making faces, see Sect. 11.2.2.2.

Secondly, (different) aggregated marginal comparisons may have different meanings (modification of Tables 11.3 and 11.4). Thirdly, as already suggested above there is room for different ways of filling the fields marked with * in Table 11.5 and also other fields in Table 11.5. For example, the last column of Table 11.5 devoted to incomparability could be modified to allow lower level evidence to play a larger role.

Fourthly, the DM could recommend all non-minimal elements of the final ranking rather than all maximal alternatives or use some other procedure discussed in Sect. 11.3.3.

11.6 Applicability of HiDAD

I now briefly discuss which other problems HiDAD may be applied to and in which cases it does not make sense to apply HiDAD.

11.6.1 Potential Further Areas of Application

In principle, in all other problems which require amalgamation the application of HiDAD may make sense, e.g., judgement aggregation (of experts) where different judgements have different levels of credibility and/or expertise, evidence amalgamation (in general), preference aggregation (in MCDM). There is no reason to think that these other potential areas of application have to be in the medical domain.

11.6.2 Restricted Scope for Application

There are cases in which the application of HiDAD is not a sensible idea, e.g., when precise quantification is possible or if the importance of the different levels of evidence is judged to vary only incrementally.

I also want to advise against the application of HiDAD in cases where there are less than a handful of levels or more than ten levels. If there are very few evidence levels, then the most important level evidence almost trumps: If $A_i >_1 A_k$ holds, then in only very few cases will A_i not be ranked higher than A_k.

If there are very many levels, then low level evidence is given too much weight in certain instances. For example, for $A_1 >_1 A_2$, $A_1 \sim_2 A_2, \ldots, A_1 \sim_{47} A_2$, $A_1 <_{48} A_2$, $A_1 <_{49} A_2$ and $A_1 <_{50} A_2$ (assuming that the DM has specified 50 levels) HiDAD ranks A_2 higher than A_1. This does not seem right since evidence on Levels 47 – 50 is much less important than evidence on Level 1. The "obvious fix" in this situation is to modify Table 11.5 by making the entries of the table to also depend on the marginal comparisons of all more important levels and not just their aggregate. I do not endorse any such "fix".

11.7 Conclusions

In this chapter, I have put forward a decision aid for ranking problems in medicine which led me to suggest answers to the three questions posed in the introduction: "What constitutes evidence in medicine?", "How does one amalgamate evidence for medical decision making?" and "How important is the most important kind of evidence?". Thr answer to the first question was that, for current purposes, evidence is all information which can on its own or jointly with other information conceivably influence the decision problem at hand.[17] The second and third questions were addressed in Sects. 11.3.2 and 11.3.3 by the aggregation of marginal comparisons.

I now discuss the claims I base this chapter on and point out some claims I do not make.

11.7.1 Claims

11.7.1.1 Normative Claims Made

The *normative* claims this approach hinges on are α) all available evidence ought to be taken into account where evidence is to be understood widely (no trumping),

[17] I do think that this definition of evidence could be of use much more generally. I shall here be content with keeping the focus on the discussed ranking problem.

β) the most appropriate framework ought to be used to tackle the ranking problem, γ) precise quantification ought to be avoided, if the DM is not able or comfortable with it and δ) overall preferences between alternatives ought to be determined from assessments that are meaningful to the DM.

This approach obeys these claims by α) taking all evidence of all kinds featuring in the specified evidence hierarchy into account and by avoiding trumping, β) the application of MCDM methodology incorporating evidence hierarchies, γ) using a qualitative approach δ) which only relies on qualitative marginal comparisons to construct the final ranking. The marginal comparisons are elicited from the DM by posing simple questions concerning the decision problem with a clear intuitive meaning.

11.7.1.2 Claims Not Made

I do not claim that any evidence hierarchy is normatively correct for all decision problems nor do I relativise this claim to the particular decision problem. The choice of an appropriate evidence hierarchy is left to the DM which entails determining the relevant kinds of evidence and ordering these by their importance.

Furthermore, no claims are made that the marginal comparisons or their aggregation are normatively correct. In Sect. 11.5, I talked about alternative ways of formalising marginal comparisons and their aggregations.

I do not claim that there exists a quantitative model which outputs the same final ranking as HiDAD.[18] The simplest such quantitative model would use scaling constants reflecting the importance of a criterion, see for example Billaut et al. (2010, p. 250) and Belton (2009). To make the use of scaling constants sensible, one has to be clear about the scale (or normalisation procedure) of a criterion otherwise the model makes no sense and is bound to give absurd results. According to Billaut et al. (2010, p. 250) this non-sensicality is "invariably taught in any basic course on MCDM". I have no idea what a proper scale in the ranking problem is and hence do not entertain the use of a qualitative model using scaling constants. To me, gerrymandering a more complex quantitative model which fits with HiDAD for the only sake of cooking up such a model does not appear like a good use of time.

HiDAD is not a comprehensive decision aid which is fully systematic and free of all human judgement: I did not give any guidelines of how to determine the levels of the evidence hierarchy nor how to rank these levels. Furthermore, there is no suggestion of how exactly to differentiate between, say, $A_k >_L A_i$ and $A_k \sim_L A_i$. Rather, HiDAD is a decision heuristic.

[18] If precise quantification were possible, then I do recommend to use these numbers. However, I supposed that precise quantification is not possible and hence went down a qualitative path. For cases in which precise quantification of the importance of criteria is feasible the reader is referred to Mussen et al. (2009), Tervonen et al. (2011).

Furthermore, HiDAD does not model all important evidential inferences in medicine. For example, it is here proposed, that items of evidence are put into categories (levels) and that interaction across levels *only* happens via the aggregation of preferences delineated in Table 11.5. In reality, new pharmacogenomic evidence showing that there are important genetic subgroups within an RCT trial arm, may lead to a reversed interpretation of the trial results and hence a reversed marginal comparison. This kind of inference is aptly depicted in Clarke et al. (2014, Fig. 1).

11.7.2 Assessing Properties of HiDAD

As with any decision aid, HiDAD is not universally applicable, see Sect. 11.6. While a restricted applicability is surely not ideal, I think that the kind of decision problems HiDAD can be applied to is important enough to warrant this approach.

As already indicated by its name, I think that compatibility (Sect. 11.4.1) is desirable. Failure of acyclicity and full equivalence of the indifference relations $\perp$ and $\approx$ (Sect. 11.4.2) cannot be blamed on HiDAD. Rather, it is a consequence of the DM's less-than-godlike evidential situation.

HiDAD is hierarchy neutral, in that HiDAD is not tied to any particular hierarchical ordering of the levels of evidence and may be applied whatever the DM deems the appropriate hierarchy at the time. I take this neutrality to be desirable. Those worried about the under-appreciation of "lower level" evidence have two ways of incorporating their thinking in HiDAD: (1) Re-arrange the (ordering of the) levels of the evidence hierarchy used. (2) On the more important levels, raise the bar for setting $A_k \gg_L A_i$.

HiDAD demands the incorporation of all evidence in a systematic way, as long as all items of evidence are of one kind which is ranked in the hierarchy of levels of evidence. Supposing that the DM can (be aided to) specify such a hierarchy, this requirement of total evidence ought to be a corner stone in evidence based medicine. I hence think that this requirement is desirable.

HiDAD outputs the final ranking which is completely determined by a fixed hierarchy of levels of evidence, the marginal comparisons and their aggregations which all are, relatively, easily articulated. This makes the entire decision making process transparent, explicit and reproducible by erasing all further room for subjective choice and arbitrariness.

Unlike GRADE, the operationalisation of marginal comparisons in HiDAD is in terms of a simple questions with intuitive implications for the decision problem. Furthermore, the number of judgements required to determine the final ranking of two alternatives given the body of available evidence and the evidence levels and their importance equals the number of levels of evidence. Since this number is (roughly) between five and ten I think that the final ranking of a pair of alternatives is not swamped by the number of human judgements.

To me, it is desirable that levels for which there is no evidence can be ignored.

Finally, I do not take issue with decision heuristics. Gerd Gigerenzer, for example, has long argued that decision heuristics make a good deal of sense and often outperform the most sophisticated approaches, the reader finds his arguments in, e.g., Gigerenzer and Gaissmaier (2011), Marewski and Gigerenzer (2012). Decision heuristics, such as GRADE and HiDAD aim at supporting decision making processes in the actual world; they were designed with the intention to outperform intuitive decisions or other formal methods. In the design process there are a number design choices to be made (GRADE: number of up- and down-grades possible, when to up- or down-grade, HiDAD: choice of marginal comparisons, their aggregations). There is no normatively correct way of making these design choices. The outputs of GRADE and HiDAD are hence not normatively correct in a strong sense.

Whether or not HiDAD is successful cannot be assessed based on purely academic consideration, it will depend on the experiences of the stakeholders and – most importantly – the patients who are treated by taking the guidelines produced by applying HiDAD into account.

Acknowledgements The idea for this chapter arose when the author was a research assistant on a project on the "Optimal design of biofuel production by microalgae" at INRA, UR0050, Laboratoire de Biotechnologie de l'Environnement (2009–2010). Progress all but ceased when the author joined the "From objective Bayesian epistemology to inductive logic" AHRC-funded project at the University of Kent. The great majority of the work was carried out after the author joined the ERC-funded project "Philosophy of Pharmacology: Safety, Statistical Standards, and Evidence Amalgamation" (grant 639276) at the LMU Munich. Currently, the author is the principal investigator of the project *Evidence and Objective Bayesian Epistemology* funded by the German Research Council. Regarding this chapter, the author benefited from a number of discussions with Seamus Bradley, Ricardo Büttner, Teddy Groves, Adam LaCaze, Laurent Lardon, Barbara Osimani, Roland Poellinger, David Teira and Jon Williamson as well as the members of the Environmental Lifecycle and Sustainability Assessment group. He would also like to thank an anonymous referee for the European Journal of Operational Research and three anonymous reviewers for this volume as well as the editors of this volume for their thoughtful comments and insights.

References

Alonso-Coello, P., Schünemann, H. J., Moberg, J., Brignardello-Petersen, R., Akl, E. A., Davoli, M., Treweek, S., Mustafa, R. A., Rada, G., Rosenbaum, S., Morelli, A., Guyatt, G. H., & Oxman, A. D. (2016). GRADE Evidence to Decision (EtD) frameworks: A systematic and transparent approach to making well informed healthcare choices. 1: Introduction. *BMJ, 353*. https://doi.org/10.1136/bmj.i2016.

Aumann, R. J. (1962). Utility theory without the completeness axiom. *Econometrica, 30*(3), 445–462.

Belton, V. (2009). Matters of weight. *Journal of Multi-Criteria Decision Analysis, 15*, 109–110.

Belton, V., & Stewart, T. J. (2002). *Multiple criteria decision analysis: An integrated approach*. Boston: Springer.

Bertamini, M., & Munafó, M. R. (2012). Bite-size science and its undesired side effects. *Perspectives on Psychological Science, 7*(1), 67–71.

Bes-Rastrollo, M., Schulze, M. B., Ruiz-Canela, M., & Martinez-Gonzalez, M. A. (2013). Financial conflicts of interest and reporting bias regarding the association between sugar-sweetened beverages and weight gain: A systematic review of systematic reviews. *PLoS Medicine, 10*(12), 1–9.

Billaut, J.-C., Bouyssou, D., & Vincke, P. (2010). Should you believe in the Shanghai ranking? An MCDM view. *Scientometrics, 84*, 237–263.

Borm, G. F., Lemmers, O., Fransen, J., & Donders, R. (2009). The evidence provided by a single trial is less reliable than its statistical analysis suggests. *Journal of Clinical Epidemiology, 62*(7), 711–715.

Boström, H., Andler, S. F., Brohede, M., & Johansson, R. (2007). On the definition of information fusion as a field of research. Technical report, University of Skövde.

Boutilier, C. (1994). Toward a logic for qualitative decision theory. In *Proceedings of KR* (Vol. 94, pp. 75–86). San Fransisco: Morgan Kaufmann.

Bouyssou, D., Marchant, T., Pirlot, M., Tsoukiàs, A., & Vincke, P. (2006). *Evaluation and decision models with multiple criteria*. New York: Springer.

Button, K. S., Ioannidis, J. P. A., Mokrysz, C., Nosek, B. A., Flint, J., Robinson, E. S. J., & Munafo, M. R. (2013). Power failure: Why small sample size undermines the reliability of neuroscience. *Nature Reviews Neuroscience, 14*, 365–376.

Canadian Task Force on the Periodic Health Examination. (1979). The periodic health examination. *Canadian Medical Association Journal, 121*(9), 1193–1254.

Carnap, R. (1947). On the application of inductive logic. *Philosophy and Phenomenological Research, 8*(1), 133–148.

Cartwright, N. (2007). Are RCTs the gold standard? *Biosocieties, 2*(1), 11–20.

Cartwright, N., & Munro, E. (2010). The limitations of randomized controlled trials in predicting effectiveness. *Journal of Evaluation in Clinical Practice, 16*(2), 260–266.

Chalmers, I., Hedges, L. V., & Cooper, H. (2002). A brief history of research synthesis. *Evaluation & the Health Professions, 25*(1), 12–37.

Chan, A.-W., & Altman, D. G. (2005). Epidemiology and reporting of randomised trials published in PubMed journals. *The Lancet, 365*(9465), 1159–1162.

Clarke, B., Gillies, D., Illari, P., Russo, F., & Williamson, J. (2013). The evidence that evidence-based medicine omits. *Preventive Medicine, 57*(6), 745–747.

Clarke, B., Gillies, D., Illari, P., Russo, F., & Williamson, J. (2014). Mechanisms and the evidence hierarchy. *Topoi, 33*, 339–360.

Cooper, N., Coyle, D., Abrams, K., Mugford, M., & Sutton, A. (2005). Use of evidence in decision models: An appraisal of health technology assessments in the UK since 1997. *Journal of Health Services Research & Policy, 10*(4), 245–250.

Doll, R., & Peto, R. (1980). Randomised controlled trials and retrospective controls. *British Medical Journal, 280*, 44.

Doyle, J., & Thomason, R. H. (1999). Background to qualitative decision theory. *AI Magazine, 20*(2), 55–68.

Dubois, D., Fargier, H., & Perny, P. (2002). On the limitations of ordinal approaches to decision-making. In D. Fensel, F. Giunchiglia, D. L. McGuinness, & M.-A. Williams (Eds.), *Proceedings of KR* (pp. 133–146). San Fransisco: Morgan Kaufmann.

Dubois, D., Fargier, H., & Prade, H. (1997). Decision-making under ordinal preferences and comparative uncertainty. In *Proceeding of UAI* (pp. 157–164).

Dubois, D., Fargier, H., Prade, H., & Sabadin, R. (2009). A survey of qualitative decision rules under uncertainty. In *Decision-making Process* (Chapter 11, pp. 435–473). London: Wiley Online Library.

Earman, J. (1992). *Bayes or bust?* MIT. Cambridge, MA

Etz, A., & Vandekerckhove, J. (2016). A bayesian perspective on the reproducibility project: Psychology. *PLoS ONE, 11*(2).

Evans, D. (2003). Hierarchy of evidence: A framework for ranking evidence evaluating healthcare interventions. *Journal of Clinical Nursing, 12*(1), 77–84.

Every-Palmer, S., & Howick, J. (2014). How evidence-based medicine is failing due to biased trials and selective publication. *Journal of Evaluation in Clinical Practice, 20*(6), 908–914.

Figueira, J., Greco, S., & Ehrgott, M. (2005). *Multiple criteria decision analysis: State of the art surveys*. Springer.
Gigerenzer, G., & Gaissmaier, W. (2011). Heuristic decision making. *Annual Review of Psychology, 62*(1), 451–482.
GRADE Working Group (2004). Grading quality of evidence and strength of recommendations. *British Medical Journal, 328*(7454), 1490–1494.
Guyatt, G., Oxman, A. D., Sultan, S., Brozek, J., Glasziou, P., Alonso-Coello, P., Atkins, D., Kunz, R., Montori, V., Jaeschke, R., Rind, D., Dahm, P., Akl, E. A., Meerpohl, J., Vist, G., Berliner, E., Norris, S., Falck-Ytter, Y., & Schünemann, H. J. (2013). GRADE guidelines: 11. Making an overall rating of confidence in effect estimates for a single outcome and for all outcomes. *Journal of Clinical Epidemiology, 66*(2), 151–157.
Guyatt, G. H., Oxman, A. D., Schünemann, H. J., Tugwell, P., & Knottnerus, A. (2011). GRADE guidelines: A new series of articles in the journal of clinical epidemiology. *Journal of Clinical Epidemiology, 64*, 380–382.
Guyatt, G. H., Oxman, A. D., Vist, G. E., Kunz, R., Falck-Ytter, Y., Alonso-Coello, P., & Schünemann, H. J. (2008). GRADE: An emerging consensus on rating quality of evidence and strength of recommendations. *British Medical Journal, 336*(7650), 924–926.
Horton, R. (2004). Vioxx, the implosion of Merck, and aftershocks at the FDA. *The Lancet, 364*(9450), 1995–1996.
Howick, J. H. (2011). *The philosophy of evidence-based medicine*. Chichester: Blackwell.
Howick, J., et al. (2011). *The Oxford 2011 levels of evidence*. Oxford: Oxford Centre for Evidence Based Medicine. http://www.cebm.net/wp-content/uploads/2014/06/CEBM-Levels-of-Evidence-2.1.pdf.
Jüni, P., Nartey, L., Reichenbach, S., Sterchi, R., Dieppe, P. A., & Egger, M. (2004). Risk of cardiovascular events and rofecoxib: Cumulative meta-analysis. *The Lancet, 364*(9450), 2021–2029.
Keeney, R. L., & Raiffa, H. (1993). *Decisions with multiple objectives*. Cambridge: Cambridge Books. Cambridge University Press.
Kelly, M. P., & Moore, T. A. (2012). The judgement process in evidence-based medicine and health technology assessment. *Social Theory & Health, 10*(1), 1–19.
Kelly, T. (2015). Evidence. In E. N. Zalta (Ed.), *The Stanford Encyclopedia of Philosophy* (Fall 2014 ed.). Stanford: Metaphysics Research Lab, Stanford University. https://plato.stanford.edu/cgi-bin/encyclopedia/archinfo.cgi?entry=evidence
Khangura, S., Polisena, J., Clifford, T. J., Farrah, K., & Kamel, C. (2014). Rapid review: An emerging approach to evidence synthesis in health technology assessment. *International Journal of Technology Assessment in Health Care, 30*(1), 1–8.
Krumholz, H. M., Ross, J. S., Presler, A. H., & Egilman, D. S. (2007). What have we learnt from vioxx? *British Medical Journal, 334*(7585), 120–123.
LaCaze, A. (2009). Evidence-based medicine must be .... *Journal of Medicine and Philosophy, 34*(5), 509–527.
Landes, J., Osimani, B., & Poellinger, R. (2018). Epistemology of causal inference in pharmacology. *European Journal for Philosophy of Science, 8*, 3–49.
Lehtinen, A. (2013). On the impossibility of amalgamating evidence. *Journal for General Philosophy of Science, 44*(1), 101–110.
Marewski, J. N., & Gigerenzer, G. (2012). Heuristic decision making in medicine. *Dialogues in Clinical Neuroscience, 14*, 77–89.
McGauran, N., Wieseler, B., Kreis, J., Schuler, Y.-B., Kolsch, H., & Kaiser, T. (2010). Reporting bias in medical research – A narrative review. *Trials, 11*(37), 1–15.
Mussen, F., Salek, S., & Walker, S. (2009). *Benefit-risk appraisal of medicines*. Chichester: Wiley.
Onakpoya, I. J., Heneghan, C. J., & Aronson, J. K. (2016). Worldwide withdrawal of medicinal products because of adverse drug reactions: A systematic review and analysis. *Critical Reviews in Toxicology, 46*, 477–489.
Osimani, B. (2014a). Hunting side effects and explaining them: Should we reverse evidence hierarchies upside down? *Topoi, 33*(2), 295–312.

Osimani, B. (2014b). Safety vs. efficacy assessment of pharmaceuticals: Epistemological rationales and methods. *Preventive Medicine Reports, 1*, 9–13.
Price, K. L., Amy Xia, H., Lakshminarayanan, M., Madigan, D., Manner, D., Scott, J., Stamey, J. D., & Thompson, L. (2014). Bayesian methods for design and analysis of safety trials. *Pharmaceutical Statistics, 13*(1), 13–24.
Reiss, J. (2015). A pragmatist theory of evidence. *Philosophy of Science, 82*(3), 341–362.
Revicki, D. A., & Frank, L. (1999). Pharmacoeconomic evaluation in the real world. *PharmacoEconomics, 15*(5), 423–434.
Russo, F., & Williamson, J. (2007). Interpreting causality in the health sciences. *International Studies in the Philosophy of Science, 21*(2), 157–170.
Sackett, D. L., Straus, S. E., Richardson, W. S., & Haynes, R. B. (2000). *Evidence-based medicine: How to practice and teach EBM* (2nd ed.). New York: Churchill Livingstone.
Savage, L. J. (1954). *The foundations of statistics*. New York: Dover Publications.
Shafer, G. (1976). *A mathematical theory of evidence*. Princeton: Princeton University Press.
Shafer, G., & Srivastava, R. (1990). The bayesian and belief-function formalisms: A general perspective for auditing. *Auditing: A Journal of Practice & Theory, 9*, 110–137.
Skyrms, B. (1990). *The dynamics of rational deliberation*. Cambridge, MA: Harvard University Press.
Solomon, M. (2011). Just a paradigm: Evidence-based medicine in epistemological context. *European Journal for Philosophy of Science, 1*(3), 451–466.
Stegenga, J. (2013). An impossibility theorem for amalgamating evidence. *Synthese, 190*(12), 2391–2411.
Stegenga, J. (2014). Down with the hierarchies. *Topoi, 33*(2), 313–322.
Tan, S.-W., & Pearl, J. (1994). Qualitative decision theory. In *Proceedings of AAAI* (vol. 2, pp. 928–933). Palo Alto: AAAI Press.
Tervonen, T., van Valkenhoef, G., Buskens, E., Hillege, H. L., & Postmus, D. (2011). A stochastic multicriteria model for evidence-based decision making in drug benefit-risk analysis. *Statistics in Medicine, 30*(12), 1419–1428.
Thomas, J., Newman, M., & Oliver, S. (2013). Rapid evidence assessments of research to inform social policy: taking stock and moving forward. *Evidence & Policy: A Journal of Research, Debate and Practice, 9*(1), 5–27.
Troffaes, M. C. M., & de Cooman, G. (2014). *Lower previsions*. Chichester: Wiley.
Twinning, W. (2011). Moving beyond law: Interdisciplinarity and the study of evidence. In P. Dawid, W. Twinning, & M. Vasilaki (Eds.), *Evidence, inference and enquiry* (chapter 4, pp. 73–118). Oxford University Press.
Upshur, R. (1995). Looking for rules in a world of exceptions: Reflections on evidence-based practice. *Perspectives in Biology and Medicine, 48*(4), 477–489.
Urfalino, P. (2012). Reasons and preferences in medicine evaluation committees. In H. Landemore & J. Elster (Eds.), *Collective wisdom* (pp. 173–202). Cambridge: Cambridge University Press.
van Valkenhoef, G., Tervonen, T., Zwinkels, T., de Brock, B., & Hillege, H. (2013). ADDIS: A decision support system for evidence-based medicine. *Decision Support Systems, 55*(2), 459–475.
Vandenbroucke, J. P. (2008). Observational research, randomised trials, and two views of medical science. *PLoS Medicine, 5*(3), e67.
Williamson, J. (2015). Deliberation, judgement and the nature of evidence. *Economics and Philosophy, 31*, 27–65.
Worrall, J. (2002). What evidence in evidence-based medicine? *Philosophy of Science, 69*(3), 316–330.
Worrall, J. (2007a). Evidence in medicine and evidence-based medicine. *Philosophy Compass, 2*(6), 981–1022.
Worrall, J. (2007b). Why there's no cause to randomize. *British Journal for the Philosophy of Science, 58*(3), 451–488.
Worrall, J. (2010). Evidence: Philosophy of science meets medicine. *Journal of Evaluation in Clinical Practice, 16*(2), 356–362.

Chapter 12
Assessing Drug Safety Assessment: Metformin Associated Lactic Acidosis

Adam LaCaze and Karl Winckel

Abstract This chapter examines two approaches to assessing drug safety evidence. Method-focused approaches, such as Evidence-based Medicine, are the dominant approaches to evidence evaluation in medicine. On this approach, a body of evidence is evaluated according to the method used to generate the evidence. A second approach to evidence evaluation focuses on causal assessment. One such approach is Russo and Williamson's account of epistemic causality. A key insight of Russo and Williamson's approach is that causal assessment consists in seeking evidence of difference-making and evidence of mechanisms. Evidence regarding metformin-associated lactic acidosis is used to compare the advice provided by method-focused and causal assessment approaches to evidence. We argue that the context of assessing drug efficacy and safety are more different than is typically appreciated and that these differences emphasise the importance of taking a causal assessment approach to assessing drug safety evidence.

12.1 Introduction

The methodological challenges of assessing the safety of medicines are under-appreciated. The strengths or otherwise of randomized trials for assessing the benefits of medicines has been discussed extensively, but detailed discussion of the systematic assessment of drug harms is largely confined to the pharmacoepidemiology and clinical pharmacology literature. There is a tendency to assume that the

A. LaCaze (✉)
School of Pharmacy, The University of Queensland, Brisbane, QLD, Australia
e-mail: a.lacaze@uq.edu.au

K. Winckel
School of Pharmacy, The University of Queensland, Brisbane, QLD, Australia

Princess Alexandra Hospital, Brisbane, QLD, Australia
e-mail: k.winckel@uq.edu.au

A. LaCaze, B. Osimani (eds.), *Uncertainty in Pharmacology*, Boston Studies in the Philosophy and History of Science 338,
https://doi.org/10.1007/978-3-030-29179-2_12

standards of evidence developed for the assessment of efficacy apply equally to the assessment of safety. This is a mistake—a simplification that obscures important distinctions and exceptions. This chapter describes the causal assessment of clinical drug safety with a particular focus on identifying what makes the assessment of drug safety different from efficacy.

Drug safety is as important to decisions about how the drug can be used as drug efficacy. As such, there needs to be a clear framework for the evaluation of causal claims regarding drug safety. To date, however, the assessment of drug safety has received considerably less philosophical and methodological attention.

There are exceptions. Vandenbroucke and Psaty (2008) identify the limitations of randomized trials for assessing drug harms and emphasize the importance of non-randomized, observational studies to provide evidence on drug harms in a way that complements the evidence provided by randomized trials. Vandenbroucke (2008) focuses on the distinction between the *discovery* and *evaluation* of drug effects. Vandenbroucke highlights the importance of the biological sciences, case series and observational studies for the discovery and explanation of drug harms. In contrast to efficacy, the discovery of drug harms—especially rare or unexpected harms—remains important well after the drug is approved for clinical use. Osimani (2013, 2014a,b), Osimani and Mignini (2015), and Landes et al. (2018) provide philosophical discussion regarding drug safety assessment. Osimani (2014a) emphasizes the role of a variety of sources of evidence for assessing drug safety and the need for different epistemological justifications for these sources of evidence. For instance, the justification provided for interpreting randomized trials are largely deductive (especially for the assessment of efficacy), whereas safety assessment typically relies on inductive arguments incorporating different sources of evidence.

There is further work to be done. Vandenbroucke (2008) leaves open the question of how to best *evaluate* drug harms. Osimani and colleagues primarily analyze the differences in efficacy and safety assessment in terms of different epistemologies—deductive approaches for the assessment of drug efficacy and inductive approaches for the assessment for drug safety. There is a need for a greater appreciation of the different epistemological warrant provided for different lines of evidence regarding drug safety, but, as Osimani (2014b, 304) recognises, this analysis only goes so far. Inductive approaches are important to the assessment of efficacy: the appropriate design and interpretation of a randomized trial requires the incorporation of the best current evidence and the application of randomized trial evidence to therapeutic decisions is clearly inductive. Likewise deductive inferences based on a suitably designed randomized trial can provide compelling evidence for a drug harm.

This chapter seeks to further examine the differences between efficacy and safety assessment. We argue that the dominant "method-focused" approach used to assess drug efficacy is not appropriate for the assessment of drug safety. The primary difference between the assessment of efficacy and safety, we suggest, is the context in which the assessment takes place. The context for safety assessment differs from that of efficacy assessment in terms of what evidence is available, the timing of the assessment in relation to market approval, and the roles that different lines of evidence contribute to the assessment. These differences undermine the application of the standards of evidence used to evaluate drug efficacy to the evaluation of drug safety. Section 12.2 clarifies terms associated with the effects of medicines and then outlines the standard approach to assessing efficacy and safety. Section 12.3 briefly introduces the Russo and Williamson (2007, 2011) account of causal assessment in the health sciences, which provides a contrast to method-focused approaches drug assessment. Russo and Williamson's account provides a framework for a discussion of a contemporary case in drug safety: metformin associated lactic acidosis. Section 12.4 reviews the evidence regarding the putative causal link between metformin and lactic acidosis. Key lessons for drug safety assessment are discussed in Sect. 12.5.

12.2 Assessing Medicines

It is important to start by clarifying the key terms. Informally: "efficacy" refers to whether the treatment 'works' under experimental conditions; "effectiveness" refers to whether the treatment 'works' when used in routine clinical care; drug "harm(s)" refer to specific deleterious outcomes caused by a drug, and drug "safety" refers to an overall assessment of the harms of the drug in the context of the drug's use. While each of these can (arguably) be considered in terms of individual-level claims (of efficacy, effectiveness, etc), our focus is on population-level claims. A problem in the use of these terms in the medical literature is that judgments of efficacy, effectiveness, harms and safety of a drug are highly context-specific and the context is too often left implicit. A drug may be efficacious for one indication and inefficacious for another. A drug may be judged sufficiently safe for one indication in one group of patients, but produce more harms than benefits for another indication or another group of patients.

One solution to this problem is to make the context more explicit in the definition of the terms. To adequately provide the context for a claim of efficacy, effectiveness, harm or safety, it is necessary to take into consideration at least the following[1]: the specific *benefit(s)* or *harm(s)* under consideration; the *indication(s)* the drug is being used for (i.e. the disease or illness); the (explicit or implied) *alternative* treatment that is being compared to the drug under consideration (e.g. untreated natural course of disease, placebo or active control); the characteristics of the patient *population* taking the drug, and the *dose regimen(s)* of the drug that is being used. More detailed definitions of efficacy, effectiveness, harm and safety take these contexts into account and are provided below.

Efficacy refers to the capacity of a drug to bring about specified outcomes that are beneficial in treating a disease or illness in a sample of participants. The

[1]There are several additional factors that may be important for interpreting claims of efficacy, effectiveness, harm and safety, especially in particular contexts, these include the manufacturing process for the drug and the social contexts within which the drug is used.

benefits of the drug are measured by comparing the specified outcomes observed in groups of patients treated with the drug under investigation as opposed to active or passive control. Patients are recruited and treated under the conditions of an experimental study. The study protocol specifies the inclusion and exclusion of participants, the measured outcomes and the dose regimen.

Effectiveness refers to the capacity of a drug to bring about specified outcomes that are beneficial in treating a disease or illness. The benefits of the drug are estimated or inferred as the comparative effect of the drug on the specified outcomes against an alternative treatment (which may be no treatment). The population under consideration in effectiveness claims is a population of "real-world" patients treated under routine clinical conditions. Unless otherwise specified this would include the patients who receive the drug in routine care at the dose regimens typically used in clinical practice.

Harm refers to the capacity of a drug to bring about a specific deleterious outcome for patients. Harms are typically assessed by comparing the frequency of the harm observed in patients exposed to the drug against patients not exposed to the drug.

Safety refers to an overall assessment of the observed harms of a drug in a specified context. The context may be experimental, clinical or both. Specific indications, patient populations, comparator treatments or doses may be specified as part of the context under consideration. "Safe" implies an *absence* of significant harms in the context under consideration.

We suggest the definitions provided are consistent with routine use of the terms in the medical literature with the benefit of being more explicit about the required context for each claim. It is particularly important to note that "effectiveness" and "safety" can be used in way that use of one of these terms implies the other. While effectiveness claims don't say anything specific about a drug's safety, such claims often carry the implication that the drug is sufficiently safe to warrant routine use for the stated indication. At minimum there is an expectation that any serious safety concerns would be stated in conjunction with the effectiveness claim (especially if that safety concern suggested cautious use of the drug). A related issue is that a drug's indication, and the drug's efficacy and/or effectiveness for that indication, are essential parts of the context to determine drug safety. What counts as a "significant harm" depends on the likely benefit of the drug. A drug that effectively treats a life-threatening condition may well be considered "safe" despite causing harms that would never be tolerated for a condition that was not life-threatening.

12.2.1 Assessing Efficacy

Randomized trials provide a clear and relatively uncontroversial *standard of evidence* for assessing the efficacy of medicines. Successful Phase III randomized trials are the standard of evidence for demonstrating the efficacy of a new medicine prior to market approval. The justification for adopting randomized trials as an appropriate standard of evidence for assessing efficacy is provided in the medical literature and epidemiology texts (see, for example, Collins and MacMahon 2007; Sackett 2005). The benefits of randomized trials in terms of assessing efficacy are relative to alternative methods for assessing treatment benefits. The main comparison that is being made is between well-conducted randomized trials and well-conducted observational studies (e.g. cohort and case-control studies). The central claim is that randomized trials have the capacity to avoid specific types of *confounding* that are difficult or impossible to avoid in observational studies. Additional benefits of random allocation when assessing efficacy include the opportunity to mask allocation, and the provision of an agreed starting point for statistical analysis (Senn 1994; Kadane and Seidenfeld 1990).[2]

Randomized trials play a similarly important role in how clinicians assess the efficacy of medicines when making decisions regarding the care of patients. Evidence-based Medicine (EBM) provides hierarchies of evidence to aid clinicians making decisions regarding treatment benefits, treatment harms, diagnostic tests, screening and prognosis (OCEBM Levels of Evidence Working Group 2011). On EBM's advice, randomized trials are the standard of evidence for clinicians assessing treatment benefits for individual therapeutic questions. Randomized trials and systematic reviews of randomized trials are placed high in the hierarchy of evidence for assessing treatment benefits (see Table 12.1). The hierarchy of evidence also informs the methodology of systematic reviews and the development of treatment guidelines (See, for example, the advice provided by GRADE and the Cochrane Collaboration: Guyatt et al. 2008; Higgins and Green 2008). Evidence from studies using methods listed higher in the hierarchy are given more weight than other sources of evidence.

Table 12.1 This is a summary of Oxford Center for Evidence-based Medicine's (2011) levels of evidence for assessing drug benefits and harms. The document provides guidance for grading evidence higher or lower based on methodological considerations and effect sizes

Level	Benefits	Harms
1	Systematic review of RCTs, n-of-1 studies	*Common*: Systematic review of RCTs, nested case-control studies, n-of-1 studies or observational study with large effect
		Rare: Systematic review of RCTs or n-of-1 study
2	RCT or observational studies with dramatic effect	Individual RCTs or (exceptionally) an observational study with dramatic effect
3	Non-randomized controlled cohort or follow-up study with sufficient numbers	
4	Case-series, case control, or historically controlled studies	
5	Mechanism-based reasoning	

[2] Arguments regarding the benefits of randomized trials when assessing the efficacy of medicines have been well-rehearsed in the literature, see LaCaze (2013) for discussion.

EBM is a method-focused approach. The key idea underpinning method-focused approaches is that the efficacy and safety of a medicine are best assessed according to the *methods* employed in providing that evidence. The advice provided by these approaches is that decision-makers should locate the best evidence—i.e. the study or studies that employed the best methods—and use that evidence to inform decisions. Method-focused approaches rely on the hierarchy of evidence for advice on best methods for different contexts. Method-focused approaches can be contrasted with approaches that seek to *amalgamate* evidence (such as approaches to causal assessment discussed in Sect. 12.3)

Randomized trials have been criticized on a number of fronts. Much of this criticism is focused on the tendency within the medical literature (especially within EBM) to oversell the benefits of randomized trials. Worrall (2007), for instance, argues against the view that randomized trials are necessary to draw causal conclusions in medicine. Cartwright (2010, 2011) highlights the limitations of randomized trials for assessing the *effectiveness* of a medicine. Vandenbroucke and Psaty (2008) highlight the important limitations on randomized trials in assessing the harms of medicines. These criticisms are important correctives to the tendency within the medical literature to conflate the role of randomized trials as a appropriate *standard of evidence* for assessing efficacy with the view that randomized trials are *sine qua non*, the *only* reliable evidence for assessing medicines.[3] It is consistent to hold the view that randomized trials are an appropriate standard for assessing efficacy while recognising both the limits of randomized trials and that additional lines of evidence are important for assessing efficacy.

12.2.2 Assessing Safety

Within the medical literature, two distinct approaches to assessing drug safety have been articulated. The dominant approach is the method-focused approach epitomized by EBM. A second approach consists of methods for assessing causality in individual case reports of suspected drug harms.

Method-focused approaches apply the arguments provided for assessing efficacy to the assessment of drug safety. Some in the literature explicitly apply the same standards to safety assessment as efficacy assessment. Collins and MacMahon (2007, 24) provide a clear statement of this view[4]:

> Without clear confirmatory evidence from large-scale randomized trials or their meta-analyses, reports of moderate treatment effects from observational studies should not be

---

[3]It is relatively easy to find quotes in the medical literature that over-reach on the benefits of randomized trials (see, for example, Straus et al. (2011, 70), and Collins and MacMahon (2007, 24)). Urbach (1985) provides an early criticism of the tendency to claim too much on behalf of randomized trials.

[4]Collins et al. (2016) reiterate this view in similar terms.

interpreted as providing good evidence of either adverse or protective effects of these agents (and, contrary to other suggestions, the absence of evidence from randomized trials does not in itself provide sufficient justification for relying on observational data).

Others in the literature make small adjustments to accommodate safety assessment while retaining the method-focused approach. Note, for example, the similarities in the advice provided for assessing benefits and harms in the Oxford Centre for Evidence Based Medicine Levels of Evidence (summarized in Table 12.1). The differences are limited to elevating "nested case-control studies" and "observational studies with large effects" to "Level 1" for assessing *common* harms. "Nested case-control studies" are case-control studies nested within prospective cohort studies or randomized trials. This design has the benefit that the cases and controls are selected from a defined population. "Observational studies with large effects" are included in "Level 1" because it is thought that if the observed effects are large, it is less likely that an unrecognized confounding factor is sufficiently influential to be responsible for the observed effect (Glasziou et al. 2007).

The inclusion of these methods in the Oxford Centre for Evidence Based Medicine Levels of Evidence appears to be a partial acknowledgement of the limitations of randomized trials in assessing drug harms. Observational studies are prone to confounding whether assessing benefits or harms.[5] And the methodological benefits of randomized trials in terms of assessing drug safety are the same as the benefits in terms of assessing efficacy—at least in theory. In practice, randomized trials are typically designed to rigorously assess efficacy rather than safety. Safety is assessed, and can be assessed well, but the primary endpoint of most trials is an efficacy endpoint. The sample is selected and the study is powered to assess this endpoint. For example, elderly patients or those with multiple co-morbidities are more likely to experience drug harms are often excluded from trials assessing efficacy. And yet, once the drug is on the market, many of these patients will receive the drug in routine clinical care. It is theoretically possible to design a randomized trial to assess the capacity of the drug to cause specific harms in the patients who receive the drug in routine clinical care, but these trials are often not conducted. Instead, safety evidence regarding "real-world" use is often collected using observational studies. As a result, there are *more* well-conducted nested case-control studies and other observational studies that are explicitly designed to assess a drug harm than there are randomized trials that are explicitly designed to assess drug harms.

A second approach to assessing drug safety is articulated in the clinical pharmacology literature. These approaches focus on systems for informing a causal

[5]Vandenbroucke and Psaty (2008, 2418) argue that some forms of confounding are less common in observational studies of unintended (harms) compared to observational studies of benefits. This point is also taken up in Vandenbroucke (2008, e67–e68) and Osimani (2014a, 299). The idea is that confounding is a larger problem for observational studies assessing drug benefits because aspects of the patient's condition is more likely to influence treatment decisions (e.g. larger doses for sicker patients). Granting this point, it is still the case that observational studies are prone to more confounding than randomized trials whether assessing harms or benefits.

assessment of drug harms in individual case reports. Naranjo et al. (1981) describes one of the best known of these systems.[6] Assessing likely drug harms in an individual patient can be challenging, especially in patients who are suffering from multiple medical conditions and taking several medications. In order to improve the consistency of these assessments, Naranjo et al. (1981) provide a scoring system based on several indicators. Included in the indicators are the temporal relationship between administration and harm; whether the harm improved on withdrawal of the drug and/or reappeared on re-exposure; the presence or absence of alternative explanations for the patient's harm; presence of the drug in the blood or relevant tissues; presence or absence of a dose-response relationship, and the patient's response on any previous exposure. The main benefit of the system outlined by Naranjo et al. (1981) is that it is an approach to evidence amalgamation. A broad range of evidence is considered and the methods employed attempt to bring this evidence together to inform an overall judgement. Each of the indicators provided by Naranjo et al. (1981) are related to one or more of "Hill's criteria". Hill (1965) put forward a list of aspects that can be considered when assessing whether a relationship between drug exposure and outcome is causal in biomedicine (see Table 12.2). None of Hill's nine criteria are individually necessary to determine causality, rather they are *indicators* of a possible causal relationship. If many or all of the indicators are present in a particular situation (and there is an absence of contradictory evidence), then there is a strong case for inferring a causal link.

Table 12.2 Hill's criteria (1965)

Criteria	Question (type of indicator)
Strength of association	Does exposure greatly increase the risk of the outcome? (difference-making)
Consistency	Has a link between exposure and outcome been repeatedly observed in different populations across time? (difference-making)
Specificity	Is the particular outcome only observed following exposure? (difference-making)
Temporality	Does exposure precede outcome? (mechanism)
Biological gradient	Does increased exposure increase the risk or severity of the outcome? Is there a dose-response relationship (mechanism)
Plausibility	Is a causal relationship between exposure and outcome theoretically plausible? (mechanism)
Coherence	Is a causal relationship between exposure and outcome coherent with evidence from different sources? (mechanism)
Experiment	Is it possible to demonstrate a change in outcomes by intervening experimentally on exposure? (difference-making)
Analogy	Is the relationship between exposure and outcome similar to a well-accepted relationship in another area? (mechanism)

[6]See also Karch and Lasagna (1977), Edwards (2012), and WHO-UMC (2014).

Naranjo et al. (1981) and similar systems employed in drug safety assessment focus on the assessment of single cases. Factors that lead to the assessment that a drug caused a harm in an individual case may say little about whether the drug is likely to cause that harm more generally.

Factors relevant to the individual case may trump more general considerations.[7] Hill's criteria have received renewed interest as a way to assist the causal assessment in the health sciences at a population level (Russo and Williamson 2007; Howick et al. 2009; Landes et al. 2018). Different accounts incorporate Hill's suggestions in different ways. We outline the framework provided by Russo and Williamson (2007, 2011) and use it to consider the evidence related to metformin associated lactic acidosis.

12.3 Causal Assessment

Russo and Williamson (2011, 568) argue that causal claims in biomedicine are assessed on the basis of mixed evidence. They characterize that evidence as being of two broad types: *difference-making* and *mechanisms*. Each of the standard philosophical accounts of causation can be characterized as providing either a difference-making or mechanistic account (Russo and Williamson 2011, 564–566). Difference-making accounts define the causal relation in terms of the difference cause C makes to effect E—probabilistic, counterfactual and manipulationist theories of causation each provide a specific account of difference-making. Mechanistic accounts define the causal relation in terms of the mechanism that links cause C to effect E. Mechanistic accounts include process theories and complex-systems theories. Within the biomedical sciences, successful well-conducted randomized trials provide compelling evidence of difference-making. And biology, immunology and associated sciences identify mechanisms of disease and inform the development of drugs that target parts of that mechanism to improve patient outcomes.

Russo and Williamson argue that none of the standard philosophical accounts of causation, whether difference-making or mechanistic, are able to provide an adequate account of the need for these two types of evidence when assessing causal claims in biomedicine.

> On the one hand, the proponent of the mechanistic theory can't account for the widespread use of probabilistic data and methods in causal analysis even where the underlying mechanism is uncontroversial, and, on the other, the proponent of the probabilistic theory can't account for the fact that mechanisms are required even when appropriate probabilistic associations are well established. (Russo and Williamson 2007, 164)

Russo and Williamson's key claim—the *Russo-Williamson Thesis*—is that in order to chart successful inferences, causal claims need to made on the basis of evidence

[7] Perhaps the patient experiences an adverse effect that is tightly linked to peak drug concentrations. This may provide good *prima facie* evidence for a causal link in this patient even in the absence of any reports of the harm in any other patients.

of difference-making and mechanisms. Russo and Williamson provide examples in support of their claim such as the association between hand-washing and death following child-birth and the association between *Helicobacter pylori* and gastrointestinal ulcers. In both cases, Russo and Williamson (2007, 163) argue that evidence of mechanisms provides essential support to the probabilistic evidence in order to establish the causal claim. Russo and Williamson link their advice to that of Hill (1965)—each of Hill's criteria provide evidence of relevance to difference-making and/or mechanisms (see Table 12.2).

The Russo-Williamson Thesis is contentious. Establishing the precise details on the requirement of mechanistic evidence to inform causal inference is a common sticking point. Is mechanistic evidence really *necessary* for causal inference? If mechanistic evidence is necessary, what *kind* of mechanistic evidence is necessary? There appears to be many cases in which there is sufficiently rigorous difference-making evidence such that a successful causal inference can be made despite the absence of a fully worked-out correct mechanism linking C and E.[8]

Illari (2011) and Gillies (2011) distinguish between strong and weak versions of the Russo-Williamson Thesis.

The weak version of the Russo-Williamson Thesis requires only that there is *some* mechanistic evidence regarding the relationship between C and E. In this context, Illari (2011, 151) distinguishes between possessing detailed evidence of *what the mechanism is* (which required for the strong version of the Russo-Williamson Thesis); evidence *that there is a mechanism* of the postulated kind (perhaps due to evidence of other mechanisms in the domain); evidence of *analogous* mechanisms, and evidence of *no* mechanism. Extending the notion of evidence of mechanisms in this way saves the Russo-Williamson Thesis from counter-examples, but leads to the worry that a broad conception of evidence of mechanisms may render the thesis trivial. The weak version of the Russo-Williams Thesis is adopted in this chapter. While not a main focus of the paper, we suggest that the application of the Russo-Williamson Thesis to a case such as metformin associated lactic acidosis demonstrates that the weak version of the Russo-Williamson Thesis is far from trivial.

Section 12.4 outlines the evidence related to metformin associated lactic acidosis (a potential harm related to metformin). The primary purpose of the case is to highlight the complexity of drug safety assessment and to provide a specific context for examining the consequence of adopting different theoretical approaches to drug safety assessment. In Sect. 12.5, we draw two lessons from this case for the assessment of drug safety. First, the limitations of method-focused approaches to drug safety assessment. There are important differences between the assessment

[8]For instance, Broadbent (2011) argues that Semmelweis established the benefits of hand-washing in a controlled trial, and this *should* have been sufficient for causal inference despite the lack of a completely worked-out correct mechanism linking hand-washing and death prior to the development of germ theory. Consider also observed drug harms, such as the effects of thalidomide on foetal development. Action on clear difference-making evidence regarding the foetal toxicity of thalidomide shouldn't be delayed until a completely worked-out mechanism is available.

of efficacy and safety that are under-recognised in method-focused approaches. Second, the importance of taking a *causal* approach to drug safety assessment. Specifically, a causal approach that provides guidance regarding the amalgamation of the different types of evidence that inform an assessment of drug safety. The method-focused approach and the causal approach to drug safety assessment provide different outputs for decision-making. An important consequence of adopting the causal approach to drug safety assessment, we argue, is that it provides outputs that better inform clinical and regulatory decisions based on the assessment of a drug's safety.

12.4 Metformin Associated Lactic Acidosis

Metformin is a biguanide drug used primarily in type 2 diabetes. Metformin's key pharmacological actions are to reduce glucose production in the liver, increase glucose uptake in skeletal muscle and alter endocrine function in the pancreas (Pernicova and Korbonits 2014). First-line treatment for type 2 diabetes is lifestyle and dietary modification. After diet and lifestyle changes have been initiated, metformin is the first-line drug treatment (The Royal Australian College of General Practitioners 2016; American Diabetes Association 2016). The United Kingdom Prospective Diabetes Study (UKPDS Study Group 1998) provides compelling evidence that metformin treatment leads to a greater reduction in mortality compared to the most commonly used alternative treatments for type 2 diabetes. Metformin has been widely used in Europe and Australia for the treatment of diabetes since the 1960s, but was not marketed in the US until 1995 due to safety concerns (Fig. 12.1 provides a timeline of key events). Phenformin, another biguanide medication, was withdrawn from the US market by the Food and Drug Administration (FDA) in 1977 due to its association with lactic acidosis (Misbin 1977). Lactic acidosis is a rare but life-threatening metabolic condition consisting in an increase in lactate and blood acidity. Lactic acidosis occurs in a very small number of diabetic patients

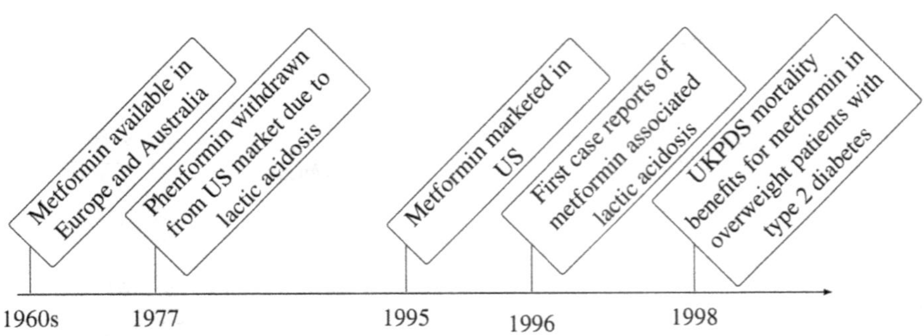

Fig. 12.1 Key milestones in the history of metformin

taking therapeutic doses of metformin, whether or not metformin is the *cause* of lactic acidosis in these patients remains controversial.

Lactic acid is a metabolite produced by physiological processes in the body. Elevated levels of lactic acid, above what can be used or removed, can be caused by tissues not getting enough oxygen (Type A lactic acidosis) or due to diseases, drugs, or errors of metabolism that lead to increased lactate production or reduced elimination despite the presence of oxygen (Type B lactic acidosis). Type A lactic acidosis is a relatively common consequence of serious acute illness. Anaemia, shock, critical illness and diabetes are among the conditions that can cause or contribute to lactic acidosis. For some patients who develop Type A lactic acidosis, the underlying condition causing lactic acidosis will be life-threatening and treatment will focus on this condition. Phenformin and metformin are considered to cause Type B lactic acidosis. The Type A/Type B distinction refers to the postulated etiology of lactic acidosis. It is not a distinction that is made clinically. Patients are often seriously unwell by the time lactic acidosis is diagnosed and this makes it difficult to judge whether the lactic acidosis is a cause of—or caused by—the serious acute illness (i.e. whether it is Type A or Type B lactic acidosis). "Metformin associated lactic acidosis" is a causally neutral term that refers to the development of lactic acidosis in a patient taking metformin.

The decision by the FDA to withdraw phenformin is summarized in the following statement.

> This action is being taken on the basis of the clear association demonstrated between the use of phenformin and the occurrence of lactic acidosis...
>
> Although the true frequency of the occurrence of lactic acidosis in diabetics using phenformin cannot be determined, the relative frequency among such diabetics is considerably higher than that being reported for any other form of treatment for diabetes...
>
> Because lactic acidosis is a frequently fatal complication, because no patient population exists in whom all risk factors for this complication can always be either identified or predicted and because of the availability of effective alternative products and other drug and nondrug methods for diabetic therapy, the Director [of the Bureau of Drugs] concludes that the risk of phenformin therapy outweighs the potential benefits that can be derived from its use. (Food and Drug Administration 1979, 20967)

Several aspects of the decision are noteworthy. First, the primary safety evidence on which the FDA acted was *difference-making*. Rates of phenformin associated lactic acidosis are higher than rates of lactic acidosis associated with alternative treatments for diabetes. Bailey (1992) estimates the incidence rate of phenformin-associated lactic acidosis to be 64 cases per 100,000 patient-years.[9] Second, the FDA withdrew phenformin despite the absence of a completely worked-out *mechanism* for phenformin-induced lactic acidosis (Misbin 1977).[10] Third, the

[9]"Patient-years" provides a measure of patient numbers given years of exposure. 64 cases per 100,000 patient-years is the number of cases expected if 100,000 patients took phenformin for 1 year.

[10]The FDA were challenged on this point by the manufacturers of phenformin. Food and Drug Administration (1979) details the response of the Administrative Law Judge who reasoned that

absence of relative *benefits* of phenformin treatment compared to the available alternatives (including no treatment) played an important role in the decision to withdraw phenformin. At the time diabetes was treated relatively conservatively and there was little evidence that any particular treatment led to superior long-term outcomes. Safety assessment is a judgement regarding drug harms in the context of how the drug is used.

Once approved in the US, metformin was closely monitored. Two letters to the *New England Journal of Medicine* describe early spontaneous reports of metformin-associated lactic acidosis to the FDA (Gueriguian et al. 1996; Misbin et al. 1998). Gueriguian et al. (1996) describe 10 reports of metformin-associated lactic acidosis and emphasize the challenges in accurately diagnosing metformin-associated lactic acidosis from the provided information and the existence of independent risk factors for lactic acidosis in the reported cases. Misbin et al. (1998) describe 47 cases of confirmed lactic acidosis associated with metformin reported to the FDA in the first year of metformin availability. Most of the cases involved patients with independent risk factors for lactic acidosis, especially heart failure and renal impairment. Four cases involved patients with no other risk factors for lactic acidosis apart from diabetes.[11] Misbin et al. (1998) estimate the reported rate of metformin-associated lactic acidosis to be approximately 5 cases per 100,000 users in 1 year, but warn that the estimate could be artificially low due to non-reporting. Following these reports the product information of metformin was modified to include heart failure as a contraindication to metformin treatment.

Spontaneous reports of adverse drug reactions have a number of limitations: they don't provide an accurate measure of incidence; they are influenced by how long the drug has been on the market; and clinician perceptions regarding the drug's possible adverse effects will influence the likelihood of reporting. Due to these limitations, the spontaneous reports of metformin associated lactic acidosis alone are unable to assess the degree to which metformin contributes to the development of lactic acidosis. Specifically: Do therapeutic doses of metformin cause lactic acidosis in diabetic patients, or is metformin merely an innocent bystander in unwell diabetic patients with lactic acidosis due another cause? The difference-making and mechanistic evidence regarding metformin-associated lactic acidosis is summarized below.

12.4.1 Evidence of Difference-Making

Brown et al. (1998), in a manufacturer funded study, used electronic hospital records to estimate the incidence of lactic acidosis in patients *prior* to the introduction of

---

since drugs can be approved on the basis of efficacy without a fully worked-out mechanism, so too can drugs be withdrawn.

[11] Misbin (2004, 1791) has since suggested that one of these four had a urinary tract infection, which may have triggered lactic acidosis.

metformin in the US. These authors estimate a rate of lactic acidosis in diabetic patients (not taking metformin) to be between 9.7–16.8 cases per 100,000 person-years. The lower figure corresponds to confirmed cases of lactic acidosis, the higher figure corresponds to possible cases. Risk factors for lactic acidosis in these cases included heart failure, hypotension and sepsis. The authors note the similarity of this figure to the estimate based on spontaneous reports to the FDA.

Stang et al. (1999) analysed dispensing and hospital records to estimate the rate of metformin-associated lactic acidosis in a Canadian province between 1980 and 1995. These data provide an incident rate of metformin-associated lactic acidosis of 9 cases per 100,000 patient-years—a figure comparable to both Misbin et al. (1998) and Brown et al. (1998). The limitations of observational studies based on administrative records are well recognized (Collins and MacMahon 2007). A key concern in this context is the possibility of under-reporting of lactic acidosis.

Cryer et al. (2005) reports a randomized trial that was conducted by the sponsor as a postmarketing safety study requested by the FDA as a condition of the approval of metformin in the US. The study was designed to assess the rate of serious adverse events in diabetic patients randomized metformin or usual care, which consisted of diet and alternative diabetes treatments.[12] Plasma lactate levels were also measured. The study was not statistically powered to identify an increase in the incidence of lactic acidosis of the magnitude that metformin is thought to cause (though it may have identified an increase of the magnitude that phenformin is estimated to cause). Patients with heart failure or renal impairment were excluded from the trial. No difference in serious adverse events were observed in participants who received metformin compared to usual care, nor was a difference in plasma metformin levels observed. There were no cases of lactic acidosis.

Salpeter et al. (2010) conducted a Cochrane systematic review to assess the incidence of metformin-associated lactic acidosis in prospective trials and cohort studies in patients with type 2 diabetes. This review pooled data from 347 studies: 209 comparative prospective trials; 125 prospective cohort studies; 13 retrospective cohort studies. The pooled studies provide a comparison between 70,490 patient-years of diabetic patients taking metformin and 55,451 patient-years of diabetic patients not taking metformin. *No* cases of lactic acidosis were observed in the included studies. This highlights the rarity of lactic acidosis, specifically in the patients that participate in clinical studies. The average age of participants in the included studies was relatively young (late 50s), and many of the studies excluded participants with renal impairment, cardiovascular disease or liver disease.

One possibility that needs to be considered is that metformin induces lactic acidosis, but the prescribing restrictions instituted by the FDA have successfully reduced this risk. However, the available data does not support this interpretation. Studies assessing metformin prescribing found that between 18.9–94% of patients received metformin despite possessing one or more formal contraindications to

[12] 'Serious adverse events' included fatal or life-threatening events, hospitalization and events that caused significant disability (Cryer et al. 2005, 540).

metformin treatment (such as renal impairment or heart failure) (Tahrani et al. 2007). Tahrani et al. (2007) also identify three observational studies that assessed outcomes in patients with diabetes and heart failure who received metformin or alternative treatments for diabetes. These patients are considered to be at high-risk of lactic acidosis, and yet patients who received metformin had lower rates of mortality than patients who did not receive metformin. These studies are subject to confounding: specifically, that diabetic patients with heart failure who receive metformin are different to those patients who do not (perhaps they are healthier in some way that the investigators were unable to control). Even with this limitation in mind, these results provide support for the apparent safety of metformin in patients with heart failure.

Despite the extensive use of metformin in a wide variety of patients there is an absence of compelling difference-making evidence that metformin increases the incidence of lactic acidosis. If therapeutic doses of metformin induces lactic acidosis, the increase in risk appears to be small even in patients considered to be at high risk of the condition.

12.4.2 *Evidence of Mechanisms*

A completely worked-out mechanism for biguanide-induced lactic acidosis is not available. There is, however, a good understanding of the physiological processes that may result in increased lactate levels and the ways in which biguanides may influence these processes. Lactate is primarily a byproduct of glycolysis. The two key processes for utilizing lactate are (i) the oxidation of lactate in the mitochondria of particular tissues and (ii) the conversion of lactate back into glucose in the liver and kidney (Defronzo et al. 2016). Both diabetes and biguanides influence these processes. In the absence of a complete mechanism for biguanide-induced lactic acidosis, mechanistic evidence assessing the association between biguanides and lactic acidosis consists in assessing steps in the mechanistic chain linking putative cause and effect. This evidence is indirect, but is important evidence for assessing the plausibility of there being a mechanism (or mechanisms) that link biguanides and lactic acidosis. Furthermore, assessing differences between phenformin and metformin in relation to aspects of the postulated mechanism provide possible explanations for the differences seen in the difference-making evidence between these two drugs.

12.4.2.1 Do Therapeutic Doses of Metformin Increase Blood Lactate Levels?

If metformin causes lactic acidosis, we might expect to see increases in blood lactate as a precursor to lactic acidosis. There is inconsistent evidence of this and it is complicated by the different methods employed to assess lactate levels. Cusi et al. (1996), Cryer et al. (2005), and Salpeter et al. (2010) found no increase in lactate levels in patients taking metformin.[13] Waters et al. (1978) conducted an observational study assessing mean lactate blood levels in patients taking different diabetes medicines: the percentage of patients exceeding the 95th percentile of the normal range of lactate blood levels were 63% of patients taking phenformin, 22% of patients taking metformin and 13% of patients taking sulfonylureas. Defronzo et al. (2016) summarizes the data and concludes that metformin likely causes a small increase in blood lactate levels, and that this occurs to a much smaller degree than the increase that occurs with phenformin.

12.4.2.2 Are There Case Reports of Metformin Associated Lactic Acidosis in Which Metformin is the Only Likely Contributor?

Most reported cases of metformin associated lactic acidosis are in patients with conditions that make them susceptible to lactic acidosis whether or not they are taking metformin. If metformin induces lactic acidosis, there should be cases of lactic acidosis in which metformin is the only likely cause. Reported cases of metformin overdose in otherwise healthy individuals fit this category. Reports of metformin overdose are infrequent, but there are enough reported cases to conclude that a metformin overdose can cause lactic acidosis (Dell'Aglio et al. 2009).

The second type of case that is relevant to this question are clear cases of Type B lactic acidosis in patients taking therapeutic doses of metformin. That is, cases in which a patient develops lactic acidosis in the absence of any conditions that may lead to tissue hypoxia. Such cases represent a very small proportion of reports of metformin-associated lactic acidosis, but do seem to occur (Misbin et al. 1998). The problem, however, is that there are no clear diagnostic criteria for distinguishing between Type A and Type B lactic acidosis. The reports themselves don't label cases of lactic acidosis as "Type A" or "Type B", rather they just mention the presence or absence of independent risk factors for lactic acidosis. The lack of diagnostic specificity combined with the infrequency of cases of metformin associated lactic acidosis in which the patient does not have any independent risk factors makes any assessment of "Type B" lactic acidosis open to challenge.

The lack of diagnostic precision also inflicts cases of phenformin associated lactic acidosis. The difference, however, is that cases of phenformin associated lactic acidosis in patients without serious acute illness represent a larger proportion of the cases. For instance, Fulop and Hoberman (1976) examine 18 cases of phenformin-associated lactic acidosis—10 of the 18 had no serious acute illness that would

[13]Cusi et al. (1996) randomized diabetic and non-diabetic patients to metformin or placebo for 15 weeks and took care to minimize changes in body weight. Cryer et al. (2005) is discussed in Sect. 12.4.1. Salpeter et al. (2010) was a systematic review comparing participants in prospective trials and cohort studies that received metformin or alternative diabetic treatments.

explain the development of lactic acidosis (contrast this with metformin cases; 4/47 in Misbin et al. (1998) and 0/49 in Lalau and Race (1999b)).

12.4.2.3 Is the Severity of Lactic Acidosis Related to Metformin Blood Concentrations?

If metformin causes lactic acidosis, we might expect that higher concentrations of metformin lead to more severe lactic acidosis. This question is a topic of debate in the literature. Lalau and Race (1999b) report 49 cases of metformin-associated lactic acidosis and compare the median metformin plasma and arterial lactate levels in patients who survived and those who died. Lactate levels were similar in both groups of patients, but metformin plasma levels were significantly *lower* in the patients who died (20.6 mmol/L v 6.3 mmol/L). As with other reports, the patients involved in these cases had independent risk factors for the development of lactic acidosis. Lalau and Race (1999b) infer that mortality in patients with metformin associated lactic acidosis is associated with the patient's underlying condition, rather than metformin or lactate levels. On the basis of this data, Lalau and Race (1999a,b) suggest that metformin may not be the cause of many cases of death in metformin associated lactic acidosis.[14]

One of the limitations of Lalau and Race (1999b) is that the timing of metformin and lactate samples is unspecified. Duong et al. (2013) seek to address this limitation in their report of 15 cases of metformin associated lactic acidosis. Metformin levels, lactate levels, renal function and pH were measured repeatedly in each of the cases. The correlation between metformin plasma levels and each of the other measures were calculated for individual patients (providing there were 5 or more samples available). The correlation between metformin plasma levels and lactate, renal function and pH were significant in the majority of patients: as metformin levels reduced, there was a tendency for lactate concentration to reduce; pH to rise; and renal function to improve (Duong et al. 2013). Duong et al. (2013, 739) draw the following inference (discussed further below):

> Correlations certainly do not prove causation but the significant correlations in several patients is consistent with metformin being a major contributor to MALA [metformin-associated lactic acidosis].

Lalau and Race (1999b) and Duong et al. (2013) identify three possible subgroups of patients that suffer from metformin-associated lactic acidosis. Group 1 includes patients with lactic acidosis without metformin accumulation. In these patients, lactic acidosis is likely due to tissue hypoxia caused by an underlying condition (and not metformin). Group 2 includes patients with lactic acidosis in the presence of metformin accumulation without other risk factors for lactic acidosis

[14] More provocatively, Lalau and Race (1999b, 383–384) suggest that metformin may be protective in cases of lactic acidosis. While the authors proffer some evidence for this claim, they make it clear that the suggestion is speculative.

apart from renal impairment. This group includes otherwise well patients who have ingested a metformin overdose; metformin is presumed to be the causative agent in these patients. Group 3 includes patients with lactic acidosis, metformin accumulation and an underlying condition (or conditions) that can induce lactic acidosis. Most cases of metformin-associated lactic acidosis in the literature are patients in Group 3.

What contribution does metformin make to the development of lactic acidosis in patients in Group 3? Duong et al. (2013, 744) postulate that lactic acidosis in these patients may be a result of a positive feedback loop. The components of the feedback loop are *diarrhoea/vomiting*, *acute renal impairment*, *increased metformin plasma levels* and *lactic acidosis*. Duong and colleagues suggest that lactic acidosis can develop when any component of this feedback loop is triggered as each component has the propensity to trigger a worsening of the other components.

For example, sepsis may lead to *diarrhoea/vomiting* which leads to dehydration and *acute renal impairment*, which will lead to *increased metformin plasma levels* and *lactic acidosis*. *Lactic acidosis* will then worsen the patient's *diarrhoea/vomiting*, which in turn will worsen dehydration and *acute renal impairment* and so on. Likewise, tissue hypoxia may lead to *lactic acidosis*, which in turn causes *diarrhoea/vomiting*, *acute renal impairment* and *increased metformin plasma levels*, which further increase lactic acid levels, and so the cycle restarts. The existence of a positive feedback loop is a compelling explanation for the available data—it also explains the difficulties involved in identifying any *single* contributor to the development of lactic acidosis in patients in Group 3.

Lalau and Race (1999b) argue that metformin is frequently (if not always) an innocent bystander in cases of metformin associated lactic acidosis on the basis that there is no correlation between metformin plasma levels and prognosis. Duong et al. (2013) argue that metformin induces lactic acidosis by playing a part in a positive feedback loop. The problem is that the data do not discriminate between these alternative mechanistic explanations. The problem for Lalau and Race (1999b) is that patients that experience metformin-associated lactic acidosis are not a homogeneous group. The likely presence of different subgroups makes any attempt to interpret correlations between metformin or lactate levels and patient outcomes in an undifferentiated group difficult. The problem for Duong et al. (2013) is that the proposed feedback loop was developed on the assumption that metformin induces lactic acidosis. Metformin is cleared from the body via the kidneys—if renal function reduces, metformin will accumulate; if renal function improves, metformin blood levels will reduce. The data provided by Duoung and colleagues is consistent with increased metformin plasma levels resulting from acute renal impairment and any one of the other components of the feedback loop or an external trigger causing lactic acidosis. It is true that the observed correlations are consistent with metformin being a major contributor to lactic acidosis, but they are also consistent with metformin *not* being a major contributor to lactic acidosis. And the question remains: if metformin *is* a significant contributor to the development of lactic acidosis, why is it so hard to find compelling difference-making evidence that metformin increases the rate of lactic acidosis?

The difference-making and mechanistic evidence regarding metformin associated lactic acidosis is summarized by the following statements.

1. Cases involving metformin overdose provide evidence that metformin can cause lactic acidosis.
2. Patients that develop lactic acidosis typically have multiple conditions and/or risk factors that are considered to contribute to the development of lactic acidosis.
3. It is plausible that these risk factors contribute to a positive feedback loop that lead to the development of lactic acidosis.
4. There is no reliable comparative evidence regarding the incidence of lactic acidosis in at-risk diabetic patients who are similar in all relevant respects but differ in relation to metformin use.

12.5 Causal Assessment Is Superior to a Method-Focused Approach to Assessing Drug Safety

It is now possible to contrast the method-focused approach to assessing the evidence regarding metformin-associated lactic acidosis with the approach to causal assessment outlined above. Sophisticated versions of each of these approaches will acknowledge many of the same points in relation to the specific evidence. There is, however, differences in what evidence is included in the assessment and how that evidence is assessed. This leads to a difference in how evidence is summarized and how that summary is used to inform decisions.

The method-focused approach focuses on key metformin-related *clinical* studies and notes the limitations of these studies. If well conducted randomized trials and systematic reviews of randomized trials were available, the method-focused approach would recommend basing decisions on these studies. In the absence of these studies the method-focused approach recommends seeking evidence from lower levels of the hierarchy of evidence for drug harms (see Table 12.1). In addition to these recommendations the method-focused approach provides the tools for assessing the methods used in the clinical studies assessing metformin-associated lactic acidosis. It is from within the method-focused approach that you can highlight the limitations of the randomized trials (e.g. lack of power) and observational studies (e.g. confounding). Thus, the overall assessment of the evidence regarding metformin-associated lactic acidosis using a method-focused approach will note the lack of compelling evidence for metformin increasing the incidence of lactic acidosis, the methodological limitations of the available studies, and recommend the need for future research.

This assessment can be seen in Salpeter et al. (2010). Salpeter et al. (2010) is a Cochrane systematic review that includes comparative clinical studies (randomized trials and observational studies) involving metformin—as such, it provides the highest level of evidence regarding metformin-associated lactic acidosis. The conclusions of this review are (i) that there is no evidence from comparative clinical studies that metformin increases the incidence of lactic acidosis and (ii) further large prospective comparative trials are necessary in patients with type 2 diabetes who are considered to be at risk of lactic acidosis (Salpeter et al. 2010, 10). This review does not include an assessment of case reports involving metformin overdose, nor does it evaluate mechanisms associated with the development of lactic acidosis while taking metformin.

The method-focused approach emphasises an assessment of the *clinical* evidence: What evidence is available, and what is the quality of this evidence? Causal assessment, by contrast, looks to amalgamate all the available evidence—clinical and non-clinical—to better understand the putative causal relationship: How does metformin cause lactic acidosis (mechanism), and what are the effects of that relationship in patients (difference-making)? It provides an assessment of the available clinical evidence, using the same tools as method-focused approaches, and incorporates this into a broader analysis of how that evidence relates to what is understood about the mechanisms at play. What is known (and uncertain) about the mechanism(s) is part of the evaluation of the available evidence as well as an important output of the assessment in its own right. Thus, the overall causal assessment (as summarised above) notes the lack of compelling clinical evidence that metformin increases the incidence of lactic acidosis in routine use, the likely risk factors for development of lactic acidosis, that a positive feedback loop involving these risk factors is likely to occur in many cases of lactic acidosis, and notes the somewhat discordant evidence regarding cases involving lactic acidosis due to metformin overdose.[15]

Causal assessment is superior method-focused approaches to assessing drug safety. This is due to the specific context in which safety assessment takes place. Specifically, the factors that support the application of the method-focused approach to the assessment of efficacy are frequently absent when assessing safety. A randomized trial provides an appropriate standard of evidence for assessing an explicitly specified claim when the trial is designed in a way that is consistent with what is understood about the mechanisms at play, the characteristics of the population receiving the drug and what is known about the pharmacokinetics and pharmacodynamics of the drug. The primary outcome of the trial needs to be well-defined and the trial needs to employ methods to ensure the outcome is measured with minimal error. This is the context within which well-conducted efficacy trials are conducted—indeed, it is the objective of pre-clinical development to provide the necessary information such that efficacy trials are conducted in this context. It is in this context that the application of the hierarchy of evidence as recommended in method-focused approaches can be defended—at least in as far as randomized trials provide an appropriate standard of evidence for assessing efficacy.[16] For any

[15]Of the papers discussed throughout the chapter, the discussion section of Duong et al. (2013) most clearly resembles the causal assessment approach.

[16]This is not to say that other lines of evidence are not relevant in the assessment of drug efficacy, nor that is particular circumstances evidence from observational studies or the basic sciences could not trump the results of a randomized trial. It is also worth noting that assessments of the *effectiveness* of a drug will typically require a broader range of evidence.

explicitly specified efficacy claim, a specifically designed randomized trial can be conducted such that it will provide superior evidence of efficacy in a defined sample of patients compared to an observational study or reasoning based solely the mechanisms provided by the biological sciences. Figure 12.2, Part A, represents the evidential inputs that typically go into the assessment of drug efficacy. A lot of evidence goes into the design of a randomized trial to assess the efficacy of a drug—once conducted, the outcomes of the trial provides evidence for the efficacy (or lack of efficacy) of the drug.

To the extent that evidence from well-conducted explicitly specified and designed randomized trials is available, method-focused approaches will often provide appropriate advice regarding whether a drug exhibits an explicitly specified effect.

However, specifically designed, well-conducted randomized trials assessing explicit safety claims are rare. This is one reason that safety assessment requires the amalgamation of different lines of evidence: difference-making evidence from studies employing different methods to address different aims and mechanistic

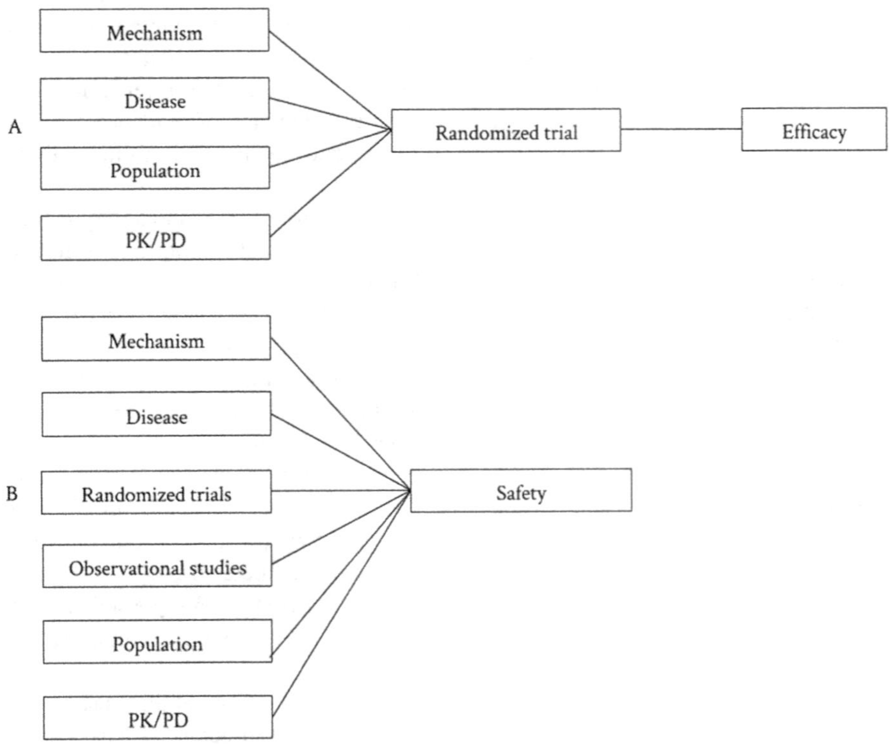

Fig. 12.2 A representation of the different evidential inputs into assessment of efficacy and safety. 'Mechanism': what is understood about the action of the drug; 'Disease': what is known about the disease process; 'Population': what is understood about the treated population and how they may vary; 'PK/PD': what is understood about the pharmacokinetics and pharmacodynamics of the drug

evidence that is frequently evolving to provide better methods to identify and explain the specific harm under investigation. The hierarchy of evidence provides poor advice in this context. When assessing metformin-associated lactic acidosis, the available evidence from the randomized trials is in no way superior to the evidence from the observational studies. The relevant randomized trials did not observe cases of lactic acidosis and were not designed to. These trials provide evidence that the rate of metformin induced lactic acidosis in relatively young and healthy patients cannot be above the lower-bound that the trials had sufficient power to detect.[17] By contrast, some of the observational studies were explicitly designed to assess the incidence of lactic acidosis in patients who receive metformin in routine clinical care. These studies have their own limitations, but the explicit focus on lactic acidosis is an important benefit.

Drawing appropriate inferences for safety assessment relies on the specific details: What patients are included? What factors led the patients to take metformin? How was lactic acidosis defined and detected? What additional monitoring was conducted? ..., and so on. Deciding what details matter depends heavily on what evidence is available regarding the mechanism(s) linking cause and effect. What is needed in this context is a framework for evidence amalgamation. Russo and Williamson's account of causal assessment provides a such a framework, emphasizing in particular the need to amalgamate evidence of difference-making and evidence of mechanisms. The lack of a single standard of evidence for the assessment of drug safety, and the need to assess and amalgamate multiple lines of evidence is represented in Fig. 12.2, Part B.

Another important difference between the assessment efficacy and safety is that safety assessments depend on a broader range of contextual factors than efficacy assessment. The efficacy assessment of one drug does not depend on the efficacy or safety of another drug, but this is not the case for safety assessment. The changes that have occurred in relation to the management of type 2 diabetes over the past 30 years are critical to the assessment of metformin associated lactic acidosis. A thorough assessment of this possible harm is important because of the evidence provided by the UKPDS study that metformin has benefits over the most commonly used alternative diabetes medications. In the absence of distinctive benefits for metformin, there would be a strong case for caution in relation to metformin associated lactic acidosis. A lower standard of evidence would be sufficient to justify clinicians (and perhaps regulators) to prefer alternative diabetes treatments. In this context, the early caution demonstrated by the FDA is justifiable. But the standard of evidence needed to shift following publication of the UKPDS study demonstrating the benefits of metformin.

Better evidence is required to justify denying patients a treatment that is superior in terms of long-term outcomes. This shift has been slow to occur, at least partly due to the dominance of the method-focused approach, which suggests waiting for

[17] Cryer et al. (2005), for example, was powered to detect an adverse event rate of 50 per 100,000 patient-years providing there was a 2.4 fold difference in event rates between metformin and control. This a rate is considerably higher than the estimate of metformin associated lactic acidosis provided by Misbin et al. (1998).

well-conducted randomized trials rather than amalgamating the available evidence. Metformin associated lactic acidosis provides a case study for the appropriate application of the precautionary principle, as discussed in Osimani (2013).[18] Considerations such as these can be incorporated into the causal assessment framework; it is less clear how this can be managed within the method-focused approach.

Despite sharing much with the analyses provided by Vandenbroucke (2008) and Osimani (2014a), there are some points of difference. Taking on the points Vandenbroucke (2008) makes regarding the difference between discovery and evaluation; our focus has been on the evaluation of drug safety. Osimani (2014a) identifies the key difference between safety and efficacy assessment as a difference in epistemologies.

On our analysis, the more important difference is in the mix of evidence that is available for causal assessment. This has consequences for the role of different lines of evidence in efficacy and safety assessment, and the types of justification that are provided for inferences relating to these lines of evidence. We haven't provided much detail regarding how evidence is assessed. Specifying an appropriate theory of evidence is an area for further work. Reiss (2015b, 33–57) provides a critical survey of the main alternatives. See Landes et al. (2018) for a framework for safety assessment based on Bayesian epistemology, and Reiss (2015a) for a pragmatist theory of evidence.

12.6 Conclusion

Drug safety assessment will always be more challenging than efficacy assessment. Presently, however, this challenge is exacerbated by the tendency to assume method-focused approaches are equally applicable to the assessment of efficacy and safety. Placing the assessment of drug effects within a more general framework of causal assessment retains what is right about the hierarchy of evidence while remedying the misapplication of method-focused approaches to drug safety. This can lead to real improvements in the care of patients. The case of metformin associated lactic acidosis is not unique. There are many scenarios in which decisions regarding the use of life saving or life changing medicines are influenced by a misapplication of method-focused approaches to the assessment of drug safety.

[18]It is also a helpful demonstration of inductive risk (Douglas 2009). In this context, the inductive risk associated with selecting the appropriate standard of evidence for assessing drug safety.

References

American Diabetes Association. (2016). Approaches to glycemic rreatment, Sec. 7. *Diabetes Care, 39*(Suppl. 1), S52–S59. ISSN: 0149-5992. https://doi.org/10.2337/dc15-S010.

Bailey, C. J. (1992). Biguanides and NIDDM. *Diabetes Care, 15*(6), 755–772.

Broadbent, A. (2011). Inferring causation in epidemiology: Mechanisms, black boxes, and contrasts. In *Causality in the sciences* (pp. 45–69). Sept 2010. Oxford: Oxford University Press. ISBN: 9780191728921. https://doi.org/10.1093/acprof:oso/9780199574131.003.0003.

Brown, J. B., Pedula, K., Barzilay, J., Herson, M. K., & Latare, P. (1998). Lactic acidosis rates in type 2 diabetes. *Diabetes Care, 21*(10), 1659–1663. http://carediabetesjournalsorg/content/21/10/1659.short.

Cartwright, N. (2010). What are randomised controlled trials good for? *Philosophical Studies, 147*(1), 59–70. https://doi.org/10.1007/s11098-009-9450-2.

Cartwright, N. (2011). Predicting what will happen when we act. What counts for war rant? *Preventive medicine, 53*(4–5), 221–224. https://doi.org/10.1016/j.ypmed.2011.08.011. http://linkinghub.elsevier.com/retrieve/pii/%20S0091743511003008

Collins, R., & MacMahon, S. (2007). In Rothwell, P. M. (Ed.), *Reliable assessment of the effects of treatments on mortality and major morbidity* (p. 326). Edinburgh: Elsevier. ISBN: 9780080447391.

Collins, R., Reith, C., Emberson, J., Armitage, J., Baigent, C., Blackwell, L., Blumenthal, R., et al. (2016). Interpretation of the evidence for the efficacy and safety of statin therapy. *The Lancet, 6736*(16). ISSN: 0140-6736. https://doi.org/10.1016/S0140-6736(16)31357-5.

Cryer, D. R., Nicholas, S. P., Henry, D. H., Mills, D. J., & Stadel, B. V. (2005). Comparative outcomes study of metformin intervention versus conventional approach the COSMIC Approach Study. *Diabetes Care, 28*(3), 539–543. ISSN: 0149-5992.

Cusi, K., Consoli, A., & Defronzo, R. A. (1996). Metabolic effects of metformin on glucose and lactate metabolism in noninsulin-dependent diabetes mellitus. *Journal of Clinical Endocrinology & Metabolism, 81*(11), 4059–4067. http://press.endocrine.org/doi/abs/10.1210/%20jcem.81.11.8923861.

Defronzo, R., Fleming, G. A., Chen, K., & Bicsak, T. A. (2016). Metformin-associated lactic acidosis: Current perspectives on causes and risk. *Metabolism: Clinical and Experimental, 65*(2): 20–29. ISSN: 15328600. https://doi.org/10.1016/jmetabol.2015.10.14.

Dell'Aglio, D. M., Perino, L. J., Kazzi, Z., Abramson, J., Schwartz, M. D., & Morgan, B. W. (2009). Acute metformin overdose: Examining serum pH, lactate level, and metformin concentrations in survivors versus nonsurvivors: A systematic review of the literature. *Annals of Emergency Medicine, 54*(6), 818–823. ISSN: 01960644. https://doi.org/10.1016/j.annemergmed.2009.04.023.

Douglas, H. (2009). *Science policy and the value-free ideal*. Pittsburgh: University of Pittsburgh Press.

Duong, J. K., Furlong, T. J., Roberts, D. M., Graham, G. G., Greenfield, J. R., Williams, K. M., & Day, R. O. (2013). The role of metformin in metformin-associated lactic acidosis (MALA): Case series and formulation of a model of pathogenesis. *Drug Safety, 36*, 733–746. ISSN: 0114-5916. https://doi.org/10.1007/s40264-013-0038-6. http://wwwncbinlm.nih.gov/pubmed/23549904.

Edwards, I. R. (2012). Considerations on causality in pharmacovigilance. *International Journal of Risk and Safety in Medicine, 24*(1), 41–54. ISSN: 09246479. https://doi.org/10.3233/JRS-2012-0552.

Food and Drug Administration. (1979). *Phenformin hydrochloride; Withdrawal of approval of new drug application; Final decision*. Technical report 58. http://wwwfdagov/ohrms/dockets/ac/98/briefingbook/1998-%203454B1%7B%5C_%7D03%7B%5C_%7DWL38.pdf

Fulop, M., & Hoberman, H. D. (1976). Phenformin-associated lactic acidosis. *Diabetes, 25*(4), 292–296.

Gillies, D. (2011). The Russo-Williamson thesis and the question of whether smoking causes heart disease. In *Causality in the Sciences* (pp. 110–125). ISBN: 9780199682676. https://doi.org/10.1093/acprof.

Glasziou, P., Chalmers, I., Rawlins, M., & Mcculloch, P. (2007). When are randomised trials unnecessary? Picking signal from noise. *BMJ (Clinical Researched), 334*(7589), 349–351. https://doi.org/10.1136/bmj.39070.527986.68.

Gueriguian, J., Green, L., Misbin, R. I., Stadel, B., Fleming, G. A., Deutsch, J. C., Santhosh-Kumar, C. R., Kolhouse, J. F., DeFronzo, R. A., & Goodman, A. M. (1996). Efficacy of metformin in non-insulin-dependent diabetes mellitus [2]. *New England Journal of Medicine, 334*(4), 269–270.

Guyatt, G. H., Oxman, A. D., Vist, G. E., Kunz, R., Falck-Ytter, Y., Alonso-Coello, P., & Schunemann, H. J. (2008). GRADE: An emerging consensus on rating quality of evidence and strength of recommendations. *BMJ, 336*(7650), 924–926. ISSN: 0959-8138. https://doi.org/10.1136/bmj.39489.470347AD.

Higgins, J. P. T., & Green, S. (Eds.). (2008). *Cochrane handbook for systematic reviews of interventions*. Chichester: John Wiley & Sons, Ltd. ISBN: 9780470712184. https://doi.org/101002/9780470712184.

Hill, A. B. (1965). Austin Bradford Hill,"The Environment and Disease: Association or Causation?" *Proceedings of the Royal Society of Medicine, 58*, 295–300.

Howick, J., Glasziou, P., & Aronson, J. K. (2009). The evolution of evidence hierarchies: What can Bradford Hill's 'guidelines for causation' contribute? *JRSM, 102*(5), 186–194. ISSN: 0141-0768. https://doi.org/10.1258/jrsm.2009.090020.

Illari, P. M. (2011). Mechanistic evidence: Disambiguating the Russo-Williamson thesis. *International Studies in the Philosophy of Science, 25*(2), 139–157. ISSN: 0269-8595. https://doi.org/10.1080/02698595.2011.574856.

Kadane, J. B., & Seidenfeld, T. (1990). Randomization in a bayesian perspective. *Journal of Statistical Planning and Inference, 25*(3), 329–345. http://www.sciencedirect.com/science/article/B6V0M45SJDGS%20B/2/a0534f4c1e0734852c3b08fdfc8961c.

Karch, F.E, & Lasagna, L. (1977). Toward the operational identification of adverse drug reactions. *Clinical Pharmacology and Therapeutics, 21*(3), 247–254.

LaCaze, A. (2013). Why randomized interventional studies. *The Journal of Medicine and Philosophy, 38*(4), 352–368. https://doi.org/10.1093/jmp/jht028.

Lalau, J. D., & Race, J. M. (1999a). Lactic acidosis in metformin therapy. *Drugs, 58*(Suppl. 1), 55–82. http://eutils.ncbi.nlm.nih.gov/entrez/eutils/elink.fcgi?%20dbfrom=pubmed%7B%5C&%.7Did=10576527%7B%5C&%7Dretmode=ref%7B%5C&%7Dcmd=prlinks.

Lalau, J. D., & Race, J. M. (1999b). Lactic acidosis in metformin treated patients. *Drug Safety, 20*(4), 377–384.

Landes, J., Osimani, B., & Poellinger, R. (2018). Epistemology of causal inference in pharmacology: Towards a framework for the assessment of harms. *European Journal for Philosophy of Science, 8*(1), 3–49. https://doi.org/10.1007/s13194-017-0169-1.

Misbin, R. I. (1977). Phenformin-associated lactic acidosis: Pathogenesis and treatment. *Annals of Internal Medicine, 87*(5), 591–595. http://eutils.ncbi.nlm.nih.gov/entrez/eutils/elinkfcgi?%20dbfrom=pubmed%7B%5C&%7Did=921092%7B%5C&%7Dretmode=ref%7B%5C&%7Dcmd=prlinks.

Misbin, R. I. (2004). The phantom of lactic acidosis due to metformin in patients with diabetes. *Diabetes Care, 27*(7), 1791–1793.

Misbin, R. I., Green, L., Stadel, B. V., Gueriguian, J. L., Gubbi, A., & Fleming, G. A. (1998). Lactic acidosis in patients with diabetes treated with metformin. *The New England Journal of Medicine, 338*(4), 265–266. https://doi.org/10.1056/NEJM199801223380415. http://eutils.ncbi.nlm.nih.gov/entrez/eutils/elinkfcgi?%20dbfrom=pubmed%7B%5C&%7Did=9441244%7B%5C&%7Dretmode=ref%7B%5C&%7Dcmd=prlinks

Naranjo, C. A., Busto, U., Sellers, E. M., Sandor, P., Ruiz, I., Roberts, E. A., Janecek, E., Domecq, C., & Greenblatt, D. J. (1981). A method for estimating the probability of adverse drug reactions. *Clinical Pharmacology & Therapeutics,*

30(2), 239–245. http://eutils.ncbi.nlm.nih.gov/entrez/eutils/elink.fcgi?%20dbfrom=pubmed%7B%5C&%7Did=7249508%7B%5C&%7Dretmode=ref%7B%5C&%7Dcmd=prlinks

OCEBM Levels of Evidence Working Group. (2011). *Oxford centre for evidence-based medicine 2011 levels of evidence*. http://www.cebm.net/ocebm-levels-of-evidence

Osimani, B. (2013). Until RCT proven? On the asymmetry of evidence requirements for risk assessment. *Journal of Evaluation in Clinical Practice, 19*(3), 454–462. https://doi.org/10.1111/jep.12039. http://eutils.ncbi.nlm.nih.gov/entrez/eutils/elink.fcgi?%20dbfrom=pubmed%7B%5C&%7Did=23692227%7B%5C&%7Dretmode=ref%7B%5C&%7Dcmd=prlinks

Osimani, B. (2014a). Hunting side effects and explaining them: Should we reverse evidence hierarchies upside down? *Topoi, 33*, 295–312. ISSN: 19516401. https://doi.org/10.1007/s11245-013-9194-7.

Osimani, B. (2014b). Safety vs. Efficacy assessment of pharmaceuticals: Epistemological rationales and methods. *Preventive Medicine Reports, 1*, 9–13. ISSN: 22113355. https://doi.org/10.1016/j.pmedr.2014.08.002.

Osimani, B., & Mignini, F. (2015). Causal assessment of pharmaceutical treatments: Why standards of evidence should not be the same for benefits and harms. *Drug Safety, 38*, 1–11. https://doi.org/10.1007/s40264-014-0249-5.

Pernicova, I., & Korbonits, M. (2014). Metformin—mode of action and clinical implications for diabetes and cancer. *Nature Reviews. Endocrinology, 10*(3), 143–156. https://doi.org/10.1038/nrendo.2013.256.

Reiss, J. (2015a). A pragmatist theory of evidence. *Philosophy of Science, 82*(3), 341–362.

Reiss, J. (2015b). *Causation, evidence and inference* (1st ed., p. 258). New York: Routledge.

Russo, F., & Williamson, J. (2007). Interpreting causality in the health sciences. *International Studies in the Philosophy of Science, 21*(2), 157–170. https://doi.org/10.1080/02698590701498084.

Russo, F., & Williamson, J. (2011). Epistemic causality and evidence-based medicine. *History and Philosophy of the Life Sciences, 33*(4), 563–581. http://eutils.ncbi.nlm.nih.gov/entrez/eutils/elink.fcgi?%20dbfrom=pubmed%7B%5C&%7Did=22662510%7B%5C&%7Dretmode=ref%7B%5C&%7Dcmd=prlinks

Sackett, D. L. (2005). The principles behind the tactics of performing therapeutic trials. In R. Brian Haynes, D. L. Sackett, G. H. Guyatt, & P. Tugwell (Eds.), *Clinical epidemiology: How to do clinical practice research* (3rd ed., pp. 173–243). Lippincott Williams & Wilkins.

Salpeter, S. R., Greyber, E., Pasternak, G. A., & Salpeter, E. E. (2010). Risk of fatal and nonfatal lactic acidosis with metformin use in type 2 diabetes mellitus. *Cochrane Database of Systematic Reviews (Online), 1*, CD002967.

Senn, S. 1994. Fisher's game with the devil. *Statistics in Medicine, 13*(3), 217–230.

Stang, M., Wysowski, D. K., & Butler-Jones, D. (1999). Incidence of lactic acidosis in metformin users. *Diabetes Care, 22*(6), 925–927. http://eutils.ncbi.nlm.nih.gov/entrez/eutils/elink.fcgi?%20dbfrom=pubmed%7B%5C&%7Did=10372243%7B%5C&%7Dretmode=ref%7B%5C&%7Dcmd=prlinks

Straus, S. E., Glasziou, P., Richardson, W. S., & Haynes, R. B. (2011). *Evidence-based medicine: How to practice and teach it* (4th ed., p. 312). Edinburgh: Churchill Livingstone. ISBN: 9780702031274.

Tahrani, A. A., Varughese, G. I., Scarpello, J. H., & Hanna, F. W. F. (2007). Metformin, heart failure, and lactic acidosis: Is metformin absolutely contraindicated? *BMJ (Clinical research ed), 335*(7618), 508–512. https://doi.org/10.1136/bmj.39255.669444.AE.

The Royal Australian College of General Practitioners. (2016). *General practice management of type 2 diabetes: 2016–18*. East Melbourne: VIC Royal Australian College of General Practitioners. www.racgp.org.au

UKPDS Study Group. (1998). Effect of intensive blood-glucose control with metformin on complications in overweight patients with type 2 diabetes (UKPDS 34). *The Lancet, 352*(9131), 854–865. http://eutils.ncbi.nlm.nih.gov/entrez/eutils/elink.fcgi?%20dbfrom=pubmed%7B%5C&%7Did=9742977%7B%5C&%7Dretmode=ref%7B%5C&%7Dcmd=prlinks

Urbach, P. (1985). Randomization and the design of experiments. *Philosophy of Science, 52*, 256–273.

Vandenbroucke, J. P. (2008). Observational research, randomised trials, and two views of medical science. *PLoS Medicine, 5*(3), e67. https://doi.org/10.1371/journal.

Vandenbroucke, J. P., & Psaty, B. M. (2008). Benefits and risks of drug treatments. *Journal of the American Medical Association, 300*(20), 2417–2419.

Waters, A. K., Morgan, D. B., & Wales, J. K. (1978). Blood lactate and pyruvate levels in diabetic patients treated with biguanides with and without sulphonylureas. *Diabetologia, 98*(14), 95–98.

WHO-UMC. (2014). *The use of the WHO-UMC system for standardised case causality assessment.* Technical report.

Worrall, J. (2007). Why there's no cause to randomize. *The British Journal for the Philosophy of Science, 58*(3), 451–488. https://doi.org/10.1093/bjps/axm024.

Chapter 13
Robust Biomarkers: Methodologically Tracking Causal Processes in Alzheimer's Measurement

Vadim Keyser and Louis Sarry

Abstract In biomedical measurement, biomarkers are used to achieve reliable prediction of, and useful causal information about, patient outcomes while minimizing complexity of measurement, resources, and invasiveness. In this paper we discuss a specific *methodological use* of clinical biomarkers in pharmacological measurement. We confront the *reliability* of clinical biomarkers that are used to gather information about clinically meaningful endpoints. Next, we present a *systematic methodology for assessing the reliability of multiple surrogate markers (and biomarkers in general)*. We propose three relevant conditions for a robust methodology for biomarkers: (R1) *Intervention-based demonstration of partial independence of modes*; (R2) *Comparison of diverging and converging results across biomarkers*; and (R3) *Information within the context of theory*. Finally, we apply our robust methodology to currently developing Alzheimer's research to make specific theoretical conclusions about *promising causal culprits as well as decoupled biomarkers and endpoints.*

13.1 Introduction

In biomedical measurement biomarkers are used to achieve reliable prediction of, and useful information about, patient outcomes while minimizing complexity of measurement, resources, and invasiveness. A biomarker is an assayable metric that discloses the status of a biological process of interest, be it pathophysiological or in response to intervention (De Gruttola et al. 2001). The greatest utility from biomarkers comes from their ability to help clinicians (and researchers) make and evaluate clinical decisions. As such, a clinical biomarker is any biomarker deemed

V. Keyser (✉) · L. Sarry
California State University, Fresno, CA, USA
e-mail: vkeyser@csufresno.edu

A. LaCaze, B. Osimani (eds.), *Uncertainty in Pharmacology*, Boston Studies in the Philosophy and History of Science 338,
https://doi.org/10.1007/978-3-030-29179-2_13

universally applicable across the patient population that discloses the state of a biological process suitable to inform clinical decision-making and its subsequent assessment.

Clinical biomarkers (CBs) are stratified in many ways. For example, by: measurement *purpose* (e.g., the extent of disease, nature of response); *what* they measure (e.g., function, biochemistry/histology, behavior, genetics, morphology); clinical use (e.g., predictive, diagnostic, prognostic); situation within a physiological model (e.g., causative, resultant, correlative); how they can be used to examine biology (e.g., interventional analysis); and composition (i.e. singular, composite, serial).[1]

In this paper we analyze the *methodological use* of CBs in pharmacological measurement: some biomarkers are used to substitute for a clinically meaningful endpoint corresponding to events and their penultimate risk factors. For example, the biomarker LDL cholesterol (LDL-C) corresponds to the clinical endpoint heart attack, with penultimate risk factors such as coronary artery stenosis, atherosclerosis with unstable fibrous cap, and severe angina pectoris. A coveted CB is supposed to provide reliable information regarding processes that cannot reasonably be interrogated directly. Such reliable biomarkers are referred to as 'surrogate markers' (SMs).[2]

The story becomes intriguing when multiple surrogate markers are combined for measurement purposes. It seems that a combination of SMs will provide more useful results than just a single SM, to the extent that that those SMs are independent and do not carry the same confounds. But there is neither a scientific nor a philosophical account that can be used to assess the reliability of multiple biomarkers. In this paper we present a *systematic methodology for assessing the reliability of multiple surrogate markers* (and biomarkers in general). To do this we draw upon the robustness analysis literature in the philosophy of science as well as the clinical use of biomarkers.

In Sect. 13.2, we present background on robustness analysis to show that the focus within the philosophy of science literature has been on converging results. At first glance, CBs seem like a perfect candidate to be incorporated into a robustness analysis focused on converging results However, in Sect. 13.3 we analyze the reliability of CBs that are used to gather information about clinically meaningful endpoints, both within single biomarkers and within combined biomarkers. We argue that even in surrogate markers that are validated, readings of a given surrogate marker may diverge from readings of nominally equivalent, or parallel, surrogate markers. For this reason, we focus on a type of robustness analyses that can account for diverging results between modes of measurement. In Sect. 13.4, we present an intervention-based systematic methodology for assessing the reliability of surrogate markers. This methodology draws on two key accounts in the robustness literature (Schupbach 2016; Keyser 2016), both of which focus on elimination and provide uses for diverging results. In Sect. 13.5, we present a case study from Alzheimer's

[1] See Mayeux (2004) and Aronson (2005) for classification of CBs.

[2] Katz (2004) describes all biomarkers as being "candidate" surrogate markers.

research. We argue that current research in Alzheimer's disease is fitting for specifying causal relations and revising theory based on diverging results of CBs.[3]

13.2 Robustness Analysis

In this discussion, we present an intervention-based systematic methodology for assessing the reliability of surrogate markers. This methodology draws on two key accounts in the robustness literature (Schupbach 2016 and Keyser 2016), both of which focus on elimination and provide uses for diverging results. Before discussing these robustness analyses, it is important to give a bit of background about the different uses of robustness analysis.

13.2.1 Robustness Analysis: Uses

Philosophers have discussed robustness analysis in the context of models[4] as well as in the context of experiment and evidence.[5] We can specify three robustness analyses that are relevant for setting up biomarker analysis. Woodward (2006) distinguishes 'inferential robustness,' 'measurement robustness,' and 'causal robustness.' The details of these analyses are important because they can be applied to show how we can differentiate modes in biomarker measurement (Sect. 13.4.1) and how we can use robustness analysis to specify causal relationships Sect. 13.4.2.

Inferential robustness is used when there is a fixed body of data (D) and a conclusion (S). This is for cases with a number of additional assumptions (Ai), with no strong scientific preference for any of them. Woodward says: "A number of writers suggest that if for each of these Ai, D supports S, this provides a strong reason for belief in S, even in the absence of information about which of these Ai is correct—S is said in this case to be robust or sturdy or insensitive to alternative assumptions Ai, given D" (2006, 219–220).[6] If S is not robust, then it is believed to be "fragile" and Woodward suggests a suspension of belief (2006, 220).

[3]The modeling work for this project was completed in 2015 and 2016 when this was still an unfolding empirical puzzle.

[4]See Wimsatt (2007); Levins (1966); Weisberg (2006); and Glymour (1980).

[5]See Horwich (2011); Hacking (1983); Franklin (1997); Sober (1989); Trout (1998); Culp (1994); Stegenga (2009, 2012).

[6]Woodward adds that inferential robustness "... is taken to show that D supports S or provides a reason for believing S" (2006, 220).

'Measurement robustness' consists of using different measurement procedures to measure the value of some quantity (Woodward 2006, 234). One of the important features of measurement robustness is that different procedures are used. This increases our confidence that we have measured the quantity "accurately" (234). Woodward says,

> The underlying rationale is usually taken to be something like this: The different measurement procedures are in some relevant sense or senses 'independent' of each other—they involve instruments of different design, operating according to different causal principles, they employ different assumptions to interpret the data they produce and so on. This suggests that while each procedure may be subject to various sources of error, they are unlikely to be subject to exactly the same kinds of error . . . " (2006, 234)

Both inferential robustness and measurement robustness increase our confidence about an invariant result. However, the mechanism in each analysis differs: inferential robustness data may be produced by a single procedure, even though the assumptions differ. In a case of measurement robustness, there may be shared assumptions whereas measurement procedures differ.[7]

Finally, 'causal robustness' consists of finding a structural relationship that is invariant over changes in intervention. For example, a regression equation is robust if the causal relationship between the independent and dependent variables remains invariant over some class of manipulations on the independent variable (Woodward 2006, 235).

Woodward's account breaks down important features of robustness analyses, namely, theoretical assumptions, procedures, and causal conditions. We can use these features as building blocks to individuate modes.

13.2.2 Individuating Modes

There are different ways to differentiate modes.[8] Recall Woodward's characterization of independence in assumptions (inferential robustness) vs. independence in physical procedures (measurement robustness). This offers a precise way of characterizing different accounts of 'independence'. Culp (1994) posits that the use of different background theories for evidential processes contributes to the success of robustness analysis. Stegenga (2009) discusses the absence of shared problematic background assumptions to individuate modes. Specifically, converging modes of

[7]For example, assumptions in kinetic theory of heat can be used to explain the function of two different thermometry procedures.

[8]Philosophers have addressed the "individuation" (independence) of modes of evidence See Franklin (1997); Sober (1989); Culp (1994); Keeley (2002); Staley (2004); Douglas (2004); Wimsatt (2007); and Stegenga (2009, 2012); Lloyd (2010); Schupbach (2016); Keyser (2016).

evidence are more likely to signal truth, if they do not share problematic background assumptions.[9]

The second approach in characterizing 'independence' draws on Woodward's general characterization of different "procedures". These may differ because of underlying physical principles or processes. For example, optical microscopes and electron microscopes are governed by different physical principles. The electron microscope uses a high voltage electron beam to form the image (transmission electron microscope), or it uses detection of low energy "secondary electrons" emitted by the surface of the object as a result of excitation by the "primary electron beam" (scanning electron microscope). In contrast, light microscopes use techniques involving light—such as interference, polarization, phase contrast, direct transmission, and fluorescence.

Physical principles are not sufficient in differentiating modes. For example, both the scanning electron microscopes and transmission electron microscopes use the same physical principles (Stegenga 2012). However, it matters in what way the physical principles are employed. Both microscopes use high-energy beams of electrons but there is a difference between how, in each of the processes, the beams interact with the samples. Similarly, a given set of biomarkers may use the same physiological pathways, but it matters in what way these biomarkers *interact with a given clinical intervention.* The physical principle account can be modified to a *physical process account.*

Woodward's (2006) specification of causal robustness is useful for this. Causal robustness focuses on invariant relationships over interventions or manipulations in conditions. So, if we observe invariance in some EHR result after we have modified recordings of social and behavior choices, we can specify under what conditions invariance occurs. Trends that survive modified conditions are robust and, to use Woodward's (2006) reasoning, indicate causal structure. We can also specify the conditions under which the robust result remains invariant—e.g., some result R occurs under admissible changes in a specified range of conditions C.

This section serves as proper background to set up the methodological puzzle of surrogate markers that we present in the subsequent section. At first glance it seems like robustness analysis and the evaluation of independent modes are straightforward in clinical biomarkers. But what we see is that *both robustness analysis and independence analysis require re-conceptualization to make adequate solutions in the complex and divergence-filled realm of biomarkers.*

[9]For example, consider data on some disease derived from electronic health record (EHR) by clustering health and lifestyle variables, and lab data on the development of the same condition from animals exposed to a specific toxin. The EHR data build on the idealized assumption that patients surveyed in the given sample, do not all share a specific hidden confound that would skew the effect. The lab data build on the idealized assumption that the animal physiology used is relevantly similar to human physiology for this type of effect. Because these assumptions are not shared, convergence of results is less likely to come from the same systematic error.

13.3 Surrogate Markers

In this section, we describe surrogate markers (SM) and the methodological difficulty of evaluating their reliability as well as the reliability of unvalidated biomarkers. First, we define the specific role of surrogate markers according to the standards in clinical practice. Second, we describe the problematic use of multiple biomarkers: whether biomarkers are validated or unvalidated, the use of multiple markers often yields discordant or diverging results.

13.3.1 Surrogate Markers and Clinical Standards

Broadly defined, a SM is any assayable metric, which substitutes for a clinically-meaningful endpoint or penultimate pathological state, relating to important patient-centered outcomes—including affective measures and physiological function or survival (De Gruttola et al. 2001). Because an SM achieves its expected value prior to the consummation of the clinical endpoint—i.e. it precedes it in time—surrogate endpoints are of great utility in contexts where the clinical endpoint is grave and best not sampled in a patient or subject population (e.g. kidney failure). Surrogate endpoints are also of great utility in contexts where the clinical endpoint is beyond our ability to easily or reliably measure it—e.g., it would require a measurement of progression over years, prohibitively expensive equipment, or a technical demand beyond current capabilities.

Before describing a problem for SMs and biomarkers in general, it is important to note that there are stringent tests to establish a marker as a 'validated surrogate maker'. The methodological goal is to have surrogate markers that provide reliable information regarding processes that cannot be interrogated directly. We characterize 'reliability' in two senses: predictive and causal-explanatory. 'Predictive reliability' means that there is a high predictive value, resultant in the diagnostic device having both high sensitivity and high specificity. 'Causal-explanatory reliability' means that the use of biomarkers provides key causal information about a given state of a clinical endpoint.

Predictive reliability There is a high predictive value resultant in both high sensitivity and high specificity. High sensitivity conditions refer to the minimization of false negatives, whereby the marker always/often achieves an expected value when a process threshold[10] or clinical endpoint is proximate. High specificity refers to the minimization of false positives, whereby the marker never/rarely achieves an expected value when a process threshold or clinical endpoint is not proximate.

[10]A process threshold is analogous to a point of no return or a penultimate risk factor that relates to some physical/structural/tissue change in the underlying workings of a system. For example, glomerulosclerosis precedes the endpoint of end-stage renal disease/kidney failure.

Causal-Explanatory reliability The use of biomarkers provides key *causal information* about a given state of a clinical endpoint. The biomarker provides information that is embeddable within available theoretical models. These models can explain the relations between biomarker, clinical endpoint, and the risk factors associated with the endpoint.

In the literature on surrogate markers, most of the focus is placed on satisfying predictive reliability by looking at the close-knit relationship between surrogate (marker or endpoint) and clinical endpoint. Statistically, surrogate endpoints constitute "response variables" for which any manipulation on the surrogate variable will yield an effect that has a relevantly similar statistical significance to the effect (of the same manipulation) on the true clinical endpoint (Prentice 1989). Buyse et al. (2000) suggest that an adequate surrogate must be "tightly correlated" with the true clinical endpoint. In addition to this, any treatment effect on the surrogate must be tightly correlated with the treatment effect on the true clinical endpoint (Buyse et al. 2000). A simple example of this is that elevated concentrations of LDL cholesterol (LDL-C) have been associated with cardiovascular risk (Gofman et al. 1950a, b). With therapeutic interventions—e.g., the use of agents which inhibit the rate-limiting enzyme of the cholesterol biosynthesis pathway (statins)—LDL levels were lowered, which induces a reduction in cardiovascular disease (Downs et al. 1998; LaRosa et al. 2005; Ridker et al. 2008).

However, it has been argued that correlation, even if extensive, is not sufficient for a given metric to be considered a SM (Fleming and DeMets 1996). In order for a surrogate marker to be validated, the correlation must *hold under distinct (causal) interventions*. That is, a given SM and its corresponding endpoint would require causal robustness. For example, a marker specific for kidney failure such as cystatin C would need to be elevated under distinct causes of kidney damage—such as, poison-induced, ischemic, mechanical, genetic, and inflammatory—be they direct or indirect.

By the recommendation of the International Conference on Harmonisation Guideline E9 *Statistical Principles for Clinical Trials,* a surrogate endpoint (and, by extension, an ideal clinical biomarker) should have biological plausibility, prognostic value for the disease outcome, and there should be a strong association between changes in the surrogate and changes in the outcome with therapeutic intervention (Cleophas et al. 2007). Ideally, the change in the surrogate will correspond to changes in the target more or less linearly[11]—or according to a very well-characterized curve—e.g. 1-to-1, monotonic, or single-concavity—across their entire quantifiable range, and without respect to the trajectory of change (i.e. no observed hysteresis), and in both natural and interventional contexts. That is, the surrogate must act as *both epidemiological marker and therapeutic responder* (Cohn 2004).

[11] Woodward's (2004) account of causation is relevant here. An interesting discussion would be to apply Woodward's operation of manipulating one variable (surrogate marker) to see changes in another (the target variable).

Lassere et al. (2007) and Lassere (2007) summarize three conditions for validated SMs:

1. serve as harbingers of an irreversible clinical burden of disease, major morbidity or mortality,
2. have been repeatedly validated in randomized controlled trials using a diversity of interventional targets, and
3. consistently meet the highest standards of predictive association within and between studies.

For example, despite the fact that low HDL-C cholesterol may be causally relevant to coronary disease, due to the fact that HDL-raising interventions have not shown effectiveness for patient outcomes, HDL is not a classified as a validated surrogate whereas LDL has sufficiently satisfied these criteria (IOM 2010, Ch. 4).[12]

13.3.2 Biomarkers in Practice: Problems with Uncertainty and Divergence.

While idealized descriptions—e.g., Prentice (1989) and Lassere et al. (2007; 2008)—aim at establishing high standards for validated SMs, in practice, SMs never reach 100% sensitivity or specificity. For example, in a meta-analysis on Alzheimer's disease, dementia was evaluated using the cerebrospinal fluid peptide $Aß_{42}$ as a biomarker (Mo et al. 2015). In this meta-analysis, 436 out of 1349 participants developed Alzheimer's dementia. According to Ritchie et al. (2014), "Individual study estimates of sensitivity were between 36% and 100% while the specificities were between 29% and 91%" (7). Not only do sensitivity and specificity range between studies for a given surrogate marker, but sometimes it is even difficult to make a proper estimate of sensitivity and specificity, because we do not have information about the true endpoint. That is, the diagnosis of the actual health status of the study participants is itself based on *uncertain measures*. Cure (2014) uses autopsy in order to tighten the range of uncertainty for Alzheimer's measures. While

[12]If a surrogate is "reasonably likely" to forecast an outcome, but such a tether is not fully conclusive based on the evidence, a surrogate may be considered *unvalidated* and used for accelerated approval of drugs and medical devices in pressing clinical situations with few alternatives. In accordance with FDA regulations (CFR Title 21 Subpart H), these unvalidated SMs must be subsequently validated (Katz 2004). Unvalidated surrogates are also used in pre-clinical or pilot trials exploring safety or reasonable likelihood alone. As the spectrum of disease far outstrips our toolkit of validated surrogates, most disease-centered biomedical literature utilizes unvalidated surrogates as sources of evidence. It is important to note that the FDA only lists four validated surrogate markers: systolic blood pressure (SBD), low density lipoprotein cholesterol (LDL) level, forced expiratory volume in 1 second (FEV1), and human immunodeficiency virus (HIV) viral load (http://www.fda.gov/AboutFDA/Innovation/ucm512503.htm).

effective for generating specificity and sensitivity, such outcome measures are rarely accessible.

Overall, there is disagreement about cut-offs for sensitivity and specificity in CBs in general. Some researchers continue to posit high standards—e.g., 90% (Brower 2011). Others argue that sensitivity and specificity alone only work for discrete, not continuous variables (Cleophas et al. 2007). Pepe et al. (2016) posits that the thresholds we accept for CBs should depend on the prevalence of the disease endpoint and also on the costs of making a mistake versus the benefits of correct identification. For example, there are unvalidated surrogates, "reasonably likely to predict clinical benefit (for a specific disease setting and class of interventions)"(Fleming and Powers 2012). There are also "correlates that serve as measures of biological activity", such as CD-4, in HIV infected patients and fever in Community Acquired Bacterial Pneumonia (Fleming and Powers 2012, our emphasis).

Divergence between values of validated markers that are presumed to be systematically connected (and assumed to 'line up' in a causal structure), may also occur. When judging the difference in quantitative values between two biomarkers, whose values are expected to be aligned, scientists may refer to theory to embed both biomarker trajectories along a series of interventions and realize that one of the biomarkers is not following its theoretically predicted trajectory. For instance, in the aforementioned example, LDL-C was categorized as a validated SM; but Otvos et al. (2011) show the *inconsistency* of LDL measurement in relation to the clinical endpoint by comparing LDL-C and LDL particle number (LDL-P) measures.

It is worth noting that pragmatic considerations are also important in judgments about converging or diverging biomarkers. Scientists may judge that a given biomarker is "similar enough" or "differs too much" based on considerations that are outside of the theoretical and methodological parameters.

The problem of SM divergence is not directly solvable by referring to already available theoretical or explanatory considerations. Even in the category of validated SMs, causal complexity prevents clean-cut relationships between SMs and endpoints. This is where causal-explanatory reliability, i.e. extracting causal information from surrogates, becomes difficult. As reviewed in Fleming and Powers (2012), a marker may:

(C1) be a parallel output of an upstream event in the causal pathway, but not tie into the downstream causal pathway of the outcome of interest;
(C2) lie only within one of many parallel causal pathways in multifactorial disease processes;
(C3) be the target of an intervention that has direct or indirect off-target effects on the outcome of interest.[13] (See Fig. 13.1)

[13]That is, the intervention may produce other causal interactions that are relevant to the outcome of interest. Additionally, combinations of (C1), (C2), and (C3) are probable in biological systems.

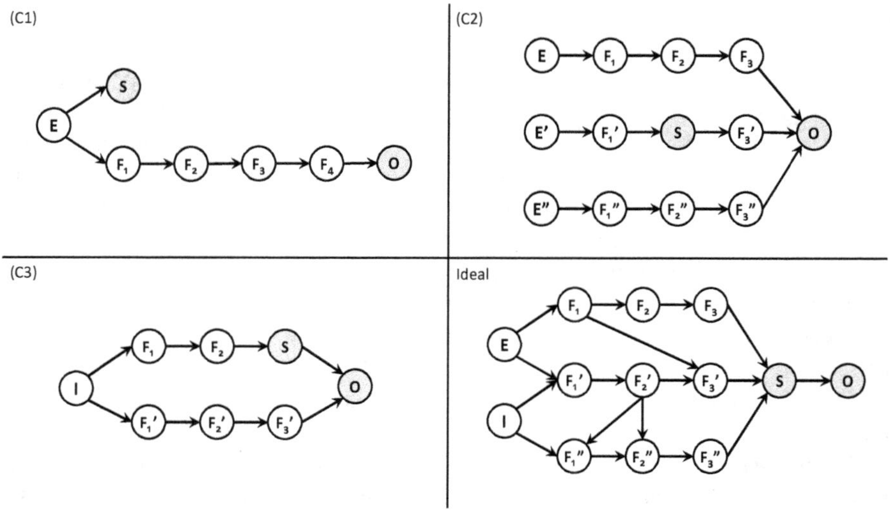

Fig. 13.1 E = event; S = surrogate; F = biological factor; O = outcome; I = intervention; C1-C3 adapted from the circumstances presented in Fleming & Powers; 'Ideal' refers to the behavior/location of an ideal surrogate within a biological web

In selecting targets, we must be cautious before concluding that the marker is located squarely within the causal pathway for our disease of interest. Even with perfect correlation and proximity, a marker may be part of a parallel process, an inert byproduct, an essential compensation mechanism, or an inessential compensation mechanism; and we can imagine that directly targeting the marker in this case might actually make for worsened outcomes. Hence, (C1)-(C3) reduce the validity of the surrogate marker.

When multiple surrogate markers are used, the methodological situation worsens.[14] Suppose that two surrogate markers diverge in results. Each marker may diverge *for any of the specified reasons (C1)-(C3)*. But in each case we do not know what is *producing the divergence*. Take the example of LDL as a surrogate for cardiovascular disease:

> Serum lipoproteins are particles that transport water-insoluble lipids such as triglycerides, cholesterol and cholesterol esters between the organs of energy transformation (liver, kidneys) and the organs of energy expenditure and storage (musculature, adipose). Low-density lipoprotein cholesterol (LDL-C), a measure of the total amount cholesterol contained within all LDL particles, is an accepted surrogate marker for cardiovascular disease risk, despite the fact that about half of hospitalized cardiovascular disease occurring in individuals with

[14]There is disagreement about combining bimarkers to make useful predictions. In Alzheimer's research, Lehmann et al. (2014) say that combining adequate markers (e.g., 80% specificity and sensitivity) improves their utility. Palmqvist et al. (2015) argue that combining markers does not improve their predictive utility—although he does not directly address higher ranges of specificity and sensitivity.

> normal or "optimal" LDL-C (< 100 mg/dL) (Sachdeva et al. 2009). Unsurprisingly, since the carriage capacity of lipoprotein particles is bound by lower and upper limits, the number of LDL particles (LDL-P) correlates strongly with LDL-C and has emerged as a strong predictor of risk.

In a large cohort, LDL-C and LDL-P values *differed* by more than 12 percentile units in half of the participants, with one-quarter of them confined to high LDL-C/low LDL-P and another quarter to low LDL-C/high LDL-P subgroups. Moreover, there seemed to be a difference in predictive value. The low LDL-C/high LDL-P subjects had: a 67% greater risk of a cardiovascular event than the high LDL-C/low LDL-P group; a 25% greater risk than the non-discordant group; and they were found to have a greater prevalence of the metabolic syndrome (MetS) and its associated biomarkers such as elevated triglycerides. Furthermore, in those with LDL-P below the 30th percentile, low LDL-C (also below the 30th percentile) increased cardiovascular event rates by 30%, while in those with LDL-P above the 30th percentile, low LDL-C increased event rate to a similar extent (Otvos et al. 2011).

In such cases where SMs diverge in results, the causal conclusions are puzzling. The primary methodological instinct is to weed out the SM with low sensitivity and specificity.[15] But in each case we do not know what is producing the divergence. We offer a methodological suggestion: *to use multiple surrogate markers in order to make causal conclusions at the theoretical level.* The solution that we will focus on for the remainder of the discussion is how to methodologically account for multiple surrogate markers with converging and diverging results.

13.4 Robustness Analysis and Divergence

In this section, we present an intervention-based methodology for assessing the reliability of surrogate markers. We do this by discussing a new method for individuating modes in Sect. 13.4.1 and then discussing the importance of converging and diverging results in biomarker readings in Sect. 13.4.2. Our intervention-based methodology will refer to two eliminative robustness analyses that use diverging results to make conclusions (Schupbach 2016 and Keyser 2016), both of which focus on elimination and indicate possible inferential uses of diverging results.

---

[15] As discussed, because sensitivity and specificity carry uncertainty, it would not be a simple case of using the "highest" scoring surrogate marker.

13.4.1 Individuating Modes; (Partially) Independent Pathways vs. Physical Principles

To use Woodward's (2006) reasoning, it is unlikely that different physical processes would share the same source for error: it is unlikely that different types of biomarkers share the same error source.[16] Standard accounts of robustness analysis rely on distinct physical principles or distinct physical processes underpinning the independence of modes of observation. [17] There are difficulties in specifying what counts as a 'different physical process' (Keyser 2016). Applied to biomarkers, there are many questions about differences in process: Do small adjustments, e.g., in only some of the conditions, constitute changes in physical process? What kinds of differences are important between biomarkers: the source, composition, and/or use? If only some of these conditions differ in two given biomarkers, are the biomarkers relevantly similar?

Keyser (2016) presents a heuristic-based account for differentiating physical modes: *the manipulation indicator*. One way to differentiate biomarkers is to look for indicators of physical independence:

> In the presence of the manipulation indicator, one measurement process can manipulate the results of other processes, and independence is compromised. For example, suppose that in (robustness analysis applied to the measurement process) you have two modes of gathering information through measurement, M1 and M2. To see if M1 and M2 are individuated modes, we can introduce a manipulation to the measurement conditions (e.g., instrumental, preparatory, etc.) behind M1 to see if it produces a change in M2. If it does, then there is successful manipulation, and we conclude that M1 and M2 are not independent. (2016, 17)

This account can be modified to be useful for individuating biomarkers. In the case of Keyser's (2016) measurement individuation, *independence is demonstrated through a manipulation indicator in upstream measurement processes*. It would not be sufficient to demonstrate independence by altering a downstream determinant just prior to or at the point of measurement output.

In using biomarkers as individuation indicators there are some difficulties. The first difficulty is that because we are using biology to probe biology, we are dealing with interconnected systems. This means that full independence of modes may be out of the question. This is where pragmatic considerations become important: modes are independent enough for the diagnostic purpose. Our account of the individuation/independence of biomarkers has a pragmatic component. However, whether partial or full, independence will also be determined using causal models that characterize a given intervention. So, we propose that partial independence is *determined* by using causal models—that characterize a given intervention, includ-

[16]Here, "error" does not refer to measurement error. In other words, we assume that the measurements reflect the actual value of the biomarker. In the case of biomarkers, "error" refers to the interference or confounding of unspecified biological variables in a physiological network.

[17]See Hacking (1983), Barad (2007), Stegenga (2009, 2012), and Keyser (2016), Woodward (2006).

ing the variables and values that constitute the model—in addition to pragmatic considerations about the purposes of a given intervention.[18]

Demonstrating (partial) independence in biomarkers involves the use of exogenous interventions that do not interact with one of the biomarkers of interest, but rather modify a process some number of "steps" (factors, voxels) removed from each of the markers.[19]

Here, pharmacological intervention is proven to be a fertile device for demonstrating independence in biomarkers. In this case, all we need to show is that, for any two biomarkers, the equation describing the curve of their relationship under natural conditions, is non-identical to the equation describing the curve of their relationship under interventional conditions that are indirect. If this is demonstrated, we have (partial) independence. In simple terms, the intervention does not affect each biomarker in the same way. The following two examples are presented for illustrative purposes: biomarkers of inflammation (CRP and ESR), and biomarkers of cognitive impairment (PET Amyloid and CSF Aβ).

CRP and ESR The acute-phase response is a systemic defense program initiated by the liver in response to rising levels of interleukin-6 (IL-6), an innate inflammatory messenger produced in response to tissue damage, infectious agents, environmental toxicants and other salient abnormalities in cellular function. During the acute-phase response, levels of several serum proteins change to assist the organism in meeting those needs, which commonly prove most adaptive under such challenges. Therefore, hallmarks of the acute phase response are commonly used as indicators of all-cause inflammation. Two such markers are C-reactive protein (CRP), a protein that approximates the function of a non-specific antibody (Marnell et al. 2005), and the erythrocyte sedimentation rate (ESR) that is essentially a measure of blood viscosity resulting from the abundance of other acute-phase proteins such as fibrinog en and α-globulins.

Despite both markers traditionally being considered interchangeable, given that they derive from the same organ under presumptively identical conditions, CRP and ESR may be discordant/divergent in 1030% of cases, with singly-elevated ESR being 6-fold more common (Costenbader et al. 2007; Colombet et al. 2010; Feldman et al. 2013; Sbong and Feldman 2014). The observation of discordance in simultaneous measurement of these markers is important in exploring their partial independence.

By introducing a pharmacological intervention we can see how each biomarker is affected. Prednisone is a synthetic glucocorticoid that reduces inflammation by strong activation of the glucocorticoid receptor in various tissues throughout the

[18]We thank an anonymous reviewer for the suggestion that 'partial independence' involves causal models.

[19]Absent of this, we run into the circularity that two metrics are partially-independent because we observe discordance and we can glean causal information from discordance among said markers because they are partially-independent.

body. Over 140 days of prednisone administration, CRP will decline monotonically, while ESR will show a backwards-J-shaped non-monotonic decrease over the same time period (McConkey et al. 1979). This illustrates that the two biomarkers respond differently to intervention and are therefore partially-independent.

PET Amyloid and CSF Aβ The drug Semagacestat is an inhibitor of the integral membrane protease complex known as γ-secretase, which is the terminal processing step in the production of the Aβ peptides and other essential protein cleavage products within neurons (and other tissues). As an intervention, Semagacestat is therefore indirect for the biomarkers under consideration.[20] Semagacestat dose-dependently reduces CSF Aβ production in humans (Imbimbo and Giardina 2011). Furthermore, similar compounds, administered to humanized transgenic mice, have been shown to prevent new plaque formation with no change in existing plaques as visualized with amyloid plaque labels (Garcia-Alloza et al. 2009). Therefore one may conclude that in the presence of these agents, PET-measured amyloid and CSF Aβ change with independent dynamics in contrast to the natural history of Alzheimer's disease.

An intervention-based account of individuation/independence can be useful for understanding robustness analysis. *Once biomarkers are shown to be partially-independent we can begin to ask questions about what we can conclude from converging and diverging results in the presence of intervention.* In the next section we discuss how robustness analysis can be used in biomarkers in order to make conclusions about causal relations.

13.4.2 Robustness and Elimination: Specifying Causal Relations.

Traditionally, robustness analysis with respect to the modeling process provides a way to illuminate the relationship between the robust consequence and certain aspects of each model in order to conclude that the inessentials of the model (the idealizations) are not producing the robust consequence. Framing robustness analysis this way, we focus on eliminating the inessentials of the model.[21] Two philosophical accounts that focus on elimination in robustness analysis are useful: Schupbach (2016) and Keyser (2016). The former focuses on using robustness

[20] Due to a significant worsening of cognitive scores and the emergence of several alarming off-target effects in the semagacestat groups, the full panel biomarker assessments were not completed in humans prior to termination of clinical trials.

[21] Orzack and Sober's (1993) criticism still looms. Perhaps it is some common core, shared by the individual modes, that is driving the robust result. See Justus (2012) for a summary of the concern: robustness analysis may only reflect shared properties of models rather than anything about the real world system (798).

analysis to eliminate competing hypotheses. The latter focuses on tracking causal relations based on multimodal comparison.

Schupbach (2016) uses Horwich's (1982) eliminative account of evidential robustness, the emphasis of which is using evidence to eliminate competing hypotheses in relation to some hypothesis (H) (2016, 314). According to Schupbach:

> What really matters for (Robustness Analysis (RA))-diversity is that the means (which may actually be quite similar in most respects) are different in just the sense required to rule out H's salient competitors. Accordingly, when seeking to increase the RA-diversity of our evidence, we search for a new way of detecting (the robust result) R that rules out some of H's still-standing competitors. (2016, 316)

According to Schupbach, new modes of evidence need not be fully independent. Rather, they just have to be "relevantly different" for the purpose of eliminating competing hypotheses (2016, 316). This fits nicely with our use of 'partial independence' in biomarkers.[22] Finding modes that are different enough is important for biomarker analysis. In Alzheimer's research, PET amyloid & CSF Aβ are biomarkers that are individuated enough within the context of intervention. We can slightly change Schupbach's account, so that robustness offers hypothesis *revision*, not just elimination. As we will see in the Alzheimer's research application in Sect. 13.5, revision gets at an important process in robustness analysis: the theoretical specification of causal details.

To make the eliminativist picture even more fine-grained, we can focus on the use of multiple modes to figure out what parts of each mode are producing causally relevant results. This is where we can use Woodward's (2006) specification of causal robustness. Causal robustness focuses on invariant relationships over manipulations. Manipulation is relevant to the case of biomarkers that undergo interventions because some relationships remain invariant while others produce diverging results. Trends that survive modified conditions are robust and, to use Woodward's (2006) reasoning, indicate causal structure. We can specify the conditions under which the robust result remains invariant—e.g., some result R occurs under admissible changes in a specified range of conditions C. However, as discussed in Sect. 13.3, many biomarkers do not maintain converging results, and results from multiple biomarkers often diverge. So we need a robustness account that provides causal information while effectively interpreting diverging as well as converging results.

Keyser (2016) offers an account of robustness analysis in measurement that consists of the comparison between independent modes of measurement along with the use of theory to make sense of that comparison. He illustrates that theory in the presence of diverging measurement results can pinpoint *what is producing the divergence*. For example, when multiple thermometers converge but others diverge,

[22] To illustrate his point, Schupbach uses the example of Perrin's modes of measuring Brownian motion. While Perrin's use of varieties of pollen are not "strongly heterogeneous" because each experiment uses a type of pollen; the experiments with varieties of pollen are different enough in order to rule out potential confounding hypotheses—such as, Brownian motion is only due to a specific type of pollen (2016, 316).

theory is used to analyze the conditions behind the divergence (Keyser 2016, 10). Keyser proposes that after initial cross-comparison of measurement methods and results, we use theory to home in on specific physical differences in thermometers—e.g., the liquid used in a given thermometer—in order to *explain* how those features produce differences in results (Keyser 2016, 10–11).[23] This differential comparison and explanation process in the presence of theory is useful for measurement in terms of specifying causal relations within the measurement set-ups, not only in terms of which conditions are *difference-makers* but also in terms of how error is "tracked"—i.e., located (Keyser 2016, 11). This is precisely why this account is useful for biomarkers: Cross-comparison allows for *locating relevant causal relations in biomarker measures*. We can refine and structure Keyser's (2016) account in order to provide useful steps for our proposed intervention-based robustness methodology.

The two methods of eliminative robustness can be used in a complementary fashion. *One focuses on eliminating alternative hypotheses (or revising hypotheses). The other can be used to further specify relations between variables within the context of theory*. We can generate three conditions for robustness analysis in biomarkers, which draw on these two methods. These conditions are useful for producing causal information.

R1) *Partial independence of modes*: Demonstrating partial physical independence in biomarkers involves an exogenous intervention that does not directly interact with one of the biomarkers of interest, but rather modifies a process some number of "steps" removed from each of the markers. Partially-independent biomarkers can serve as crosschecks on each other's results.

R2) *Comparison of diverging and converging results across biomarkers*: Partially-independent biomarkers (along with the relevant external causal conditions) are systematically juxtaposed to see: (A) where they converge and diverge in results; and(B) what conditions are responsible for the convergence and divergence. By systematically comparing partially-independent biomarkers we can track under what conditions they fail to converge in results, and under which conditions they successfully converge. Here, in addition to focusing on the robust result (the result that is invariant over biomarker measures), we *also* focus on the contextual failure of convergence in the cross-comparison of biomarkers (divergence). From both converging and diverging results in cross-comparison we get causal information about how the biomarkers work.

R3) Through (R1) and (R2), theoretical information is generated within the context of theory. By applying theory, cross-comparison provides information about causal relations. That is, through a systematic cross-comparison of the markers

[23]Keyser (2016) draws on van Fraasen's (2008) discussion of the relation between theory and measurement practice: Theory classifies what is being measured. However, as Keyser points out, this does not have to be a fully developed theory. It can even amount to a theory of how the instruments work. This may be applicable to cases of biomarker measurement where there is no overarching theory.

we can discover under what conditions certain changes are being produced, including confounding results. This kind of causal specification is useful for theory revision. That is, based on the juxtaposition of results, theories are eliminated and/or revised in order to provide adequate representations of the system being studied.

We apply (R1)-(R3) to analyze how biomarker modes can be useful for making causal conclusions in Alzheimer's research.

13.5 Robustness Analysis Applied to Alzheimer's Research

Before we apply our intervention-based robustness analysis, in Sect. 13.5.1 we briefly introduce theory and measurement in Alzheimer's disease research with particular attention to the revision of theory in the context of multiple biomarkers. In Sect. 13.5.2, we apply our methodology to Alzheimer's research in order to show its adequacy for analyzing currently developing empirical research. Following our robustness methodology conditions (R1)(R3), we show that when systematically combining different interventions on multiple biomarkers, promising causal information emerges.

13.5.1 Alzheimer's and Theoretical Revision

The theoretical and biomarker details of Alzheimer's research are important for our applied account of robustness analysis. Through the details, we show *how* theoretical models are *refined in relation to the converging and diverging results in biomarker measures.* A way to usefully apply robustness analysis to Alzheimer's research is to look at how biomarker robustness leads to specifications in theoretical robustness.

Traditionally, Alzheimer's disease was presumptively diagnosed based on clinical observation, such as cognitive testing battery biomarkers like the Mini-Mental State Exam (MMSE), and it was confirmed via autopsy by presence of the neuropathological signs of disease—e.g., amyloid plaques, neurofibrillary tangles, synaptic decay and neuronal loss. Recently, the development of bio-analytics and advanced neuroimaging has allowed these hallmarks of disease to be tracked and visualized in living subjects. Massive initiatives have helped to establish the current use of clinical biomarkers, which themselves have given rise to a multi-decade staging of the disease process (Jack and Holtzman 2013; Young et al. 2014). Notable are the refinements to the prevailing theoretical disease model, the *amyloid cascade hypothesis* (ACH) as a consequence of multiple biomarker measures.

Currently, the biomarkers in greatest use are cerebrospinal fluid (CSF) tau protein (the principal component of neurofibrillary tangles), CSF 42-amino acid amyloid-β

($A\beta_{42}$; hereafter CSF Aβ). regional brain volume as determined by structural MRI, neuronal metabolism measured by glucose analog uptake and positron emission tomography (FDG-PET), and amyloid plaques measured with PET detection of amyloid-binding dyes. CSF Aβ is the protein cleavage product believed to precipitate disease by forming neuron-damaging plaques as it aggregates, and "42 amino acids" refers to the length of this large peptide, measured by the number of basic amino acid building blocks it possesses.[24]

Based on the ACH, it was posited that CSF Aβ would be elevated in Alzheimer's dementia. Thus the discovery that levels were significantly reduced in Alzheimer's was initially met with surprise (Hampel et al. 2008). Hardy and Allsop (1991) argued that an imbalance between amyloid production and clearance (potentially caused by a number of factors) led to an accumulation of Aβ which eventually precipitated into plaques, damaged neurons and sequentially led to the downstream pathology (tangles of modified tau, neuronal stress, damage, and ultimately loss) of AD.

Youngkin (1995) advanced that a 42-amino acid (rather than a 40-amino acid) Aβ peptide was likely the culprit by. This peptide was characterized as being "stickier" in the sense of more likely to aggregate, and it was correlated better with the production/elimination imbalance, especially in genetic (early-onset) forms of AD.

Over time many potentially toxic mechanisms of Aβ were characterized, such as triggering inflammation and excitotoxicity (Behl 1997; Canevari et al. 2004; Carrillo-Mora et al. 2014)—with many, now hypothesized to result from repeating units of adhered misfolded Aβ peptides referred to as Aβ oligomers. Recently, the discussion of oligomeric forms of Aβ 42 has expanded.[25] Very soon, a gestalt shift in theoretical explanation allowed for a crucial discovery: amyloid plaques within the brain were acting as sinks for the circulating CSF Aβ and, beyond a critical concentration, Aβ would precipitate into these plaques, or, at the very least, newly produced Aβ would be sequestered in plaques to diminish steady-state levels of CSF Aβ. If the brain were imaged, one should observe the retention of amyloid-stains as CSF Aβ dropped.

This specification of the original amyloid cascade hypothesis (call it ACH_*) relied on different assumptions about the nature of Aβ and amyloid plaques. In ACH_* the assumption is not that soluble monomers (single units) are elevated

---

[24]Additional imaging techniques are being adapted and blood plasma measurements of Aβ are being developed.

[25]Currently, these are thought to be the most toxic form, but lesser or interacting toxicity of other forms of Aβ is not discounted. The current discussion revolves around the many forms and sizes of oligomers: some toxic, some not, prefibrillar forms, and fibrillary (fiber-like) forms which can form either diffuse or dense plaques. The general order of formation is asserted to be: peptides, small oligomers, larger oligomers, prefibrillar forms, fibrillary forms, diffuse plaques, dense neuritic plaques. But this formation order is in no way invariant, as there are branches, two-way streets and overlaps. It is important to note that much of this information on ACH and oligomers was hypothesized before the biomarkers were characterized. It can be argued that such theoretical explanations were not fully integrated into the theoretical model of ACH until recently with the help of biomarkers. The most current model of the theory is discussed by Selkoe and Hardy (2016).

throughout the time course of pathology; also they are not a primary agent responsible for damage.[26] Even though the importance of Aβ is invariant over both theoretical models, ACH_* provided most of the useful explanation. But it too required empirical confirmation; and this was done in relation to multiple *diverging biomarker results.*

ACH_* predicted that depressed CSF Aβ and an elevated amyloid plaque PET signal would be fully correlated over their numerical range, in that they were assumed to be dependent biomarkers. That is, PET signals should increase as CSF Aβ levels decrease and vice-versa. Indeed, they were shown to correlate on the order of 85% using threshold values (Landau et al. 2013; Zwan et al. 2013; Vos et al. 2016), but even a *15% discordance appeared to indicate missing causal specifications*. So another theoretical model is now needed. In fact, biomarker discordance, especially in the preclinical and early stages of Alzheimer's coupled with the lack of success of targeted pharmacologic agents has led many to question the accuracy of the ACH/ACH_*.

In normal subjects and those with mild cognitive impairment (MCI), PET positivity and CSF Aβ "negativity" (that is, higher values) is the most common form of discordance (Landau et al. 2013).[27] This is despite the fact that early on it was observed that, in cases of discordance, CSF Aβ decline would *precede* PET positivity (Musiek and Holtzman 2012; Zwan et al. 2013).

Such discordant/diverging results supported the current development of a new sub-model according to which:

1. either Aβ is incorporated into soluble oligomers (Lesne 2014), pre-fibrillary forms and/or diffuse fibrillary precipitates *prior to* the appearance of dye-retaining plaques—an issue of laddering (oligomer submodel, OSM, Walsh et al. 2000),
2. or potential blood-brain barrier dysfunctions increase the dilution of CSF Aβ into peripheral compartments (Erickson and Banks 2013).

The OSM produces useful explanations with new causal components neither included in ACH nor in ACH_*. On the basis of OSM, cognitive impairment can precede amyloid biomarker positivity despite later progression to Alzheimer's dementia.[28] Oligomers seem to form to some degree prior to plaques, so we could have high CSF Aβ without much plaquing.[29]

[26]The second reason had been suspected (Reviewed in Walsh and Selkoe 2007; Review that monomers may actually be protective: Giuffrida et al. 2009).

[27]Such discordance has been observed in 21% of normal individuals, 12% of MCI cases and in 6% of cases with diagnosed Alzheimer's dementia (Mattsson et al. 2014). It is worth noting that the oligomer sub-model, discussed later in the paragraph, can account for both types of discordance mentioned in this study: the larger discordance of the florbetapir$^+$ (PET positivity) /CSF Aβ(-) group and the smaller discordance of the florbetapir$^-$ (PET negativity)/CSF Aβ(+) group.

[28]This has indeed been demonstrated by Toledo et al. 2014 and Ritchie et al. 2016.

[29]Additionally, OSM can account for the fact that an individual may have cognitive decline even with low CSF Aβ and no plaques because processing enzymes that produce Aβ are also necessary

Juxtaposing OSM with ACH and ACH_* shows some robustness. Each model—ACH, ACH_*, and OSM's agree that Aβ is an important factor in Alzheimer's disease, but they do not posit the same causal relations.

However, the robust explanation, which has to do with the causal role of Aβ in Alzheimer's, doesn't result from using different models to explain a result. Rather, these models are *refined in relation to the converging and diverging results in biomarker measures*.

One could argue that there are two types of robustness occurring here: the robustness of biomarker measures (biomarker robustness) and the robustness of theoretical models (theoretical robustness). A way to usefully apply robustness analysis to Alzheimer's research is to look at how biomarker robustness leads to specifications in theoretical robustness. As we have just seen, in Alzheimer's research, diverging biomarker readings catalyzed the specification of new theoretical models. Discrepancies in multiple biomarkers in various studies facilitated the move from ACH to ACH_* and then to OSM. This can be analyzed as a case where *multiple modes (biomarkers) support theoretical revision*. As mentioned in Sect. 13.4 and illustrated in this section, this is precisely why it is important to change Schupbach's account so that robustness offers hypothesis revision, not just elimination. Relevant details of ACH_*, for example, were eliminated for OSM, but some previous aspects did not change. As we will see next, revision is an important process in robustness analysis because it is behind the theoretical specification of causal details.

13.5.2 Intervention-Based Robustness Analysis Applied to Alzheimer's Measurement

Under the amyloid cascade models discussed above (ACH, ACH_*, and OSM), reduction of cerebral Aβ is expected to reduce cognitive decline and may afford neurons a chance to repair, either by delaying debilitating dementia or restoring cognitive function. It is predicted that a 25% reduction in Aβ species may prove to be an adequate therapeutic target for delayed disease progression to be observed (Toyn 2015).

Human trials of these agents have heretofore been disappointing largely because agents proved to be poorly tolerated at biochemically active doses. At those same doses disagreement among clinical biomarker findings as well as between biomarkers and established functional endpoints were observed (Toyn 2015). Altogether, this has contributed to several calls for a rejection of the ACH, or at the very least,

---

for the production of neurotrophic and neurodifferentiation factors (Willem et al. 2006; Woo et al. 2009; De Strooper et al. 2010). Thus, low activity might lower both CSF Aβ and factors important for optimal neuronal function. This may be seen in neuroinflammation or CNS infection as well (Krut et al. 2012).

its role as the cardinal focal point of disease research. Though difficult to interpret resultant to biomarker discordance, fragmentary positive findings have shifted focus toward more upstream interventions as desirable targets in antibody trials (Toyn 2015).

Presently, promising research has emerged: The PRIME trial (Sevigny et al. 2016) is a Phase 1b clinical trial (a dose-escalation trial in a limited number of patients) conducted across 33 locations within the United States testing the efficacy of Aducanumab. Aducanumab specifically targets Aβ fibrils as well a toxic Aβ oligomers including decamers (arc-, disc- or pore-shaped composites consisting of 10 regularly repeated Aβ monomers adhered together) (Ratner 2015). Binding studies reveal that Aducanumab does, in fact, bind fibrils in both diffuse deposits and compact plaques and that chimeric mouse-adapted versions of the drug precipitate Aβ oligomers in vitro by clumping them together into large complexes.[30] Aducanumab binds CAA deposits (see above) with less prominence and has 10,000-fold target selectivity for these composite forms over Aβ monomers.

Subjects were divided into five arms, to receive monthly IV infusions of either placebo or 1, 3, 6 or 10 mg/kg bodyweight aducanumab over a one-year period. After 12 months, the 10 mg/kg group had PET dye retention reduced 2530%, almost to the cutoff level for admission into the trial. Dose- and time-dependent reductions in PET positivity were seen across the intervention arms.

Despite being neither designed nor powered to assess efficacy with respect to cognitive outcomes, there was also a dose- and duration-dependent slowing of cognitive decline (but not reversal or improvement) in the Clinical Dementia Rating-Sum of Boxes (CDR-SB) and Mini-Mental State Examination (MMSE) scores. The 10 mg/kg group retained significantly more cognitive function than the placebo group and had effectively unchanged cognitive performance over the 12 months. Placebo-exposed subjects had no more PET-detected amyloid accumulation over the yearlong trial but continued to experience cognitive decline, pointing to one or more unidentified variables, presumably Aβ oligomers. In the 10 mg/kg group, despite continued reductions in PET amyloid positivity from 6 to 12 months, CDR-SB scores did not continue to improve. This indicates that changes in plaque abundance and cognitive function are divorced over their dynamic ranges.[31]

Following our robustness methodology conditions (R1)(R3), when systematically combining different interventions on multiple biomarkers, promising causal information emerges. We discuss this by directly comparing bapineuzumab and solanezumab interventions, along with aspects of the aducanumab intervention.

[30] In simpler terms, scientists replaced the "tail", also known as the constant region of the antibody with that from a mouse so that the mouse's immune system wouldn't have an immune response to the human antibody. This shows that it is the specific antigen-binding region from the screen, which interacts with Aβ oligomers.

[31] While awaiting Phase III efficacy trials, which were begun immediately upon consolidation of positive findings, a lingering question (Lee et al. 2006) poses whether plaque destabilization could actually lead to increased exposure of neurons to toxic forms of amyloid as many structural models indicate that plaque-oligomer interconversion could be bidirectional.

Marker ‖ Condition	Disease	Bapineuzumab	Solanezumab	Aducanumab
PET Amyloid Dye Retention	↑	↓	=	↓
CSF Aβ	↓	=	↓	NM
CSF Tau	↑	↓	=	NM
Plasma Aβ	↑	NM	↑	NM
Regional Brain Volume	↓	NM	NM	NM
Neuronal Metabolism	↓	NM	NM	NM
Cognitive Battery	↓	=	↑†	↑

Fig. 13.2 Arrows indicate increase or decrease relative to unaffected population (column 2) or relative to control or baseline (columns 3, 4, 5); '=' indicates no difference relative to the control; 'NM' indicates Not Measured; '†' In a subset of participants. Crenezumab, another humanized mouse monoclonal antibody which targets Aβ aggregates, has a profile strikingly similar to solanezumab

Bapineuzumab is a humanized mouse monoclonal antibody that targets the N-terminal (front end) region of fibrillary (the principal configuration within plaques) and monomeric Aβ. In *APOE-ε4* gene carriers and non-carriers, bapineuzumab reduced Aβ PET positivity and decreased the neurodegeneration biomarker CSF tau, but did not alter levels of CSF Aβ (Toyn 2015). Subjects exposed to active treatment initially showed better retention of MMSE scores in the highest dose group but failed in late-stage clinical trials to improve cognition or activities of daily living versus control (Ratner 2015).

In contrast to bapineuzumab, solanezumab, a humanized mouse monoclonal antibody targeting the central region of Aβ monomers and small oligomers, did not alter levels of plaque assayed by Aβ PET, nor did it alter CSF tau. In addition, it led to an increase in plasma Aβ and a reduction in CSF Aβ (recall, this reduction is a marker for poor cognitive outcome in the *absence* of intervention) (Toyn 2015; Ratner 2015; Selkoe and Hardy 2016).[32]

Putting these results side by side (see Fig. 13.2), we see that there is divergence in relation to multiple biomarkers:

1. levels of plaque assayed by Aβ PET;
2. CSF tau; and
3. CSF Aβ.

Bapineuzumab reduced (1) and (2), but did not alter levels of (3).
Solanezumab did not alter (1) and (2) but led to a reduction in (3).[33]

[32] Solanezumab is currently undergoing pooled subgroup re-evaluation from two Phase III trials after small but significant positive findings (34% deceleration of cognitive decline over 18 months versus placebo) were observed in Alzheimer's Disease Assessment Scale (ADAS) cognitive domain scores in those with mild impairment (Toyn 2015; Ratner 2015; Selkoe and Hardy 2016). Results of a follow-up Phase III study are in the offing (http://www.alzforum.org/therapeutics).

[33] As well as an increase in a factor not measured in the bapineuzumab study: plasma Aβ.

Methodologically, there are two important points here.

1. There is a contrasting picture between three biomarkers in relation to two interventions (R1 and R2). The divergence and convergence is systematic in that one intervention shows changes in two biomarkers and no change in the final one, and the other intervention shows stability in two biomarkers and a change in the final one. Furthermore, there is a difference in the outcome of cognitive improvement. So, the two different interventions not only produce differences in biomarker measures, but they also produce differences in cognitive results.
2. Within the context of theory, specifically OSM, this systematic divergence and convergence leads to useful causal conclusions (R3). Using robustness analysis, we specify multiple *promising conclusions*, which can be explored in further research:

Conclusion 1) Oligomers are a more promising causal target for preserving cognitive function (C1):

Following the antibody trials, solanezumab as well as aducanumab target oligomers as part of their specificity. Solaneuzumab results suggest improvement in cognition in mild cases based on the post-hoc pooling analysis. Aducanumab seems to improve cognition based on the Phase 1b results relative to control/placebo. Solaneuzumab and aducanumab target oligomers, but without a contrasting intervention that does not target oligomers, we cannot see if oligomer intervention is *necessary* for cognition improvement. Bapineuzumab provides such a contrast. It does not target oligomers and shows no evidence for improving cognition. The *convergence between solaneuzumab and aducanumab compared to the divergence of bapineuzumab can be made sense of in a theoretical context.* In this case OSM makes sense of how the targeting occurs in the case of the different interventions in order to conclude which interventions are and are not targeting oligomers. For example, OSM explains that aducanumab targets Aβ fibrils as well a toxic Aβ oligomers including decamers; solaneuzumab targets the central region of Aβ monomers and small oligomers; and bapineuzumab targets the N-terminal region of fibrillary (the principal configuration within plaques) and monomeric Aβ. From this we understand that solaneuzumab and aducanumab target oligomers and bapineuzumab does not. The converging results of cognition improvement in the oligomer-targeting interventions compared to the diverging results of the non-oligomer-targeting intervention supports causal specification that oligomers modification is necessary for cognitive improvement. That is, without manipulation of oligomers, cognitive improvement will not occur.

Conclusion 2) CSF tau may not be a *complete marker* of neurodegeneration (C2):

It is assumed that cognition reflects neurodegeneration (Terry et al. 1991; Savva et al. 2009). In bapineuzumab, CSF tau declined (presumably less or at lower rate of neurodegeneration) but cognition did not change. In solanezumab, CSF tau did not change (no change in rate of neurodegeneration) but cognition improved relative to control.

Neurodegeneration and cognition are not correlated *in the way that we assume*. For example, there could be extra variables related to rate or duration of damage in upstream or downstream processes that are causally linked to cognition. This option would imply that neurodegeneration and cognitive decline do not change in lockstep due to variable complexity. If tau is a marker for neurodegeneration, and tau and cognition do not correlate, then neurodegeneration and cognition do not correlate. The lack of correlation between tau and cognition is supported by the shared lack of convergence for tau and cognitive markers in bapineuzumab and solanezumab interventions. However, we need to further explore the first conjunct in the antecedent, which is that tau is a maker for neurodegeneration.

Conclusion 3) PET positivity is not a surrogate for oligomers (C3).

Sevigny et al. (2016) refers to PET positivity as a "surrogate" of oligomers. After targeting oligomers, which cannot be measured directly in vivo, as well as plaques, Sevigny et al. (2016) found that PET-detected plaques decreased. This does not point to a clear relationship between PET positivity and oligomers because the latter were not directly measured. One theoretical option is that there is a transition of monomers to oligomers to plaques that is unidirectional, and reducing plaques will pull oligomers into new plaques (thus reducing them).[34]

However, even without direct oligomer measurements, we can use robustness analysis to question Sevigny's theoretical model. If it is the case that oligomers are causally connected with cognition[35] and solanezumab actually targeted oligomers,[36] then PET positivity cannot be a surrogate of oligomers, as Sevigny claims. The reason for this is that, in the solanezumab intervention, while oligomers were targeted which produced a change in cognition, *PET positivity did not change*. In addition to this, the bapineuzumab intervention resulted in a *PET positivity decline* but cognition did not change. Because of the support from C1, which posits a close connection between oligomers and cognition, we can argue that no cognition change indicates that there is no oligomer change. *Both* interventions converge on PET positivity change not being correlated with cognition change, even though oligomer change is. This implies that PET positivity is not a surrogate for oligomers.[37]

[34]This has the corollary that plaques are being pulled out faster than oligomers can form new plaques, and monomers cannot form new oligomers as fast as oligomers go into plaques. This is similar to LeChatelier's principle. This could also be turned into a "bidirectionality" theoretical model.

[35]See C1 for support.

[36]See background on solanezumab earlier in the section for support.

[37]In a stronger form of robustness analysis, we can use the convergence of results to eliminate Sevigny's theoretical model. Alternatively, auxiliary modifications can be made positing a more complex causal relationship between PET positivity, oligomer change, and cognition. But given the analysis in C1 and C3 thus far, we can at least cast doubt on Sevigny's theoretical model.

13.6 Concluding Remarks

We have presented an intervention-based robustness methodology that specifies how to check for partial independence in biomarkers, how divergence as well as convergence is useful for figuring out causal information related to biomarkers and endpoints, and how theory revision works in relation to biomarker analysis. The aim of this discussion was for the purpose of philosophical methodological analysis, but the application of our account results in important consequences for empirical research. When applied to Alzheimer's research, our robustness methodology produces specific conclusions about *promising causal culprits as well as decoupled biomarkers and endpoints (C1)-(C3)*. By transforming robustness analysis into an eliminative account, we have made it useful for dealing with complex and discordant/diverging results in biomarker measurement.

References

ALZFORUM. (n.d.). *Networking for a cure*. http://www.alzforum.org/therapeutics

Aronson, J. K. (2005). Biomarkers and surrogate endpoints. *British Journal of Clinical Pharmacology, 59*(5), 491–494. https://doi.org/10.1111/j.1365-2125.2005.02435.x.

Barad, K. (2007). Meeting the universe halfway. In *Meeting the universe halfway* (pp. 39–70). Duke University Press. https://doi.org/10.1215/9780822388128-002

Behl, C. (1997). Amyloid β-protein toxicity and oxidative stress in Alzheimers disease. *Cell and Tissue Research, 290*(3), 471–480. https://doi.org/10.1007/s004410050955.

Brower, V. (2011). Biomarkers: Portents of malignancy. *Nature, 471*(7339), S19–S20. https://doi.org/10.1038/471s19a.

Buyse, M., Molenberghs, G., Burzykowski, T., Renard, D., & Geys, H. (2000). The validation of surrogate endpoints in meta-analyses of randomized experiments. *Biostatistics, 1*(1), 49–67. https://doi.org/10.1093/biostatistics/1.1.49.

Canevari, L., Abramov, A. Y., & Duchen, M. R. (2004). Toxicity of amyloid β peptide: Tales of calcium, mitochondria, and oxidative stress. *Neurochemical Research, 29*(3), 637–650. https://doi.org/10.1023/b:nere.0000014834.06405.af.

Carrillo-Mora, P., Luna, R., & Colín-Barenque, L. (2014). Amyloid beta: Multiple mechanisms of toxicity and only some protective effects? *Oxidative Medicine and Cellular Longevity, 2014*, 1–15. https://doi.org/10.1155/2014/795375.

Cleophas, T., Zwinderman, A., & Chaib, A. (2007). Novel procedures for validating surrogate endpoints in clinical trials. *Current Clinical Pharmacology, 2*(2), 123–128. https://doi.org/10.2174/157488407780598126.

Cohn, J. N. (2004). Introduction to surrogate markers. *Circulation, 109*(25_suppl_1), IV–20–IV–21. https://doi.org/10.1161/01.cir.0000133441.05780.1d.

Colombet, I., Pouchot, J., Kronz, V., Hanras, X., Capron, L., Durieux, P., & Wyplosz, B. (2010). Agreement between erythrocyte sedimentation rate and c-reactive protein in hospital practice. *The American Journal of Medicine, 123*(9), 863.e7–863.e13. https://doi.org/10.1016/j.amjmed.2010.04.021.

Costenbader, K. H., Chibnik, L. B., & Schur, P. H. (2007). Discordance between erythrocyte sedimentation rate and c-reactive protein measurements: Clinical significance. *Clinical and Experimental Rheumatology, 25*(5), 746–749.

Culp, S. (1994). Defending robustness: The bacterial mesosome as a test case. *PSA: Proceedings of the Biennial Meeting of the Philosophy of Science Association, 1994*(1), 46–57. https://doi.org/10.1086/psaprocbienmeetp.1994.1.193010.

Cummings, J. L., Dubois, B., Molinuevo, J. L., & Scheltens, P. (2013). International work group criteria for the diagnosis of Alzheimer disease. *Medical Clinics of North America, 97*(3), 363–368. https://doi.org/10.1016/j.mcna.2013.01.001.

Cure, S., Abrams, K., Belger, M., Happich, M., & others. (2014). Systematic literature review and meta-analysis of diagnostic test accuracy in Alzheimer's disease and other dementia using autopsy as standard of truth. *Journal of Alzheimer's Disease, 42*(1), 169–182.

De Gruttola, V. G., Clax, P., DeMets, D. L., Downing, G. J., Ellenberg, S. S., Friedman, L., Gail, M. H., Prentice, R., Wittes, J., & Zeger, S. L. (2001). Considerations in the evaluation of surrogate endpoints in clinical trials. *Controlled Clinical Trials, 22*(5), 485–502. https://doi.org/10.1016/s0197-2456(01)00153-2.

De Strooper, B., Vassar, R., & Golde, T. (2010). The secretases: Enzymes with therapeutic potential in Alzheimer disease. *Nature Reviews Neurology, 6*(2), 99–107. https://doi.org/10.1038/nrneurol.2009.218.

Douglas, H. (2004). The irreducible complexity of objectivity. *Synthese, 138*(3), 453–473. https://doi.org/10.1023/b:synt.0000016451.18182.91.

Downs, J. R., Clearfield, M., Weis, S., Whitney, E., Shapiro, D. R., Beere, P. A., Langendorfer, A., et al. (1998). Primary prevention of acute coronary events with Lovastatin in men and women with average cholesterol levels. *JAMA, 279*(20), 1615. https://doi.org/10.1001/jama.279.20.1615.

Erickson, M. A., & Banks, W. A. (2013). Blood-brain barrier dysfunction as a cause and consequence of Alzheimer's disease. *Journal of Cerebral Blood Flow & Metabolism, 33*(10), 1500–1513. https://doi.org/10.1038/jcbfm.2013.135.

Feldman, M., Aziz, B., Kang, G. N., Opondo, M. A., Belz, R. K., & Sellers, C. (2013). C-reactive protein and erythrocyte sedimentation rate discordance: Frequency and causes in adults. *Translational Research, 161*(1), 37–43. https://doi.org/10.1016/j.trsl.2012.07.006.

Fleming, T. R., & DeMets, D. L. (1996). Surrogate end points in clinical trials: Are we being misled? *Annals of Internal Medicine, 125*, 605–613. https://doi.org/10.7326/0003-4819-125-7-199610010-00011.

Fleming, T. R., & Powers, J. H. (2012). Biomarkers and surrogate endpoints in clinical trials. *Statistics in Medicine, 31*(25), 2973–2984. https://doi.org/10.1002/sim.5403.

Food and Drug Administration. (2017). *FDA facts: Biomarkers and surrogate endpoints*. http://www.fda.gov/AboutFDA/Innovation/ucm512503.htm

Food and Drug Administration. (n.d.). *FDA Title 21 of US CFR*. https://www.accessdata.fda.gov/scripts/cdrh/cfdocs/cfcfr/CFRSearch.cfm?CFRPart=314&showFR=1&subpartNode=21:5.0.1.1.4.8

Franklin, A. (1997). Calibration. *Perspecties on Science, 5*, 31–80.

Garcia-Alloza, M., Subramanian, M., Thyssen, D., Borrelli, L. A., Fauq, A., Das, P., Golde, T. E., Hyman, B. T., & Bacskai, B. J. (2009). Existing plaques and neuritic abnormalities in APP:PS1 mice are not affected by administration of the gamma-secretase inhibitor LY-411575. *Molecular Neurodegeneration, 4*(1), 19. https://doi.org/10.1186/1750-1326-4-19.

Giuffrida, M. L., Caraci, F., Pignataro, B., Cataldo, S., De Bona, P., Bruno, V., Molinaro, G., et al. (2009). β-Amyloid monomers are neuroprotective. *Journal of Neuroscience, 29*(34), 10582–10587. https://doi.org/10.1523/jneurosci.1736-09.2009.

Glymour, C. (1980). *Theory and evidence*. Princeton: Princeton University Press.

Gofman, J. W., Jones, H. B., Lindgren, F. T., Lyon, T. P., Elliott, H. A., & Strisower, B. (1950a). Blood lipids and human atherosclerosis. *Circulation, 2*(2), 161–178. https://doi.org/10.1161/01.cir.2.2.161.

Gofman, J. W., Lindgren, F., Elliott, H., Mantz, W., Hewitt, J., Strisower, B., Herring, V., & Lyon, T. P. (1950b). The role of lipids and lipoproteins in atherosclerosis. *Science, 111*(2877), 166–186. https://doi.org/10.1126/science.111.2877.166.

Hacking, I. (1983). *Representing and intervening*. Cambridge: Cambridge University Press. https://doi.org/10.1017/cbo9780511814563.

Hampel, H., Bürger, K., Teipel, S. J., Bokde, A. L. W., Zetterberg, H., & Blennow, K. (2008). Core candidate neurochemical and imaging biomarkers of Alzheimer's disease. *Alzheimers & Dementia, 4*(1), 38–48. https://doi.org/10.1016/j.jalz.2007.08.006.

Hardy, J., & Allsop, D. (1991). Amyloid deposition as the central event in the aetiology of alzheimers disease. *Trends in Pharmacological Sciences, 12*(January), 383–388. https://doi.org/10.1016/0165-6147(91)90609-v.

Horwich, P. (2011). *Probability and evidence*. Cambridge University Press. https://doi.org/10.1017/cbo9781316494219.

Imbimbo, B. P., & Giardina, G. A. M. (2011). γ-secretase inhibitors and modulators for the treatment of Alzheimers disease: Disappointments and hopes. *Current Topics in Medicinal Chemistry, 11*(12), 1555–1570. https://doi.org/10.2174/156802611795860942.

Institute of Medicine. (2010). *Evaluation of biomarkers and surrogate endpoints in chronic disease*. Washington, DC: National Academies Press. https://doi.org/10.17226/12869.

Jack, C. R., & Holtzman, D. M. (2013). Biomarker modeling of Alzheimer's disease. *Neuron, 80*(6), 1347–1358. https://doi.org/10.1016/j.neuron.2013.12.003.

Justus, J. (2012). The elusive basis of inferential robustness. *Philosophy of Science, 79*(5), 795–807. https://doi.org/10.1086/667902.

Katz, R. (2004). Biomarkers and surrogate markers: An FDA perspective. *NeuroRX, 1*(2), 189–195. https://doi.org/10.1602/neurorx.1.2.189.

Keeley, B. L. (2002). Making sense of the senses. *Journal of Philosophy, 99*(1), 5–28. https://doi.org/10.5840/jphil20029915.. Edited by John Smylie.

Keyser, V. (2016). *A new theory of robust measurement*. http://www.apaonline.org/members/group_content_view.asp?group=110424&id=476093

Krut, J. J., Zetterberg, H., Blennow, K., Cinque, P., Hagberg, L., Price, R. W., Studahl, M., & Gisslén, M. (2012). Cerebrospinal fluid Alzheimers biomarker profiles in CNS infections. *Journal of Neurology, 260*(2), 620–626. https://doi.org/10.1007/s00415-012-6688-y.

Landau, S. M., Lu, M., Joshi, A. D., Pontecorvo, M., Mintun, M. A., Trojanowski, J. Q., Shaw, L. M., Jagust, W. J., et al. (2013). Comparing positron emission tomography imaging and cerebrospinal fluid measurements of β-amyloid. *Annals of Neurology, 74*(6), 826–836. https://doi.org/10.1002/ana.23908.

LaRosa, J. C., Grundy, S. M., Waters, D. D., Shear, C., Barter, P., Fruchart, J.-C., Gotto, A. M., et al. (2005). Intensive lipid lowering with atorvastatin in patients with stable coronary disease. *New England Journal of Medicine, 352*(14), 1425–1435. https://doi.org/10.1056/nejmoa050461.

Lassere, M. N. (2007). The biomarker-surrogacy evaluation schema: A review of the biomarker-surrogate literature and a proposal for a criterion-based, quantitative, multidimensional hierarchical levels of evidence schema for evaluating the status of biomarkers as surrogate endpoints. *Statistical Methods in Medical Research, 17*(3), 303–340. https://doi.org/10.1177/0962280207082719.

Lassere, M. N., Johnson, K. R., Boers, M., Tugwell, P., Brooks, P., Simon, L., Strand, V., et al. (2007). Definitions and validation criteria for biomarkers and surrogate endpoints: Development and testing of a quantitative hierarchical levels of evidence schema. *The Journal of Rheumatology, 34*(3), 607–615.

Lee, H.-g., Zhu, X., Nunomura, A., Perry, G., & Smith, M. A. (2006). Amyloid-β vaccination: Testing the amyloid hypothesis? *The American Journal of Pathology, 169*(3), 738–739. https://doi.org/10.2353/ajpath.2006.060633.

Lehmann, S., Dumurgier, J., Schraen, S., Wallon, D., Blanc, F., Magnin, E., Bombois, S., et al. (2014). A diagnostic scale for Alzheimer's disease based on cerebrospinal fluid biomarker profiles. *Alzheimers Research & Therapy, 6*(3), 38. https://doi.org/10.1186/alzrt267.

Lesne, S. (2014). Toxic oligomer species of amyloid-β in Alzheimers disease, a timing issue. *Swiss Medical Weekly*, November. https://doi.org/10.4414/smw.2014.14021.

Levins, R. (1966). The strategy of model building in population biology. *American Scientist, 54*(4), 421–431.

Lloyd, E. A. (2010). Confirmation and robustness of climate models. *Philosophy of Science, 77*(5), 971–984. https://doi.org/10.1086/657427.

Marnell, L., Mold, C., & Du Clos, T. W. (2005). C-reactive protein: Ligands, receptors and role in inflammation. *Clinical Immunology, 117*(2), 104–111. https://doi.org/10.1016/j.clim.2005.08.004.

Mattsson, N., Insel, P. S., Donohue, M., Landau, S., Jagust, W. J., Shaw, L. M., Trojanowski, J. Q., Zetterberg, H., Blennow, K., & Weiner, M. W. (2014). Independent information from cerebrospinal fluid amyloid-β and florbetapir imaging in Alzheimers disease. *Brain, 138*(3), 772–783. https://doi.org/10.1093/brain/awu367.

Mayeux, R. (2004). Biomarkers: Potential uses and limitations. *NeuroRX, 1*(2), 182–188. https://doi.org/10.1602/neurorx.1.2.182.

McConkey, B., Davies, P., Crockson, R. A., Crockson, A. P., Butler, M., Constable, T. J., & Amos, R. S. (1979). Effects of gold, dapsone, and prednisone on serum c-reactive protein and haptoglobin and the erythrocyte sedimentation rate in rheumatoid arthritis. *Annals of the Rheumatic Diseases, 38*(2), 141–144. https://doi.org/10.1136/ard.38.2.141.

Mo, J.-A., Lim, J.-H., Sul, A.-R., Lee, M., Youn, Y. C., & Kim, H.-J. (2015). Cerebrospinal fluid β-Amyloid142 levels in the differential diagnosis of Alzheimer's disease systematic review and meta-analysis. *PLOS One, 10*(2), e0116802. https://doi.org/10.1371/journal.pone.0116802. Edited by Rosanna Squitti.

Musiek, E. S., & Holtzman, D. M. (2012). Origins of Alzheimer's disease. *Current Opinion in Neurology, 25*(6), 715–720. https://doi.org/10.1097/wco.0b013e32835a30f4.

Orzack, S. H., & Sober, E. (1993). A critical assessment of Levinss the strategy of model building in population biology (1966). *The Quarterly Review of Biology, 68*(4), 533–546. https://doi.org/10.1086/418301.

Otvos, J. D., Mora, S., Shalaurova, I., Greenland, P., Mackey, R. H., & Goff, D. C. (2011). Clinical implications of discordance between low-density lipoprotein cholesterol and particle number. *Journal of Clinical Lipidology, 5*(2), 105–113. https://doi.org/10.1016/j.jacl.2011.02.001.

Palmqvist, S., Zetterberg, H., Mattsson, N., Johansson, P., Minthon, L., Blennow, K., Olsson, M., & Hansson, O. (2015). Detailed comparison of amyloid PET and CSF biomarkers for identifying early Alzheimer disease. *Neurology, 85*(14), 1240–1249. https://doi.org/10.1212/wnl.0000000000001991.

Pepe, M. S., Janes, H., Li, C. I., Bossuyt, P. M., Feng, Z., & Hilden, J. (2016). Early-phase studies of biomarkers: What target sensitivity and specificity values might confer clinical utility? *Clinical Chemistry, 62*(5), 737–742. https://doi.org/10.1373/clinchem.2015.252163.

Prentice, R. L. (1989). Surrogate endpoints in clinical trials: Definition and operational criteria. *Statistics in Medicine, 8*(4), 431–440. https://doi.org/10.1002/sim.4780080407.

Qu, Y. (2013). Statistical evaluation of surrogate markers: Validity, efficiency, and sensitivity. *Clinical Trials: Journal of the Society for Clinical Trials, 10*(5), 693–695. https://doi.org/10.1177/1740774513499652.

Ratner, M. (2015). Biogens early Alzheimers data raise hopes, some eyebrows. *Nature Biotechnology, 33*(5), 438–438. https://doi.org/10.1038/nbt0515-438.

Reiman, E. M. (2016). Attack on amyloid-β protein. *Nature, 537*(7618), 36–37. https://doi.org/10.1038/537036a.

Ridker, P. M., Danielson, E., Fonseca, F. A. H., Genest, J., Gotto, A. M., Kastelein, J. J. P., Koenig, W., et al. (2008). Rosuvastatin to prevent vascular events in men and women with elevated c-reactive protein. *New England Journal of Medicine, 359*(21), 2195–2207. https://doi.org/10.1056/nejmoa0807646.

Ritchie, C., Smailagic, N., Noel-Storr, A. H., Takwoingi, Y., Flicker, L., Mason, S. E., & McShane, R. (2014). Plasma and cerebrospinal fluid amyloid beta for the diagnosis of Alzheimer's disease dementia and other dementias in people with mild cognitive impairment (MCI). *The Cochrane Database of Systematic Reviews, 6*. CD008782, https://doi.org/10.1002/14651858.CD008782.pub4.

Ritchie, K., Carrière, I., Berr, C., Amieva, H., Dartigues, J.-F., Ancelin, M.-L., & Ritchie, C. W. (2016). *The clinical picture of Alzheimer's disease in the decade before diagnosis* (Vol. 77, pp. e305–e311). *The Journal of Clinical Psychiatry*. https://doi.org/10.4088/jcp.15m09989.

Sachdeva, A., Cannon, C. P., Deedwania, P. C., LaBresh, K. A., Smith, S. C., Dai, D., Hernandez, A., & Fonarow, G. C. (2009). Lipid levels in patients hospitalized with coronary Artery disease: An analysis of 136,905 hospitalizations in get with the guidelines. *American Heart Journal, 157*(1), 111–117.e2. https://doi.org/10.1016/j.ahj.2008.08.010.

Savva, G. M., Wharton, S. B., Ince, P. G., Forster, G., Matthews, F. E., & Brayne, C. (2009). Age, neuropathology, and dementia. *New England Journal of Medicine, 360*(22), 2302–2309. https://doi.org/10.1056/nejmoa0806142.

Sbong, S., & Feldman, M. (2014). Frequency and causes of c-reactive protein and erythrocyte sedimentation rate disagreements in adults. *International Journal of Rheumatic Diseases, 18*(1), 29–32. https://doi.org/10.1111/1756-185x.12537.

Schneider, L. S., Kennedy, R. E., & Cutter, G. R. (2010). Requiring an amyloid-β1-42 biomarker for prodromal Alzheimers disease or Mild cognitive impairment does not lead to more efficient clinical trials. *Alzheimers & Dementia, 6*(5), 367–377. https://doi.org/10.1016/j.jalz.2010.07.004.

Schupbach, J. N. (2016). Robustness analysis as explanatory reasoning. *The British Journal for the Philosophy of Science*, axw008. https://doi.org/10.1093/bjps/axw008

Selkoe, D. J., & Hardy, J. (2016). The amyloid hypothesis of Alzheimers disease at 25 years. *EMBO Molecular Medicine, 8*(6), 595–608. https://doi.org/10.15252/emmm.201606210.

Sevigny, J., Chiao, P., Bussière, T., Weinreb, P. H., Williams, L., Maier, M., Dunstan, R., et al. (2016). The antibody aducanumab reduces aβ plaques in Alzheimer's disease. *Nature, 537*(7618), 50–56. https://doi.org/10.1038/nature19323.

Sober, E. (1989). Independent evidence about a common cause. *Philosophy of Science, 56*(2), 275–287. https://doi.org/10.1086/289487.

Sperling, R. A., Jack, C. R., Black, S. E., Frosch, M. P., Greenberg, S. M., Hyman, B. T., Scheltens, P., et al. (2011). Amyloid-related imaging abnormalities in amyloid-modifying therapeutic trials: Recommendations from the Alzheimer's association research roundtable workgroup. *Alzheimers & Dementia, 7*(4), 367–385. https://doi.org/10.1016/j.jalz.2011.05.2351.

Staley, K. W. (2004). Robust evidence and secure evidence claims. *Philosophy of Science, 71*(4), 467–488. https://doi.org/10.1086/423748.

Stegenga, J. (2009). Robustness, discordance, and relevance. *Philosophy of Science, 76*(5), 650–661. https://doi.org/10.1086/605819.

Stegenga, J. (2012). Rerum concordia discors: Robustness and discordant multimodal evidence. In *Characterizing the robustness of science* (pp. 207–226). Dordrecht: Springer. https://doi.org/10.1007/978-94-007-2759-5_9.

Terry, R. D., Masliah, E., Salmon, D. P., Butters, N., DeTeresa, R., Hill, R., Hansen, L. A., & Katzman, R. (1991). Physical basis of cognitive Alterations in Alzheimers disease: Synapse loss is the major correlate of cognitive impairment. *Annals of Neurology, 30*(4), 572–580. https://doi.org/10.1002/ana.410300410.

Toledo, J. B., Weiner, M. W., Wolk, D. A., Da, X., Chen, K., Arnold, S. E., Jagust, W., et al. (2014). Neuronal injury biomarkers and prognosis in ADNI subjects with normal cognition. *Acta Neuropathologica Communications, 2*(1). https://doi.org/10.1186/2051-5960-2-26.

Toyn, J. (2015). What lessons can be learned from failed Alzheimer's disease trials? *Expert Review of Clinical Pharmacology, 8*(3), 267–269. https://doi.org/10.1586/17512433.2015.1034690.

Trout, J. D. (1998). *Measuring the intentional world*. Oxford University Press. https://doi.org/10.1093/0195107667.001.0001.

Van Fraassen, B. C. (2008). *Scientific representation: Paradoxes of perspective*. Oxford: Oxford University Press.

Vos, S. J. B., Gordon, B. A., Yi, S., Visser, P. J., Holtzman, D. M., Morris, J. C., Fagan, A. M., & Benzinger, T. L. S. (2016). NIA-AA staging of preclinical Alzheimer disease: Discordance and concordance of CSF and imaging biomarkers. *Neurobiology of Aging, 44*(August), 1–8. https://doi.org/10.1016/j.neurobiolaging.2016.03.025.

Walsh, D. M., & Selkoe, D. J. (2007). Aβ oligomers – a decade of discovery. *Journal of Neurochemistry, 101*(5), 1172–1184. https://doi.org/10.1111/j.1471-4159.2006.04426.x.

Walsh, D. M., Tseng, B. P., Rydel, R. E., Podlisny, M. B., & Selkoe, D. J. (2000). The oligomerization of amyloid β-protein begins intracellularly in cells derived from human brain. *Biochemistry, 39*(35), 10831–10839. https://doi.org/10.1021/bi001048s.

Weisberg, M. (2006). Robustness analysis. *Philosophy of Science, 73*(5), 730–742. https://doi.org/10.1086/518628.

Willem, M., Garratt, A. N., Novak, B., Citron, M., Kaufmann, S., Rittger, A., DeStrooper, B., Saftig, P., Birchmeier, C., & Haass, C. (2006). Control of peripheral nerve myelination by the -secretase BACE1. *Science, 314*(5799), 664–666. https://doi.org/10.1126/science.1132341.

Wimsatt, W. C. (2007). *Re-engineering philosophy for limited beings: Piecewise approximations to reality*. Cambridge, MA: Harvard University Press.

Wolz, R., Schwarz, A. J., Gray, K. R., Yu, P., & Hill, D. L. G. (2016). Enrichment of clinical trials in MCI due to AD using markers of amyloid and neurodegeneration. *Neurology, 87*(12), 1235–1241. https://doi.org/10.1212/wnl.0000000000003126.

Woo, H.-N., Park, J.-S., Gwon, A.-R., Arumugam, T. V., & Jo, D.-G. (2009). Alzheimer's disease and notch signaling. *Biochemical and Biophysical Research Communications, 390*(4), 1093–1097. https://doi.org/10.1016/j.bbrc.2009.10.093.

Woodward, J. (2004). *Making things happen: A counterfactual theory of causal explanation*. Oxford: Oxford University Press. https://doi.org/10.1093/0195155270.003.0005.

Woodward, J. (2006). Some varieties of robustness. *Journal of Economic Methodology, 13*(2), 219–240. https://doi.org/10.1080/13501780600733376.

Young, A. L., Oxtoby, N. P., Daga, P., Cash, D. M., Fox, N. C., Ourselin, S., Schott, J. M., & Alexander, D. C. (2014). A data-driven model of biomarker changes in sporadic Alzheimers disease. *Brain, 137*(9), 2564–2577. https://doi.org/10.1093/brain/awu176.

Younkin, Steven G. 1995. Evidence that aβ42 is the real culprit in Alzheimers disease. *Annals of Neurology* 37 (3): 287–288. https://doi.org/10.1002/ana.410370303.

Zwan, M., van Harten, A., Ossenkoppele, R., Bouwman, F., Teunissen, C., Adriaanse, S., Lammertsma, A., Scheltens, P., van Berckel, B. N. M., & Van der Flier, W. (2013). Concordance between CSF biomarkers and [11c]PIB PET in a memory clinic population. *Alzheimers & Dementia, 9*(4), P830. https://doi.org/10.1016/j.jalz.2013.04.476.

Chapter 14
Modelling Individual Response to Treatment and Its Uncertainty: A Review of Statistical Methods and Challenges for Future Research

Ulrich Mansmann and Anne-Laure Boulesteix

Abstract Clinicians often have to make treatment decisions based on the likelihood that individual patients will receive benefit. This information has to be derived from appropriate studies and provided in a way that easily transforms from the statistical model into clinical practice. This chapter discusses four methodological strategies to provide information for treatment decisions: subgroup analysis, regression models, models for *potential outcome*, and prediction models. Based on how statistical techniques are able to provide information for stratified, precision, or individualized medicine we present the challenges that exist in analysis, reporting and applying those results. We also identify relevant ethical issues. There is still a substantial gap between formal statistical results on patient treatment interaction and ways in which they can be presented to clinicians as easy to use tools in clinical decision-making. At the moment, most energy is spent in the discovery of biomarkers which help to grasp patient treatment interaction. Less energy is spent in corresponding validation studies and the process of translating statistical results into clinical practice. Furthermore, clinical studies that provide evidence for successful translation of predictive biomarkers into clinical practice are needed.

14.1 Introduction

The problem of tailored knowledge for decision making regarding an appropriate choice of treatment has two origins: one is epistemological (inherent difficulties in getting such information) and one relates to the translation of statistical data into meaningful information for the doctor and patient.

U. Mansmann · A.-L. Boulesteix (✉)
Institute for Medical Informatics, Biometry and Epidemiology, Ludwig-Maximilians Universität München, Munich, Germany
e-mail: mansmann@ibe.med.uni-muenchen.de; boulesteix@ibe.med.uni-muenchen.de

A. LaCaze, B. Osimani (eds.), *Uncertainty in Pharmacology*, Boston Studies in the Philosophy and History of Science 338,
https://doi.org/10.1007/978-3-030-29179-2_14

Treatments are not equally useful to all patients, a phenomenon recognized in the statistical literature as *patient treatment interaction* (Gail and Simon 1985). If the individual effect of a treatment was completely unpredictable for a given patient population, medical doctors could do no better than prescribing to all patients the treatment that is *on average* best. In general, this knowledge is provided by results from randomized controlled trials summarized in systematic reviews. But, doctors want to treat *individual* not *average* patients (Rothwell et al. 2005). Especially costly innovative treatments in oncology request an effective use in patients, which potentially respond to them with high probability. Predicting response is also relevant for *already known* substances that failed during the clinical drug development process which did not prove sufficient overall success. Reanalysis of trial data is hoped to help in the discovery of patient subgroups with sufficient and economically relevant success for the substances that failed as *blockbusters*.

It turns out that the response of patients to treatments can often be partly predicted based on patients' characteristics also called *biomarkers* – or, in statistical terms, "covariates". The term *biomarker* is generic and denotes any measurable indicator of some biological state or health condition: age, sex, and parameters of clinical chemistry, imaging, or each form of molecular (e.g., genetic) information. Biomarkers can fulfill different tasks like elucidating prognostic, diagnostic or treatment related information. A biomarker related to the effect of a treatment is called *predictive*.

Using appropriate modelling strategies and estimating the corresponding parameters based on patient data available from clinical trials, one can obtain rules to predict the individual response to therapy of future patients (Vickers et al. 2007; Casella 2008). Uncertainty regarding a patient's response to treatment remains, but is reduced by use of available information. Such information can be derived from registries and clinical trials. In this paper, we focus on data provided by randomized controlled trials (RCT). Data from RCTs is of higher quality compared to data from registries. It is collected with much more care. Further, data generated by experimental designs in RCTs is less confounded compared to the observational data in registries.

In the eyes of the epistemologist: to find the probability of an individual treatment response, one must first incorporate the case into a suitable reference class. Typically, the clinician defines the reference class via biomarkers: specific biomarker profiles will define a group of individuals with comparable probabilities for outcome. This is a pragmatic frequentist framing of this specific reference class problem. But is this reference class (which carries a lot of features inherited from the trial design and conduct) the appropriate one which will help future clinician's to perform unbiased treatment decisions? Is the future patient with a specific biomarker measurement like the patients with the same biomarker measure but recruited into the RCT whose data is used for deriving the treatment response probabilities? In statistical terms: Is the probability of treatment response well calibrated with regard to the probabilities used in the decision rule?

In this paper, we first give a brief overview of standard statistical approaches for modelling individual treatment effects based on treatment-covariate interactions in the context of personalized medicine. We thereby particularly focus on the use of empirical data to estimate parameters and to select an appropriate model. Secondly, we outline how this information is communicated to make clinical decisions and how ethical aspects are supported by statistical methodology. Thirdly, we explain why there is a need for further developments of statistical methods in this context based on examples from recent clinical trials. The paper ends with an epistemological summary.

14.2 Overview of Standard Statistical Approaches

14.2.1 Settings

Suppose that there are two available treatments for a particular condition, for example, the *standard treatment* (denoted as *ST*) and an *innovative treatment* (denoted as *IT*). In a randomized clinical trial, *ST* and *IT* are compared with respect to the so-called "primary endpoint", which we will denote by *Y*. In this case, *Y* may be a (continuous) blood value, but also a binary outcome such as dead vs. alive within a well-defined time period or an event time like overall survival (OS) or progression free survival (PFS). For simplicity, in this section, *Y* is assumed as continuous unless otherwise stated and high values of *Y* are assumed to indicate a good outcome.

The standard analysis of data from two-armed RCTs consists of computing a difference (often the absolute difference or the log odds ratio, the log relative risk, the log hazards ratio) between the outcomes in both treatment groups. This difference is often called *treatment effect*. It is of interest to derive its confidence interval and to test the null-hypothesis that there is no difference between both groups. In the case of odds ratios, relative risks, or hazard ratios, the statements are derived on the log-transformed scale and typically transformed back into the standard scale.

The standard analyses focus on the expected treatment effect over the entire patient population of interest. They address the question of how the (mean) health status of a population changes if the standard is replaced by the innovative treatment. In other words, they allow inference on the expected difference between *Y* under both treatments for a patient randomly selected from the population of the patients of interest. In terms of expectation (denoted as *E[.]* in the sequel) the null-hypothesis is defined by $E[Y_{IT}\text{-}Y_{ST}] = 0$. Note, however, that the substance matter null-hypothesis is that there is no difference in the distribution of *Y* between the two groups. Focusing on the expectations is only one of several conceivable approaches. Alternatives exist, for example approaches based on stochastic dominance.

No matter which approach is taken to compare the distributions of *Y* in the two groups, clinicians often have to make treatment decisions based on the likelihood that individual patients will receive benefit. The most important thing about a treatment is that it is effective in the individual who is being treated, not merely that it ought to be effective on the basis of the overall result of an RCT or a systematic review (Rothwell et al. 2005).

This likelihood is not provided by the classical RCT result. Because of the variability of Y, an expected positive difference does not imply that the difference would be positive for all (or most of the) patients. A caregiver is not interested in the expected treatment effect in the whole population but in the expected difference for the patient sitting in front of him/her. In this context, an important question is: How it is possible to relate biomarker information to expected treatment effects?

Let Y_{ST} and Y_{IT} denote the observed endpoints under standard and innovative treatment, respectively. For simplicity we consider the difference Y_{IT} -Y_{ST}, where larger values of the difference imply superiority of the innovative treatment. Relating biomarker information to expected treatment effects means exploring whether the difference Y_{IT} -Y_{ST} is associated with these patients' characteristics. In terms of expectations, the question is whether there are characteristics X such that the conditional expectation given X, $E[Y_{IT} - Y_{ST}|X]$, depends on X and does not equal $E[Y_{IT} - Y_{ST}]$? If yes, and if the caregiver knew $E[Y_{IT} - Y_{ST}|x_0]$ for his patient with observed characteristics x_0, she/he would choose the standard treatment if $E[Y_{IT} - Y_{ST}|x_0] < 0$ and the innovative treatment if $E[Y_{IT} - Y_{ST}|x_0] > 0$, no matter the sign of $E[Y_{IT} - Y_{ST}]$.

Thus, a crucial challenge in medical statistics is to estimate $E[Y_{IT} - Y_{ST} \mid X]$ from the available data, where X are available patients' characteristics (predictive biomarkers). There are two popular approaches to this question: subgroup analyses and regression modelling with treatment-covariate interactions. These two approaches are briefly reviewed in Sects. 14.2.2 and 14.2.3, respectively. Section 14.2.4 introduces the *potential outcome* approach as a more refined strategy to conceptualize *individual treatment effects*. Section 14.2.5 gives a short account to prediction rules.

14.2.2 Subgroup Analyses

Subgroup analyses for differential treatment effects in continuous outcome Y are typically performed in the cases where the available patients' characteristics define categories (called *subgroups*) of patients. For example, let us consider $X = $ (SEX, SMOKER) where SEX and SMOKER are two binary variables with categories *male* (SEX $= 0$) vs. *female* (SEX $= 1$) and *non-smoker* (SMOKER $= 0$) vs. *smoker* (SMOKER $= 1$). These variables define a total of 2 x 2 $= 4$ subgroups. Estimating $E[Y_{IT} - Y_{ST} \mid X]$ here simply means estimating $E[Y_{IT} - Y_{ST}]$ within each of these four subgroups. When the outcome Y is continuous, this is typically done by fitting, for each subgroup, a linear regression model of the form:

$$Y = \beta_0 + \beta_G \cdot G + \varepsilon,$$

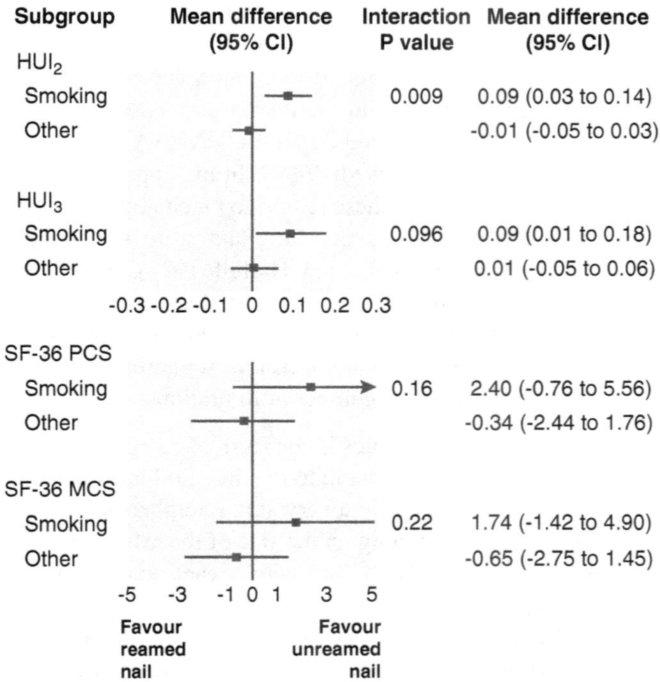

Fig. 14.1 Effect of reamed vs unreamed nailing on Health Utility Index (HUI, 2a) and Short Form-36 (SF-36, 2b) in subgroups of smoking and other tibial fracture patients. PCS = physical component summary. MCS = mental component summary (Sun et al. 2010)

where G stands for the treatment. In case of two treatment options G is a binary covariate ($G = 0$ if ST, $G = 1$ if IT), ε is a standard normally distributed independent error term and β_0 and β_G are coefficients that have to be estimated from the data at hand in each subgroup. Under this model, β_G is the treatment effect $E[Y_{IT} - Y_{ST}]$. The estimation is done by least squares estimation (Rao et al. 2008), i.e. by finding the values of β_0 and β_G that minimize the sum of the squared errors made by predicting Y as $\beta_0 + \beta_G \cdot G$. The estimate of β_G estimates $E[Y_{IT} - Y_{ST}]$ in the considered subgroup. It is also standard to provide subgroup specific confidence intervals for the estimated treatment effect.

The following Fig. 14.1 shows two subgroup effects observed for four continuous endpoints, which are subscales of the Health Utility Index (HUI) and the Short Form-36 (SF-36) in an orthopedic trial of *reamed versus unreamed nailing of tibial shaft fractures* (Bhandari et al. 2008). The subgroups are defined by *Smoking* and *Other*. The figure represents mean effects in both subgroups as forest plot (Cuzick 2005). The figure also presents p-values of an interaction test, which will be explained in the following section.

A procedure similar to linear regression, called logistic regression, exists for binary endpoints when the treatment effect is expressed by the odds ratio (OR) or the log-transformed odds ratio (logOR) (Collett 2002). For endpoints consisting of a time-to-event whereby the event (e.g., death) did not occur for all patients and may be (right-) censored, the standard regression approach is known as Cox proportional hazards regression focusing on the hazards ratio (HR) or log(HR) as the measure for a treatment effect (Collett 2014). Relative risks are determined using Poisson regression (Cameron and Trivedi 1998). In the appendix we offer a short introduction into the practical use of these regression techniques.

Subgroup analyses are often done post hoc and were not predefined in the protocols of the specific RCTs (Senn and Harrell 1997). Therefore, subgroup analyses may be informative, but also sometimes misleading. There is a search for practical and effective criteria that help to gauge the likelihood that a subgroup effect is real (Sun et al. 2010). Beside this issue of lack of reliability, the straightforward subgroup analyses approach suffers a number of limitations:

- It is applicable only for characteristics in the form of categories.
- Each subgroup has to be large enough to allow reliable estimation of the coefficients, which implies that only a very small number of characteristics can be considered in practice (depending on the size of the available dataset).
- It yields an estimation of $E[Y_{IT} - Y_{ST}]$ within each subgroup, but does not tell us whether $E[Y_{IT} - Y_{ST}]$ is significantly different between the subgroups, which may also be of importance. For example, each subgroup may show a significant difference even if the effect is not the same in each subgroup. The other way around, there may be a significant difference in one subgroup and no significant difference in the other subgroup, even if the true effect is the same in both subgroups. The presentation of subgroup results in a forest plot allows to compare confidence intervals of subgroup effects and to derive first information on differences between subgroups
- Subgroup analyses are mostly explorative analyses, performed in a post-hoc approach.

In order to strengthen the external validity of subgroup findings, confirmation has to be performed by looking at similar but independent studies (called meta-analyses within a systematic review) (Kasenda et al. 2014). At the one hand, meta-analyses reduce the probability of random error by producing a reduced CI. On the other hand, the systematic review assesses differences of methodological quality and variability in results between studies may increase the confidence in the external validity of such aggregate results.

A more general approach addressing the three first points consists of estimating $E[Y_{IT} - Y_{ST} \mid X]$ within a regression model through interactions between treatment and covariates, as briefly reviewed in Sect. 14.2.3.

14.2.3 Regression Modelling with Treatment-Covariate Interactions

Central to this approach is one regression model that incorporates the covariates (without forming specific groups) and the treatment indicator. In the case of a single covariate X, the simplest model is given as

$$Y = \beta_0 + \beta_G \cdot G + \beta_X \cdot X + \beta_{GX} \cdot G \cdot X + \varepsilon,$$

where the product term $\beta_{GX} \cdot G \cdot X$ is called an (covariate-treatment) *interaction term.* In the simple case of a binary X, such as sex taking values 0 (male) and 1 (female), the expectation of Y under standard treatment ($G = 0$) is β_0 for males and $\beta_0 + \beta_X$ for females, while under the innovative treatment ($G = 1$) it is $\beta_0 + \beta_G$ for males and $\beta_0 + \beta_G + \beta_X + \beta_{GX}$ for females. Turning to $E[Y_{IT} - Y_{ST}]$, we see that it is β_G for males and $\beta_G + \beta_{GX}$ for females. This form of regression modelling thus takes into account that the treatment effect may be different depending on X. More generally (i.e. for X covariates that are not necessarily binary), we have

$$E\left[Y_{IT} - Y_{ST} \mid X\right] = \beta_G + \beta_{GX} \cdot X$$

or, even more generally (to take into account that the effect of the continuous covariate X is not necessarily linear):

$$E\left[Y_{IT} - Y_{ST} \mid X\right] = \beta_G + f_{GX}\ (X),$$

where $f_{GX}\ (X)$ *is* an appropriate transformation of X estimated, for example, by splines (Tian et al. 2014) or fractional polynomials (Royston and Sauerbrei 2008); see also Sect. 14.3.3.

If the interaction coefficient β_{GX} is non-zero, the difference between the expectations of Y under standard and innovative treatments depends on X. The covariate X is then called *predictive* in contrast with a *prognostic* covariate which has an effect on Y that does not depend on G ($\beta_X \neq 0$ but $\beta_{GX} = 0$). Thus, this modelling approach quantifies so called *individual treatment effects* where only the individual feature of interest is given by X. This modelling approach allows the discovery of (or definition of) subgroups with a differential treatment effect. Part of the individual variability of the difference between the treatment effects is captured by X. In other words, the knowledge of X allows us to refine the prediction of $E[Y_{IT} - Y_{ST}]$ for a given patient and thus reduce the uncertainty of his/her response to the treatment. Estimation of the model parameters β_0, β_G, β_X, β_{GX} is again performed by minimization of the sum of squared errors. The inclusion of further covariates in the model is straightforward.

In Sect. 14.2.1 we also mentioned the non-linear regression techniques (logistic, Poisson-, and Cox-regression). They also allow for similar reasoning regarding the coefficients β_0, β_G, β_X, β_{GX}. Importantly, for all four regression approaches, it is possible to obtain confidence intervals for the estimated coefficients and to test that the null-hypothesis equals zero.

The following Fig. 14.2 gives a graphical representation of a regression result based on a Cox regression and representing the treatment biomarker interaction in terms of a hazard ratio. It shows the association between imputed anti-circumsporozoite antibody titers and the hazard ratio for clinical malaria episodes among children who received the RTS,S/AS01E vaccine versus no vaccination (standard), according to a Cox regression model with cubic splines and with a baseline titer of 1.0 enzyme-linked immunosorbent assay unit (EU) per milliliter as the reference. The dotted lines indicate the 95% confidence interval. There was no significant variation in risk between 1 EU per milliliter and 1000 EU per milliliter (i.e., the confidence intervals include a hazard ratio of 1.0); at values above 1000 EU per milliliter, however, there was a reduced risk of clinical malaria with increasing antibody titers (Olotu et al. 2013). The "upside down" J-curve beyond the no-effect point clearly shows that there is heterogeneity of protection effect across the range

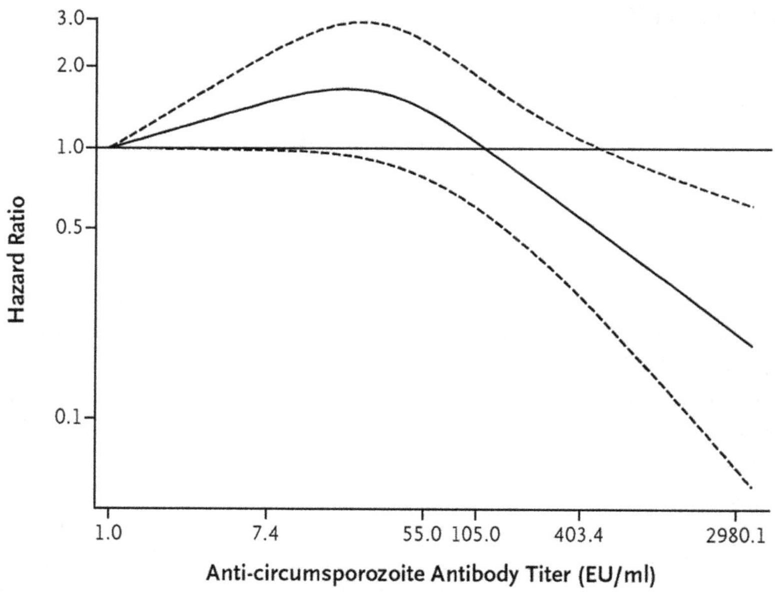

Fig. 14.2 The association between imputed anti-circumsporozoite antibody titers and the hazard ratio for clinical malaria episodes among children who received the RTS,S/AS01E vaccine, according to a Cox regression model with cubic splines and with a baseline titer of 1.0 enzyme-linked immunosorbent assay unit (EU) per milliliter as the reference. The dotted lines indicate the 95% confidence interval. There was no significant variation in risk between 1 EU per milliliter and 1000 EU per milliliter (i.e., the confidence intervals include a hazard ratio of 1.0); at values above 1000 EU per milliliter, however, there was a reduced risk of clinical malaria with increasing antibody titers (Olotu et al. 2013)

of anti-circumsporozoite antibodies titer. It represents the interaction term between biomarker and treatment and is formally given by $exp(f_{GX}(X))$.

However, the benchmark line at the no-effect point (i.e., a hazard ratio of 1) leads to false interpretation. The interpretation of the heterogeneous protection effect would be correct if this benchmark line was moved to a horizontal line at the overall treatment effect level. Looking at $E[Y_{IT}\text{-}Y_{ST}|X] = \beta_G + f_{GX}(X)$, the outcome under *IT* is worse compared to the outcome under *ST* if $\beta_G + f_{GX}(X) < 0$ or $exp(f_{GX}(X)) < exp(-\beta_G)$, A horizontal line at $exp(-\beta_G)$ would increase the interpretability of the figure. Its intersection with $exp(f_{GX}(X))$ defines the region of biomarker values, which indicate that the patient has specific advantage under the innovative treatment *IT*.

Although the authors indicate that there is reduced risk of clinical malaria with increasing antibody titers at values above 1000 enzyme-linked immunosorbent assay unit (EU) per milliliter in the legend, a visual display of the threshold of protection change would be very useful for clinicians. However, one may be more optimistic regarding the decision for the placement of the threshold by positioning it where the fitted regression curve reaches the no-effect point (around 105 EU/ml).

14.2.4 Individual Treatment Effects and Potential Outcome

The regression framework adheres to the idea of an average treatment effect (ATE), which is inferred for a subgroup. It approximates *individual treatment effects* (ITEs) by the *average treatment effect* (ATE) within the group of all individuals with a specific biomarker value *X*. When we look at *individual treatment effects,* we have to change perspectives and use a new theoretical concept. The *potential outcome framework* is recognized as a good starting point to study ITEs in RCTs, to elucidate their common distribution, and to gain methodological advantages like improving precision of estimates. The regression framework offers marginal distributions of both treatment outcomes conditioned on the biomarker *X* while the potential outcomes approach allows inferring a bivariate distribution as well as correlations between both outcomes.

The Neyman *potential outcomes framework* is based on the idea that every observational unit has different potential outcomes depending on their "assignment" to a treatment (Neyman 1923). In the case of two assignments (standard treatment vs. innovative treatment), it is impossible to see more than one potential outcome at once: the second potential outcome is always missing, which is described as the "*fundamental problem of causal inference*". A randomized controlled trial works by assigning people randomly to treatments (in this paper, *ST* and *IT*). Because the assignment was random, the groups are *in distribution* identical with respect to all patients' characteristics, and the difference in outcomes is only caused by the difference between the treatments. The random assignment mechanism is the explanation as to why some units received the *IT* and others *ST*. Rubin, together with a number of other contributors developed this approach into a formal framework for assessing causation in randomized and observational data (Rubin 1974).

Formally, for each individual the *potential outcome* is given by a bivariate random variable (Y^{IT}, Y^{ST}). This binary random variable is not observable since two effects of two simultaneous treatments cannot be measured. We use the upper indices for treatment groups to stress the issue that a *potential outcome* is only observable after a specific treatment is assigned to a subject. Thus Y^{IT} is the *potential outcome* that can be measured as Y_{IT} after having assigned the innovative treatment (*IT*) to this person. The specific distribution of *potential outcomes* within a population is a two dimensional distribution. For simplicity, assume that

$$\left(Y^{IT}, Y^{ST}\right) \sim N\left(\mu, \Sigma\right),$$

where $N(.)$ denotes the bivariate normal distribution, $\mu = (\mu^{IT}, \mu^{ST})$ denotes the expectation of Y in the *IT* and *ST* groups, respectively. The covariance matrix Σ is determined by the within-group variances σ^{IT}, σ^{ST}, and the correlation ρ. In a classical RCT, it is possible to estimate μ^{IT} and σ^{IT} as well as μ^{ST} and σ^{ST}, but not ρ.

Introducing predictive biomarkers X, as well as structural assumptions (like the assumption of a normal distribution) into the analysis, it is possible to elucidate the entire *potential outcome* setting (Y^{IT}, Y^{ST}, X). Here the variable X is defined by $X = a + b\,(Y^{IN}\text{-}Y^{ST}) + \varepsilon$ with $\varepsilon \sim N(0,\tau^2)$ and the error term ε is independent of the *potential outcomes* (Y^{IT}, Y^{ST}). The idea behind the model is that the biomarker is a linear function of the individual treatment effect and therefore carries the relevant information. It is also assumed that the measurement error regarding the biomarker is independent of the *potential outcomes*. This model is discussed for the normal case in a technical report by Laubender, R. P., and Mansmann, U. (2014) and for the setting of binary outcomes by Huang et al. (2012). There are three papers from the mid-1950s that formulated this problem in the language of psychometrical test theory (Lord 1955a; b) and provided an exact and explicit ML-estimation procedure (Anderson 1957). In principle, the solution to the Gaussian structural *potential outcome* problem was known in the literature but not recognized by statisticians. At the beginning of the new millennium, Garry L. Gadbury tried to estimate the correlation ρ for the Gaussian *potential outcome* model with biomarkers (Gadbury and Iyer 2000; Gadbury et al. 2001, 2004). In the 1950's, Lord (1955a, b) framed this problem in terms of an calibration approach: How can we transform the score of test 1 into the score of test 2 (which was not taken) or vice versa. Lord used a simple test 3 taken by all students as the basis for the calibration. Interpreting test 3 as the *biomarker* and test 1 and test 2 as treatment outcome allows the transfer of Lord's idea into the therapeutic potential outcome problem. Furthermore, during the 1970's the connection of potential outcomes with the concept of missing values evolved since Rubin's contribution to both. Starting from the 1990's, potential outcomes approaches have been used to gain a deeper understanding for confounding in observational studies. Only recently, the concept of individual treatment effect

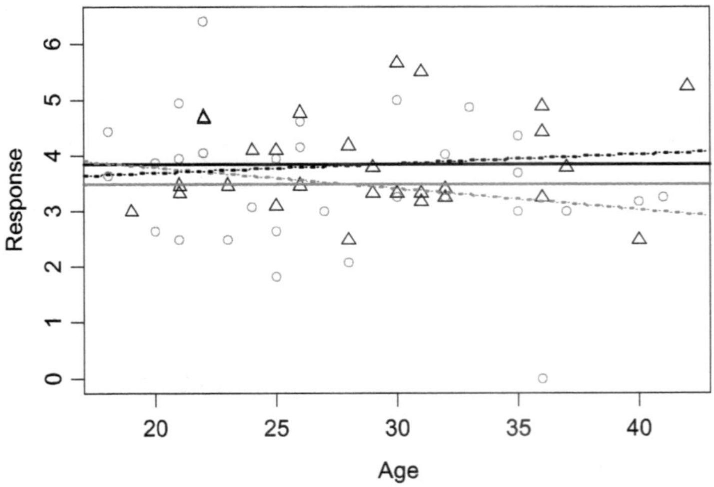

Fig. 14.3 A total of 59 tumor patients with epileptic seizures are randomized on two treatments: anti-epileptic treatment Progabide + Chemotherapy (IT – red) and Placebo + Chemotherapy (ST – black). The outcome is the asinh-transformed total sum of epileptic seizures over the 8 weeks of treatment

which is immanent in the potential outcome approach was operationalized and applied to RCT data.

Having statistical tools to estimate the full model for (Y^{IT}, Y^{ST}, X) allows for presenting estimates for $E[Y^{IT} - Y^{ST}|X]$ together with a distribution of individual treatment effects $Y^{IT} - Y^{ST}$ conditional on X. Here, the expression $P[Y^{IT} - Y^{ST} > 0| X]$ may be of specific interest. Garry L. Gadbury used simple data sets to illustrate his ideas. His data from Thall and Vail (1990) are shown in the following figure where the biomarker (X) is given by age (Fig. 14.3).

The dots belong to the patients in the IT group. The triangles belong to the patients under standard treatment. The outcome is the asinh-transformed total sum of epileptic seizures over the 8 weeks of treatment. Lower values represent a better outcome (lower number of epileptic seizures). The corresponding solid horizontal lines represent the mean treatment effects. The dotted lines represent the corresponding linear relation between biomarker and treatment effect. Based on standard regression analyses per treatment group, the randomization, and the independence between both treatment groups, one can determine the biomarker treatment response relationship and use it to quantify $P_1[Y_{IT} < Y_{ST} \mid X]$. Using the ML-estimation approach as outlined by Anderson (1957) and Laubender, R. P., and Mansmann, U. (2014) one can estimate the parameters of the three-dimensional Gaussian structural *potential outcome* model (Y^{IT}, Y^{ST}, X). This allows for the calculation of a second version $P_2[Y^{IT} < Y^{ST} \mid X]$ based on the *potential outcome* model. A comparison between both approaches is given in Fig. 14.4, which illustrates that the curve $P_2[Y^{IT} < Y^{ST} \mid X]$ developed from the *potential outcome* perspective is steeper compared to the curve derived from standard regression

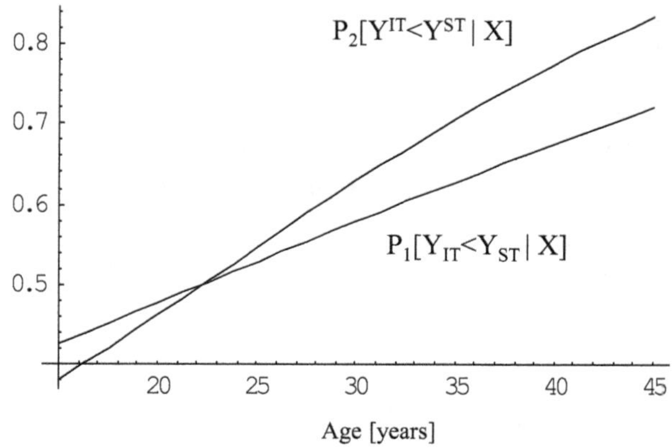

Fig. 14.4 Biomarker induced success probabilities for the innovative treatment derived from the Gadbury data using the standard regression approach ($P_1[Y^{IT} < Y^{ST}|X]$) and the potential *outcome* approach ($P_2[Y_{IT} < Y_{ST}|X]$)

analysis $P_1[Y_{IT} < Y_{ST}|X]$. It illustrates that the *potential outcome* approach has the potential to provide sharper decision rules. A mathematical proof of this statement can be based on the fact that the potential outcomes (Y^{IT}, Y^{ST}) are correlated for each biomarker value X.

Figure 14.4 does not show confidence bands for the estimates of the quantities P_1 and P_2. It also gives no information on the range of biomarkers where clear evidence supports the decision to treat the patients using the innovative treatment (*IT*). Corresponding statistical methods are provided by Laubender, R. P., and Mansmann, U. (2014).

14.2.5 Statistical Challenges: Some Examples

The regression methods presented so far are particularly easy. In many cases, however, they may be too simplistic to reflect the complexity of the dependence between patients' characteristics and effect of the therapies (*functional form problem*).

For example, in the case of a continuous X, the assumption of linearity may be critical as already mentioned in Sect. 14.2.3. Why do we include the term $\beta GX \cdot G \cdot X$ in the model and not, say, $\beta_{GX} \cdot G \cdot X^2$ or $\beta_{GX} \cdot G \cdot I(X > 3)$? (Thereby $I(.)$ is the so-called indicator function equaling 1 if the condition given between the brackets holds and 0 otherwise). Beyond the simple linear form, there are plenty of possible dependence patterns. The choice of one of these patterns is denoted as model selection in statistics and is anything but easy. Currently, a plethora of methods are being proposed in the literature to select an adequate form for the dependence of Y, G and X. They can roughly be split into two categories of approaches. The first category encompasses modelling approaches, i.e. procedures assuming that the data follow a specified stochastic model, whose parameters have to be estimated, for example by maximizing the likelihood. The regression approach sketched in Sect. 14.2.3 belongs to this category. The choice of the specific stochastic model is a crucial issue. The second category consists of data-driven algorithms that do not assume any particular model but rather constructs the function connecting Y with G and X - without making initial conjectures on its form – directly from the data, often in a step-by-step manner. In his paper, Breiman extensively discusses the differences between these two approaches (Breiman 2001). The evaluation of the performances of the numerous existing methods in various scenarios is an active research topic.

Another issue arising from the increasing availability of large-scale molecular data collected by high-throughput technology is the potentially high dimension of X. In our example we considered only one or two covariates affecting the effects of treatments. In some cases, however, a larger – sometimes even huge – number of potential predictive covariates are available in the dataset. For example, genome-wide gene expression data may be available and it is conceivable that some of the gene expression levels may act as predictive biomarkers. It is impossible to fit models of the type of those outlined in Sect. 14.2 if there are a high number of covariates. How can functions connecting Y with G and X be derived from data in such a situation? How can we test whether a (large) set of covariates as a whole modifies the treatment effect? Existing approaches are still in their infancy; see for example the very recent articles providing first comparison studies on testing large sets of covariates for interactions (Callegaro et al. 2016) and regression modelling for the purpose of prediction (Ternès et al. 2016). Methodological statistical research has to be conducted to address these issues.

Furthermore, methods reviewed in Sects. 14.2.2 (subgroup analyses), 14.2.3 (regression modelling with treatment-covariate interaction) and 14.2.4 (*potential outcome* approach) may be used to derive prediction rules for Y given X and G. By a prediction rule, we mean a rule that takes the covariates X, as well as the assigned treatment G, as an input and then outputs a prediction for Y. Going one step further, one could alternatively define a prediction rule as a rule taking X as the input and returning the "best treatment" as the output. Note that the term "best" has to be defined carefully and that several definition are conceivable. These rules may be useful in clinical practice to support treatment choice for individual patients; they suggest a therapeutic course of action. In this way, prediction rules turn into decision aids. They can "change clinical behavior and reduce unnecessary costs while maintaining quality of care and patient satisfaction (McGinn 2000)."

14.3 Challenges: Reporting, Statistics, and Ethics

14.3.1 Good Reporting Challenges

Section 14.2 gave a short introduction to the statistical issues related to treatment of individual patients. It focuses on regression models that combine biomarker information and treatment assignment with a clinical outcome. It was argued that differential treatment effects within subgroups identified via biomarker measurements provide key information for treatment decisions. It was also mentioned that the regression framework is helpful for the complete range of clinically relevant outcomes (continuous measurements, binary endpoints, and time-to-event data). It was argued that the interaction effect between treatment assignment and biomarkers provides essential information on differential treatment effects within subgroups. The relevant questions are now: Do we inform the clinician sufficiently if the paper just reports the interaction effect? Is it appropriate to report its confidence interval or just a p-value? To discuss this issue we start with continuous outcomes and their differences between the treatment groups describing the treatment effect and will look in a second step at binary and time-to-event data.

In practice, caregivers are not always interested in $E[Y_{IT} - Y_{ST} \mid X]$ only. It is a mean of the distribution that is relevant for the group of patients expressing feature X. The true interest of a clinician consists in the complete distribution. A particular feature of the distribution is $P[Y_{IT} - Y_{ST} > 0|X]$, i.e. the probability that the innovative treatment leads to a better outcome than the standard treatment. If the variability of the distribution is large (in statistical terms, if the variance of ε is large), $E[Y_{IT} - Y_{ST} \mid X]$ may be large even if $P[Y_{IT} - Y_{ST} > 0|X]$ is close to 0.5, which would obviously mitigate the informative value of $E[Y_{IT} - Y_{ST}|X]$. Since a clinician often has to weigh advantages of treatment effects against disadvantages of the treatment, like adverse events, it may be the case that accurately predicting $P[Y_{IT} - Y_{ST} > 0|X]$ is more relevant than accurately predicting $E[Y_{IT} - Y_{ST}|X]$. In a more general sense it may be relevant for a clinician to determine $P[Y_{IT} - Y_{ST} > \delta|X]$ for a specific advantage of amount δ, which may depend on the patient under consideration. Therefore, a good estimate for the distribution of $Y_{IT} - Y_{ST}$ is of high relevance. Considering the complexity of the needed information, the issue of transparent reporting of prediction models is a challenging issue and a focal point of research on reporting methods (Collins et al. 2015). Note that, even if this information is given in a paper, the clinician will have problems deriving the information she/he needs to treat a patient (Collins et al. 2015). In this context, statistical training of clinicians is crucial.

Based on the simple linear regression framework introduced in Sect. 14.2.3, it follows that $E[Y_{IT}-Y_{ST}| X] = \beta_G + \beta_{GX}.X$. Therefore, a mean advantage of the innovative treatment will occur if $E[Y_{IT}-Y_{ST}| X] > 0$, *consequently* $\beta_G + \beta_{GX}.X > 0$ *or* $-\beta_G/\beta_{GX.} < X$. It is of interest to see for which X the equipoise $\beta_G + \beta_{GX}.X = 0$ is reached or where the line $\beta_{GX}.X$ crosses the horizontal line $-\beta_G$. Unfortunately, in medical publications, this fact is most often not communicated to clinicians in such a way that she/he can use it as input into a shared decision making process with the patient. Shen, Y.-M., Le, L. D., Wilson, R., and Mansmann, U. (2017) found that some articles that report the interaction between treatment and continuous biomarkers also report the interaction $f_{GX}(X) = \beta_{GX}.X$ or more generally the non-linear interaction term $f_{GX}(X)$ as a graph, but do not show the horizontal line at $-\beta_G$ to indicate the boundary of biomarker value regions with potential disadvantage or advantage for the innovative treatment. Most often, these graphs only show the horizontal 0-line, which separates negative and positive treatment-biomarker interaction effects, but do not indicate regions of disadvantage or advantage for the innovative treatment. Therefore, current reporting practice needs improvement as far as relevant information on treatment-biomarker interactions is concerned.

As a good example, we present a figure from the reanalysis of the study MRC RE01, which compared two treatments: interferon-α and medroxyprogesterone acetate (MPA) (Royston et al. 2004). Royston and colleagues used fractional polynomials to investigate the updated data from this trial for the possible interaction of treatment with biomarkers. In the spirit of hypothesis generation, they studied ten biomarkers, of which white cell count (WCC) was found to influence the effectiveness of interferon-α treatment. In patients treated with MPA, there was no effect of WCC, whereas, in patients treated with interferon-α, the risk of dying increased significantly with WCC level. Their results are presented in the following figure.

The appendix B of the Royston et al. paper reports the results of the multivariate analysis (Royston et al. 2004). The given formulae allow quantifying the advantage of an interferon-α treated patient compared to a MPA treated patient given a specific biomarker WCC measurement (and other covariates kept constant). The corresponding log-hazard ratio is $\log HR_{\text{Interferon/MPA}} = -0.94 + (1.34\text{–}0.19)\cdot X^2 - (0.33 + 0.0089)\cdot X^3$ where $X = \text{WCC}/10$. The $HR_{\text{Interferon/MPA}}(\text{WCC})$ derived from this expression together with its 95% confidence limits is presented in Fig. 14.5. The upper 95% confidence bound of the $HR_{\text{Interferon/MPA}}$ function suggests that there is evidence that WCC values below 8 provide better response under interferon-α compared to MPA. Looking simply at the estimate $HR_{\text{Interferon/MPA}}$, the figure

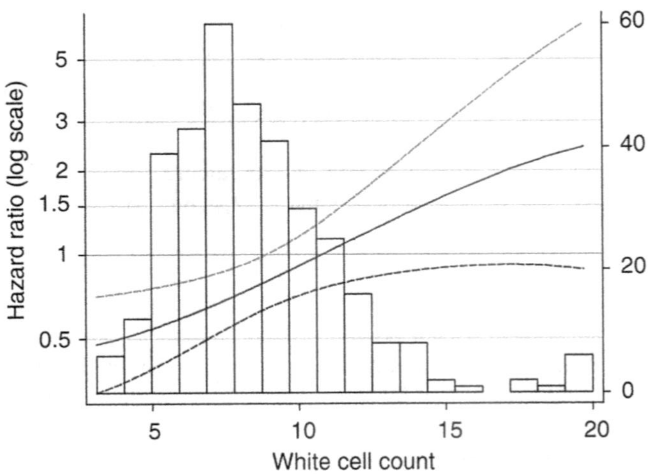

Fig. 14.5 Dependence on white cell count (x $10^9\ l^{-1}$) of the effect of treatment with interferon-a compared with MPA, estimated from fractional polynomial model. Solid line: hazard ratio; dashed lines: 95% CI for hazard ratio. The horizontal line at hazard ratio ¼ 1 denotes equivalence of treatment effects. Values beneath this line indicate that interferon-α is more effective than MPA, and vice versa. A bar chart showing the distribution of white cell count values for all patients in the study is superimposed (Royston et al. 2004)

suggests that WCC values above 10 results in a better response under MPA compared to interferon-α. In their abstract, the authors propose a subgrouping of patients on WCC levels and estimated a hazard ratio of 0.53 in favor of interferon in patients with WCC $< 6.5 \times 10^9$ whereas, for patients with WCC $> 10 \times 10^9$ the risk appears to be similar between the treatment groups, or even slightly higher in the interferon group. Here the authors report the estimates without taking their uncertainties into account.

Above, we stressed the fact that a good estimate of $P[(Y_{IT}, Y_{ST})|X]$ is necessary to derive relevant quantitative information for the decision between two optional treatments. In a typical RCT setting with biomarker measurements, the randomization can be used to strengthen the assumption of a comparable distribution of biomarker values between the treatment groups. Thus, $P[(Y_{IT}, Y_{ST})|X]$ can be calculated from the marginal distributions (Y_{IT}, X) and (Y_{ST}, X) assuming independence between the two. Engelhardt et al. (2016) show how non-parametric density estimators can be used to achieve this goal. Figures that show estimates of the probability of a specific advantage implied by the innovative treatment $P[Y_{IT}\text{-}Y_{ST} > \delta|X]$ together with the uncertainty related to this estimate may be very helpful for the translation of treatment-biomarker interaction into clinical practice.

Estimates of $P[Y_{IT} - Y_{ST} > 0|X]$ can also be provided for binary, as well as, time-to-event outcome. For the binary setting $P[Y_{IT} - Y_{ST} > 0|X] = P[Y_{IT} = 1, Y_{ST} = 0|X]$, using the independence information between the two arms of a RCT implies

$$\begin{aligned} P[Y_{IT} = 1, Y_{ST} = 0|X] &= P[Y_{IT} = 1|X] \cdot P[Y_{ST} = 0|X] \\ &= exp\{\beta_0 + \beta_G + \beta_G + \beta_{GX} \cdot X\} / \\ &\quad [(1 + exp\{\beta_0 + \beta_G + \beta_X \cdot X + \beta_{GX} \cdot X\}) \\ &\quad \cdot (1 + exp\{\beta_0 + \beta_X \cdot X\})] . \end{aligned}$$

Similarly, for time-to-event data it is possible to calculate the probability that the event (e.g., death) under the innovative treatment (T_{IT}) happens after the event under the standard (T_{ST}):

$$P[T_{IT} < T_{ST} \mid X] = exp\{\beta_G + \beta_{GX}.X\} / (1 + exp\{\beta_G + \beta_{GX}.X\}) .$$

These formulae demonstrate that results from regression models, as described in Sect. 14.2.3., provide the input for quantitative information useful in decision making.

Using the last formula, the results of Royston et al. could also be presented in an alternative way. Figure 14.6 shows the probability that a person treated by interferon-α dies before a person (with comparable features) who is treated with MPA. A horizontal line shows the value of indifference (0.5). The curve $P[T_{IT} < T_{ST} \mid X]$ crosses the 0.5 line at 11. If a patient shows WCC values above 11, the probability to die earlier under interferon-α when compared with MPA is above 0.5. If a patient shows WCC values below 11, the probability to die earlier under

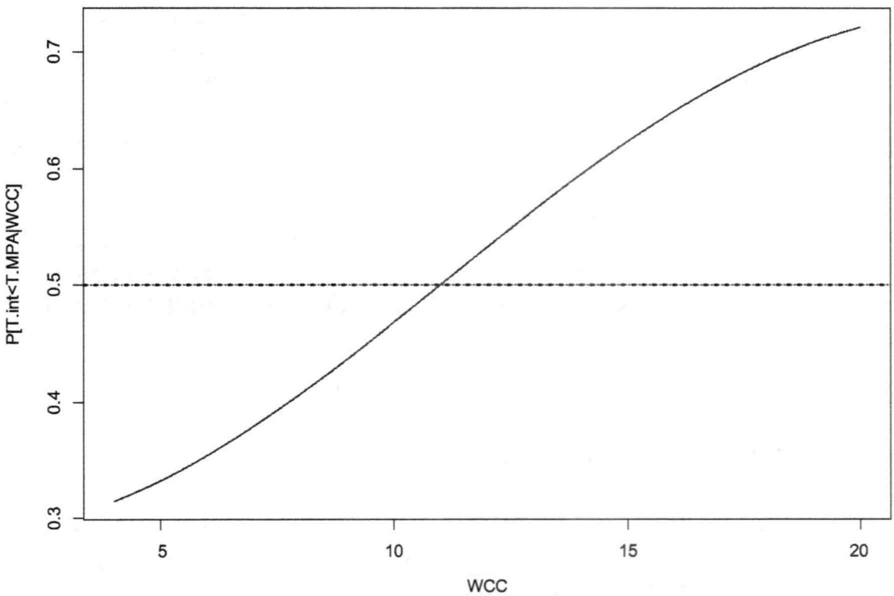

Fig. 14.6 $P[T_{IT} < T_{ST}|X] = exp\{\beta_G + \beta_{GX}.X\}/(1 + exp\{\beta_G + \beta_{GX}.X\})$ applied to the data of Royston et al. Increasing WCC results in a higher probability to die earlier under Interferon-α treatment compared to MPA treatment

interferon-α earlier when compared with MPA is below 0.5. Unfortunately, the distribution of survival under both treatments is not clear. Confidence intervals can be derived from a confidence interval for $\beta_G + \beta_{GX}X$ and using the inverse logit transformation. The confidence interval for $\beta_G + \beta_{GX}{\cdot}X$ is a straightforward consequence of the covariance matrix of the regression coefficients.

In our review Shen, Y.-M., Le, L. D., Wilson, R., and Mansmann, U. 2017 of clinical reports on treatment-biomarker interaction effects we did not find any paper that tried to provide the information in a way that allowed a straightforward translation into the practical clinical use of the results. In our opinion, plots that show the relationship between X and $P[Y_{IT} \text{-} Y_{ST} > \delta|X]$, together with information on the uncertainties of this estimate, would be helpful for translating complex statistical results on treatment-biomarker interaction into clinical decision making. Implementing a structural modelling approach to perform a *potential outcome analysis* that estimates conditional distributions for individual treatment responses $(Y^{IT},Y^{ST})|X$ allows to infer $P[Y^{IT}\text{-}Y^{ST} > \delta|X]$, which is more precise than $P[Y_{IT}\text{-}Y_{ST} > \delta|X]$ by using additional information on the possible correlation between the components of (Y^{IT},Y^{ST},X). This, again, stresses the importance of the *potential outcome* framework in studying treatment-biomarker interaction.

In summary, good information that supports clinical decision-making is, to date, often not straightforward in medical literature. Significant interaction effects do trigger inadequate qualitative judgments. Comparison of different biomarkers for a problem of interest is mostly impossible, because just comparing the extent of interaction effects is misleading. This impairs the choice of the biomarker that may be most useful for the decision making. Which one has the best discriminative power for treatment decision under focus? Especially the last issue relates to new forms of ROC curves that are discussed (Huang et al. 2012; Shen, Y.-M., Le, L. D., Wilson, R., and Mansmann, U. 2017 Engelhardt et al. 2016).

It is quite straightforward to transform the statistical models in algorithms that provide a quantitative measure for $P[Y_{IT}\text{-}Y_{ST} > \delta \mid X]$ or $P[Y^{IT}\text{-}Y^{ST} > \delta|X]$. It is, however, not straightforward to create an easy to use interface between the clinician and the algorithm. But research is under its way to explore this field (Gentles et al. 2013).

14.3.2 Simple Risk Calculations

In general, for a set of dichotomous/categorical treatment-selection biomarkers, a modified forest plot can represent heterogeneity of treatment effects within subpopulations (Collett 2002). The forest plot shows for each subgroup its specific treatment effect. The results are easily communicated in terms of absolute risk changes if the absolute risk difference is also reported. In case of relative risks (RR), odds ratios (OR) or hazard ratio (HRs), changes of absolute risks can be communicated if a baseline risk is known. Figure 14.7 shows the differential effect between carotid endarterectomy (*IT*) versus medical treatment alone (*ST*) on 5-year risk of stroke and operative death depending on the grade of stenosis.

Assuming that the baseline risk is $p_{ST} = 0.1$ (5 year overall survival rate under standard) and the effect between an innovative treatment and the standard is 0.7, the following can be concluded (1) if the effect is given by a relative risk $RR_{IN/ST} = 0.7$: 5 year overall survival rate under the *IN* is $p_{IN} = 0.1 \cdot 0.7 = 0.07$; (2) if the effect is given by an odds ratio $OR_{IN/ST} = 0.7$: $odds_{ST} = 0.1/0.9 = 0.11$, $odds_{IN} = odds_{ST} \cdot OR_{IN/ST} = 0.11 \cdot 0.7 = 0.077$, $p_{IN} = 0.077/1.077 = 0.072$; (3) if the contrast is given by a hazard ratio $HR_{IN/ST} = 0.7$: $p_{IN} = p_{ST}{}^{HR} = 0.1^{0.7} = 0.200$. In the case

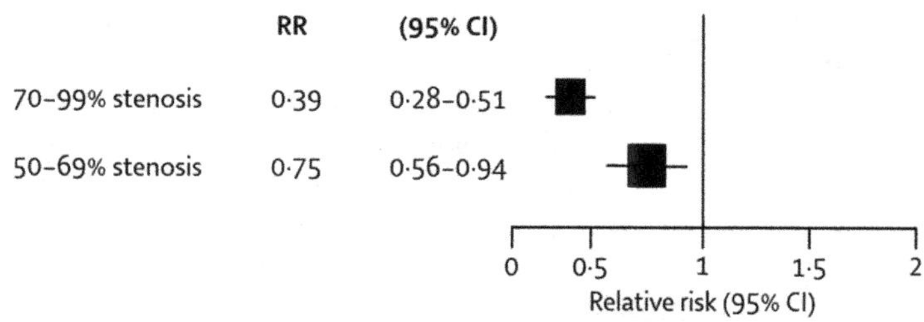

Fig. 14.7 Effect of carotid endarterectomy for ≥70% and 50–69% symptomatic stenosis on 5-year risk of stroke and operative death Data taken from RCTs of endarterectomy versus medical treatment alone (Rothwell et al. 2003)

of a given $HR_{IN/ST}$ it is also possible to quantify the probability that the event under *IN* (T_{IN}) will happen before the event under *ST* (T_{ST}): $P(T_{IN} < T_{ST}) = HR_{IN/ST}/(1+HR_{IN/ST})$. In our example this results in $P(T_{IN} < T_{ST}) = 0.7/1.7 = 0.41$.

14.3.3 Ethical Aspects in Connection with Statistics

This section starts with a sketch of a complex setting of how to treat a patient with Multiple Sclerosis (MS). It emphasizes the uncertainty in which a doctor has to operate and needs to interact with the patient to reach an optimal treatment decision.

Multiple sclerosis (MS) is a chronic-inflammatory degenerative disease of the central nervous system. In Germany, MS affects more than 200,000 individuals and is the main cause of non-traumatic disability in young adults. During the early course of MS, 11 approved drugs offer efficient treatment options. Additional drugs are approaching approval. These drugs differ with respect to biological mechanisms, efficacy and adverse effects. Therapy initiation and selection of a particular treatment currently relies on a few parameters (e.g. number and severity of relapses, number of lesions identified by MRI of brain and spinal cord). Treatment is adjusted by trial and error until sufficient control of disease activity is achieved with acceptable adverse reactions. Thus a patient may be exposed to ineffective drugs, unnecessary side effects, and discontinuation of medication for years. It is necessary to find ways to early tailored therapies, improving patient-related outcomes, minimizing adverse effects, improving compliance and reducing costs. In order to develop predictors relevant to therapy decisions, a wide variety of clinical and molecular parameters are, in principle, available at clinical sites: (a) case data (e.g., flare-up severity and frequency), (b) MRI-derived parameters (e.g., number and localization of lesions), (c) cerebrospinal-fluid data (e.g., oligoclonal bands, other biomarkers), (d) genomic data (e.g., HLA type, genomic variants), (e) electrophysiological data (e.g., evoked potentials), (f) data from optical coherence tomography, and (g) biomarker data (e.g., serological data, proteome, transcriptome). There is a high interest in the MS-community to develop and validate prediction models leading towards a decision support system with regard to effective and safe treatment of MS-patient.

In order to understand the relevance of a clinical translation of statistical information on treatment-biomarker to the place where treatment decisions are made, we give a short introduction into shared decision making. Shared decision-making is increasingly advocated as an ideal model of treatment decision-making in the medical encounter. Its key characteristics are (1) that at least two participants—physician and patient be involved; (2) that both parties share information; (3) that both parties take steps to build a consensus about the preferred treatment; and (4) that an agreement is reached on the treatment to implement.

Good quantitative information is an essential part of shared decision making. How can a doctor and a patient, without specific skills in statistics, understand extent and probability of a potential advantage or damage, which may happen if a specific treatment will be given? What is an informed consent on which the decision can be built on? How to make a shared decision? Individual risk calculation needs individual risk communication. It may be wise to establish, together with the patient, a series of questions relevant for his/her treatment to be planned. A formal risk-benefit calculation does not replace the information procedure to reach the patient's consent for the planned intervention or treatment. A risk calculation complements this discussion.

The appropriate interpretation of the quantitative output of a risk-benefit calculation needs a good understanding of probabilities. Using prediction models and corresponding algorithms will be important. The doctor must know the clinical background of the statistical models used to calculate the numbers and must make a judgment whether the patient belongs to the study population behind the calculated numbers. The doctor must be able to explain the probabilities calculated in a way that can be understood by the patient. This requires specific training for the doctor (Wegwarth 2012; Zikmund-Fisher, B. J., Fagerlin, A., and Ubel, P. A. 2010; Zikmund-Fisher, B. J. (2012). The probabilities should not be given in percent, like 20%, but presented in absolute numbers, like 20 out of 100 (Bodemer et al. 2014). Clinical decisions must be based on absolute risks, not on relative risk measures like relative risk, hazard ratio, or odds ratio. These do not quantify an actual individual risk.

Given the situation as described in it is obvious that the formal requirements for shared decision making between doctor and patients are not always fulfilled. Also the doctor is left alone because all these myriads of papers on treatment-biomarker interaction do not provide sensible information for an informed clinical decision making. The papers do not even give useful information for a qualitative judgement: Regions of biomarker values with a high likelihood of an advantage by the innovative treatment cannot be easily derived and, for the most part, the provided quantitative and graphical information is misleading.

14.4 Discussion

Doctors have to treat patients in the best way possible. To this end they have roughly speaking three options: (1) using individual experience; (2) using evidence based guidelines; (3) using objective individual decision making based on risk calculation. Options (2) and (3) rely on data derived statements and tools. The term *objective* in option (3) relates to processes that allow a transparent description and documentation of the decision process. Often, option (3) is the only choice, since guidelines and experience may not be adapted to innovative treatments.

Finding best treatment options for an individual patient is often termed individualized, precision, or stratified medicine. The overview of methods presented in Sect. 14.2 suggests subtle terminological differences between the three terms. The subgroup analyses described in Sect. 14.2.2 provide the basis for the stratified medicine: Classification of a patient in specific categories leads to specific treatment recommendations. The regression approaches presented in Sect. 14.2.3 fit the ideas of precision medicine by including continuous biomarkers and the fine-tuning of response predictions. The idea of individualized medicine is represented by the potential outcome setting as presented in Sect. 14.2.4 when individual treatment effects are inferred.

Good treatment decisions need evidence from good data and trial results, which are considered as relevant by clinicians. Traditionally, this is achieved by the complex evidence synthesis of systematic reviews of randomized controlled trials (RCTs), which are used to build clinical guidelines. Thereby, clinical knowledge is created by impressive machinery out of an immense body of available papers: systematic review, meta-analysis, techniques of evidence synthesis, health technology assessment, consensus processes, etc. This approach basically aggregates individual studies to create reliable clinical knowledge. From a very formal data specific perspective – individual data are held separate, only statistical results are combined.

An alternative to this paper based approach is a computational thinking, which aggregates clinical experience in computational algorithms using a large database of well-documented clinical experience. This process does not aggregate results of individual papers but creates large data sets that allow the development and validation of a variety of prediction tools. Prediction tools may be the carrier of knowledge by giving clear advice on which actions may be helpful for an individual patient.

Finally, we point out that the authors of a study whose main output is a prediction rule (as defined in Sect. 14.2.5) should ensure that the prediction rule is available to and applicable by potential users. "Available" means that all required information is contained in the paper, in supplementary materials or in the form of a companion tool. In the case of a regression model, for example, the needed information defining the prediction rule consists of the model equation and the estimated values of the regression coefficients (including the intercept/baseline). This information can simply be made available in a paper. For more complex prediction rules, however, it may be necessary that the authors make a software tool available to the readers of the paper. "Applicable" means that the targeted audience (typically, medical practitioners) can make use of the prediction rule for future patients without too much time and effort. There are very simple prediction rules that can be explained "in words" in a paper (for example: if the number of risk factors exceeds 6, use therapy A, otherwise use therapy B). However, most prediction rules are more difficult to apply. The implementation within easy-to-use software or in the form of an online tool is often indispensable. Besides developing and validating prediction rules for clinical practice, the implementation of such instruments in the clinical work-up may be a challenge.

14.5 Epistemological Summary

The paper's focus was how to provide knowledge to the clinicians which enables good patient-oriented treatment decisions. It is written from a Frequentist perspective. Frequentism implies that, in the observer-independent reality, true probabilities are unique, but there is considerable room for multiple perspectives regarding the definition of replicable experiments, collectives or reference sets. The concept of biomarkers (X) is used to define the patient's reference class on which corresponding quantitative information regarding differential treatment response is provided.

This information is formalized by the conditional means and conditional probabilities of differential treatment effects: $E[Y_{IT} - Y_{ST} \mid X]$ and $P[Y_{IT} - Y_{ST} \mid X]$. We presented three different statistical strategies to estimate these quantities of interest: (1) *Subgroup analysis* as the basis of *stratified medicine* with very broad reference classes, (2) *regression analyses* with interaction terms as the basis of *precision medicine* where reference classes were define by single numeric values of biomarkers, and (3) *potential outcome* approaches as a basis of the *individualized medicine* where individual treatment effects could be derived. It was discussed that the extent of theoretical assumption increases when one progresses from (1) to (3). We use the potential outcome approach in the setting of a randomized experiment which avoids a lot of epistemological issues which arise in general epidemiological research (Vandenbroucke et al. 2016).

We did not enter the topic of model diagnoses to establish the adequacy of all assumptions used. Statistical techniques for model diagnosis do exist, but were not discussed. Data from randomized clinical trials (RCT) was used to generate the quantitative estimates. Theoretically, in RCT randomization assures that patients can be exchanged between the treatment groups conditional on the biomarker value X. Thus, other features of the patient (except the biomarker value) are irrelevant for the differential treatment effect. This also is a brave assumption. Therefore, the epistemological value of the statistical strategy and the data base used may not be perfect.

As a consequence, validation studies have to be implemented to assess the clinical value of the derived quantities $E[Y_{IT} - Y_{ST} \mid X]$ and $P[Y_{IT} - Y_{ST} \mid X]$. But it is not enough to simply assess the calibration of $E[Y_{IT} - Y_{ST} \mid X]$ and $P[Y_{IT} - Y_{ST} \mid X]$ for new future patients. Additionally, we stressed that implementing statistical concepts in clinical decision making is not trivial. Therefore, the only way to assess the usefulness of the body of knowledge on differential treatment effects for real patients would be a randomized controlled trial which splits patients into a *standard group* with no treatment decision and the *intervention group* where the statistical information is implemented in a clinical decision procedure and used. If such a trial shows superiority of the intervention compared to standard evidence would exist that biomarker driven treatment decisions are useful.

Formally, the derivation of $E[Y_{IT} - Y_{ST} \mid X]$ and $P[Y_{IT} - Y_{ST} \mid X]$ provides an hypothesis which needs to be tested (by the just mentioned RCT). There is a lively discussion if registries or other kinds of observational data are able to assess the usefulness of such decision procedures. This issue is not settled yet. Also, there is an industry of exploratory studies which claim to provide better biomarkers for better decisions. Due diligence is often overlooked in this committed competition to create good knowledge for treatment decision making.

A.1 Statistical Excursus: Cox Regression and Logistic Regression

The Cox proportional Hazards model quantifies the effect of a treatment (G) and a biomarker (X) on an event time T by specifying the survival probability. $P[T > t|X, G]$ represents the probability to survive the time t given biomarker measurement X and treatment G. The proportional hazards model is given by

$$P\left[T > t \mid X, G\right] = exp\left[-\int_0^t \alpha_0(s) \bullet exp\left\{ \beta_G \cdot G + \beta_X \cdot X + \beta_{GX} \cdot G \cdot X\right\} ds\right].$$

Here $\alpha_0(s)$ describes an unspecific baseline hazard and $\exp\{ \beta_G \cdot G + \beta_X \cdot X + \beta_{GX} \cdot G \cdot X\}$ quantifies the hazards ratio between subjects treated by *IN* $(G = 1)$ versus subjects treated by *ST* $(G = 0)$ and presenting a biomarker measurement of X. In a more general sense the model can also handle nonlinear influence of the biomarker on outcome by

$$P\left[T > t \mid X, G\right] = exp\left[-\int_0^t \alpha_0(s) \bullet exp\left\{ \beta_G \cdot G + f_X(X) + f_{GX}\ (X)\right\} ds\right].$$

Here $f_{GX}(X) = 0$ for $G = 0$ and $f_{GX}(X) = h_X(X)$ for $G = 1$. Using the survival function of the baseline group $P[T > t \mid X = 0,\ G = 0] = P_0[T > t]$, the above expression is equivalent to

$$P\left[T > t \mid X, G\right] = P_0\left[T > t\right] exp\left\{\beta_G \cdot G + f_X\ (X) + f_{GX}\ (X)\right\}.$$

The term $\beta_G \cdot G + f_X(X) + f_{GX}\ (X)$ or $\beta_G \cdot G + \beta_X \cdot X + \beta_{GX} \cdot G \cdot X$ are called linear predictors and written as $LP(X,G)$. Therefore the above formulae translate to

$$P\left[T > t \mid X, G\right] = exp\left[-\int_0^t \alpha_0(s) \bullet exp\left\{ LP\left(X, G\right)\right\} ds\right]$$

and

$$P\left[T > t \mid X, G\right] = P_0\left[T > t\right] exp\left(LP\left(X, G\right)\right)\Big\}.$$

The logistic regression quantifies the effect of a biomarker measured by X and treatment defined by G on a binary outcome Y using the model:

$$P\left[Y=1\mid X,G\right]=exp\left\{LP\left(X,G\right)\right\}/\left(1+exp\left\{LP\left(X,G\right)\right\}\right).$$

The linear predictor is different from the linear predictor of the Cox regression because it also contains a baseline term β_0:

$$LP\left(X,G\right)=\beta_0+\beta_G\cdot G+f_X\left(X\right)+f_{GX}\ \left(X\right)$$

or

$$LP\left(X,G\right)=\beta_0+\beta_G\cdot G+\beta_X\cdot X+\beta_{GX}\cdot G\cdot X.$$

The value $exp\{LP(X,\ 1)\}$ quantifies the odds ratio (OR) of a person showing a biomarker measurement of value X and being treated by the innovative treatment versus a person showing a biomarker measurement of value $X = 0$ and being treated by the standard. The odds ratio for a person with biomarker measurement $X = x_0$ between both treatments is

$$exp\left\{LP\left(x_0,1\right)-LP\left(x_0,0\right)\right\}=exp\left\{h_X\left(X\right)\right\}.$$

References

Anderson, T. W. (1957). Maximum likelihood estimates for a multivariate Normal distribution when some observations are missing. *Journal of the American Statistical Association, 52*(278), 200–203. https://doi.org/10.1080/01621459.1957.10501379.

Bhandari, M., Guyatt, G., Walter, S. D., Tornetta, P., Schemitsch, E. H., Swiontkowski, M., & Sanders, D. (2008). Randomized trial of reamed and Unreamed intramedullary nailing of Tibial shaft fractures. *The Journal of Bone and Joint Surgery-American Volume, 90*(12), 2567–2578. https://doi.org/10.2106/jbjs.g.01694.

Bodemer, N., Meder, B., & Gigerenzer, G. (2014). Communicating relative risk changes with baseline risk. *Medical Decision Making, 34*(5), 615–626. https://doi.org/10.1177/0272989x14526305.

Breiman, L. (2001). Statistical modeling: The two cultures (with comments and a rejoinder by the author). *Statistical Science, 16*(3), 199–231. https://doi.org/10.1214/ss/1009213726.

Callegaro, A., Spiessens, B., Dizier, B., Montoya, F. U., & van Houwelingen, H. C. (2016). Testing interaction between treatment and high-dimensional covariates in randomized clinical trials. *Biometrical Journal, 59*(4), 672–684. https://doi.org/10.1002/bimj.201500194.

Cameron, A. C., & Trivedi, P. K. (1998). *Regression analysis of count data*. New York: Cambridge University Press. https://doi.org/10.1017/cbo9780511814365.

Casella, G. (2008). *Statistical design*. New York: Springer. https://doi.org/10.1007/978-0-387-75965-4.

Collett, D. (2002). *Modelling binary data* (2nd ed.). London: Chapman Hall/CRC Press.

Collett, D. (2014). *Modelling survival data in medical research*. Boca Raton: Chapman Hall/CRC Press.

Collins, G. S., Reitsma, J. B., Altman, D. G., & Moons, K. G. M. (2015). Transparent reporting of a multivariable prediction model for individual prognosis or diagnosis (TRIPOD). *Annals of Internal Medicine, 162*(10), 735–736. https://doi.org/10.7326/l15-5093-2.

Cuzick, J. (2005). Forest plots and the interpretation of subgroups. *The Lancet, 365*(9467), 1308. https://doi.org/10.1016/s0140-6736(05)61026-4.

Engelhardt, A., Shen, Y. M., & Mansmann, U. (2016). Constructing an ROC curve to assess a treatment-predictive continuous biomarker. *Studies in Health Technology and Informatics, 228*, 745–749.

Gadbury, G. L., & Iyer, H. K. (2000). Unit-treatment interaction and its practical consequences. *Biometrics, 56*(3), 882–885. https://doi.org/10.1111/j.0006-341x.2000.00882.x.

Gadbury, G. L., Iyer, H. K., & Allison, D. B. (2001). Evaluating subject-treatment interaction when comparing two treatments. *Journal of Biopharmaceutical Statistics, 11*(4), 313–333. https://doi.org/10.1081/BIP-120008851.

Gadbury, G. L., Iyer, H. K., & Albert, J. M. (2004). Individual treatment effects in randomized trials with binary outcomes. *Journal of Statistical Planning and Inference, 121*(2), 163–174. https://doi.org/10.1016/s0378-3758(03)00115-0.

Gail, M., & Simon, R. (1985). Testing for qualitative interactions between treatment effects and patient subsets. *Biometrics, 41*(2), 361–372. https://doi.org/10.2307/2530862.

Gentles, S. J., Stacey, D., Bennett, C., Alshurafa, M., & Walter, S. D. (2013). Factors explaining the heterogeneity of effects of patient decision aids on knowledge of outcome probabilities: A systematic review sub-analysis. *Systematic Reviews, 2*(1). https://doi.org/10.1186/2046-4053-2-95.

Huang, Y., Gilbert, P. B., & Janes, H. (2012). Assessing treatment-selection markers using a potential outcomes framework. *Biometrics, 68*(3), 687–696. https://doi.org/10.1111/j.1541-0420.2011.01722.x.

Kasenda, B., Schandelmaier, S., Sun, X., von Elm, E., You, J., Blumle, A., Tomonaga, Y., et al. (2014). Subgroup analyses in randomised controlled trials: Cohort study on trial protocols and journal publications. *BMJ, 349*(Jul16 1), g4539–g4539. https://doi.org/10.1136/bmj.g4539.

Laubender, R. P., & Mansmann, U. (2014). Estimating individual treatment effects from responses and a predictive biomarker in a parallel group RCT. Department of Statistics: Technical Reports 176. https://epub.ub.uni-muenchen.de/22207/1/TR176.pdf.

Lord, F. M. (1955a). Equating test ScoresA maximum likelihood solution. *Psychometrika, 20*(3), 193–200. https://doi.org/10.1007/bf02289016.

Lord, F. M. (1955b). Estimation of parameters from incomplete data. *Journal of the American Statistical Association, 50*(271), 870. https://doi.org/10.2307/2281171.

McGinn, T. G., Guyatt, G. H., Wyer, P. C., Naylor, C. D., Stiell, I. G., Richardson, W. S., & for the Evidence-Based Medicine Working Group. (2000). Users guides to the medical literature. *JAMA, 284*(1), 79. https://doi.org/10.1001/jama.284.1.79.

Neyman, J. (1923). Sur Les Applications de La Theorie Des Probabilites Aux Experiences Agricoles: Essai Des Principes. *Statistical Science, 5*, 463–472. (Dabrowska DM and Speed TP, Translators).

Olotu, A., Fegan, G., Wambua, J., Nyangweso, G., Awuondo, K. O., Leach, A., Lievens, M., et al. (2013). Four-year efficacy of RTS,S/AS01E and its interaction with malaria exposure. *New England Journal of Medicine, 368*(12), 1111–1120. https://doi.org/10.1056/nejmoa1207564.

Rao, C. R., Toutenburg, H., & Shalabh, H. (2008). *Linear models and generalizations: Least squares and alternatives* (3rd ed.). Berlin: Springer Series in Statistics.

Rothwell, P. M., Eliasziw, M., Gutnikov, S. A., Fox, A. J., Taylor, D. W., Mayberg, M. R., Warlow, C. P., & Barnett, H. J. M. (2003). Analysis of pooled data from the randomised controlled trials of endarterectomy for symptomatic carotid stenosis. *The Lancet, 361*(9352), 107–116. https://doi.org/10.1016/s0140-6736(03)12228-3.

Rothwell, P. M., Mehta, Z., Howard, S. C., Gutnikov, S. A., & Warlow, C. P. (2005). From subgroups to individuals: General principles and the example of carotid endarterectomy. *The Lancet, 365*(9455), 256–265. https://doi.org/10.1016/s0140-6736(05)17746-0.

Royston, P., & Sauerbrei, W. (2008). *Multivariable model-building: A pragmatic approach to regression anaylsis based on fractional polynomials for modelling continuous variables* (Vol. 777). New York: Wiley.

Royston, P., Sauerbrei, W., & Ritchie, A. (2004). Is treatment with interferon-α effective in all patients with metastatic renal carcinoma? A new approach to the investigation of interactions. *British Journal of Cancer, 90*(4), 794–799. https://doi.org/10.1038/sj.bjc.6601622.

Rubin, D. B. (1974). Estimating causal effects of treatments in randomized and nonrandomized studies. *Journal of Educational Psychology, 66*(5), 688–701. https://doi.org/10.1037/h0037350.

Senn, F., & Harrell, S. (1997). On wisdom after the event. *Journal of Clinical Epidemiology, 50*(7), 749–751. https://doi.org/10.1016/s0895-4356(97)00023-1.

Shen, Y.-M., Le, L. D., Wilson, R., & Mansmann, U. (2017). Graphical presentation of patient-treatment interaction elucidated by continuous biomarkers. *Methods of Information in Medicine, 56*(01), 13–27. https://doi.org/10.3414/me16-01-0019.

Sun, X., Briel, M., Walter, S. D., & Guyatt, G. H. (2010). Is a subgroup effect believable? Updating criteria to evaluate the credibility of subgroup analyses. *BMJ, 340*, c117–c117. https://doi.org/10.1136/bmj.c117.

Ternès, N., Rotolo, F., Heinze, G., & Michiels, S. (2016). Identification of biomarker-by-treatment interactions in randomized clinical trials with survival outcomes and high-dimensional spaces. *Biometrical Journal, 59*(4), 685–701. https://doi.org/10.1002/bimj.201500234.

Thall, P. F., & Vail, S. C. (1990). Some covariance models for longitudinal count data with Overdispersion. *Biometrics, 46*(3), 657–671. https://doi.org/10.2307/2532086.

Tian, L., Alizadeh, A. A., Gentles, A. J., & Tibshirani, R. (2014). A simple method for estimating interactions between a treatment and a large number of covariates. *Journal of the American Statistical Association, 109*(508), 1517–1532. https://doi.org/10.1080/01621459.2014.951443.

Vandenbroucke, J. P., Broadbent, A., & Pearce, N. (2016). Causality and causal inference in epidemiology: The need for a pluralistic approach. *International Journal of Epidemiology, 45*(6), 1776–1786. https://doi.org/10.1093/ije/dyv341.

Vickers, A. J., Kattan, M. W., & Sargent, D. J. (2007). Method for evaluating prediction models that apply the results of randomized trials to individual patients. *Trials, 8*(1). https://doi.org/10.1186/1745-6215-8-14.

Wegwarth, O. (2012). Do physicians understand cancer screening statistics? A national survey of primary care physicians in the United States. *Annals of Internal Medicine, 156*(5), 340–349. https://doi.org/10.7326/0003-4819-156-5-201203060-00005.

Zikmund-Fisher, B. J. (2012). The right tool is what they need, not what we have. *Medical Care Research and Review, 70*(1_suppl), 37S–S. https://doi.org/10.1177/1077558712458541.

Zikmund-Fisher, B. J., Fagerlin, A., & Ubel, P. A. (2010). Risky feelings: Why a 6% risk of Cancer does not always feel like 6%. *Patient Education and Counseling, 81*(December), S87–S93. https://doi.org/10.1016/j.pec.2010.07.041.

Chapter 15
Epistemic Gains and Epistemic Games: Reliability and Higher Order Evidence in Medicine and Pharmacology

Barbara Osimani

Abstract In this paper I analyse the dissent around evidence standards in medicine and pharmacology as a result of distinct ways to address epistemic losses in our game with nature and the scientific ecosystem: an "elitist" and a "pluralist" approach. The former is focused on reliability as minimisation of random and systematic error, and is grounded on a categorical approach to causal assessment, whereas the latter is more focused on the high context-sensitivity of causation in medicine and in the soft sciences in general, and favours probabilistic approaches to scientific inference, as better equipped for defeasibility of causal inference in such domains. I then present a system for probabilistic causal assessment from heterogenous evidence that makes justice of concerns from both positions, while also incorporating "higher order evidence" (evidence/information about the evidence itself) in hypothesis confirmation.

15.1 Introduction

Medical science distinguishes itself with respect to other sciences and technologies by the joint interaction of following phenomena:

1. Epistemic uncertainty with respect to the real state of nature and of the outcome of interventions is generally higher than for other natural sciences (Joffe 2011),

Funding by the European Research Council through grant 639276 (PhilPharm) is grateful acknowledged.

B. Osimani (✉)
Department of Biomedical Sciences and Public Health, Polytechnic University of the Marche, Ancona, Italy

Munich Center for Mathematical Philosophy, Munchen, LMU, Munich, Germany
e-mail: b.osimani@staff.univpm.it

A. LaCaze, B. Osimani (eds.), *Uncertainty in Pharmacology*, Boston Studies in the Philosophy and History of Science 338,
https://doi.org/10.1007/978-3-030-29179-2_15

indeed among the natural sciences, medicine is the closest one to the social sciences and the humanities;

2. Medical products are so called "credence products", that is products for which the consumer (medical community, patients, and public health system in a more general sense) cannot evaluate the quality prior (and often not even after) consumption. Such state of affairs is therefore characterized by information asymmetry, which further interacts with the decision-makers' risk attitude.
3. medical choices are high-stakes decisions both in terms of physical, psychological, existential, and financial costs (Osimani 2012),
4. information asymmetry and high sensitivity of medical decisions, can be strategically exploited by producers of medical knowledge having vested interests in the research outputs and dissemination (such strategic behaviour may of course also evolve in time) (Teira 2011; Teira and Reiss 2013; Holman 2015);
5. medical decisions are intertwined with pressing ethical dilemmas touching different anthropological dimensions (Papa 2014; Sgreccia 2007; Pessina 2009; Beauchamp 2011; Faden and Beauchamp 1986; Scheu 2003);

This picture is especially vivid in pharmacology; the complex network of interests (financial issues, concern for reputation etc.), as well as legal rights and duties which frame the scientific and social ecosystem in which pharmacology is embedded make it a unique blend of science and technology. Indeed, it is manifest that in scientific domains characterised by vested interests, the production and evaluation of evidence is embedded in a strategic game, where agents obviously try to maximise their payoff. This state of affairs strongly emphasises the role of reliability (in its various aspects) as a decisive dimension of evidence.

This paper focuses on the roots of methodological dissent in medicine as a dissent related to how reliability is conceptualised and warranted in contending schools of thoughts.

The paper is organised as follows: in Sect. 15.2, I present the debate around standards of evidence in medicine and the role played by evidence from randomized controlled trials vs. evidence of biological mechanisms or other kinds of sources. Section 15.3 presents the two poles on which such debate rests: the "elitist game" and the "pluralist game". These two distinct approaches focus on different aspects of reliability as a result of weighting the costs of different kinds of errors differently. In particular, elitists strive to maximise internal validity by minimising random and systematic error, while the pluralists are rather concerned by the high context-sensitivity of causes in the biological realm (and in the soft sciences in general), hence they are more concerned about the stability of causal knowledge across populations and domains (failure of external validity and false predictions). In Sect. 15.4, I illustrate a meta-level perspective where the *structure* and *organisation* of the entire body of evidence brings its own contribution to hypothesis confirmation. In this perspective, epistemic dimensions of evidence itself, such as its reliability, consistency/coherence and (in)dependence also contribute to hypothesis confirmation in an interactive way, and these are connected with research organisation issues: how the scientific ecosystem is modelled and interacts with the broader social system. This perspective is concretely translated in a multilayer framework which models probabilistic causal inference through evidence synthesis elaborated by De Pretis et al. (2019). This framework, henceforth "E-Synthesis", reconciles various concerns around causal inference in medicine and pharmacology, by allowing various kinds of evidence to jointly contribute to hypothesis confirmation, as well as to incorporate higher order evidence, such as evidence about the reliability of the sources and about its relevance with respect to the target population, hence to account both for elitists' and pluralists' concerns.

From a philosophical point of view, "E-Synthesis" provides a solution that accommodates the intuitions underpinning apparently conflicting concerns,[1] and has the virtue of being able to be adapted to various theoretical stances on causality (counterfactual vs. process theories vs. regularity vs. inferentialist theories of causality) (see Poellinger 2018). Section 15.5 presents how "E-Synthesis" addresses and solves philosophical issues around causal inference in medicine; in particular, the privilege accorded to randomised controlled trials, the debate around evidence hierarchies, the epistemic status of evidence for biological mechanisms, and the role of higher order evidence (evidence about evidence itself) in hypothesis confirmation.

15.2 Isolating Causes vs. Causes in Interaction: The Two Contending Paradigms

The philosophical and methodological debate concerning evidence in medicine can be roughly made sense of as a sophisticated elaboration of arguments in favour of (or against) two main paradigms for scientific inference: a categorical approach – inherited from frequentist statistics – and a probabilistic approach (where hypothesis confirmation comes in degrees) – inherited from an inductive/Bayesian approach to scientific inference.[2] Nancy Cartwright (2007a) for instance speaks about "clinching" and "vouching" methods as distinctive ways in which causal inference may be carried out in medicine and in the social sciences: vouchers, other than clinchers, do not force any conclusion, but rather suggest or support it more or less strongly.

- *Clinchers.* In principle, clinchers deductively force their conclusions and deliver an acceptance/rejection verdict on the hypothesis under investigation, on the basis of a "modus tollens" reasoning:

[1] This is possible also because the topology of our framework reveals that these conflicts relate to different levels or dimensions of causal inference and therefore can be deflated.

[2] These two approaches intersect with other epistemic stances such as empiricism vs. methodological pluralism, as well as various programs for causal inference from statistical data, however the above mentioned dichotomy can be analyzed relatively independently from these perspectives.

$$\frac{H \supset E, \neg E}{\neg H}$$

where H stays for hypothesis and E for evidence. However in experimental settings the syllogism is applied probabilistically, since the evidence $\neg E$ consists in an event that is not impossible under the assumption of H holding, but rather only very improbable. That is why $\neg E$ only invites the following statement, also called "Fisher disjunction": either something very improbable happened, but H is nevertheless true, or H is false. However, in practice hypotheses are either rejected or not, and because of this categorical approach, clinchers are characterised by a greater inductive risk, which explains why they are focused on the *probability of erroneous conclusions* (Mayo and Spanos 2006; Sprenger 2016). More importantly, the threshold for hypothesis acceptance/rejection is based explicitly on the degree of reliability that one wishes to obtain: that is, on significance levels, which are substantially based on the size of random error considered to be tolerable (for the purpose at hand). In the standard approach of hypothesis testing, the hypothesis under investigation regards the individual contribution of a putative cause to its effect (possibly moderated by some prognostic factors) and such hypothesis is contrasted with the so called null hypothesis of the cause under investigation making no difference to the observed effect. For an observed positive effect, the hypothesis space consists of three general alternatives: (1) either the treatment causes the observed effect; (2) or the effect is due to chance; (3) or the effect is due to some alternative confounding factors. Reliability theory, and methodology in general, define the instruments that should be used in order to exclude such alternative hypotheses as safely as possible: Large sample sizes help to exclude chance, by the law of large numbers (hence, the larger the sample sizes, the lower the random error, ceteris paribus); study design helps to evaluate the internal validity of the results, i.e. the extent to which alternative causes can be excluded. Therefore, clinchers are studies that are reliable both in the sense that they exclude chance (random error), and in the sense that they exclude bias or confounding (systematic error) to a large extent. In sum, clinchers maximise accuracy and internal validity.

- *Vouchers* are methods whose results cannot force any conclusion, but only suggest it. They allow only defeasible inference, but they are flexible enough to incorporate new possibly conflicting evidence in the inferential framework, without necessary eliminating old beliefs. This is a particularly appropriate approach when causes are highly context-sensitive (such in biology or the social sciences). Reliability refers here rather to whether a causal link established for some population may also hold in other contexts, and on how interacting causes modulate the causal effect.

In medical research the two approaches are reflected in two different perspectives with respect to the evaluation and use of evidence coming from different sources/methods.

The so called "Evidence Based Medicine" paradigm has emphasized the selection of "best evidence", meant as accurate and internally valid evidence, and therefore has privileged some methods rather than others, on grounds of their higher reliability warrant. The contending view criticizes this paradigm for being uselessly and harmfully monistic and for leaning on misleading heuristics (such as evidence hierarchies).[3]

In particular, critics of the EBM approach have emphasized its shortcoming in dealing with external validity issues (Cartwright 2012; Clarke et al. 2014), which is indeed nothing else than the straightforward consequence of context sensitivity of causation, and in not paying due attention to the ontology of the phenomena being investigated (Cartwright 2007d; Anjum 2012). Furthermore, EBM is blamed for unknowingly bringing back through the window what they threw back from the door, i.e. "subjectivity" (Stegenga 2011), and for not keeping their epistemic promises (Worrall 2007b), or for running against their intended goals by not adequately distinguishing between harm vs. benefit assessment (Osimani 2014), and not being cost-effective in using the available evidence (De Pretis et al. 2019; Russo and Williamson 2007).[4]

The debate has been mainly carried on by way of illustrating case studies where one or the other approach fails to deliver the optimal information for making clinical decisions or policies. For instance, Worrall (2007b,c) illustrates the case of Extracorporeal Membrane Oxygenation for newborn babies affected by persistent pulmonary hypertension. Notwithstanding high success rate of the treatment and knowledge of the mechanisms explaining that success, efficacy was considered not to be established until proven by randomized controlled trials. This obviously exposed the newborns in the no-treatment arm to almost certain, and, given the available treatment, avoidable death. Since the newborns in the treatment arm had instead a much higher chance to survive, this was a dramatic case of lack of equipose. By pointing to the neglected sources of knowledge (mechanisms and past success record in uncontrolled case series) and by downsizing the reliability of RCTs, Worrall explains why "there is no cause to randomise" (see Worrall 2008).

Cartwright exemplifies her theory of extrapolation, by reporting about an integrated nutrition program which, while fully successful in India, completely failed in Bangladesh, and imputes this failure to a mismatch between the causal structure and social mechanisms working in the original vs. the target population (Cartwright 2012).

Also Russo and Williamson (2007) and Clarke et al. (2013, 2014) provide a rich list of case studies to support their view that evidence of different kinds are co-

[3]Also, a parallel issue concerns the strong preference accorded to the so called Potential Outcome Approach (POA) vs. more pluralistic views of causal inference and causality itself (see: Vandenbroucke et al. (2016)).

[4]Not to mention the fact that, traditionally, epistemology has considered varied evidence as more confirmatory than repetitive data (see Osimani and Landes (2020) for a discussion of the "Variety of Evidence Thesis".)

supportive and provide a stronger basis for causal assessment, by providing distinct types of information.

Instead Howick and colleagues (2013) warn that knowledge of mechanims need always to be complemented by randomised controlled trials, while the reverse does not hold. The basis for this position is that biological pathways are extremely complicated and rich of feedback loops and backup mechanisms, which may lead to surprising outcomes at the phenotypic level. Although RCTs provide black-box evidence only, they are the only reliable basis for causal inference, in that they deliver clear information about input-output data at the level of interest, the clinical one.[5]

At its root such dissent is originated by the two sides playing different games with nature and within the scientific ecosystem. EBM plays an "elitist" game where evidence is put to test before entering the court: a certain threshold for reliability is established before the evidence is considered at all. Hence reliability is the gatekeeper and has precedence over other possible epistemic values such as the principle of total evidence, the precautionary principle, and external validity. Furthermore the notion of reliability developed with the EBM framework is method-specific; it has been constructed around a specific statistical school of thought: frequentist hypothesis testing.

Opponents of the EBM approach play a different game: they are mainly concerned by context-sensitivity of causation, and, more generally, by defeasibility of scientific inferences; consequently they appeal to the Principle of Total Evidence, as an essential desideratum in non-monotonic reasoning, as opposed to the hypothetico-deductive paradigm; furthermore, they are also worried as to the extent to which such evidence may be used for extrapolation and prediction. Hence for them also relevance of the evidence, and external validity play an important role. In both games, there is a considerable amount of uncertainty, but dissent arises as to how such uncertainty should be managed and weighted,[6] and consequently, as to the costs and utilities of each research strategy, or more generally, of any regulatory standards.[7]

---

[5]They cite the example of the drug "Torcetrapib", as a case where even perfect knowledge of its functioning mechanisms did not enable its producers to avoid excess death rates in the treatment arm of the third-phase trials, and therefore denial of approval. However, this failure was in fact due to lack of knowledge about (the mechanisms leading to) the side-effect, for which knowledge of the mechanisms for the intended effect is not necessarily helpful.

[6]In the statistical counterparts of these games (frequentist vs. Bayesian statistics) probability as a measure of uncertainty is attached to the probability of error (in the long run) in one game, whereas in the other it is attached to the hypothesis itself.

[7]See also Podolsky and Powers (2015) for a critique on a recent shift in FDA evidential standards, from what I call an "elitist" to a "pluralist" view. The considerations underpinning such critiques are cast exactly in terms of the costs and benefits of each regulatory approach. Analogously Osimani (2014) and Osimani and Mignini (2015) insist that evidential standards should not be the same for assessing efficacy and harm, exactly on these epistemic and strategic grounds.

15.3 The Elitist and the Pluralist Game

The above games are associated with different epistemic goals and value specific epistemic virtues over others: it is this different preference ordering that might help us understand the dissent. Let's look a little bit closer at these preference orderings and the related payoffs and losses. In the EBM framework reliability (in both senses) is highly valued: the inferential game is seen as a game against nature, chance and biased scientists: and truth is what remains after all other factors have been eliminated. The pluralist stance is more concerned about complexity of causal phenomena and about combining a plurality of evidence to find truth in the web of interactions.

The general idea here is that whereas EBM puts efforts in decontextualising nature's signals from noise and is more focused on science distorting them (through unreliable instruments), its opponents are rather worried that nature's signals are embedded in a symphony and cannot be interpreted in isolation. Metaphorically speaking, the point is whether one considers context as noise to be eliminated, or as music which gives meaning to the individual note.

15.3.1 The Elitist Game

Clinching methods are elitist methods which value evidence for the degree to which they exclude random and systematic error. Random error refers to the "chance" variability around an otherwise reliable measurement. This can be amended by averaging the results of various replications of the study. Instead, systematic error refers to possible confounders, which systematically distort the results through replications.

The different strategies advocated by so called Evidence Based Medicine have been developed in order to secure studies from two kinds of error[8]:

1. random error;
2. systematic error (this can be generated by bias, confounding or both).

The strategies adopted to address these two kinds of error are orthogonal to each other, and are partly a consequence of the statistical approach entrenched in the EBM paradigm, that is, classical frequentist statistics.

[8]The whole enterprise of the EBM paradigm can indeed be seen as the tireless effort to systematize a set of techniques to track and possibly minimize random and systematic error.

15.3.1.1 Random Error

Random error is dealt with straightforwardly; indeed it is the basis of the entire statistical machinery of the frequentist approach: the level of tolerated *random error* is inscribed in the very methodological procedure by establishing it beforehand, and *deciding on that basis, whether the hypothesis will be accepted or rejected*, given the observed data. Hence, it is the desired "reliability level" ("significance level" and "power") that determines the ultimate fate of the hypothesis under investigation.

In this first sense, belief in the reliability of the result can be enhanced in two related ways: either by pooling results of various studies, thought to be homogeneous enough as to the sampled populations – so called meta-analyses – or by performing "identical" replications of the original study.

As much as the confidence interval may be narrow or the number of consistent replications high, these indicators of accuracy could never set you free from another kind of error, namely systematic error; that is the error arising from the in principle "identical" studies measuring the effect of a confounder instead of the investigated variable. Pure statistical considerations cannot address the problem of systematic error, which is indeed rather approached by *study design*.

It is as if we are dealing with three possible explanations when examining a given scientific result: (1) chance, (2) the investigated causal factor, (3) other causal factors (confounders, bias). Consistent replications reduce the probability of chance producing the results, but cannot do anything in discriminating between 2 and 3.

15.3.1.2 Systematic Error

The entire EBM paradigm can be seen as an effort to foster quality of evidence by both insisting on maximisation of accuracy and minimisation of systematic error (or maximisation of "internal validity"). This effort is concretised in the development of evidence hierarchies that rank studies according to their design. Hence, one finds randomised controlled trials at the top level of the ranking (and higher than that, systematic reviews and meta-analyses of RCTs); followed by comparative observational studies (such as cohort and retrospective – or, historical – studies), and below these, uncontrolled observational studies (that is, case series, and single case studies). Evidence concerning cellular mechanisms or, generally any data below the phenotypic level is ranked lowest.[9]

Internal validity refers to the neutralisation of possibly confounding factors, as candidate alternative explanations for the observed effect. These may be alternative or interactive causes (also known as prognostic and predictive factors). The preferred means to counteract this sort of problem is to use intervention and randomization, so as to avoid (self-)selection bias and have as much as possible

[9]However the reason for ranking this data below the phenotypic level lowest, is due to issues of external rather than internal validity (see Howick (2011)).

balanced groups in the experiment arms. This aims to guarantee that the observed effect is due to the treatment, and only to it. However, also random error is taken into account; hence, larger RCTs are ranked higher in the hierarchy than smaller ones, and meta-analyses of several studies are ranked higher than individual studies.

Below the standard design ranking of EBM hierarchies (with related epistemic goal) are:

1. *random vs. non-random sampling*: this criterion is important both for the *representativeness* of the sampled population (of subjects or studies) and for the external validity of the results. A randomised controlled study, whose treatment and control arms are perfectly balanced, but whose subjects have not been sampled randomly from the target population will be internally valid, but not representative of the population under study (and will suffer from low external validity if the results need to be applied to that population). Indeed samples for randomised clinical trials are almost never sampled randomly from the original population. However, systematic reviews of meta-analyses aims to achieve this same goal, in constructing a population of studies which is as representative as possible of the sampled population.
2. *experimental vs. observational design* (RCTs vs. cohort studies): experimental interventions are the best warrant of internal validity in that they severe the cause under investigation from other potential confounding factors, which could alternatively explain the observed difference in the two arms of the study;
3. *controlled vs. uncontrolled design* (cohort – or any kind of comparative studies, vs series of clinical cases with no control group of "untreated" patient). Control serves the purpose of comparing whether the same causal effect is observed also in the absence of the investigated cause; that is, to verify relevance of the putative cause to the investigated effect.

Hence, (1) control serves the goal of verifying (causal) *relevance*; (2) intervention through random allocation has the purpose of testing *causal sufficiency*; (3) and random sampling is a way to establish *causal necessity*:

1. the statistical *non-spurious* difference between outcomes in the exposed and non-exposed groups provides evidence that the presence of the putative cause makes a difference with respect to the investigated outcome;
2. random allocation ensures that no latent variable confounds the experimental results, and thus ensures that the set of causes taken into account is complete – although they are unknown. Causal sufficiency[10] is warranted by the very fact that the possible influence of all confounders (known and unknown) is neutralized through random allocation.
3. the sample population for RCTs is very rarely sampled randomly from a source population. However, the ranking of systematic reviews of meta-analyses over

[10]In the causal search literature Pearl (2000) and Spirtes et al. (2000), causal sufficiency is a fundamental assumption which grounds the algorithmic search, and undermines it if it fails to hold.

simple meta-analyses responds to the recognition that the latter may be biased by cherry-picking the included studies. Systematic reviews ensure that the sampled population *of studies* is non-biased and therefore representative of the source population. This warrants that the "causal law" inferred from the RCTs is at least valid in that specific population (the one from which the study population has been sampled from).

15.3.2 The Pluralist Game

Advocates of a pluralist methodology oppose the idea that *RCTs have a privileged role* in establishing causation (Worrall 2007a,b) as well as the view that *difference-making is necessary or sufficient* to establish causation (Cartwright 2007a). Objections to this paradigm have been raised on the following grounds:

1. *internal/external validity trade-off*: the causal structure underpinning the study results in the sample population, may not be equivalent to the one which characterises the causal dynamics in the target population, hence even studies that are strongly internally valid may fail to provide the right evidence for other kinds of populations. Indeed a sort of "inverse proportionality" relationship is posited between internal and external validity: the stricter the study protocol (inclusion/exclusion criteria, mode of administration, etc), the more likely that the study result will be internally valid, but also that its results will lead to wrong inferences in relation to real-life settings (Cartwright 2012)[11];
2. the putative privilege of randomised evidence is ill-founded (Worrall 2007b,c): that is, randomised controlled trials do not even provide any guarantee of internally valid results;

As a constructive response to these criticisms, a conciliatory position has been proposed, arguing for the adoption of a pluralistic approach to causal inference. According to this view, neither statistical evidence, nor evidence on mechanisms underpinning it are per se sufficient to establish causation, however they are both necessary. This position has been elaborated in Clarke et al. (2013, 2014), where it is advanced that different sorts of evidence may have complementary roles in supporting causal hypotheses. In particular, evidence about difference making helps de-masking causes which might be canceled out by back-up compensatory mechanisms in the organ system, whereas evidence about mechanisms is needed in order to design and interpret statistical studies – the upshot of this reasoning is that different kinds of evidence may systematically support each other and jointly (dis)confirm the causal claim under investigation.

[11] These criticisms have kindled a series of defenses of RCTs on various grounds: see Senn (2003), Papineau (1994), La et al. (2012), Teira (2011), Teira and Reiss (2013) and Osimani (2014) for an overview on the debate.

De Pretis et al. (2019) have extended this approach to various indicators of causality and relaxed the requirement that any of them be necessary, by allowing for probabilistic causal assessment, rather than categorical ones.[12]

15.3.3 Context-Sensitivity of Causality and Causal Modulation

The insistence on mechanisms and external validity from proponents of the pluralist view on medical evidence straightforwardly derives from the intuition that causation in this scientific domain is highly context sensitive. However, what does context-sensitivity exactly mean? Roughly speaking, context sensitivity refers to the joint interaction of various causes in bringing about a given effect and in modulating its intensity. Hence, subjects in studies can be conceived as vectors consisting of possible combinations of variables (at different levels: molecular, cellular, organic), that jointly contribute to the occurrence and the modulation of the effect in the presence (vs. absence) of the investigated cause.

Leonid Hanin (2017) makes this point very vividly when he explains the irreproducibility of trial results by drawing on various sources of (uncontrollable) variation in clinical research. Speaking about predictors of metastatic recurrence after breast cancer surgery, Hanin points out that:

> Whether or not a metastases will escape from dormancy in a particular patient depends not only on the effects of treatment, functioning of the immune system, concentrations of circulating angiogenesis promoters and inhibitors, and other internal factors; exhacerbation of the disease may also be triggered by intercurrent sporadic external events such as surgery unrelated to breast cancer, infection, trauma, radiation, stress, etc. Another highly significant prognostic factor is the intrinsic aggressiveness of the disease; however, its reliable assessment at early stages of the disease has proven to be far elusive. Thus the most critical determinants of the trial outcome are largely unobservable and/or unpredictable.

To make things worse, an additional level of opacity should be added to the picture (my emphasis):

> In practice the above observable prognostic factors are substituted with less informative observable surrogates such as (1) age at trial entry; (2) stage and historical grade of of the disease at surgery; (3) localization and size of the primary tumor; (4) whether or not the tumor invaded surrounding tissues; (5) the extent of nodal involvement; (6) menopausal status; (7) estrogen and progesterone receptor status; (8) presence of specific mutations in BRCA1 or BRCA2 genes; (9) family history of breast cancer; and (10) individual history of other malignancies. *Even this rough and incomplete set of surrogate clinical variables creates a large number of categories of women in both arms of the trial with potentially very different characteristics of survival. Importantly, randomization won't*

[12]This stance can be made more general by drawing on the "Variety of Evidence Thesis" (VET) which, stated in its more general form, claims that *ceteris paribus*, the more varied the evidence, the higher the confirmatory support provided to the hypothesis which explains it. Taken at face value this claim seems to favour the pluralistic methodology approach over the "evidence elitist" view adopted in the EBM paradigm. See also Osimani and Landes (2020).

> *eliminate the observable and hidden heterogeneity; it will only reduce the difference in the extent of heterogeneity between treatment and control arms.* The aforementioned inter-subject heterogeneity is quite typical of clinical trials (as opposed to in vitro experiments with cell lines or studies on animal models with tightly controlled inter-subject variation). Thus, individual responses of subjects in both arms of a trial cannot even approximately be viewed as homogeneous, let alone distributionally identical.

(Hidden) mediators and moderators may also non-additively contribute to modulating the causal effect through joint interaction. Human beings are not gas molecules; their reactions to the same treatment may be largely dependent on contingent factors (variation of individual response in different circumstances: intra-subject, or individual variation), and on various systematic mediating and interacting causes. Sample heterogeneity may be epistemically inscrutable since mediators and moderators may (and generally do) act jointly and at different organ levels (genetic/genomic, proteomic, cellular), and also emerge dynamically within one's own clinical history and social environment.

In "The Cement of the Universe" (1974) John Mackie offers a deterministic reading of so called probabilistic causation by advancing the concept of INUS condition, which provides a way to formalise the context-sensitive nature of causation. In his proposal, causes come in sets of components, (such as e.g. the presence of oxygen for a spark to start a fire), and an INUS condition is an insufficient but necessary component of an unnecessary but sufficient causal set. For instance, A is an INUS condition for E means that it is part of at least one conjunctive set standing in a biconditional relationship with E:

$$(A \wedge B) \vee (A \wedge C) \Leftrightarrow E$$

The "causal sets" can be equated to possible worlds which explain the occurrence of E in the presence of various concurring causal factors. For instance where the following holds:

$$(A \wedge B \wedge C) \vee (A \wedge D \wedge F) \vee (B \wedge D \wedge F) \Leftrightarrow E$$

the same effect E can be caused by different causal sets, such as for instance (ABC), or (ADF), or (BDF). Hence in a study which investigates the causal effect of A with respect to E (for instance a certain medical treatment with respect to a given clinical outcome) such effect will be the result of the proportion of people having also the characteristics B and C or D and F over all other subjects (since neither ACF nor ABF, nor ACD nor ABD are sufficient causal sets for E). Furthermore, there will be cases where, notwithstanding the absence of A, E will nevertheless occur (that is, when subjects present the characteristics BDF jointly).[13]

[13]Since "causal strength" is generally measured by the "effect size", that is, the proportion of subjects in the sample who show the effect E in the treatment vs. control group – for instance through relative risk ratio or odds ratio measures – causal strength can be both the result of a relative context-insensitivity of the treatment investigations (or better, the fact that it contributes

By drawing on this conceptualization of context-sensitivity, Cartwright (2012), underpins her criticisms of RCTs as gold standard in medicine, on account of the high context-sensitivity of causation in the soft sciences. Provided that the law underpinning the observed effect in relation to treatment X is formalized as follows:

$$L : Y_c = a + \beta X + W, \tag{15.1}$$

Where Y is the effect variable, X the cause variable, a is a constant, β a coefficient and W a random error summarising the effect of hidden latent variables; then, the average value of Y in the treatment group will be measured by:

$$\langle Y(u)/X(u) = x'\rangle = \langle a(u)/X(u) = x'\rangle + \tag{15.2}$$

$$\langle \beta(u)/X(u) = x'\rangle x + \tag{15.3}$$

$$\langle W(u)/X(u) = x'\rangle . \tag{15.4}$$

and, consequently, the treatment effect will be measured by the following equation:

$$T =_{def} = \langle a(u)/X(u) = x'\rangle - \langle a(u)/X(u) = x\rangle + \tag{15.5}$$

$$\langle \beta(u)/X(u) = x'\rangle x - \langle \beta(u)/X(u) = x\rangle x + \tag{15.6}$$

$$\langle W(u)/X(u) = x'\rangle - \langle W(u)/X(u) = x\rangle . \tag{15.7}$$

Since randomization is meant to warrant that the expectation of $a(u)$, $\beta(u)$ and $W(u)$ are the same whatever value X assumes (that is, that X is probabilistically independent from a, β and W), the first and last two terms in the equation cancel out; hence, the measure of the treatment effect given by an RCT results from the following formula:

$$T =_{def} \langle \beta(u)\rangle (x - x'). \tag{15.8}$$

The critical issue raised by Cartwright is that β represents all the combinations of factors that determine not only whether X contributes to Y (if β is 0, then this contribution is null); but also how (positively or negatively) and to what extent. By drawing on the notion of causes as INUS condition, Cartwright then considers the possibility that β is a disjunction of sets of interacting factors: $\beta = f_1(z_{11}, \ldots, z_{1n}) + \ldots + f_m(z_{m1}, \ldots, z_{mp})$.

to the effect in many causal sets), or to its intrinsic force (as measured for instance by a steep dose-response curve). This intrinsic ambiguity prompted the substitution of the Bradford-Hill indicator "strength of the association" with three different indicators: probabilistic dependence, dose-response, and rate of growth in (De Pretis et al. 2019, Section 3). See Sect. 15.4 below.

This translates in an irreducible context-sensitivity component of causation, which cannot be fully accounted for by classical statistical methods of causal inference.If the main information provided by experimental evidence is about whether (and to what extent) a causal link between two phenomena exists, the information provided by detailed case series, or basic science (molecular studies; in vitro or, so called, "in silico" evidence) may contribute to a greater extent to the identification of specific subclasses in which the effect follows the treatment to a higher or lower degree, that is information about causal interactions.[14] Analogous information may also come from adjustment (stratification) and subgroup analyses in controlled studies. However, the more context-sensitive the treatment under investigation, the larger the sample sizes need to be in order to obtain such information. What is needed here is the acknowledgement that the two approaches are focused on complementary (not opposite!) goals: the "elitist approach" is concerned about establishing causal links rather independently from the task of identifying mediators and moderators, and to minimize false positives resulting from random or systematic error. Reliability of the evidence consists in the isolation of the causal link under investigation from possible interferers, and is a function of the process through which it is collected. Instead, the "pluralist approach" emphasises the irreducible context-sensitive nature of causation in medicine (and in the social sciences); therefore reliability here is about whether the acquired causal knowledge is stable enough to be applied to other contexts (prediction, external validity). Any source of information is valued as a useful basis to identify such interacting co-factors.

In the following I present a multilayer approach to causal inference in pharmacology, where both objectives are considered, and accounted for in a unifying framework.

[14]The importance to predict the variability of the effect as a function of the joint interaction of possible co-factors hence casts a new light on sources of evidence, such as case reports and case series – which standard hierarchies rank as low, because it is not "controlled" and hence cannot provide high warrant of internal validity – but whose specific epistemic import is no lower, indeed very high, in that they can provide us with valuable information about the various scenarios in which a given treatment may (or may not) induce its effect in different degrees. So, very detailed case reports facilitate inference about co-factors (prognostic factors and mediators) possibly influencing whether and to what extent, the effect size occurs in a specific population. "In silico" evidence comprehends a huge class of methodologies which can be broadly subdivided into systems biology approaches to computational modelling and simulation and machine learning techniques for knowledge extraction or pattern recognition.

15.4 E-Synthesis: A Probabilistic Causal Inference with Heterogeneous and Higher Order Evidence

The so called "reproducibility crisis" (see for instance the "Reproducibility Project: Psychology" by the Open Science Collaboration) caused some stir among psychologists, methodologists as well as scientists. The crisis extends well beyond psychology and also invests medical research (Begley and Ellis 2012; Ioannidis 2005; Prinz et al. 2011).

While some analysts have provided formal confirmation for the plausibility of such explanations (Etz and Vandekerckhove 2016; Marsman et al. 2017), and some downplay the whole issue (Senn 2002), others have further insisted on the problem of noisy data and suggested that "to resolve the replication crisis in science we may need to consider each individual study in the context of an implicit meta-analysis" (Gelman 2015). By "meta-analysis" Andrew Gelman does not mean here the standard data-pooling methods developed for instance within the Cochrane Library initiative, where statistics from sufficiently homogeneous studies are averaged so as to obtain more accurate measures of effect sizes, but rather a global approach to evidence evaluation, which takes into account not only the prima facie supportive strength any individual study gives to the investigated hypothesis (and its alternatives), but also other "higher-order" dimensions related to the entire body of evidence:

> One direction for statistical analysis that appeals to me is Bayesian inference, an approach in which data are combined with prior information *(in this case, the prior expectation that newly studied effects tend to be small, which leads us to downwardly adjust large estimated effects in light of the high probability that they could be coming largely from noise)*. (Gelman 2015, p.35) (my emphasis).

Therefore, higher order evidence may consist of consistent replications across repeated measurements, knowledge about the behaviour of measurement instruments under certain conditions, or more generally, robustness of results across methods and sources, and coherence of data with the established knowledge as well as with available theories. This evidence may interact with estimation about the reliability of the source, as a function of various social factors, such as financial interests, concern for liability and regulatory issues, scientific and commercial reputation.

So far, there has been a division of labour among philosophers of science, epistemologists, and methodologists regarding these different layers. Philosophy of science and methodology (statistics) has focused on the prima facie relationship between evidence and hypothesis, and related problems (measurement error, evidential support, theory ladeness, underdetermination, internal and external validity) (Carnap 1956; Fisher 1955; Hacking 2006; Haack 2011; Lenhard 2006; Hoyningen-Huene 2013). More generally, classical and formal epistemology have paid attention to the formal warrants for knowledge justification (Audi 1993; BonJour 2009; Swinburne 2001). Social (formal) epistemology instead has concentrated its efforts on the complex framework of interests and constraints which mold the scientific ecosystem (Goldman 1999; Longino 1990; Mayo-Wilson et al. 2011).

With respect to medicine, Miriam Solomon (2015) has emphasised the social construction of medical knowledge, and David Teira has thoroughly examined the impact of the social environment (regulatory framework, economic interests, etc.) on the development of methods for evidence evaluation (see Teira, this volume, and Teira and Reiss (2013) and Teira (2011))[15]; Bennett Holman frames the evolution of regulatory tools for drug approval and monitoring as an epistemically asymmetric race of arms (Holman 2015).

In the present section I present a multilayer approach to modelling causal inference for pharmaceutical harm, that keeps track of the interactions between these different dimensions of evidence in a unitary framework (see Fig. 15.1):

1. a basic level of evidential support to the hypothesis at hand (and various evidence aggregation/amalgamation techniques, see De Pretis et al. (2019)). This constitutes the traditional focus of philosophy of science with its various approaches to scientific inference (nomological-deductive, inductive, abductive, etc.) and of statistical inference;
2. higher order level of epistemic dimensions related to the entire body of evidence: consistency/coherence of items of evidence, (in)dependence structure; reliability, relevance – this level also pertains to various meta-inferential patterns such as the "Non-Alternative-Argument" (Dawid et al. (2015)); this domain has been mainly investigated in Bayesian and formal epistemology. However, also standard statistical techniques developed to detect publication bias or other kinds of biases focus on these aspects: (Krauth et al. 2013; Rising et al. 2008; Wood et al. 2008; Lundh et al. 2017; Lundh and Bero 2017);
3. a further level comprehending information/knowledge/evidence related to such meta-epistemic dimensions themselves. This information/knowledge/evidence relates to a social epistemology level and can be more or less straightforwardly inferred from it: incentives/deterrents for bias (such as financial interests, reputation etc.), social ontology of the research domain, and nudging studies.

The philosophical foundations of such a framework have been presented in De Pretis et al. (2019). The framework consists in a Bayesian epistemic network, borrowed from Bovens and Hartmann (2003), which formalises scientific inference probabilistically. The ancestor consists in the investigated causal hypothesis, while its direct descendants are the epistemic consequences of such hypothesis, namely, indicators of causality resulting from an elaboration of Bradford-Hill guidelines for causation (Hill 1965). Concrete data is represented by reports of study results attached to the relevant causal indicator. A reliability and a relevance node input into each report node in order to weight the evidence by its level of reliability and stability with respect to the context of application (target population), see

[15]Teira analyses for instance the role played by concerns about impartiality of research in the historical establishment of randomised controlled trials as gold standards for drug approval.

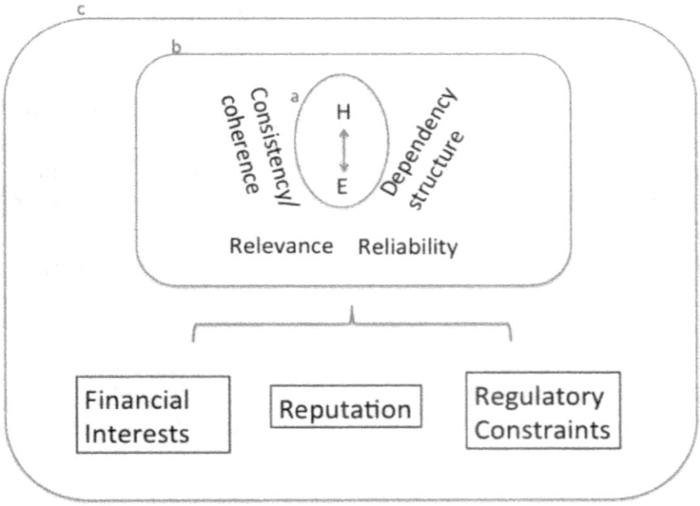

Fig. 15.1 A multilayer approach to modelling probabilistic causal inference through evidence synthesis: a. issues related to the evidential support provided by the evidence to the hypothesis at hand as discussed both in the statistical literature as well as in the philosophy of science; b. issues related to meta-evidential dimensions: consistency of studies, structure of the body of evidence in terms of mutual dependence of observations, reliability of the pieces of evidence and their relevance with respect to the target group; c. social dimensions of the pharmaceutical ecosystem (funding structures, reputational concerns, regulatory constraints), generally discussed in the social epistemology literature

Fig. 15.2. By breaking down the evidential line between pieces of evidence and causality into a two-stage process mediated by causal indicators, E-Synthesis helps disentangle philosophical issues related to the conceptualisation of causality from those related to causal inference and diagnostics.[16] At least some of the disputes among philosophers and methodologists may be solved by keeping these levels as distinct (see 15.5).

Figure 15.3 illustrates the structure of the causal inference problem: the set of dependencies and independencies determines a hierarchical structure where causal indicators – difference making (Δ), probabilistic dependency (PD), dose-response relationship (DR), rate of growth (RoG), evidence of mechanism (M), information about time asymmetries (T), etc.[17] lie on the same level, whereas study reports, possibly deriving from different methods (observational, such as cohort or

[16] This is also in analogy with Bogen and Woodward's distinction between data and phenomena (Bogen and Woodward 1988).

[17] These are derived from the epidemiological/causal literature (in particular from Bradford Hill guidelines); see De Pretis et al. (2019).

retrospective studies) or experimental (such as Randomised Controlled Studies) lie below this level.[18]

To each report a reliability node is attached: this is intended to capture the degree of systematic error (confounding and bias) estimated to affect the source report, as well as a relevance node, referring to the representativeness of the study sample for the purpose of causal inference with respect to the target population. Random error is directly represented by the likelihood function mapping reports to abstract indicators.

The graphical form provides an illustration of the epistemic dimensions at stake and thereby provides greater insight into some methodological issues by offering a mathematical explanation of their dynamics. This sort of representation allows one to single out in the mathematical formulae the specific role played by each epistemic dimension in the inferential dynamics; e.g., the role of reliability with respect to the propagation of confirmation in connection with replication of studies and with heterogenous sources of evidence. This framework has several advantages[19]:

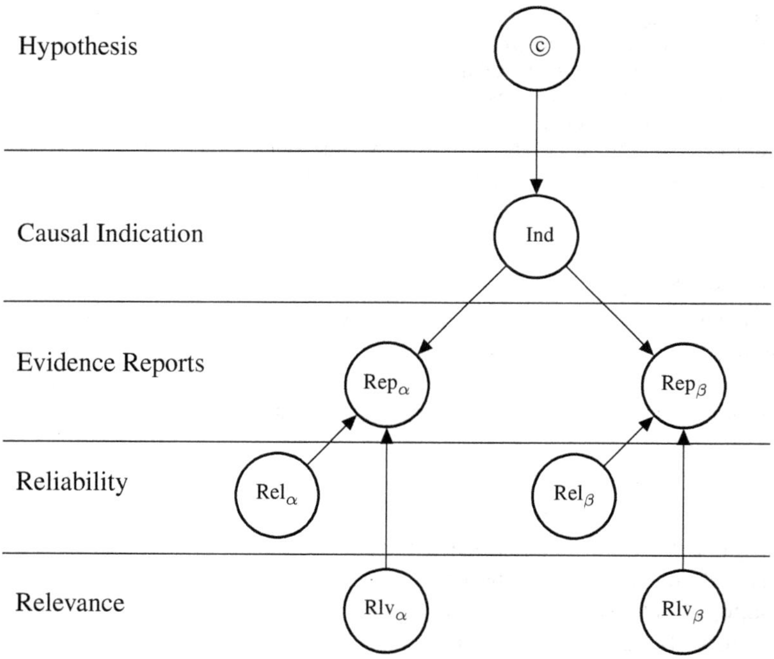

Fig. 15.2 Graph structure of the Bayesian network for two reports and epistemic categories

[18]The frameworks also allows modeling and simulation studies to play a relevant confirmatory role, especially with regard to the underlying dynamics underpinning the phenotypic causal effect. See also Osimani and Poellinger (2020).

[19]Please refer to De Pretis et al. (2019) for basics and details of the framework.

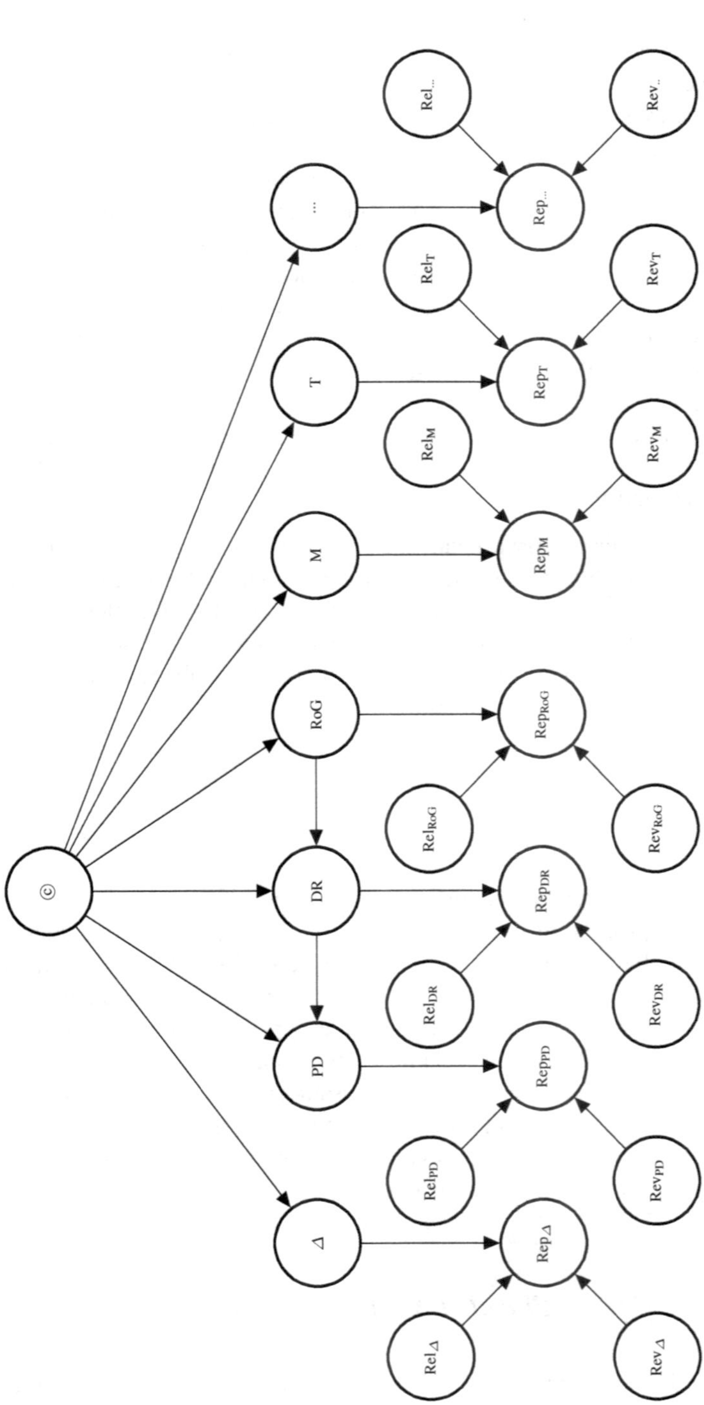

Fig. 15.3 Graph of the Bayesian network with one different report for every causal indicator variable. The dots indicate that there might be further indicators of causality not considered here

1. it identifies possible indicators of causality on the basis of the methodological and philosophical literature on causality, evidence, and causal inference;
2. embeds them in a topological framework of probabilistic dependencies and independencies grounded in assumptions regarding their reciprocal epistemic interconnections;
3. weakly orders some of these probabilistic dependencies as a function of their inferential strength with respect to the confirmation of causal hypotheses.
4. it easily accommodates many intuitions already expressed by philosophers of medicine regarding pluralistic approaches to evidence evaluation;
5. it lends itself to explicitly track the interaction of several dimensions of evidence, such as coherence and reliability;
6. it allows for a pluralistic but at the same time systematic approach to evidence amalgamation;
7. it tracks the different role of cross-validation through heterogeneous methods or sources vs. the confirmatory contribution of exact replication (see also Osimani and Landes (2020));

I present here how this framework can accommodate intuitions coming from both sides of the dispute, by considering five specific issues: the EBM vs. pluralist approach to causal inference, evidence hierarchies, causal holism, relevance (external validity), and reliability. These issues also show how our framework provides a higher order perspective on these debates by effectively embedding these various epistemic dimensions in a concrete topology.

15.5 Discussion

By breaking down the evidential line between pieces of evidence and causality into a two-stage process mediated by causal indicators, E-Synthesis helps disentangle philosophical issues related to the conceptualisation of causality from those related to causal inference and diagnostics. Furthermore, the framework aims to probabilistic causal assessment, and therefore bypasses problems generated by accounts that aim to establish causal claims categorically (through the identification of necessary and sufficient conditions for causation). More importantly, separate nodes for relevance and reliability are embedded in the network; this contributes to deflate much discussion in the methodological literature concerning the trade-off between these two dimensions of evidence quality.

15.5.1 The EBM vs. Pluralist Approach to Causal Inference

Translated within the E-Synthesis framework, the implicit assumptions underpinning the EBM viewpoint is that ideal RCTs (that is internally valid ones) are perfect indicators of difference making, and difference making is a perfect indicator of causality,[20] whereas other indicators only weakly support the hypothesis of causation. This can be represented in logical terms as an entailment relationship:

$$RCT \supset \Delta \supset ©.$$

As a consequence, EBM focuses on the level of evidence that goes through the Δ indicator, and concentrates its efforts on having as reliable as possible evidence for such indicator.

The contending view is that different indicators may have complementary epistemic roles in supporting the hypothesis of causality. However, to this view, held for instance by Clarke et al. (2014), Howick counters that: "There are many cases where patient-relevant effects of medical therapies have been established by comparative clinical studies alone." (Howick 2011, p. 939).

The debate is jeopardised by conflating the two entailment relationships RCT $\supset \Delta$, and $\Delta \supset$ © into one; that is, what is discussed is whether it can be justifiably held that RCTs provide perfect information for causality (directly): i.e. whether

$$RCT \supset ©,$$

rather than either whether RCT $\supset \Delta$, or $\Delta \supset$ ©.

If we let the epistemic net of E-Synthesis represent this discussion, we can take Howick to hold the view that:

$$P(© \mid \Delta) = 1,$$

and that ideal RCTs provide strong evidence for Δ:

$$P(\Delta \mid RCT) \approx 1,$$

hence offering strong enough evidence to establish the causal claim; while for the contenders $P(© \mid \Delta)$ is too small to establish the causal claim. For them, also having mechanistic evidence is required to establish the causal claim:

$$P(© \mid \Delta \& M) = 1.$$

E-Synthesis allows for RCT $\supset \Delta \supset$ © to hold, but the entailment relationship between RCTs and Δ is one between ideal RCTs and Δ. Hence in any concrete case $P(\Delta|\text{RCT}) < 1$, and this leaves room for other kinds of evidence to also contribute to hypothesis confirmation. On the other side, by dropping any necessity

[20] This directly derives from the potential outcome approach underpinning RCT methodology. See Holland et al. (1985), Rubin (2005), and Vandenbroucke et al. (2016) for a critical appraisal of this approach.

or sufficiency requirements for causal inference, this approach relaxes the theory that both evidence of association and of underlying biological mechanisms is necessary to establish causation. This approach is therefore much more flexible and responds both to the EBM intuition underpinning the privileged role assigned to RCTs, as well as to the pluralist intuition that various kinds of evidence are contributory to causal assessment.

15.5.2 Evidence Hierarchies

In relation to evidence hierarchies – which is a strong point of contention among philosophers of medicine and methodologists – E-Synthesis poses a set of inequalities regarding the evidential strength of various causal indicators (see De Pretis et al. (2019, p.32–33)), which nicely parallels the rankings proposed in the EBM paradigm:

$$P(©|\Delta) > P(©|DR), P(©|RoG) > P(©|PD) \tag{15.9}$$

What differentiates E-Synthesis from standard evidence rankings however is that these have predominantly been formalised as lexicographic decision rules. This means that higher-level studies trump lower-level ones: when two studies of different levels deliver contradictory findings, then the one higher in the evidence hierarchy is considered more reliable and allows one to discard the lower level one.[21] E-Synthesis incapsulates the rationale for ranking evidence (in the inequalities across probabilistic dependencies between causation and various indicators), but at the same time allows one to take into account all evidence, and to act accordingly, as soon as the probability of the causal hypothesis goes above the threshold established by the other dimensions of the decision (utility of withdrawing/not withdrawing the drug, conditional on the probability of it causing the suspected harm) (see De Pretis et al. 2019, Section 2.2.).[22]

15.5.3 Causal Holism

Methodological pluralists such as Cartwright (2011; 2007b), and Stegenga (2011), among others, express concerns against the privileged role of RCTs also on grounds

[21] A somewhat unwanted consequence of this "take the best" approach is that it has become commonplace to assume an uncommitted attitude towards observed associations least they are "proved" by gold standard evidence (see the still ongoing debate on the possible causal association between paracetamol and asthma: Osimani (2014)).

[22] This also complies with the precautionary principle in risk assessment and with how decisions should be made in health settings: Osimani (2007, 2013) and Osimani et al. (2011).

that classical 'linear' approaches to causal inference cannot do justice to the complexity of causal phenomena in the biological and social sciences, characterized by nonlinear causation and causal interactions.[23]

Strictly speaking this sort of criticism does not deny that:

$$\Delta \supset ©,$$

but only denies the reverse:

$$© \supset \Delta.$$

Since in the causal graph literature the defining features for causality jointly entail that $\Delta \nLeftrightarrow ©$ but not the reverse, this criticism misses the point. We contribute to deflate the debate by not collapsing Δ and © into a single node, therefore allowing causation to be holistic and therefore not reducible to difference-making, and at the same time letting difference-making immediately imply causation: in E-Synthesis when a difference-making relationship between two events or variables holds, then this is a sufficient – although not necessary – condition for causality. This can be characterized in logical terms as an entailment relationship: $\Delta \supset ©$. Hence, in E-Synthesis, the probability of a causal relationship, given a genuine difference making relationship is 1: $P(© \mid \Delta) = 1$. The inverse entailment though, $© \supset \Delta$, does not hold: knowledge of © does not necessitate the existence difference-making – e.g., in cases of "holistic causation".[24]

15.5.4 Evidence of Mechanisms and Relevance

Pluralists accord to evidence of mechanisms a preeminent role in establishing external validity and extrapolation, in that it helps evaluate whether the cause under investigation will work in a similar "context" also in the target population (Russo and Williamson 2007; Cartwright 2007a; Cartwright and Stegenga 2011). The present framework however, formally distinguishes the role of evidence of mechanisms for the purpose of causal assessment, from its role for the purpose of establishing external validity by associating the latter to the relevance node RLV. This allows us to explicitly distinguish the different kinds of inductive risk involved in the inference: (1) from statistical dirty data to causal indicator(s), and then to causality, in a specific population, (2) the extrapolation of a causal link established

[23] In the same line, also modular conceptualization of causes such as the ones implied in the causal graph methodology developed by Pearl (2000) and Glymour Spirtes et al. (2000) and colleagues (see also Woodward (2003)), are under attack for failing to recognize that causes may be holistic and therefore may be not adequately captured by a difference making account.

[24] This responds to concerns expressed among others by Cartwright (2007c), Mumford and Anjum (2011), Anjum (2012), and Kerry et al. (2012).

in a given population/model to another population/model. More importantly, adding a relevance node to the evidence reports allow for even highly reliable RCTs to play a low evidential support if they are not considered to be relevant to the target population.

15.5.5 Reliability and Higher Order Evidence

By introducing a reliability node *Rel*, and thereby breaking up the different dimensions of evidence (strength, relevance, reliability) E-Synthesis allows them to be explicitly tracked in the body of evidence. This makes it possible to parcel out the strength of evidence from the method with which it was obtained.[25] With this, E-synthesis provides a higher order perspective on evidential support by effectively embedding these various epistemic dimensions in a concrete topology. Indeed, the framework presented here also provides a fruitful platform for integrating insights developed in the philosophy of science around such topics as the role of replication in assessing the reliability of evidence (Open Science Collaboration 2015; Meehl 1990; Lamal 1990; Hempel 1968; Platt 1964), as well as the confirmatory role of explanatory power (McGrew 2003; Crupi et al. 2013; Cohen 2016; Lipton 2003) and coherence (Dietrich and Moretti 2005; Moretti 2007; Wheeler and Scheines 2013; Fitelson 2003; Bovens and Hartmann 2003). This paper focuses on inference *within one model*, rooting in *one hypothesis*, but E-Synthesis allows for going beyond the network's limits and for embedding it in an even larger network to trace the hypothesis' relation with other potentially concurring hypotheses. The mechanics of Bayesian epistemology are flexible enough to permit such an augmentation for the purposes of tracing further inference patterns. The framework is currently being developed into a concrete tool for evidence amalgamation (see Landes et al. (2018)) and possibly into a software.

Acknowledgements This research was supported by the European Research Council, ERC Starting Grant GA 639276: "Philosophy of Pharmacology: Safety, Statistical Standards, and Evidence Amalgamation". I thank my colleagues from the Munich Center for Mathematical Philosophy, and various audiences to which different parts of this talk were presented:

I also thank previous colleagues and students at the MCMP who provided precious comments and feedback on earlier versions of the paper. I am also grateful for discussions on the topics mentioned in the paper with Rani Anjum, Jeffrey Aronson, Bengt Autzen, Seamus Bradley, Lorenzo Casini, Vincenzo Crupi, Ralph Edwards, Branden Fitelson, Stephan Hartmann, Erik Nyberg, Elena Rocca, Jan Sprenger, Daniel Steel, and Momme von Sydow. Finally, I thank Adam LaCaze, Scott Podolsky, David Teira for reading and commenting on earlier drafts of the paper, with the usual disclaimer that any errors are my own responsibility.

[25] Osimani and Landes (2020) investigates the various concepts of reliability involved in such considerations.

References

Anjum, R. L., & Mumford, S. (2012). Causal dispositionalism, Chap. 7. In Bird, A., Ellis, B., Sankey, H. (Eds.), *Properties, powers and structure* (pp. 101–118). New York: Routledge.

Audi, R. (1993). *The structure of justification*. Cambridge: Cambridge University Press.

Beauchamp, T. L. (2011). Informed consent: Its history meaning, and present challenges. *Cambridge Quarterly of Healthcare Ethics, 20*(04), 515–523.

Begley, C. G., & Ellis, L. M. (2012). Drug development: Raise standards for preclinical cancer research. *Nature, 483*(7391), 531–533.

Bogen, J., & Woodward, J. (1988). Saving the phenomena. *The Philosophical Review, 97*(3), 303–352.

BonJour, L. (2009). *Epistemology: Classic problems and contemporary responses*. Lanham: Rowman & Littlefield Publishers.

Bovens, L., & Hartmann, S. (2003). *Bayesian epistemology*. Oxford: Oxford University Press.

Carnap, R. (1956). The methodological character of theoretical concepts. Indianapolis: Bobbs-Merrill.

Cartwright, N. (2007a). Are RCTs the gold standard? *Biosocieties, 2*, 11–20. https://doi.org/10.1017/S1745855207005029.

Cartwright, N. (2007b). Are RCTs the gold standard? *BioSocieties, 2*(1), 11–20. https://doi.org/10.1017/S1745855207005029.

Cartwright, N. (2007c). *Causal powers: What are they? Why do we need them? What can be done with them and what cannot?* Technical report, contingency and dissent in science technical report 04/07. http://www.lse.ac.uk/CPNSS/research/concludedResearchProjects/ContingencyDissentInScience/DP/CausalPowersMonographCartwrightPrint%20Numbers%20Corrected.pdf.

Cartwright, N. (2007d). *Hunting causes and using them: Approaches in philosophy and economics*. Cambridge/New York: Cambridge University Press.

Cartwright, N. (2012). Presidential address: Will this policy work for you? Predicting effectiveness better: How philosophy helps. *Philosophy of Science, 79*(5), 973–989. https://doi.org/10.1086/668041.

Cartwright, N., & Stegenga, J. (2011). A theory of evidence for evidence-based policy. In Dawid, P., Twining, W., Vasilaki, M. (Eds.), *Evidence inference and enquiry* (Chapter 11, pp. 291–322). Oxford: Oxford University Press.

Clarke, B., Gillies, D., Illari, P., Russo, F., & Williamson, J. (2013). The evidence that evidence-based medicine omits. *Preventive Medicine, 57*(6), 745–747. https://doi.org/10.1016/j.ypmed.2012.10.020.

Clarke, B., Gillies, D., Illari, P., Russo, F., & Williamson, J. (2014). Mechanisms and the evidence hierarchy. *Topoi, 33*, 339–360. https://doi.org/10.1007/s11245-013-9220-9.

Cohen, M. P. (2016). On three measures of explanatory power with axiomatic representations. Early view. *British Journal for the Philosophy of Science, 67*(4), 1077–1089. https://doi.org/10.1093/bjps/axv017.

Crupi, V., Chater, N., & Tentori, K. (2013). New axioms for probability and likelihood ratio measures. *British Journal for the Philosophy of Science, 64*(1), 189–204. https://doi.org/10.1093/bjps/axs018.

Dawid, R., Hartmann, S., & Sprenger, J. (2015). The no alternatives argument. *British Journal for the Philosophy of Science, 66*(1), 213–234. https://doi.org/10.1093/bjps/axt045.

De Pretis, F., Landes, J., & Osimani, B. (2019). E-synthesis: A Bayesian framework for causal assessment in Pharmacosurveilance. *Accepted in Frontiers in Pharmacology*.

Dietrich, F., & Moretti, L. (2005). On coherent sets and the transmission of confirmation. *Philosophy of Science, 72*(3), 403–424. https://doi.org/10.1086/498471.

Etz, A., & Vandekerckhove, J. (2016). A Bayesian perspective on the reproducibility project: Psychology. *PLoS One, 11*(2), e0149794.

Faden, R. R., & Beauchamp, T. L. (1986). *A history and theory of informed consent*. New York: Oxford University Press.
Fisher, R. (1955). Statistical methods and scientific induction. *Journal of the Royal Statistical Society Series B (Methodological), 17*, 69–78.
Fitelson, B. (2003). A probabilistic theory of coherence. *Analysis, 63*(279), 194–199. https://doi.org/10.1111/1467-8284.00420.
Gelman, A. (2015). Working through some issues. *Significance, 12*(3), 33–35. https://doi.org/10.1111/j.1740-9713.2015.00828.x.
Goldman, A. I. (1999). *Knowledge in a social world* (Vol. 281). Oxford/New York: Clarendon Press Oxford.
Haack, S. (2011). *Defending science-within reason: Between scientism and cynicism*. New York: Prometheus Books.
Hacking, I. (2006). *The emergence of probability: A philosophical study of early ideas about probability induction and statistical inference*. Cambridge: Cambridge University Press.
Hanin, L. (2017). Why statistical inference from clinical trials is likely to generate false and irreproducible results. *BMC Medical Research Methodology, 17*(1), 127. https://doi.org/10.1186/s12874-017-0399-0.
Hempel, C. G. (1968). Maximal specificity and lawlikeness in probabilistic explanation. *Philosophy of Science, 35*(2), 116–133. http://www.jstor.org/stable/186482.
Hill, A. B. (1965). The environment and disease: Association or causation? *Proceedings of the Royal Society of Medicine, 58*(5), 295–300.
Holland, P. W., Glymour, C., & Granger, C. (1985). Statistics and causal inference. *ETS Research Report Series, 1985*(2), i–72.
Holman, B. (2015). *The fundamental antagonism: Science and commerce in medical epistemology*. PhD Dissertation. Irvine: University of California.
Howick, J. (2011). Exposing the Vanities – and a qualified defense – of mechanistic reasoning in health care decision making. *Philosophy of Science, 78*(5), 926–940. https://doi.org/10.1086/662561.
Howick, J., Glasziou, P., & Aronson, J. K. (2013). Problems with using mechanisms to solve the problem of extrapolation. *Theoretical Medicine and Bioethics, 34*(4), 275–291.
Hoyningen-Huene, P. (2013). *Systematicity: The nature of science*. New York: Oxford University Press.
Ioannidis, J. P. A. (2005). Why most published research findings are false. *PLoS Medicine, 2*(8), e124.
Joffe, M. (2011). Causality and evidence discovery in epidemiology. In D. Dieks, W. J. Gonzalez, S. Hartmann, T. Uebel, & M. Weber (Eds.), *Explanation, prediction, and confirmation* (pp. 153–166). Dordrecht: Springer Netherlands. ISBN: 978-94-007-1180-8. https://doi.org/1.1007/978-94-071180-8_11.
Kerry, R., Eriksen, T. E., Lie, S. A. N., Mumford, S. D., & Anjum, R. L. (2012). Causation and evidence-based practice: An ontological review. *Journal of Evaluation in Clinical Practice, 18*(5), 1006–1012. https://doi.org/10.1111/j.1365-2753.2012.01908.x.
Krauth, D., Woodruff, T. J., & Bero, L. (2013). Instruments for assessing risk of bias and other methodological criteria of published animal studies: A systematic review. *Environmental Health Perspectives, 121*(9), 985.
LaCaze, A., Djulbegovic, B., & Senn, S. (2012). What does randomisation achieve? *Evidence-Based Medicine, 17*(1), 1–2. https://doi.org/10.1136/ebm.2011.100061.
Lamal, P. A. (1990). On the importance of replication. *Journal of Social Behavior and Personality, 5*(4), 31–35.
Landes, J., Osimani, B., & Poellinger, R. (2018). Epistemology of causal inference in pharmacology. *European Journal for Philosophy of Science, 8*(1), 3–49.
Lenhard, J. (2006). Models and statistical inference: The controversy between Fisher and Neyman–Pearson. *The British Journal for the Philosophy of Science, 57*(1), 69–91.
Lipton, P. (2003). *Inference to the best explanation*. London: Routledge.

Longino, H. E. (1990). *Science as social knowledge: Values and objectivity in scientific inquiry*. Princeton: Princeton University Press.
Lundh, A., & Bero, L. (2017). The ties that bind. *British Medical Journal, 356*. https://doi.org/10.1136/bmj.j176.
Lundh, A., Lexchin, J., Mintzes, B., Schroll, J. B., & Bero, L. (2017). Industry sponsorship and research outcome. *Cochrane Library, 2*, Art. No.: MR000033. https://doi.org/10.1002/14651858.MR000033.pub3.
Marsman, M., Schoönbrodt, F. D., Morey, R. D., Yao, Y., Gelman, A., & Wagenmakers, E.-J. (2017). A Bayesian bird's eye view of 'Replications of important results in social psychology'. *Royal Society Open Science, 4*(1), 160426.
Mayo, D. G., & Spanos, A. (2006). Severe testing as a basic concept in a Neyman-Pearson philosophy of induction. *British Journal for the Philosophy of Science, 57*(2), 323–357. https://doi.org/10.1093/bjps/axl003.
Mayo-Wilson, C., Zollman, K. J. S., & Danks, D. (2011). The independence thesis: When individual and social epistemology diverge. *Philosophy of Science, 78*(4), 653–677.
McGrew, T. (2003). Confirmation, heuristics, and explanatory reasoning. *British Journal for the Philosophy of Science, 54*(4), 553–567. http://www.jstor.org/stable/3541678.
Meehl, P. E. (1990). Appraising and amending theories: The strategy of lakatosian defense and two principles that warrant it. *Psychological Inquiry, 1*(2), 108–141. https://doi.org/10.1207/s15327965pli0102%5C_1.
Moretti, L. (2007). Ways in which coherence is confirmation conducive. *Synthese, 157*(3), 309–319. https://doi.org/10.1007/s11229-006-90575
Mumford, S., & Anjum, R. L. (2011). *Getting causes from powers*. Oxford: Oxford University Press.
Open Science Collaboration. (2015). Estimating the reproducibility of psychological science. *Science, 349*(6251), aac4716. https://doi.org/10.1126/science.aac4716.
Osimani, B. (2007). *Probabilistic information and decision making in the health context. The package leaflet as a basis for informed consent* (1st edn). Lugano: USI, Università della Svizzera italiana. http://doc.rero.ch/record/28759?ln=fr.
Osimani, B. (2012). Risk information processing and rational ignoring in the health context. *The Journal of Socio-Economics, 41*(2), 169–179.
Osimani, B. (2013). The precautionary principle in the pharmaceutical domain: A philosophical enquiry into probabilistic reasoning and risk aversion. *Health, Risk & Society, 15*(2), 123–143.
Osimani, B. (2014). Hunting side effects and explaining them: Should we reverse evidence hierarchies upside down? *Topoi, 33*(2), 295–312. https://doi.org/10.1007/s11245-013-9194-7.
Osimani, B., & Mignini, F. (2015). Causal assessment of pharmaceutical treatments: Why standards of evidence should not be the same for benefits and harms? *Drug Safety, 38*(1), 1–11. ISSN: 1179–1942. https://doi.org/10.1007/s40264-014-0249-5.
Osimani, B. & Landes, J. (2020). *Varieties of Error and Varieties of Evidence in Scientific Inference*. (Accepted).
Osimani, B., & Poellinger, R. (2020). A protocol for model validation and causal inference from computer simulation. In M. Bertolaso & F. Sterpetti (Eds.), *A critical reflection on automated science. Will science remain human*. Heidelberg: Springer Nature. (forthcoming).
Osimani, B., Russo, F., & Williamson, J. (2011). Scientific evidence and the law: An objective Bayesian formalization of the precautionary principle in pharmaceutical regulation. *The Journal of Philosophy Science & Law, 11*(2), 1–24.
Papa, A. (2014). *L'identità esposta. La cura come questione filosofica*. Milano: Vita e Pensiero.
Papineau, D. (1994). The virtues of randomization. *The British Journal for the Philosophy of Science, 45*(2), 437–450.
Pearl, J. (2000). *Causality: Models, reasoning and inference* (1st ed.). Cambridge: Cambridge University Press.
Pessina, A. (2009). Biopolitica e Persona. *Medicina e Morale, 2*, 239–253. http://hdl.handle.net/10807/4748.
Platt, J. R. (1964). Strong inference. *Science, 146*(3642), 347–353.

Podolsky, S. H., & Powers, J. H. (2015). Regulating antibiotics in an era of resistance: The historical basis and continued need for adequate and well-controlled investigations regulating antibiotics in an era of resistance. *Annals of Internal Medicine, 163*(5), 386–388.

Poellinger, R. (2018). On the ramifications of theory choice in causal assessment. Indicators of causation and their conceptual relationships. Submitted.

Prinz, F., Schlange, T., & Asadullah, K. (2011). Believe it or not: How much can we rely on published data on potential drug targets? *Nature Reviews Drug Discovery, 10*(9), 712–712.

Rising, K., Bacchetti, P., & Bero, L. (2008). Reporting bias in drug trials submitted to the Food and Drug Administration: Review of publication and presentation. *PLoS Medicine, 5*(11), e217.

Rubin, D. B. (2005). Causal inference using potential outcomes. *Journal of the American Statistical Association, 100*, 322–331.

Russo, F., & Williamson, J. (2007). Interpreting causality in the health sciences. *International Studies in the Philosophy of Science, 21*(2), 157–170. https://doi.org/10.1080/02698590701498084.

Scheu, G. (2003). *In dubio pro securitate: Contergan, Hepatitis-/AIDS-Blutprodukte, Spongiformer Humaner Wahn und kein Ende?* (Vol. 42). Baden-Baden: Nomos.

Senn, S. (2002). A comment on replication, p-values and evidence SN Good-man, statistics in medicine 1992; 11: 875–879. *Statistics in Medicine, 21*(16), 2437–2444.

Senn, S. (2003). *Dicing with death: Chance risk and health.* Cambridge: Cambridge University Press.

Sgreccia, E. (2007). *Manuale di Bioetica* (3rd ed.). Milano: Vita e Pensiero.

Solomon, M. (2015). *Making medical knowledge.* Oxford: Oxford University Press.

Spirtes, P., Glymour, C. N., & Scheines, R. (2000). *Causation, prediction, and search.* Cambridge: MIT press.

Sprenger, J. (2016). Bayesianism vs. Frequentism in statistical inference (Chap. 18). Oxford University Press.

Stegenga, J. (2011). Is meta-analysis the platinum standard of evidence? *Studies in History and Philosophy of Science Part C: Studies in History and Philosophy of Biological and Biomedical Sciences, 42*(4), 497–507. https://doi.org/10.1016/j.shpsc.2011.07.003.

Stegenga, J. (2014). Down with the hierarchies. *Topoi, 33*(2), 313–322. https://doi.org/10.1007/s11245-013-9189-4.

Swinburne, R. (2001). *Epistemic justification*. Oxford: Oxford University Press.

Teira, D. (2011). Frequentist versus Bayesian clinical trials. In Gifford, F. (Ed.), *Handbook of philosophy of medicine* (pp. 255–298). Amsterdam: Elsevier.

Teira, D., & Reiss, J. (2013). Causality, impartiality and evidence-based policy. In H.-K. Chao, S.-T. Chen, & R. L. Millstein (Eds.), *Mechanism and causality in biology and economics* (pp. 207–224). Dordrecht: Springer. https://doi.org/10.1007/978-94-007-2454-9_11.

Vandenbroucke, J. P., Broadbent, A., & Pearce, N. (2016). Causality and causal inference in epidemiology: The need for a pluralistic approach. *International Journal of Epidemiology, 45*, 1776–1786. https://doi.org/10.1093/ije/dyv341.

Wheeler, G., & Scheines, R. (2013). Coherence and confirmation through causation. *Mind, 122*(485), 135–170. https://doi.org/10.1093/mind/fzt019.

Wood, L., Egger, M., Gluud, L. L., Schulz, K. F., Juüni, P., Altman, D. G., Gluud, C., Martin, R. M., Wood, A. J. G., & Sterne, J. A. C. (2008). Empirical evidence of bias in treatment effect estimates in controlled trials with different interventions and outcomes: Meta-epidemiological study. *BMJ, 336*(7644), 601–605.

Woodward, J. (2003). *Making things happen: A theory of causal explanation (Oxford studies in the philosophy of science).* New York: Oxford University Press. ISBN: 9780195189537.

Worrall, J. (2007a). Do we need some large, simple randomized trials in medicine? *EPSA Philosophical issues in Science*, 289–301. https://doi.org/10.1007/9789048132522_27.

Worrall, J. (2007b). Evidence in medicine and evidence-based medicine. *Philosophy Compass, 2*(6), 981–1022. https://doi.org/10.1111/j1747.9991.2007.00106.x.

Worrall, J. (2007c). Why there's no cause to randomize. *British Journal for the Philosophy of Science, 58*(3), 451–88. https://doi.org/10.1093/.bjps/axm024.

Worrall, J. (2008). Evidence and ethics in medicine. *Perspectives in Biology and Medicine, 51*(3), 418–431. https://doi.org/10.1353/pbm.0.0040.

Part III
Decisions

Chapter 16
Values in Pharmacology

Sven Ove Hansson

Abstract Examples from pharmacology are often used in discussions on the role of values in science. This contribution provides an account of the various types of values that have an influence in pharmacology. Furthermore, a criterion is proposed to distinguish between legitimate and illegitimate influence of non-epistemic values in pharmacological science. According to this proposal, practical considerations may not legitimately lower the demands of evidence for scientific claims, but they may legitimately raise them. It is argued that the latter type of influence from values is common in pharmacology. This is because the demands of evidence for the safety and efficacy of drugs for human use typically exceed the level of evidence usually required when nothing more than the risk of erroneous scientific beliefs is at stake.

Keywords Enhancement · Epistemic values · Intrinsic values · Pharmacology · Risk-benefit assessment · Science-policy interface · Values in science

16.1 Introduction

The role of values in science is one of the most debated issues in the philosophy of science. It is also a prominent topic in many other academic contexts. Is scientific knowledge value-free? If not, can it be? And should it be? Some discussants promote a value-free ideal according to which science should be neutral in all value issues, whereas others consider such an ideal to be impossible and consequently promote the imbuement of science with the values they consider to be appropriate. Most working scientists side with the former view, at least implicitly through their

---

S. O. Hansson (✉)
Division of Philosophy, Royal Institute of Technology (KTH), Stockholm, Sweden

Department of Learning, Informatics, Management and Ethics, Karolinska Institutet, Solna, Sweden
e-mail: soh@kth.se

A. LaCaze, B. Osimani (eds.), *Uncertainty in Pharmacology*, Boston Studies in the Philosophy and History of Science 338,
https://doi.org/10.1007/978-3-030-29179-2_16

adherence to methodologies constructed to minimize the impact of personal value commitments on research outcomes.

Discussions on values in science have been heated, and they have often looked more like trench warfare than fruitful dialogues. One reason for this may be that these discussions have usually referred to a too small range of values. The focus has mostly been on controversial moral and political values, but there are also other types of values that can have an influence on science. Some moral values are virtually uncontroversial in modern societies; one example is the value standpoint that more efficacious pain reduction in paediatric care is a worthwhile goal in pharmacological research. Moreover, there are non-moral values that can have an impact on science, most importantly epistemic values, i.e. values of knowledge per se.[1] Some examples of epistemic values are truth, error avoidance, simplicity, and explanatory power. All these types of values need to be taken into account in a comprehensive discussion of values in science, and we need to open up for the possibility that different types of values may have to be treated differently in terms of their potential influence in science.

Examples from pharmacology have figured prominently in the debate on science and values. In this contribution I will go one step further and use pharmacology and its clinical applications as a prime example for a new account of the role of values in science. By considering the full spectrum of pharmacology, from basic biochemical studies of drug action to the application of clinical pharmacology in the treatment of individual patients, our study will have access to a microcosm of science. It contains research aimed at knowledge just for its own sake, research directly aimed at practical applications and – importantly – various forms of intermediates and combinations between these two extremes. This will give us an opportunity to see what is common but also what differs in the value aspects of these different types of scientific endeavours.

Section 16.2 will provide some basic concepts for the characterization of values in pharmacology. Section 16.3 is devoted to an overview of the moral values that have major impact in pharmacology and its applications. In Sect. 16.4 the new model of the science-value relationship is presented, and its relevance for pharmacology is exhibited. Section 16.5 briefly concludes.

A couple of omissions should be explicitly stated: The ethics of animal experimentation will not be discussed, and the same applies to veterinary pharmacology and to the ethics of experiments on patients and other human subjects. Value issues in the toxicology of non-pharmacological substances will not be treated here (but are extensively discussed in Hansson 2008, 2009; Hansson and Rudén 2008).

[1]A third category is aesthetic values, which have a particularly important role in mathematics. (Dreyfus and Eisenberg 1986; Breitenbach 2015) They do not seem to have much of an impact in pharmacology.

16.2 Some Concepts for Value Characterization

A considerable number of conceptual tools for the characterization and classification of values have been developed in moral philosophy and value theory. Some of them are directly useful for the characterization of values in pharmacology, but others need to be adjusted in order to be sufficiently helpful for that purpose.

16.2.1 Intrinsic and Extrinsic Values

Values are commonly classified as either *intrinsic* or *extrinsic* (Zimmerman 2015). Something has intrinsic (or ultimate) value if and only if it has value in itself, or for its own sake. To have extrinsic (or instrumental) value means to be valuable by contributing to something else that is valuable. For instance, it is usually assumed that a reduced risk of coronary heart disease is intrinsically valuable. In contrast, reduced blood cholesterol is only extrinsically valuable, i.e. valuable to the extent that it contributes to something else that is valued for its own sake (such as reduced risk of coronary heart disease). In pharmacology, extrinsic value is usually causal in nature; it amounts to a capacity to cause something else that has intrinsic value (usually improved health).

To determine whether something is intrinsically valuable is in principle a "pure" value issue, whereas to determine whether it is extrinsically valuable is a "mixed" issue that requires consideration of both facts and values. In our example, the claim that reduced blood cholesterol is (extrinsically) valuable depends both on the factual claim that such a reduction lessens the probability of coronary heart disease and on the value claim such a decrease is intrinsically valuable. Needless to say, it is the former (factual) claim that is controversial and in need of careful investigation. In what follows the focus will be on intrinsic values, since they provide more clear examples of the philosophical value issues. However, it should be kept in mind that pharmacology is largely concerned with establishing and characterizing causal connections between the administration of a drug and various aspects the patient's health and well-being. It is through these causal connections that the intrinsic values of health effects give rise to various extrinsic values related to pharmacology.

Intrinsic values can be of different kinds. In pharmacology, the major types of value are epistemic and moral value. Epistemic values dominate in the basic study of drug mechanisms, but they have an important role also in the more applied parts of the discipline. We certainly want physicians to base their decisions on pharmacological treatments on scientifically well-supported recommendations, rather than using rule-of-thumb practices that are not well understood. Moral values have their strongest impact in the clinical applications of pharmacology, but they can have an impact also in basic pharmacological science. For instance, an account of a drug's kinetic properties will be considered more valuable if it relates to variables that can be used in clinical optimization of dosage than if its variables lack clinical correlates. The following two subsections will be devoted to some useful distinctions referring to moral (ethical) and epistemic values respectively.

16.2.2 Moral Values

In moral philosophy it is common to distinguish between a person's self-regarding and her other-regarding interests and values. This distinction is based on the sensible assumption that a person's own interests are at the centre of most of her concerns and constitute a separate category to be contrasted with those referring to the interests of other people (or non-human concerns). In medicine, the centre of a physician's or nurse's concerns is (hopefully) not her own interests but those of the patient. We should therefore replace the distinction between self-regarding and other-regarding interests by that between *patient-regarding* and *other-regarding interests*. Here, the latter concept refers to the interests of persons other than the patient, including prominently those of her family and friends, healthcare personnel, and society at large. This includes considerations of justice and equality in access to pharmaceutical drugs and to healthcare in general.[2]

The physician–patient (nurse–patient, etc.) relationship requires that the physician (nurse, etc.) acts in the interest of the patient. However, this requirement is restricted to the patient's health-related interests. It is therefore important to distinguish between *health-related values* and *non-health-related* values. The distinction is simple in principle, but drawing the line in practice is often quite difficult, as we will see below.[3]

In moral and political philosophy, it is often important to distinguish between on the one hand the outcomes of some action or decision and on the other hand the principles and procedures guiding the attainment of those outcomes. This is a rather intricate distinction (Hansson 2014), but fortunately it can be operationalized in a fairly straightforward way in a healthcare context. We can distinguish between those health-related values that concern the *value of treatment outcomes* and those that are *procedural*, i.e. refer to the decision process leading up to the treatments in question. The latter category includes the values of informed consent and respect for the rights of patients.[4]

Finally, there is one distinction that is not usually made in moral philosophy but is nevertheless highly important in many applied contexts, including both science and healthcare: the distinction between *controversial and uncontroversial values*. The application of that distinction depends on what values are prevalent in the surrounding society. For example, it is fairly uncontroversial that if a drug prevents

[2]We will return to the distinction between patient-regarding and other-regarding interests in Sect. 16.3.5.

[3]We will return to this distinction in Sect. 16.3.2.

[4]This distinction will be further discussed in Sect. 16.3.4.

or delays the onset of Alzheimer's disease, then that is a valuable effect. However, a drug making soldiers less reluctant to kill enemies would be highly controversial in most societies.[5]

16.2.3 Epistemic and Non-Epistemic Values

Values enter science in at least three major ways: They influence (1) our choices what to investigate, (2) our decisions how we investigate it, and (3) the conclusions we draw from the outcomes of our investigations. The first two types of value influence are fairly uncontroversial. The first can be exemplified by the focus that pharmacologists put on substances that can potentially have a positive effect on human health, rather than on substances with other, perhaps biologically highly interesting, effects on the human body. This is of course the result of a generally accepted, and quite uncontroversial, influence of values on science. The second type, influence of values on how investigations are conducted, is also often quite uncontroversial. For instance, both human and animal experiments in pharmacology are restricted by ethical guidelines and principles. These restrictions are based on (moral) values. Although there are often different opinions on the more detailed ethical requirements on the conduct of research, no one seems to dispute that moral values have a legitimate role in this context. In contrast, the third type of value influence is highly controversial. Claims have been made that values should have no influence at all on what conclusions scientists draw from their investigations, but this has been countered with claims that such an influence should be allowed, at least within certain limits.

The so-called traditional view, which rejects value influence on scientific conclusions and assertions, was expressed by the American sociologist Robert K. Merton (1910–2003) in his influential characterization of the ethos of science. (Merton 1942) He summarized it in the form of four institutional imperatives. The first of these, *universalism*, asserts that all scientific truth claims should be evaluated according to pre-established, impersonal criteria. This implies that the acceptance or rejection of a scientific assertion has to be independent of the personal or social qualities of their protagonists. Furthermore:

> Objectivity precludes particularism. The circumstance that scientifically verified formulations refer to objective sequences and correlations militates against all efforts to impose particularistic criteria of validity. (Merton 1942, p. 118)

According to the second imperative, *communality* (infelicitously called "communism" by Merton), the findings of science belong to the community, rather than being owned by individuals or groups. His third imperative, *disinterestedness*, requires the presence of institutional controls to curb the effects of whatever

[5] See the last two paragraphs of Sect. 16.3.1 for an application of this distinction.

personal or ideological motives an individual scientist may have. Finally, the fourth imperative, *organized skepticism*, supports detached scrutiny of beliefs that are fostered by other institutions in society. The requirement that the acceptance or rejection of scientific claims should be value-independent is an overarching principle in Merton's account, and most clearly expressed in his explication of the first imperative. But notably, Merton does not require science to be "value-free" in a more general sense. To the contrary, the four imperatives are closely related to values, and it is for value-based reasons that decisions on acceptance or rejection of scientific claims should be independent of values.

However, already in the 1940s other views on the role of values in science were aired. In an article published in 1948, the American philosopher Charles West Churchman (1913–2004) argued that when deciding what to believe, i.e. choosing among hypotheses, we must evaluate the consequences of different such hypotheses, and for this we need to have some sort of weights determining the values of different outcomes. Therefore, he said, "the theory of inference in science demands the use of ethical judgments", and consequently our evaluation of hypotheses is "a problem of ethics". (Churchman 1948, p. 265) As an example, he noted that "in determining the efficiency of a drug for preventing a disease", conflict may arise between economic criteria and criteria concerning health or longevity. (p. 261) However, he did not explain this example more concretely, and his whole discussion was rather abstract. Five years later, his colleague Richard S. Rudner (1921–1979) published a paper in the same journal, where he developed these ideas with much more concreteness. His main argument was that a decision whether or not to accept a scientific hypothesis must take into account not only the available empirical evidence but also the seriousness of the two possible types of mistakes: accepting an incorrect hypothesis and rejecting a correct one. He, too used a pharmacological example:[6]

> Thus, to take a crude but easily manag[e]able example, if the hypothesis under consideration were to the effect that a toxic ingredient of a drug was not present in lethal quantity, we would require a relatively high degree of confirmation or confidence before accepting the hypothesis – for the consequences of making a mistake here are exceedingly grave by our moral standards. On the other hand, if say, our hypothesis stated that, on the basis of a sample, a certain lot of machine stamped belt buckles was not defective, the degree of confidence we should require would be relatively not so high. How sure we need to be before we accept a hypothesis will depend on how serious a mistake would be. (Rudner 1953, p. 2)

Rudner's article became the starting-point of a debate that led to important new ideas on the relationship between science and values. In a paper published in

[6]A third, highly influential, early participant in this debate also used a pharmacological example. Richard Jeffrey (1956) pointed out that a physician who decides whether or not to inoculate a child with a polio vaccine may very well come to another conclusion than a veterinarian who decides whether or not to use the same vaccine on a monkey. To accept or reject once and for all a hypothesis about the vaccine is in his view "to introduce an unnecessary conflict between the interests of the physician and the veterinarian." (ibid, p. 245) Jeffrey proposed that scientists should avoid such conflicts by refraining from truth claims and instead assign probabilities to the statements under consideration.

1960, Isaac Levi (1930–2018) described Rudner's article as "the most explicit and sweeping attack against the value-neutrality thesis", by which he meant the thesis that a scientist is "prohibited by the canons of scientific inference from taking his attitudes, preferences, temperament, and values into account when assessing the correctness of his inference". (Levi 1960, pp. 346, 345) In defence of that thesis, Levi maintained that "[e]ven in the case of decision problems where practical objectives are involved, it often seems appropriate to distinguish between acting on the basis of a hypothesis relative to that objective and accepting the hypothesis as true". (p. 351) Although he conceded Rudner's point that scientists need to have criteria for the acceptance and rejection of scientific claims, he did not concur with Rudner's conclusion that we therefore have to give up the value-neutrality thesis.

> When a scientist commits himself to certain "scientific" standards of inference, he does, in a sense, commit himself to certain normative principles. He is obligated to accept the validity of certain types of inference and to deny the validity of others. *The values that determine minimum probabilities may be part of this commitment*." (Levi 1960, p. 356)

In an article published the same year, Carl Hempel (1905–1997) expressed very similar ideas but with another, perhaps more clarifying, terminology. He proposed that in decisions on what conclusions to draw from the available evidence, scientists should not be influenced by *moral* values. However, they should take into account "the value or disvalue which the different outcomes have from the point of view of pure scientific research rather than the practical advantages or disadvantages that might result from the application of an accepted hypothesis". He called these "purely scientific, or epistemic, utilities" (Hempel 1960, p. 465).

> [T]he utilities should reflect the value or disvalue which the different outcomes have from the point of view of pure scientific research rather than the practical advantages or disadvantages that might result from the application of an accepted hypothesis, according as the latter is true or false. Let me refer to the kind of utilities thus vaguely characterized as *purely scientific, or epistemic, utilities*. (Hempel 1960, p. 465)

Later he called them "epistemic values" and maintained that epistemic values should reflect the usefulness of truth, simplicity, explanatory power and other desiderata of scientific theories. (Hempel 1981, p. 398) The notion of epistemic value has been further developed by other authors, including Levi (1962), Harsanyi (1983), and Feleppa (1981). In this literature, there has been a strong focus on two epistemic values, namely:

1. *Truth attainment*: believing in as many true statements as possible, and
2. *Error avoidance*: believing in as few erroneous statements as possible

These two values will often run into conflict. If we want to avoid holding false beliefs, then we should apply very strict standards of evidence. However, doing so will prevent us from believing in many statements for which we do not have conclusive evidence but which are nevertheless true. In science, a balance will always have to be struck between the two values of error avoidance and truth attainment, since we cannot obtain both to a maximal degree.[7] Scientific practice gives high (but not absolute) priority to error avoidance. In other words, it is regarded as a worse mistake in science to consider something to be scientifically established that is in fact wrong than to consider something as not scientifically established that is in fact true.[8] (There is a close analogy here with judicial principles for criminal proceedings. Sentencing someone for a crime he did not commit is always considered to be a serious mistake, but not sentencing someone for a crime he committed may be quite a reasonable decision if the evidence is inconclusive.)

The distinction between the two types of error has often been used in discussions on uncertainty in the environmental sciences. For instance, even if there is uncertainty whether a chemical substance has a certain toxic effect, it may be sensible to act *as if* that is the case. The major reasons for doing so can be expressed in terms of a comparison between how bad it would be to treat it as non-toxic if it is toxic and how bad it would be to treat it as toxic if it is non-toxic. (Cranor 1993; Shrader-Frechette 2008) Similarly, when discussing climate policy, it is highly relevant to compare the consequences of overestimating the pace of global warming to the consequences of underestimating it. (Intemann 2015) The values involved in these discussions are clearly non-epistemic. For instance, we may choose to take precautionary measures against a suspected carcinogen although we do not know for sure whether the substance has the suspected carcinogenic effect. The reason why we do so is the high priority that we assign to avoiding malignant diseases, obviously for moral rather than epistemic reasons. Generally speaking, proposals to exclude non-epistemic values from deliberations on scientific assent have much more plausibility in pure science than in applied sciences where various practical consequences are at stake.

Pharmacology has a strong focus on practical issues with considerable moral impact, namely health issues. Therefore, non-epistemic values have important roles in pharmacology. In particular the type of reasoning just referred to, based on the two types of error, is highly relevant in pharmacology, where it can be used to clarify how we weigh different types of potential drug effects against each other.[9] For claims that a drug has some desired effect, a high emphasis on error avoidance at the expense of truth attainment leads to a considerable risk that we

[7]Maximal error avoidance would amount to refraining from having any opinions at all at least in empirical matters. In order to maximize truth attainment one would have to have an opinion about virtually everything, including issues in which that opinion would have to be an uninformed guess.

[8]One aspect of the priority given to error avoidance is operationalized in the use of statistical significance levels in the evaluation of observations. See Hansson (1995) and Shrader-Frechette (2008).

[9]Many philosophically interesting discussions on the science-value relationship have limited relevance for pharmacology, since they focus on deliberations in pure science where the non-epistemic values are much less relevant than in pharmacology. The reader interested in a more general overview over the science-value discussion is referred to (Douglas 2009; Elliott 2017; Kincaid et al. 2007; Lacey 1999). There is also an interesting parallel discussion in general epistemology, see for instance (Lihoreau and Rebuschi 2014; Stanley 2005).

miss out on using drugs with beneficial effects. On the other hand, if we pay less attention to error avoidance in relation to such claims, then we run a greater risk of acclaiming drugs with little or no positive effect. For claims that a drug has some undesirable side effect, a strong focus on error-avoidance will lead to a considerable risk that we condone a drug with unacceptable side effects, whereas the opposite focus will increase the risk that we refrain from using an effective drug due to ungrounded or exaggerated fear of negative side effects. Pharmacology provides excellent examples of moral and epistemic values bearing on the same issues and decisions.

16.3 Ethical Values in Pharmacology

The epistemic values in pharmacology are essentially the same as in other empirical sciences. Moral values in science are more area-specific, and those of pharmacology are closely related to the values in clinical medicine that are studied in medical ethics, namely values referring to (professional activities aiming at) the maintenance and restoration of health and the amelioration of illness. It is the purpose of this section to clarify the nature of these values and the conflicts and controversies that they can give rise to in pharmacology and its applications.

16.3.1 Weighing Risks and Benefits

Unfortunately, positive effects from pharmacological treatments can seldom if ever be obtained without non-negligible risks of negative side effects. All decisions on drug use will have to be based on the weighing of positive and negative effects against each other. Is hair loss an acceptable side effect of a drug against acne? Is hypertension a tolerable side effect of an anti-depressive drug? How great a risk of a life-threatening anaphylactic shock is worth taking in order to achieve how great a chance of curing otitis media? How great a risk should we accept if the disease is instead a potentially life-threatening pneumonia? How great a risk of a drug-induced neoplasm should we accept from a chemotherapeutic agent used to fight a potentially deadly cancer? The weighing of positive versus negative effects is an essential component of clinical drug treatment, and its influence permeates all of pharmacology, including its most basic parts in which potential beneficial and detrimental effects on the human body are sought, identified, and characterized.

It is important to recognize that this weighing is very different from the more well known form of weighing that is prescribed by utilitarian moral theory. (Hansson 2004) According to utilitarian moral theory, all morally relevant consequences of an action should be weighed against each other, irrespectively of whom they accrue to. If this principle were to be applied in clinical decisions on drug use, then consequences for others than the patient would be assigned the same weights as corresponding consequences for the patient. For instance, the positive consequences for caregivers of a pacifying medication could weigh heavier than the negative consequences for the patient of the drug in question. In sharp contrast to this utilitarian way of thinking, the tradition in clinical medicine is to weigh the advantages and disadvantages (only) *for the patient* against each other. (Other-regarding considerations have a marginal role in a few cases, to which we will return in Sect. 16.3.5.)

The adherence of the healthcare professions to this patient-centred value base is the reason why their professional ethics excludes the administration of treatments intended to harm the treated person. That is why the World Medical Association has repeatedly declared that it is "unethical for physicians to participate in capital punishment, in any way, or during any step of the execution process, including its planning and the instruction and/or training of persons to perform executions" (World Medical Association 2008). Due to the basic allegiance of pharmacology to the purposes of medicine, this should apply equally to all forms of pharmacological research and practice.

There is no general algorithm or other mechanical method for weighing the benefits and detriments for the patient of a treatment against each other. That weighing has to be based on our intuitive judgments of what is important in life. Fortunately, as far as most health issues are concerned, such judgments are made in fairly similar ways by the vast majority of human beings. In other words, we seem to value most health conditions in roughly the same way. This is one of the reasons why most diseases can be treated according to standardized treatment schedules. In the usual treatment schedules for a disease, the first-line treatment is the one with the best balance of therapeutic effects vs. negative side effects according to how the vast majority of patients appraise these effects, and the second-line treatment is the one that comes second in terms of this balance, etc. Such standardized treatment recommendations would not be possible unless these judgments were made in concurring ways by most patients.

However, there are exceptions to this uniformity. Some treatment effects are judged differently by different people. For instance, depending on their life situations and life plans, patients differ widely in their appraisals of risks of infertility (Loren et al. 2013). There are also subcommunities in our societies that reject what most people, including the medical community, would describe as therapeutic improvements. The "fat is beautiful" movement denies that obesity is a disease requiring treatment and medical attention (Zimdars 2015). Segments of the Dwarf community have reacted against the introduction of therapies against their condition, seeing this as a threat to the future existence of their way of life (Berreby 1996). By far the strongest such counter-reaction is the negative reaction of large parts of the Deaf World towards cochlear implant surgery in prelingually deaf children (Crouch 1997; Lane and Bahan 1998; Hansson 2015). In all cases where the evaluation of clinically relevant physical and psychological conditions is controversial, there is a need for extended ethical discussions with a strong participation of persons having or risking these conditions.

16.3.2 Enhancements

We noted in Sect. 16.2.2 that the values to be counted in clinical decision-making are limited, not only to the individual patient's interests, but also – and even stricter – to his or her health-related interests. However, the delimitation of what is health-related is far from clear. This is particularly important in neuropharmacology, since neuroactive drugs can be used not only to cure disease but also to improve functions or capabilities that are already within the range of the normal. Hence, drugs developed for the treatment of narcolepsy are in use in armed forces as wakefulness drugs (Westcott 2005). Drugs against depression are used for mood elevation by persons with no psychiatric diagnosis, and drugs against erectile dysfunction are used for pleasure. Drugs developed to prevent cognitive deterioration in Alzheimer's disease seem to be capable of improving cognitive functioning in healthy persons (Wolpe 2002; Repantis et al. 2010). Although it has rightly been pointed out that both media reporting and the bioethical literature tend to grossly exaggerate the positive effects in healthy persons of currently available prescription drugs (Partridge et al. 2011), this is an issue in need of attention, and it can potentially become more urgent as new drugs are developed.

Some authors have argued that virtually all forms of enhancement should be rejected since they transcend the traditional task of medicine, which is to treat and prevent disease, not to improve human abilities in general. "[T]he goals of medicine concern not all human suffering, but only that suffering connected with a malady" (Miller et al. 2000). However, there are at least two problems with this standpoint. First, the distinction between disease and health or normality is not as clear as it may first seem.[10] Disease is not a biologically well-defined concept but one that depends largely on social values. Some conditions that were previously regarded as variations within normality are now regarded as diseases and receive pharmacological treatment. Attention-deficit hyperactivity disorder (ADHD) is an example of this. It should also be mentioned that medical tradition strongly endorses enhancements that increase resistance against disease in healthy people whose resistance is already normal, in particular through vaccination. Other ways to improve the body's resistance against disease would probably find acceptance relatively easily. In societies with widespread attempts to reduce disease risks by other means, such as physical exercise and dietary modifications, it is difficult to see how a line could be drawn that excludes pharmaceuticals reducing the risk of serious diseases in individuals with a normal initial risk level.

Secondly, it is unclear how standards of normality should relate to age. For instance, presbyopia and tooth wear are normal features of aging, but we do not hesitate to treat these conditions. It has been proposed that pharmacological treatment could be used not only against specific dementia disease but also against normal age-related decline in cognitive functionings (Zimerman et al. 2013).

[10] On the concept of disease, see Humber and Almeder (1997), Nesse (2001), and Temple et al. (2001).

Suppose that a drug treatment becomes available that mitigates or cures age-related cognitive decline. It is a good guess that – perhaps after some initial hesitation – our attitude to such a treatment would be the same as to age-related conditions of the eyes and teeth. It is also difficult to see how such drugs could, in open societies, be kept away from younger persons, would they also benefit from them.

The difficulties in specifying the disease/normality limit create problems for the closely related distinction between health-related and non-health-related interests, referred to above, that is traditionally used to demarcate those values that medicine and the sciences that support it, including pharmacology, are supposed to serve. However, the lack of such a limit does not necessarily mean that all forms of enhancement would have to be endorsed. There may be other arguments against some of them, arguments that do not depend on the distinction between disease and normality. For instance, just as in the treatment of diseases, drug use for enhancement will expectedly have side effects. The less serious a disease is, the smaller risks of side effects can be accepted in drugs (or other treatments) against it. Arguably, tolerance for side effects should be even smaller when there is no disease to be treated.

16.3.3 An Asymmetrical Weighing?

We have already noted that the weighing of advantages and disadvantages of a treatment differs from standard utilitarianism in two important ways: It is limited to the advantages and disadvantages for the *individual patient*, and furthermore, it is limited to *health-related* advantages and disadvantages (although health relatedness is, as we have just seen, far from well defined). There may also be a third difference, namely an asymmetry in how pros and cons are counted.

Classical utilitarianism treats gains and losses symmetrically. If we apply the same approach to pharmacological treatments, then a small preponderance of positive over negative health effects would be sufficient to recommend a treatment. This can be illustrated with an admittedly oversimplified example: Suppose that a patient runs a risk of 1 in 200 in dying due to respiratory arrest, and the only available treatment is a drug that would eliminate that risk but instead incur a risk of 1 in 300 of dying due to cardiac arrest. With a symmetric weighing criterion the patient should be offered this treatment.

It is far from certain, however, that such a treatment would in actual fact be offered to patients. The age-old maxim *primum non nocere* can be interpreted as saying that a higher priority should be given to avoid harming a patient's health condition than to improve it. According to this view, a small preponderance of health benefits over health detriments is not sufficient to justify a treatment. There must be a *large preponderance* of benefits over detriments in order for the treatment to be justified.

It remains to determine empirically whether such a criterion of large preponderance is really applied in clinical practice. It does not seem to be clearly expressed in ethical codes or other general guidelines for pharmacological treatment or for medical treatments in general. There may be differences between different diseases and treatments, for instance due to the severity of the conditions and the availability of alternatives.

In the terms of moral philosophy, the asymmetric approach can be justified with reference to the classical distinction between acts and omissions (Woollard and Howard-Snyder 2016). It is a common moral view that doing harm is worse than allowing harm. (In a medical context, this view is particularly prominent in the discussion on euthanasia.) Harms arising as side effects of a drug are harms of action, where harms resulting from a choice not to offer the drug are harms of omission. It could be argued that this is a valid reason to give higher weight to the avoidance of side effects than to the attainment of intended positive effects.

16.3.4 Procedural Values and the Double Requirement

Informed consent is a cornerstone of healthcare ethics. It can be spelt out as the requirement that the patient who receives a treatment must have been provided with all the information (s)he needs to decide whether she wants to receive it, and has then freely, under no pressure or duress, agreed to the treatment. The requirement for informed consent to medical treatments is based on two important procedural values, namely that the patient should be well informed about the available treatment options and that she should receive a treatment of her own choice.

It is important to observe that standard medical ethics requires *both* a positive outcome of a professionally performed clinical risk-benefit assessment, *and* the patient's informed consent. One of them is not enough (with the exception of cases when the patient is unconscious, incompetent, or for some other reason unable to give informed consent). Both the outcome-related and the procedural values have to be satisfied. This means that the administration of a treatment that does more harm than good to the patient is unethical even if the patient consents to it. It is also unethical to administer a highly beneficial drug without the patient's informed consent (provided that she is able to give it).

These two requirements combine to make quackery unethical. Homeopathy is an unusually clear example. Homeopaths administer drugs with no physiological effect whatsoever, contrary to the standard requirement of a specific positive treatment effect. Furthermore, patients are routinely given incorrect information about the drugs; according to one study patients taking homeopathic drugs were not even aware that these products are highly diluted (Holt and Gilbert 2009; Hansson 2013).

16.3.5 Other-Regarding and Social Values

As we have already noted, medical ethics requires that the interests of the individual patient determine the outcome of treatment decisions. The scope left for considering the interests of others is strictly limited. The major case in which it is allowed to have some influence is the treatment of a communicable disease that poses a threat to others. In this case, many legislations have provisions for involuntary interventions, and the administration of such interventions is not considered to be a breach of medical ethics. It should be noted that such measures are justified by the health interests of other persons, not their interests in general. These cases are extremely rare. In the vast majority of cases, the treatment of an individual patient is required to be based on values concerning the health-related interests of that patient.

However, important decisions about healthcare and its use of pharmaceuticals are also made in other contexts than individual treatment decisions. In all countries, political decisions are made on access to healthcare, priorities between different groups of patients, the availability of different drugs and other treatments, their costs to patients, etc. In addition to the values discussed above, such decisions will have to be based on values concerned with social justice and individual rights. To the extent that such decisions are based on medical values, treatment opportunities will be distributed according to medical needs.

16.4 Pharmacology and the Science-Value Issue

There are strong reasons why the uninhibited introduction of values into science would be highly problematic. We may wish something to be true, but that is certainly no reason to hold it to be a scientifically established fact. The role of science in society is to provide the ultimate reality check. And surely, pharmacology provides important reality checks for healthcare practices. Like all sciences, it has to be protected against wishful thinking in order to fulfil its task. But on the other hand, there are important values, such as those furthered in medical ethics, which have to be incorporated in pharmacological assessments.

This means that the *science-value problem* that was presented in Sect. 16.2.3 is highly relevant for pharmacology. On the one hand, we want science to answer our questions about practically relevant facts in a straightforward and useful way. On the other hand, we want to preserve the integrity of science as a purely fact-based source of knowledge. The first of these demands implies that the criteria of scientific assent can legitimately be influenced by what is practically at stake; for instance our demands on evidence concerning a potential side effect of a drug should be different if the side effect is intermittent itches than if it is an anaphylaxis. However, the second demand requires to the contrary that the practical consequences of acting as if something is true should have no influence on whether we hold it to be scientifically established.

Several solutions to the science-value problem have been proposed. To make a long story short I will introduce here a solution that I have recently proposed in more detail elsewhere (Hansson 2017), and show how it can be used to solve the value problems in pharmacological science.

16.4.1 Two Types of Decisions

To begin with we need to make a distinction that has not always been made clearly enough, namely that between scientific assent and science-based practical decision-making. Scientific assent to a statement means that the statement is provisionally taken to true. In everyday scientific parlance, we refer to such statements as "true" or "known", but it is important to keep in mind that scientific assent is always provisional. We do not take such statements for incorrigible, only for provisionally not to be doubted. Although we do not actively doubt them, we reserve the right to do so at some later point in time, should we learn something new that gives us reason for a reappraisal. The provisional nature of scientific assent is one of the major reasons why science is more successful than its competitors. For instance, homeopathy is still based on principles from the late eighteenth century that were thoroughly refuted by discoveries in chemistry in the early nineteenth century. In contrast, the basis of scientific pharmacology has repeatedly been improved and revised throughout this period, largely in response to a long series of discoveries in chemistry and other fields.

Scientific data form the basis of decisions on scientific assent.[11] In many disciplines, not least pharmacology, the same data are also used for practical decisions. For instance, suppose that new data indicates that a drug has a previously unknown side effect, namely a potentially deadly liver disease. Then pharmacologists have to make up their mind in an issue of scientific assent, namely whether it should be taken for (provisionally) accepted that the drug has the effect in question, or whether that should be considered to be uncertain. In addition, regulators and clinicians will have make another decision, based on that same information, namely whether the drug should continue to be used as before, or subject to some kind of restriction. It is essential to realize that these are two different decisions, although they are based on the same information. In this particular case, it would not be irrational for regulators and clinicians to reduce the use of the substance even if the scientific issue has not been settled.

The criteria for scientific assent are strict. Those who claim the existence of a previously unknown phenomenon – such as a previously unknown pharmacological

[11] It is important to note that these are decisions, in spite of being made in a rather disorderly way by the community of scientists in the respective fields. They are *epistemic decisions*, decisions about what to believe, and should be distinguished from *practical decisions*, decisions about what to do.

effect of some chemical substance – have the burden of proof, and their task is to prove the claim beyond reasonable doubt. There are two major reasons why we apply strict standards of proof to these decisions. First, there are *intrascientific* reasons, i.e. reasons referring to the scientific process itself. The progress of science can be seriously delayed by detours into blind alleys. Before we take something (provisionally) for granted and build on it in future investigations, we need to make sure that we have very good reasons for believing it to be true. Secondly, we also have *extrascientific* reasons for these strict standards. The claims that have been accepted in science are used as a repertory of information that is sufficiently substantiated to be relied on in a variety of practical contexts. They can only serve that purpose if the standards of acceptance are high.

16.4.2 A Baseline that We Should Not Go Below

In this and the following three subsections I will propose a solution to the science-value problem, i.e. an answer to the question whether and in that case to what extent we should let extra-scientific values influence our criteria of scientific assent.[12] Obviously, such extrascientific reasons can be morally good or bad. In what follows I will only discuss reasons that are morally good or at least acceptable. Typical such reasons are access to efficient treatments and avoidance of negative side effects. Morally dubious or inadmissible reasons such as profitability for pharmaceutical companies may have influence in practice but will not be discussed here.

The proposed solution to the science-value problem has three major components. The first of these is *an epistemic baseline for the evidential requirements of scientific assent. This baseline reflects the high (but not absolute) priority given in science to the epistemic value of error-avoidance over that of truth attainment. It does not take practical considerations into account. That it is a bottom line means that we should never go below it.* For reasons already mentioned, its puts fairly high demands on the evidence for assenting to a claim. This means that it places the burden of evidence on those proposing a new theory or claiming the existence of a previously unknown phenomenon.

The epistemic baseline cannot be defined in a precise (e.g. statistical) manner, but the phrase "beyond reasonable doubt" that is considered precise enough for legal purposes is arguably also precise enough to describe the criterion of scientific assent.[13] In the latter context, it represents a standard based on the "purely

[12]It is proposed as an idealized but on the whole largely correct picture of how practical values influence scientific assent. It is an idealization in both senses of the word (Hansson 2006), i.e. it is *both* a simplification for the sake of clarity of how science actually works *and* a description of how science works at its best. Therefore, like much of the philosophy of science, it bridges over a purely descriptive and a strictly normative approach to science.

[13]The notion of a well-defined level of evidence, such as this baseline, is obviously an idealization, analogous the legal fiction of a well-defined limit between reasonable and unreasonable doubt.

epistemic" value of error-avoidance. Therefore, the baseline can be further specified as *the lower limit for when the evidence in favour of a statement is so strong that it is not reasonable to doubt the statement if only epistemic considerations are taken into account*. It follows that in basic science, for instance biochemical studies of drug mechanisms, we can usually stick to the baseline. In practically applied science, for instance clinical pharmacology, we may have to deviate from the baseline in order to accommodate practical values (such as the importance of avoiding particularly serious side effects, or the urgency of introducing a new antibiotic for life-threatening infections with multi-resistant bacteria).

The baseline marks a level that it is illicit to go below, since doing so would assign scientific status to claims that are too uncertain to deserve that status. On the other hand, the baseline criterion does not exclude deviations in the other direction, i.e. it does not prevent us from demanding more evidence before we assent to a statement than what is required to reach the baseline.

16.4.3 Raising the Demands for Practical Reasons

There are practical decisions in which we may wish to apply evidence requirements that are higher than the baseline. Consider the following example:

> A new drug against hypertension has been developed. According to the available evidence, it has exactly the same effects on blood pressure as the currently used first-line therapy, and the side effects are also the same. However, it is 5% cheaper than the currently used drug.

Obviously, if decision-makers were absolutely sure that the new drug has the properties just described, then they would replace the old drug by the new one. The money saved could be used better elsewhere. But it can reasonably be assumed that regulators and clinical decision-makers would require extraordinarily strong evidence for an exchange like this. The properties of the new drug are presumably less known than those of the old one from which there are many years of experience. Even if the evidence on the properties of the new drug is sufficient for scientific assent (i.e. above the epistemic baseline), it may be appropriate to require further corroborating evidence before a decision is made to replace the old drug by the new one. This is therefore a case in which clinical and regulatory considerations may lead to higher demands on the evidence than what would be required in purely science-driven research on drug mechanisms.

I propose that in a case like this, when the criterion for the practical decision is on a level well above the epistemic baseline, the evidence criterion for scientific assent will in practice follow it, so that the two coincide. This will have the advantage of simplicity; we will not have to distinguish between what is scientifically established and what we can assume to be true in practical deliberations. Importantly, since we set the requirements for scientific assent higher than the epistemic baseline, the effect is that an issue will be considered as open although the baseline would allow us to consider it to be settled, not the other way around (which would have happened if we had instead lowered the requirement for scientific assent below the baseline). We will therefore lower the risk of importing an erroneous claim into science (instead as increasing it as we would have done in the other case). Therefore, this adjustment of the criterion of scientific assent does no damage to the high priority that we have given to error avoidance in order to secure the integrity of science.

It is therefore no surprise that we do not hear regulators or clinical decision-makers say: “There is conclusive scientific evidence that the new, cheaper drug is equally efficient and safe as the old one, but we do not believe it.” Such a message would have been confusing since we normally assume that what is considered to be scientifically established can be used as a safe ground for practical decisions. Instead, they will simply say that the available scientific evidence is not sufficient.

Generalizing this, the second component of the proposed model is the following practice: *When moral or other practical values give us reason to apply a higher evidential standard than the epistemic baseline, then we adjust the criterion of scientific assent to that same, higher level.*

16.4.4 Dealing with Requirements to Go Below the Baseline

There are also practical decisions in which we have moral reasons to apply criteria that are lower than the epistemic baseline. Let us consider another hypothetical example:

> Two drugs are used alternatively as first-line treatments against acne vulgaris. From the viewpoint of clinical science they are equal both in terms of the intended effect and the side effects. However, new studies on a usually reliable animal model indicate that one of the two drugs (but not the other) can cause serious damage to the retina. According to the leading experts, this effect is probable but has not been scientifically established.

In the choice between these two drugs, there is a strong justification for deviating downwards from the epistemic baseline, i.e. in the direction opposite to that in the previous example. Presumably, we would use only the drug not suspected for retinal damage as a first line treatment until further research has settled the issue. However, there are also strong reasons not to lower the criterion of scientific assent along with that of the practical decision. Doing so would mean that we take it as given that the substance causes retinal damage. Such an assumption would lead us to refrain from further investigations aimed at finding out whether the substance is in fact retinotoxic. This could potentially prevent us from discovering that the substance does not have that effect in humans. It could also lead us astray in attempts to understand the relationships between retinotoxicity and chemical structure, thereby preventing us from making correct predictions about other substances. As this example shows, lowering the requirements for scientific assent could have negative consequences, not only for purely scientific investigations, but also for practical decisions. To avoid this we should retain the high evidential requirements for scientific assent in spite of adopting lower requirements for the practical decision.

As far as I can see, this is also how such a case would normally be treated in practice. In cases like this, scientists, clinicians and government officials all tend to say (or at least imply) that the evidence for the negative effect is not strong enough to be scientifically conclusive but yet strong enough for decisions to be made that protect against the effect, should it be real (Hansson 2008, 2018; Wiener and Rogers 2002). The third component of this proposal is the following practice: *When moral or other practical values give us reason to apply an evidential standard that is lower than the epistemic baseline, then we do so but without adjusting the criterion of scientific assent. Instead, we introduce a separate standard for the purpose in question, and distinguish it from the requirements for scientific assent.*

16.4.5 A Two-Branched Strategy

In summary, all this amounts to a two-branched strategy for science-based decision-making in which we have moral or other practical reasons to deviate from the epistemic baseline. When the deviation goes upwards, then the criterion of scientific assent is also adjusted so that the scientific and the practical criteria will coincide. However, if the deviation goes downwards, then the criterion of scientific assent is left unchanged.

Above, this two-branched model has been introduced in terms of exceptions from a "purely epistemic" baseline standard. However, in areas of science that have extensive connections with practical decision-making, such impact of practical values on scientific assent may be the rule rather than the exception. This applies to pharmacological science to a high degree, since it aims to provide knowledge that is helpful for practical decisions on the use of pharmaceutical drugs. In practice, evidential criteria differ widely between different such decisions. Decision-makers who apply a precautionary approach are prepared to restrict the use of a drug based on less than conclusive evidence that is has a serious side effect (provided, of course, that reasonable treatment alternatives are available). A single, well-conducted study on patients taking the drug can be sufficient. Much more evidence is required for statements to the effect that the drug has the intended effects or that it has no serious side effects.[14] For such statements, extensive evidence from each of the phases of drug testing must be available, and this usually includes several independent, large clinical trials. Only if such evidence is available will the medical community consider the claim to be beyond reasonable doubt for both scientific and clinical

[14]Strictly speaking, the latter type of statement is inaccessible to direct investigation. For purely statistical reasons, the best that we can obtain is evidence strongly indicating that the frequency of serious side effects is below some low but non-zero number (Hansson 1995, 2002).

purposes. This means that the evidential criteria are higher than the epistemic baseline that is applied in general in science.

The most important feature that distinguishes this account of the science-value relationship from previous models is the distinction that it makes between movements upwards and downwards from the basic criteria of scientific assent. The epistemic strategy that is based on that distinction has the crucial advantage of maintaining the integrity of science as a source of reliable knowledge. This is achieved by always staying above a bottom line constructed to keep the likelihood of accepting false statements at a manageable, low level. On the other hand, this model allows for value-based deviations from the purely epistemic criteria (i.e. from the epistemic baseline) in all cases when such deviations do not threaten the integrity of science, namely the cases when the evidential requirements are strengthened rather than weakened by the deviation in question.

16.5 Conclusion

This chapter has provided a set of distinctions for the characterization of values, an overview over values in pharmacology that makes use of these distinctions, and finally an account of what influence practical values can legitimately have in pharmacological science. The gist of that account is that we must distinguish between influence that would lower the evidential demands for scientific judgments and influence that would raise them. The former kind of influence must be resisted, in other words values should not be allowed to lower the criteria for what we consider to be scientifically demonstrated. In contrast, impact from values can legitimately raise the evidential demands in science. In fact this happens routinely in pharmacological science, since the demands of evidence for the safety and efficacy of drugs for human use typically exceed the level of evidence usually required when nothing more is at stake than the risk of being mistaken in a purely scientific issue.

Acknowledgement I would like to thank Adam LaCaze and two anonymous referees for very useful comments on an earlier version of this chapter.

References

Berreby, D. (1996). Up with people: Dwarves meet identity politics. *New Republic, 214*(18), 14–19.

Breitenbach, A. (2015). Beauty in proofs: Kant on aesthetics in mathematics. *European Journal of Philosophy, 23*(4), 955–977.

Churchman, C. W. (1948). Statistics, pragmatics, induction. *Philosophy of Science, 15*(3), 249–268.

Cranor, C. F. (1993). *Regulating toxic substances: A philosophy of science and the law*. USA: Oxford University Press.

Crouch, R. A. (1997). Letting the deaf be deaf. Reconsidering the use of cochlear implants in prelingually deaf children. *Hastings Center Report, 27*, 14–21.

Douglas, H. E. (2009). *Science, policy, and the value-free ideal*. Pittsburgh: University of Pittsburgh Press.

Dreyfus, T., & Eisenberg, T. (1986). On the aesthetics of mathematical thought. *For the Learning of Mathematics: An International Journal of Mathematics Education, 6*(1), 2–10.

Elliott, K. C. (2017). *A tapestry of values: An introduction to values in science*. New York: Oxford University Press.

Feleppa, R. (1981). Epistemic utility and theory acceptance: Comments on Hempel. *Synthese, 46*, 413–420.

Hansson, S. O. (1995). The detection level. *Regulatory Toxicology and Pharmacology, 22*, 103–109.

Hansson, S. O. (2002). Replacing the no effect level (NOEL) with bounded effect levels (OBEL and LEBEL). *Statistics in Medicine, 21*, 3071–3078.

Hansson, S. O. (2004). Weighing risks and benefits. *Topoi, 23*, 145–152.

Hansson, S. O. (2006). Levi's ideals. In E. J. Olsson (Ed.), *Knowledge and inquiry. Essays on the pragmatism of Isaac Levi* (pp. 241–247). Cambridge University Press.

Hansson, S. O. (2008). Regulating BFRs from science to policy. *Chemosphere, 73*, 144–147.

Hansson, S. O. (2009). Should we protect the most sensitive people? *Journal of Radiological Protection, 29*, 211–218.

Hansson, S. O. (2013). Homeopathy and consumers' right to know. *Journal of Internal Medicine, 274*, 493.

Hansson, S. O. (2014). The moral Oracle's test. *Ethical Theory and Moral Practice, 17*, 643–651.

Hansson, S. O. (2015). Ethical implications of sensory prostheses. In J. Clausen & N. Levy (Eds.), *Handbook of neuroethics* (pp. 785–797). Dordrecht: Springer.

Hansson, S. O. (2017). How values can influence science without threatening its integrity. In H. Leitgeb, I. Niiniluoto, P. Seppälä, & E. Sober (Eds.), *Logic, methodology and philosophy of science – proceedings of the 15th international congress* (pp. 207–221). College Publications.

Hansson, S. O. (2018). The precautionary principle. In N. Möller, S. O. Hansson, J.-E. Holmberg, & C. Rollenhagen (Eds.), *Handbook of safety principles* (pp. 258–283). Hoboken: Wiley.

Hansson, S. O., & Rudén, C. (2008). A risk-neutral default for chemical risk management. *American Journal of Industrial Medicine, 51*, 964–967.

Harsanyi, J. C. (1983). Bayesian decision theory, subjective and objective probabilities, and acceptance of empirical hypotheses. *Synthese, 57*, 341–365.

Hempel, C. G. (1960). Inductive inconsistencies. *Synthese, 12*, 439–469.

Hempel, C. G. (1981). Turns in the evolution of the problem of induction. *Synthese, 46*, 389–404.

Holt, S., & Gilbert, A. (2009). Beliefs about homeopathy among patients presenting at GP surgeries. *New Zealand Medical Journal, 122*, 94–95.

Humber, J. M., & Almeder, R. F. (Eds.). (1997). *What is disease?* Totowa: Humana.

Intemann, K. (2015). Distinguishing between legitimate and illegitimate values in climate modeling. *European Journal for Philosophy of Science, 5*(2), 217–232.

Jeffrey, R. C. (1956). Valuation and acceptance of scientific hypotheses. *Philosophy of Science, 23*, 237–246.

Kincaid, H., Dupré, J., & Wylie, A. (Eds.). (2007). *Value-free science?: Ideals and illusions*. New York: Oxford University Press.

Lacey, H. (1999). *Is science value free?: Values and scientific understanding*. London: Routledge.

Lane, H., & Bahan, B. (1998). Ethics of cochlear implantation in young children: A review and reply from a deaf-world perspective. *Otolaryngology and Head and Neck Surgery, 119*, 297–313.

Levi, I. (1960). Must the scientist make value judgments? *Journal of Philosophy, 57*, 345–357.

Levi, I. (1962). On the seriousness of mistakes. *Philosophy of Science, 29*, 47–65.

Lihoreau, F., & Rebuschi, M. (Eds.). (2014). *Epistemology, context and formalism*. Cham: Springer.

Loren, A. W., Mangu, P. B., Beck, L. N., Brennan, L., Magdalinski, A. J., Partridge, A. H., Gwendolyn, Q., Hamish Wallace, W., & Oktay, K. (2013). Fertility preservation for patients with cancer: American society of clinical oncology clinical practice guideline update. *Journal of Clinical Oncology, 31*(19), 2500–2510.

Merton, R. K. (1942). A note on science and democracy. *Journal of Legal and Political Sociology, 1*, 115–126.

Miller, F. G., Brody, H., & Chung, K. C. (2000). Cosmetic surgery and the internal morality of medicine. *Cambridge Quarterly of Healthcare Ethics, 9*, 353–364.

Nesse, R. M. (2001). On the difficulty of defining disease: A Darwinian perspective. *Medicine, Health Care and Philosophy, 4*(1), 37–46.

Partridge, B. J., Bell, S. K., Lucke, J. C., Yeates, S., & Hall, W. D. (2011). Smart drugs "as common as coffee": Media hype about neuroenhancement. *PloS One, 6*(11), e28416.

Repantis, D., Schlattmann, P., Laisney, O., & Heuser, I. (2010). Modafinil and methylphenidate for neuroenhancement in healthy individuals: A systematic review. *Pharmacological Research, 62*(3), 187–206.

Rudner, R. (1953). The scientist qua scientist makes value judgments. *Philosophy of Science, 20*, 1–6.

Shrader-Frechette, K. (2008). Statistical significance in biology: Neither necessary nor sufficient for hypothesis acceptance. *Biological Theory, 3*(1), 12–16.

Stanley, J. (2005). *Knowledge and practical interest*. Oxford: Clarendon Press.

Temple, L. K. F., McLeod, R. S., Gallinger, S., & Wright, J. G. (2001). Defining disease in the genomics era. *Science, 293*(5531), 807–808.

Westcott, K. J. (2005). Modafinil, sleep deprivation, and cognitive function in military and medical settings. *Military Medicine, 170*(4), 333–335.

Wiener, J. B., & Rogers, M. D. (2002). Comparing precaution in the United States and Europe. *Journal of Risk Research, 5*, 317–349.

Wolpe, P. R. (2002). Treatment, enhancement, and the ethics of neurotherapeutics. *Brain and Cognition, 50*, 387–395.

Woollard, F., & Howard-Snyder F. (2016). Doing vs. allowing harm. In E. N. Zalta (Ed.), *Stanford encyclopedia of philosophy*. http://plato.stanford.edu/archives/sum2016/entries/doing-allowing/

World Medical Association. (2008). *WMA resolution on physician participation in capital punishment.* Adopted by the 34th World Medical Assembly, Lisbon, Portugal, September/October 1981 and amended by the 52nd WMA General Assembly, Edinburgh, Scotland, October 2000 and the 59th WMA General Assembly, Seoul, Korea, October 2008. Available on http://www.wma.net

Zimdars, M. (2015). Fat Acceptance TV?: Rethinking Reality Television With TLC's Big Sexy and the Carnivalesque. *Popular Communication, 13*(3), 232–246.

Zimerman, M., Nitsch, M., Giraux, P., Gerloff, C., Cohen, L. G., & Hummel, F. C. (2013). Neuroenhancement of the aging brain: Restoring skill acquisition in old subjects. *Annals of Neurology, 73*(1), 10–15.

Zimmerman, M. J. (2015). Intrinsic vs. extrinsic value. In E. N. Zalta (Ed.), Stanford encyclopedia of philosophy (Spring 2015 ed.). http://plato.stanford.edu/archives/spr2015/entries/value-intrinsic-extrinsic/

Chapter 17
Humbug, the Council of Pharmacy and Chemistry, and the Origin of "The Blind Test" of Therapeutic Efficacy

Bennett Holman

Abstract This paper examines an early attempt to purge the medical field of commercial exploitation and promote rational therapeutics. The historical study of the origin of the double-blind randomized controlled trial as a standard of therapeutic efficacy reveals that concerns which seem entirely modern, are actually deeply entrenched in the history of medical research. Specifically, those whose primary concern was with ending the use of inefficacious, and sometimes positively dangerous drugs, found themselves in an ideological battle with those who benefitted financially from their use.

17.1 A Practical Scientific System

> At first sight it seems disheartening to find that physicians are so easily humbugged. Yet when it is remembered that it is impracticable for physicians either to analyze such products themselves or to go to the expense of having chemists do it for them, it is evident that the fault lies not so much with the physicians as with the conditions that make the exploitation of such frauds possible. Puckner and Warren (1910, p. 1705)

The US government's first attempt to regulate medical practice was not carried out by the Food and Drug Administration, or indeed any part of (what is now) the Department of Health, Education and Welfare. The first attempt to regulate medical practice was instigated by the United States Post Office. The government's loss of the ensuing Supreme Court case complicated subsequent federal attempts

Forthcoming: In *Uncertainty in Pharmacology: Epistemology, Methods and Decisions.* (Eds. B. Osimani and A. LaCaze). Springer: Dordrecht

B. Holman (✉)
Underwood International College, Yonsei University, Seoul, South Korea

Faculty of Humanities, University of Johannesburg, Johannesburg, South Africa
e-mail: bholman@yonsei.ac.kr

A. LaCaze, B. Osimani (eds.), *Uncertainty in Pharmacology*, Boston Studies in the Philosophy and History of Science 338,
https://doi.org/10.1007/978-3-030-29179-2_17

to constrain medical practice for the next half of a century (Hart 1956). While it appeared that the case would hinge on whether government had the right to prevent an ineffective treatment from being offered, the court ruled on whether efficacy (or in this case lack thereof) was something that could be established.[1]

With the advent of a functioning postal system came the ability to bilk people that lived too far away to seek restitution through normal legal remedies. In the interest of providing some measure of protection to citizens with more money than sense, congress gave the Postmaster General the ability to prevent the delivery of mail to someone obtaining "money through the mails by means of false or fraudulent pretenses, representations or promises" (17 Stat. 1872, 322–323). When the Postmaster issued a fraud order, all subsequent victims would have their mail returned with an official warning –fraudulent— stamped across the envelope.

While originally the law was applied to financial frauds, towards the end of the nineteenth century the Postmaster General began issuing fraud orders on medical scams, including one on J.H. Kelly and his school of Magnetic Healing. Through spiritual revelation, Kelly claimed to have discovered that physical illness was caused by a disordered mind. Moreover, he promised that if he properly aligned a patient's thoughts, the patient gained the ability to cure themselves of nearly any illness. Conveniently, Kelly had the ability to guide a patient's thinking from any distance. The patient merely need send in their fee. In exchange they were appointed a time during which, having cleared their mind, they would receive the "practical, scientific system" telepathically transmitted to them from Kelly's home in Nevada. By the turn of the century Kelly was receiving roughly 3000 letters a day, netting between $30,000 and $50,000 a month (roughly $1.25–$2 million adjusted for inflation). In an attempt to put an end to Kelly's racket, a fraud order was placed on his mail. This action was appealed by Kelly and the case was ultimately heard by the high court.

In the case, the Supreme Court ruled that when the Nevada postmaster, J. M. McAnnulty, halted Kelly's mail, he had done so on the basis of a fact not capable of being proven. Writing for the majority, Justice Peckam reasoned that the mind unquestionably had some effect on the body. He further noted that the effectiveness of nearly any method of treating disease was accepted by some doctors and disputed by others,[2] thus:

> "One person may believe it of far greater efficacy than another, but surely it cannot be said that it is fraud for one person to contend the mind has an effect upon the body and its

[1]This section relies heavily on the excellent historical work of James Harvey Young (1967).

[2]It is important to note that at the time, there was no consensus regarding the correct approach to medicine. Homeopaths and eclectic practitioners (essentially those who espoused standard medical knowledge, except in therapeutics where they advocated herbal remedies over heavy drugging and bleeding) were operating medical schools and made up over 20% of practicing physicians. Other unorthodox approaches were common though with fewer adherents. For an excellent discussion of the interaction between "regular physicians" and the "irregulars" during this time see Starr (1982), esp. Chapter 3. For a historical account that emphasizes the abilities of those within medical schools to exercise conceptual authority over "legitimate medicine" see Rothstein (1987).

> physical condition greater than even [what] a vast majority of intelligent people would be willing to admit or believe... We may not believe in the efficacy to the extent claimed and we may have no sympathy with them in such claims, and yet their effectiveness is but a matter of opinion in any court... There is no exact standard of absolute truth by which to approve... [an] assertion false and a fraud. (American School of Magnetic Healing and J. H. Kelly vs. J. M. McAnnulty, 187 U.S. 94)

The court's ruling served as benchmark for subsequent debates about therapeutic efficacy. It was the McAnnulty decision that the Honorable Oliver Wendell Homes relied on when he ruled that the Pure Food and Drugs Act of 1906 did not give the Bureau of Chemistry (the forerunner of the FDA) the right to judge claims of efficacy and it was the wording of Justice Peckam that was echoed by the congress in the Sherley Amendment that gave the right back to the Bureau of Chemistry if they could prove that the defendant did not believe their own claims.[3]

In retrospect what is most striking about the McAnnulty ruling is the assertion that there is no standard by which to judge a therapeutic claim true or false. Indeed, one of the hallmarks of modern medical practice is the widespread belief that we now have just such a standard. In this chapter, I will argue that at least for a number of influential researchers, a central motivation for methodological sophistication was the need to develop a means to "tame the market." Specifically, I will defend the claim that in response to a growing trend of manufacturers making questionable claims of therapeutic efficacy, medical reformers developed and advocated "the blind test" as means of disproving commercially convenient falsehoods.

Given the extent to which changing market structures shaped early twentieth century medical reform movements, Sect. 17.2 provides an overview of the development of the pharmaceutical market in nineteenth century. In Sect. 17.3, I discuss the Council of Pharmacy and Chemistry (CPC), a prominent group of medical reformers that attempted to combat the commercial domination of the therapeutics market. In Sect. 17.4 I discuss one part of this effort, the establishment of the Committee on Therapeutics, which among other projects conducted what might be considered the first modern double-blind randomize controlled trial. I conclude with the reminder that though there are modern concerns that the pharmaceutical industry is corrupting medical research, such concerns are not new. The RCT was born in an attempt to reign in the market; if industry has loosened these bonds, it should be seen as a need to reestablish control rather than some modern failing or fall from grace.

---

[3]The 1906 Pure Food and Drug act required accurate labeling (as well as requiring disclosure of certain dangerous or addictive ingredients such as cocaine). In writing for the majority, that the law did not pertain to efficacy claims, Holmes argued that it would be an abuse of power if congress began "establishing criteria in regions where opinions were far apart." Because the ruling meant that hundreds of similar cases would have to be dropped against rank fraud, President Taft called for immediate action from congress. The amendment (introduced by Kentucky Congressman Swagar Sherley) declared that a product was misbranded if a claim regarding a curative or therapeutic effect was "false and fraudulent." The requirement that the claim was false *and* fraudulent was an extremely strong requirement. As long as the defendant believed his own claims, nothing could be done. Such a requirement encouraged manufacturers to find doctors that would attest to their product's efficacy in order to provide themselves legal protection (Young 1967, pp. 48–50).

17.2 Fleecing the Sick and Miserable

When the American Revolution disrupted drug importation from England, a burgeoning American industry devoted to producing tinctures, elixirs and panaceas took root. Collectively these were referred to as patent medications although very few of them had American patents. Indeed, very often the appeal of nostrums like Curforhedake Brane-Fude (pronounced: Cure-for-headache brain-food), was a secret formula (Young 1967). Such a claim could no longer be made if the ingredients were specified in a patent application.[4] During the 1800s, the spirit of self-reliance that dominated the American ethos, pervaded medical treatment as well.[5] Of course, given the purges, caustics, and bleedings that a patient might endure if they were to see a "regular" physician, patients may have been better off using Buffalo Lithia Water or Dr. Munyon's Kidney Cure. Though the first was shown to be indistinguishable from the contents of the Potomac and the latter was literally a sugar pill, at least they did no harm.[6]

Though there were few truly effective treatments there was no shortage of supposed cures marketed with the same scientific patter. Layman and doctor alike were continuously bombarded by mixtures of long-forgotten substances, revived from the therapeutic graveyard by proprietary manufacturers and rebranded as another new "German-style synthetic." Each new product was accompanied to market with testimonials of therapeutic efficacy.

In contrast to the heavily marketed patent medications, were the "ethical" manufacturers. Makers of "ethical" drugs eschewed private ownership of medical knowledge and thus published their formulae. Ethical drug manufacturers advertised only to doctors, generally sold in bulk to pharmacists, and sought to win sales within

---

[4]For example, "The Nature's Creation Co. Discovery" advertised a purported cure for tuberculosis that contained "At least one ingredient that the medical world knows nothing about . . . no chemist has ever been able to determine what it is." Discovery Co. was a prime example of the kind of patent medicine vendors the AMA would seek to discredit. Indeed there was special antipathy for those who defrauded patients with terminal illnesses such as tuberculosis or cancer. In this case, like many others, the AMA simply tracked down the graves of those who had recently testified to being cured (Burrow 1963, pp. 119–124).

[5]As Starr (1982) points out, self-reliance was reinforced by economic considerations. Given the low population density on the frontier, retrieving the nearest doctor could add a day or two of lost labor to the cost of treatment. The cost of a physician (who also charged by the mile) thus far outstripped the panaceas populating the store shelves (c.f. Burrow 1963, pp. 107–108). Yet there may have been other reasons for the success of patent medicines patronized by morally upright citizen. The high alcohol content of some (e.g. the 82 proof Celerina) allowed the outwardly abstemious an avenue to secure alcoholic substances under the guise of good health. Additionally, the cocaine and opiates in several products created many unintentional addicts.

[6]For a fuller account of regular physicians see Young (1967, esp. ch. 2); the distinction between regular and irregular physicians is discussed in (Starr 1982, pp. 93–112). The "cures" mentioned can be found discussed in CPC (1914a); and Edmunds (1909). Lithia water was a paragon case of the worthlessness of testimonials and the unreliability of clinical experience.

the medical community via reputations of integrity (Gabriel 2009; 2014a).[7] In addition to wide range of basic ingredients, ethical companies began to market their brand of the handful of advances appeared over the nineteenth century: the isolation of morphine in 1816; the development of disinfectants such as iodin, potassium iodid, and bromids in the first half of the century; anethetics such as ether (1846); carbolic acids for sterilization (1867); and antipyretics (fever-reducers) in the last few decades before 1900.[8]

To more critical minds, the discovery of effective treatments placed in relief the worthlessness of contemporary *materia medica* (i.e. the embodiment of doctors' collective medical knowledge of therapeutics). During the 1860s and 70s the best American doctors trained in Europe and many came back committed to rational therapeutics, Claude Bernard's doctrine that medicine should be driven by a scientific understanding of physiology and illness (Warner 2003). Ambassadors of the doctrine held that much of contemporary practice was either worthless or interfered with patients' natural ability to heal. To those holding on to traditional healing practices it seemed nothing more than "therapeutic nihilism".[9] Nevertheless, rational therapeutics began to find a foothold in the medical colleges of the Northeast, such as Harvard whose dean Oliver Wendell Holmes (father of the court Justice) avowed: "I firmly believe that if the whole of *materia medica,* as now used, could be sunk to the bottom of the sea, it would be all the better for mankind, — and

[7]It might be more correct to say the preceding description is the ideal ethical drug manufacturer. In the late nineteenth century a new hybrid entity emerged, a *proprietary* medicine (derogatively referred to as a nostrum) was any drug advertised only to doctors, but that was protected against free competition by secrecy (i.e. undisclosed ingredients or manufacturing process), patent, or trademarked name. As Gabriel notes, the best proprietary drugs were made by innovative pharmaceutical companies trying to find ways to ethically introduce products to markets in a profitable manner within a system that considered patenting a medical innovation unethical and introducing new products to market beyond the remit of reputable manufacturers. During the early twentieth century, a repeated refrain of reformers would be that many proprietary drugs that doctors relied on were no different in ethical behavior than the quackish patent drugs that most doctors decried (e.g. CPC, 1915b).

[8]See Edmunds (1909) for a view of the nineteenth century from the beginning of the twentieth.

[9]Though Bernard is often credited as the founder of rational therapeutics, the origins of "nihilistic" attitude that found its way to the US trace back to Germany, as exemplified by Skoda's emphasis on the importance of diagnosis and post-mortem analysis: "We can diagnose disease, describe it and get a grasp of it, but we dare not expect by any means to cure it." (Quoted in Barker (1900, p. 153) cf. Meltzer (1909); Temkin (1964)). Such sentiments existed in France and the US before Bernard, stimulated by the numerical method of Louis. For example, in an 1835 address on spontaneous recovery from illness, Harvard professor Jacob Bigelow claimed "the unbiased opinion of most medical men of sound judgment and long experience is made up, that the amount of death and disaster in the world would be less, if all disease were left to itself, than it now is under the multiform, reckless, and contradictory practice, good and bad, with which practitioners of adverse denominations carry on their differences at the expense of their patients" (Bigelow 1858, p. 41).

all the worse for the fishes" (Holmes 1860, as quoted in Parascandola 1992, p. 8).[10] The eminent historian of nineteenth century patent medicine, James Harvey Young described Holmes as the leading critic of the industry's "dishonest therapeutic claims, faked statistics, fabricated testimonials, and hazardous ingredients" (Young 1967, p. 28; Warner 1978).[11]

But even as doctors became wary of therapeutic claims, patent medicine manufactures were themselves becoming more effective in representing their interests. Groups such as the Proprietary Medicine Manufacturers and Dealer Association (PMMDA), organized in 1881, were not only effective at lobbying congress, but in managing threats to their collective interests by manipulating the public discourse.[12] Most notable was the "red clause" that began to appear in advertising contracts between PMMDA members and periodicals. The attention grabbing red print of the clause read: "It is mutually agreed that this contract is void, if any law is enacted by your state restricting or prohibiting the manufacture or sale of proprietary medicines." Contracts would also be cancelled if the paper ran negative stories regarding patent medications. As the progressive movement began, popular press was littered with advertising for, but almost no criticism of, patent medication (Adams 1907, p 123–142). Indeed, periodicals and drugs quickly formed a close alliance. Patent medications provided large percentages (often over half) of a newspaper's gross revenue and with the red clause in contracts, the bribes paid out by the head of the PMMDA to influence public discourse and legislation dropped from $75,000 to $6000 per year.[13]

While rival medical approaches threatened the dominant medical establishment, the patent medicine industry was the primary rival for medical authority (Simmons 1907). Not only did they offer prepared concoctions in place of the tailored mixtures

[10]While this phrase is oft quoted, "as now used" is frequently omitted, yet it underscores that even drugs like morphine and quinine while being effective in some cases were all too often prescribed in inappropriate dosages, or in circumstances where they did no good.

[11]As Scott Podolsky pointed out to me in personal communication, the title of Young's earlier book, *The Toadstool Millionaires*, derives directly from a Holmes' epithet for the mendacious and avaricious hucksters that preyed upon the desperate, who themselves were not above his reproach: "Another portion of the blame rests with the public itself which insists on being poisoned. Somebody buys all the quack medicines that build palaces for the mushroom, say rather, the toadstool millionaires" (Holmes 1892, p. 186).

[12]For example, during the civil war a 4% tax had been place on proprietary medication over the complaints of many individual manufacturers. Two years after the formation of the PMMDA, the group accomplished what its individual members had failed to do for 20 years and repeal the tax (Young 1967, p. 24).

[13]This collusion was uncovered by Samuel Hopkins Adams in one of the more famous episodes of muckraking and published as part of an eleven part series in *Colliers* from October 1905 through September 1906. These were later collected and published in 1907 by the AMA as *The Great American Fraud*. Along with Upton Sinclair's *The Jungle*, these articles are widely credited with stirring the public outrage that lead to the passage of the 1906 legislation establishing the Bureau of Chemistry and a "truth in labeling" requirement for medical products. This was also the beginning of what would become the FDA (Young 1967, pp. 29–32; Hilts 2003, pp. 46–55; Starr 1982, pp. 129–131).

created by a doctor, they also frequently included polemic tracts within their packaging deriding the medical profession. Customers were encouraged to write in to patent medicine proprietors for medical advice. Doctors were portrayed as a cabal of obscurantists, opposing each person's natural right to self-medicate (Starr 1982, pp. 127–144).

In summary, those who wished to improve medical treatment faced the following difficulties: The primary competition was better organized, financed, and politically connected. Patent drug manufacturers also had significantly greater influence over public discourse through their alliance with the press. In such an environment, testimonials of therapeutic efficacy were regarded as legitimate, authoritative, and persuasive.[14] Judging by their market share, the claims of patent medicine manufacturers, via the newly emerging advertising industry held sway in the majority of treatment decisions (Marks 1997, pp. 18–20; cf. Young 1967, pp. 21–22; 143–155). Those in the American Medical Association (AMA) that sought to set themselves up as a rival source of authority, not only faced a public ethos of self-reliance that was resistant to interference, but a decentralized, fractious profession, that had considerable dissention within its own ranks.

17.3 The Turbidity of Fermentation

Though previous efforts had been made by pharmacists to establish standards of purity (Dowling 1973), the AMA's Council on Pharmacy and Chemistry (CPC), represents the first organized attempt to regulate the pharmaceutical market, including judgments of efficacy (for an analysis of why the council emerged in 1905 and not earlier see Historian's Appendix A). The establishment of the council also represents the beginning of the movement to rest the authority to determine therapeutic efficacy away from the practicing physician. Indeed, perhaps the most telling fact about the CPC was its composition; seven of the fifteen original members lacked a medical degree and none were actively engaged in treating the sick (Marks 1997, p. 24).[15] The council's expertise lay not at the bedside, but at the bench. It was chaired by George Simmons (the secretary and general manager of the AMA, and editor of JAMA), and composed of professors of pharmacology, pharmacy, bacteriology, and chemistry, complemented with government officials from the Bureau of Chemistry.[16] In retrospect, one can note that these men represented the

[14]Besides their ubiquity, one good indication of the influence of testimonials is the repeated efforts made to discredit them (e.g., Sollmann 1908, pp. 21–4; 28–30; Adams 1907, p. 4).

[15]Of the non-original members that served before September 1946, nearly 80% (39/49) would hold an M.D. For credentials of CPC members and short descriptions of the original members see Smith (1947, pp. 865–7).

[16]Government officials included the inveterate pure food advocate and soon to be father of the FDA, Harvey W. Wiley. For early links between the council and government see: Marks 1997 ch 3; Burrow 1963, pp. 76–82, 117–118).

cutting edge of changes that would fundamentally alter medical practice. The rise of the clinical pharmacology and emergence of effective chemotherapy would soon dominate botanicals; however at the time, there was only the hope that disputes in therapeutic practice were "the turbidity of fermentation, rather than that of purification" (Sollmann 1908, p. 5).

In its first meeting the CPC adopted ten rules that would serve to guide the acceptability of drugs. First and foremost, manufacturers were prohibited from advertising to the public. Additionally, manufacturers were required to disclose ingredients (percentage, purity, and strength) and prohibited from making "unwarranted, exaggerated or misleading statements as to therapeutic value" (Smith 1947, p. 870). While a company was under no compulsion to submit their products for evaluation, advertising in JAMA was restricted to drugs approved by the council and the council could and did review drugs unbidden.[17]

This power was amplified as state journals (and even some lay periodicals) began restricting their advertisements to approved products (Smith 1947, p. 877). In this regard, the power of the council lay in the ability to restrict the flow of information (i.e. advertisement) to doctors. Further, the restriction on lay advertising made the council's approval a mandatory passage point for ethical drugs.[18] It should be noted that the restriction of advertising had very real economic costs, which may have led a number of smaller journals to resist following JAMA's lead.[19]

[17]Of the first 670 products reviewed, 90 were investigated on the initiative of the council (Sollmann 1908, p. 11).

[18]Starr (1982, pp. 129–134) argues that it was crucial that muckrakers joined forces with the AMA and that the public began to rely more heavily on doctors. However, while it is certainly the case that the AMA was helped by the public outcry that followed publication of *The Jungle* and exposés in *Colliers* and *Ladies Home Journal,* such outrage did not entail reliance on doctors. Beyond keeping poisonous substances off the market, the 1906 legislation was in fact inconsistent with the intent of the Council. As Carpenter describes it: "The set of specific beliefs described by the 'right of self-medication' emerged with greater cultural force during the antebellum period and continued in force through the early 1900s. At its core, claims for the primacy or priviledged status of self-medication depended upon beliefs in consumer autonomy and judgment. Votaries of self-medication frequently voiced support for stronger labeling requirements for drug manufacturers, in part out of the belief that 'the intelligent layman' needed maximal information to render an informed pharmaceutical purchacing decision." (Carpenter 2010, p. 79). Whereas the truth in labeling laws were intended to facilitate an individual's right to self-medicate, the AMA vigorously fought against self-medication and tried to inject itself between manufacturers and patients.

[19]JAMA reported turning down $25,000 in advertising revenue in the first year that the policy was in action in a year where they had just under $100,000 in ad revenue. Nevertheless they made up for the loss with major increases in subscriptions and accordingly charging more for the ads which were accepted (JAMA 1907, p. 1962).

A more active form of influence lay in the regular space allotted to the council in JAMA and other AMA publications.[20] The council's findings became a regular feature in a JAMA section originally titled "Pharmacology", before being changed in 1911 to the apposite "Propaganda for Reform."[21] The drugs that were approved were compiled into "New and Nonofficial Remedies" which the council published on an annual basis beginning in 1907.[22] In addition to publishing their therapeutic investigations, the council also used the column tactically against threats to rational therapeutics. For example, in 1905 the CPC published a list of all the members of the Proprietary Association of America, a widely influential advocacy group for patent medication manufacturers, reviled by advocates of rational therapeutics. The CPC also published the subsequent resignations of any company that wished to clear their names in the eyes of the profession, which most did (CPC 1906).

Though the extent to which the council had an immediate impact, on day-to-day practice was unclear (especially outside the northeast), was unclear, the council at least made it possible for the doctor to prescribe "rationally" by "putting the science of therapeutics and pharmacology on a sounder and more scientific basis than ever before."[23] It worked prodigiously, first sending samples out to allied universities for chemical analysis, before constructing its own laboratories to conduct the work in house.

Within the first 3 years the Council had reviewed 600 drugs, accepting less than half (Sollmann 1908, p. 11).[24] Perhaps more important than the rejection rate, is the fact that only a small fraction the drugs accepted had conformed to the council's ideals prior to the existence of the council and the enforcement of their rules (Sollmann 1908, p. 43). After spending 3 years reviewing the vast majority of available medicines, the council tasked Torald Sollmann with the job of summarizing for the medical community—in a twenty-two part series—

[20]The importance of JAMA as an official organ of the council is underscored by the Die Arzneimittelkommission des Kongresses fur innere Medizin. The German group, formed in 1911 and modeled on the CPC, had no official organ but originally believed the medical press would voluntarily and enthusiastically accept the recommendations of the council. With a single exception, they were sadly mistaken (CPC 1914b).

[21]This shift marked the inclusion of patent medication into the purview of the council (Burrow 1963, p. 110).

[22]Moreover, the reach of the Council extended outside the US as correspondences emerged with medical organizations in European, Asian, and South American countries, including regular space in the influential London-based *Lancet* (Smith 1947, p. 877).

[23]The quote comes from an address by George Simmons to the Southern Medical Association in 1914 and is related in (Smith 1947, p. 866).

[24]As council members lived in far flung parts of the nation communication among members was coordinated through the weekly circulated Bulletin. A submission was originally considered by a single member then discussed within one of the four committees (chemistry, pharmacy, pharmacology, and therapeutics) and then finally presented for a vote by the entire council which needed a 3/4th vote for any definitive action. See Simmons 1907, p. 1650 and Edsall (1910) for a description of approval procedure.

the accomplishments, ethos, and broader aims of the CPC.[25] He condensed the council's purpose to "the general advancement of the sane study of scientific therapeutics" (Sollmann 1908, p. 44).

While it operated, the CPC was one of the loudest advocates for rational therapeutics: the prescription of known substance for a specific disease based on scientific validation of its therapeutic effect. In its early years, the council perceived that it was too large a "task to select those [treatments] which could be recommended as good enough to use; but it was a relatively simple task to pick out those which were so bad that should certainly *not* be used. (Sollman 1908, p. 8, italics in original). In the majority of cases that the council considered it was sufficient to determine the chemical composition of a substance and judge whether the therapeutic claims were plausible based on the known pharmacodynamic effects. For example, in their report on Budwell's Emulsionof Cod-Liver Oil, the council wrote:

> In the advertising circular statements regarding the various ingredients of Budwell's Emulsion are quoted from obsolete text books. These statements, for the most part, do not represent modern opinions on the subject. For instance, the circular praises the action of guaiacol as eliminated directly by the lungs, thus exerting a beneficial local effect and causing bacilli to diminish in numbers or to disappear. All of this is directly contradicted in authoritative modern publications on pharmacology, which hold that the excretion of guaiacol by the lungs is infinitesimal and its action on bacilli is nil. (CPC 1915a, p. 685)

Thus, for the majority of cases, the council sought to eliminate rank fraud on the basis of the growing body of work in clinical pharmacology (cf. Marks 1997). As such, it typically sufficed to demonstrate that the claims of manufacturers contained blatant falsehoods. However, though it was not typically employed in the council's decision making process, Sollmann and members of the Committee on Therapeutic

[25] Sollmann was a founding member of the CPC and early advocate for the scientific determination of drug efficacy. Like John Able, Sollmann's post-graduate education at Strassburg Pharmakologische Institut, provided an alternative vision of medical science, and the two were instrumental in establishing the discipline of pharmacology in America. Though he never had the public recognition of men like Cushny and Hatcher, Sollmann was highly involved behind the scenes in formative organizations linking laboratory science to medical practice such the American Society for Pharmacology and Experimental Therapeutics (Parascandola 1992, p. 65), The Biochemical Society and the Pharmacological Society (Sollmann 1961, p. 2). He was the third chairman (1938–1960) of the CPC and was a member from its founding in 1905 until he was removed in 1960 by a commercial-friendly coup that had taken power of the AMA leadership (Dowling 1973, p. 162). Sollmann's passion was a driving force behind the ideals of the council. For example, upon discovering his former student and fellow council member Robert Hatcher had contracted with Coca-cola to summarize the known properties of caffeine, Sollmann's reprimand caused Hatcher to feel "sicker of this thing than I have ever been of anything I can recall" (Hatcher (1910), quoted in Swann (1988, p. 33). An AMA historian described Sollmann as: "kindly fair, determined to get the facts and see that justice was done, eager to learn of the new, possessed of a brilliant and discerning mind" (Smith 1947, p. 867). One might note that his belief that doctors had a moral obligation to form true beliefs was a hallmark of the council's propaganda for reform (c.f. Hatcher 1916).

Research were developing an alternative means of assessing therapeutic efficacy: the blind test.

17.4 The Crucial Proof of the Pudding Is in the Eating Therof

In support of their claims of therapeutic efficacy, it was not uncommon for companies to send in testimonials from doctors averring to the therapeutic value of the good. Accordingly, in cases where the council rejected therapeutic claims on the basis of laboratory-based pharmacological evidence, the council, *ipso facto,* rejected and opposed the clinical experience of daily practitioners. Capitalizing on this, manufacturers (and medical journals that catered to the interests of manufacturers) promoted the image that the council was both self-interested and removed from the practical concerns of physicians (CPC 1915c; The Medical Brief 1910; Medical Standard 1911).[26]

While laboratory tests were sufficient to identify rank fraud, there was a large and growing gray area of therapeutic claims that were suspect, but not blatantly false (Puckner and Hilpert 1908). To deal with such issues the council sought to develop the state of clinical research by founding its Committee on Therapeutic Research. The committee engaged in a number of tasks pertinent clinical practice (e.g. literature reviews, work on drug metabolism, investigations of mechanisms of drug action) with the goal to "bridge the chasm between science and its application" (Sollmann 1916, p. 1439). The most intellectually ambitious projects involved modeling their ideal for clinical research, what Sollmann (1917) referred to as "the blind test."

The first such test conducted under the auspices of the committee was not motivated by a burning therapeutic question, but to definitively refute commercial humbug. Sodium salicylates were then in use to treat rheumatic fever and while they could be cheaply and easily synthesized, commercial manufacturers of "natural salicylates" (i.e. sodium salicylate derived from natural oils) were promoting the view that their products had superior therapeutic properties over the synthetic versions. In principle an improvement on therapeutic quality or a reduction of toxicity would represent a significant therapeutic advance given the widespread use of salicylates and the slim margin between the effective and the toxic dose (Waddell 1911).[27] This being said, the council considered the claims of increased efficacy

---

[26]For a discussion of manufacturers response to the council see Boyle (2007, 2013).

[27]This justification comes from the introduction of this article written by Sollmann. He there quotes a price difference of $0.45 vs $7.20 per pound for synthetic vs. natural salicylates. The concern with cost efficiency of was a frequent refrain of the council who objected to doctors facilitating commercial exploitation of the lay public (for an earlier example see: Adams 1907).

to be a "superstition [that] had arisen in the profession" (Sollmann 1916, p. 1439) rather than a live hypothesis.

Prior to the clinical trial the Committee appointed members to run animal trials (Waddell 1911), summarize the existing literature (Eggleston 1912), and to assess whether there was any legitimacy to the concerns that cheapest brands of salicylate were adulterated—they were not (Hilpert 1913). In so doing, they found no evidence that would support the claims of superiority made by the manufacturers. The council's campaign against "natural salicylates" culminated in a large multi-site "cooperative clinical investigation." Eighty-two doctors "of recognized standing" were recruited to run patients in the trial and to each was sent ten boxes identified only by a serial number. The doctors were informed that the purpose of the trial was to determine if there was a difference between natural and synthetic salicylate, that each of the boxes contained one or the other, but the doctors were kept ignorant of which was which "in order to place the evidence above any possibility of prejudice or bias" (Hewlett 1913, p. 319). Though they ran into problems with the completeness of some of the reports, they accepted the reports of 230 patients from 27 doctors. On the basis of the resulting data they concluded there was no difference in either the therapeutic efficacy of the toxicity of the two substances.

Of note methodologically, the trial was double-blind—neither the patient nor the doctor knew which substances were being administered. Moreover, it seems that to protect any ordering effects they created two groups, one in which boxes 3, 5, 6, 8, & 9 contained the natural salicylate and a second group in which those boxes contained the synthetic ones. As a result, the treatments were randomized. The use of synthetic salicylate serves as the (active) control by which to compare the experimental substance (natural salicylate). While double-blinding, randomization, and control, all had been components of medical trials before (for further details, see Historian's Appendix B), the salicylate trial may be the first time they were all united in a single modern medical trial. With the exception of statistical analysis of data, the trial contains all of the components of the modern double-blind randomized clinical trial.

Unlike contemporaneous "controlled trials," the testing of synthetic salicylate was recognizably modern in multiple regards. Though the council was too early to benefit from Fisher's formal development of randomization, they appreciated the crucial methodological function played by both random assignment and the large number of patients enrolled in the trial: "Since each observer received an equal number of natural and synthetic powders and used them indiscriminately, the results form a reliable basis for comparison between the two sets of powders for the variations in administration, type of case, etc. tend to distribute themselves equally between the two classes of salts" (Hewlett 1913, p. 320). Moreover, though the council did not use a placebo in the clinical trial, they were used in some contemporary laboratory analyses (CPC 1913) and the possibility of using them in clinical trials was clearly articulated in cases where efficacy (rather than equivalence/superiority) was to be determined:

> In [the blind test], the patient, or a series of patients, is given the preparation which is to be tested, and another preparation which is inactive, and the observer aims to distinguish the two preparations by their effects on the patient. Surely if the drug has any actions at all it will be possible to select correctly in a decided majority of the administrations. The same principle can be applied in distinguishing the superiority of one preparation over another. In this case, the two preparations would be given alternately to different patients, and the observer would try to distinguish them by their effects. Here again, if one drug is really superior or otherwise different from another, to a practically important extent, the observer will surely be able to make the distinction. This method is really the only one that avoids the pitfalls of clinical observation; it is the only method that makes the results purely objective, really independent of the bias of the observer and the patient. It is the only method, therefore, which determines whether it was really the pudding that was eaten and not some other dessert. (Sollmann 1917, p. 199)

The council is also recognizably modern in their justification for the use of placebos to correct for a positive effect. In numerous places the council cites the possibility of "psychic effects" producing better clinical outcomes (e.g. CPC 1910). Accordingly, contra Kaptchuk (1998), it is clear that when the council speaks of bias, it is not just that they wanted to control for the natural course of the illness. Thus, not only did the council employ the same experimental techniques, they applied them for recognizably modern reasons (randomization is a possible exception, see Historians Appendix B).

17.5 A Plea and a Program for Work

The Salicylate trial is the earliest, but by no means the only randomized controlled trial the council conducted (Bastedo 1915; CPC 1917). They were however, typically engaged in other work. Their small number of trials seems to be more of a proof of concept and model for a new type of clinical investigation rather than a shift from the chemical analyses upon which the council regularly made their judgments. Rather than carry out such trials themselves, such trials stood as a standard that could be invoked in evaluating clinical evidence and indeed, such a standard was realized at times by the more reputable pharmaceutical houses (Gabriel 2014b). There is also some indication that the CPC may have simply found centrally organizing and running large trials in far flung locations to be a challenge. At the very least they subsequently discouraged others from trying to do so, instead advocating close supervision by experienced investigators (Sollmann 1930). Ultimately, it would be Harry Gold, a student of council member Robert Hatcher, who would make a career promoting such trials (Kaptchuk 1998).

In the hundred years since the trial on Salicylates, the "blind test" has been adopted as the basis of the regulatory policy in the modern FDA and been promoted as the gold standard of medical evidence by the evidence-based medicine movement. In stark contrast to the McAnulty decision that ruled the efficacy of any substance was a matter of opinion in any court, the Supreme Court ruled in 1970 that such a standard for judging efficacy now existed. Specifically, they ruled that two randomized controlled trials were necessary to fulfill the legal mandate of the 1962 amendments to the Federal Food, Drug and Cosmetics Act that required companies to demonstrate safety and efficacy by "adequate and controlled investigations" (Carpenter 2010).

Despite the prevalence of the blind test, there are now concerns that the pharmaceutical industry has "hijacked" the evidence-based medicine movement and the function of the blind test along with it (Ioannidis 2016). Perhaps this is true, but to construe such actions as surprising relies on the idea that there is some natural direction that pure science would take and from which it has been deflected by vested interests. What the history above suggests is that vested interests have long been a part of medical research. If they are not a natural part of medical science, they are at least a stable and entrenched aspect of it. Most importantly, some of the most important developments in medical epistemology have stemmed not from a disinterested search for truth, but in an active and uphill battle against commercially profitable humbug. While it may no longer be sufficient to curtail commercial exploitation, not least amongst these developments was the double-blind randomized clinical trial.

Historian's Appendix A: Non Sibi Sed Medicinae

A historical question remains as to the timing of the emergence of the CPC especially given the fact that many of the pressures that led to the founding were present long before 1905. To wit, when the AMA was first established in 1847, it distinguished between patent and ethical medicines. In 1849, the AMA resolved to create a board to evaluate drugs and demanded that companies forgo public advertising. Despite concerns over drug adulteration and quack cures, the AMA lacked the funds to carry out such measures (Smith 1947, p. 865).

Similarly unsuccessful demands on drug companies surfaced periodically (1850, 1879, and 1900). For example, in 1900 the AMA demanded that patent medications manufacturers who wished to continue advertising in JAMA divulge their ingredients (JAMA 1900).[1] In addition, several state societies had been pressing JAMA editors to uphold resolutions passed by AMA membership regarding clean advertising. In an 1895 case brought by the state society of Pennsylvania in the personage of Dr. Solis-Cohen, the judicial council of the AMA found that the editors of JAMA had placed the financial needs of the journal over its scientific ideals when determining the acceptability of advertising copy. Despite the reprimand, little

[1]In 1900, the JAMA journal editors began a minor campaign against proprietary medicine in a reoccurring series "Relation of Pharmacy to the Medical Profession" (published weekly from April 21 to July 14th). See especially the July seventh contribution for an ethical drug manufacturer's argument for trade names and secret formulae; namely that such secrets prevent pharmacists from compounding inferior imitations. The rejoinder by the editorial staff upheld the disclosure of ingredients, arguing that doctors must know the composition of a drug is to prescribe it rationally.

changed and the "greed for advertising patronage [which] leads the editor only too often to prostitute his pen" continued unabated (Forshay (1900) as cited by Dykstra 1955, p. 410).[2] At the turn of the century the AMA had neither the organizational unanimity nor the financial resources to carry out these plans (Starr 1982, p. 129; cf. Burrow 1963, pp. 27–36; Warner 1991).[3]

Perhaps the most critical change was the 1901 revision of the AMA constitution. Previously, local societies were independent of state societies and societies could send one representative per ten members to annual meetings. As a result, meetings became so ponderously packed that business became difficult to conduct. The new constitution nested each of the local associations within the state institution, all federated under the AMA. In 1900 only 7% of doctors were members of the AMA and fully 70% belonged to no association at any level; within 10 years, half of the country's doctors had joined. More importantly they were paying dues. Moreover, the constitution created a representative House of Delegates capped at 150 members to conduct business (Starr 1982, pp. 102–112).

In addition to reorganization, in 1902 the AMA joined forces with pharmacists to study issues relating to drug purity and therapeutic claims. The joint committee called for a national agency to regulate the drug market (JAMA 1903).[4] In 1904, the state delegation from Michigan petitioned the house of delegates to do what the federal government would not and form a commission of impartial experts to assess the composition and effects of drugs, so as to provide American doctors with a source of information besides the drug detailer and company published literature (JAMA 1904, p. 1576; c.f. Simmons 1907, p. 1646; Burrow 1970, pp. 112–114).[5]

Though the House of Delegates demurred, a small group of reformed minded doctors began corresponding amongst themselves and presented a detailed plan to the AMA Board of Trustees. The board agreed that needs of the profession were too important to be left to the dithering of the House and created the Council on Pharmacy and Chemistry (CPC) on their own authority, February 5th 1905. The Council held its first official meeting on the 11th and adopted the motto "Non sibi sed medicinae (Not for ourselves but for medicine)." The work of the CPC was financed through The Board of Trustees and all council members served without remuneration (Smith 1947, pp. 866–869). The concentration of power created by

---

[2]Dykstra contains an elongate discussion of the ways in which advertising could corrupt a journal by tying advertising contracts to the publication of favorable scientific articles and editorials in a journal. In general, I have found Dykstra to be an excellent source on the conditions relevant to the formation of the council, but I have also found at least one case of blatant plagiarism (compare to Simmons 1907), thus quotes or citations made to Dykstra may indeed be of other provenance.

[3]Dowling (1973, pp. 154–5) notes that some progress had been made on advertising, but the reform was minor in comparison to the changes that would be brought about after 1905.

[4]While pharmacists and doctors both spoke out against secret formulae and adulterated products, pharmacists and doctors didn't walk in lockstep. For instance, pharmacists found doctors campaign against patent medicines and self-medication to be selfishly designed to force everyone to see a doctor for any complaint (Dykstra 1955, p. 405).

[5]The proposal was discussed and the then rejected (see p. 1579, 1640, 1644 of the same issue).

the hierarchical structure would allow the council to take actions that a democratic organization may have disapproved of and more decentralized organization would have lacked the power to enforce.

Historian's Appendix B: A Brief History of Randomization, Blinding, and Control

The published article of the Salicylate trial gives neither an indication of the rationale undergirding the experimental design nor the intellectual background for their methodological decisions. Nevertheless, it worth briefly summarizing and locating the Salicylate trial in terms of what preceded it. The interested reader may seek more sustained treatments on the topic in the articles cited below.

Randomization: Many philosophers will no doubt be familiar with Ian Hacking's (1988) work on the origins of randomization in experimental methodology. In this work, Hacking traces back many of the methodological issues surrounding randomized experiments to the process of establishing a reliable means of assessing psychic ability (or the lack thereof) in the nineteenth century. Hacking argues that though Fisher's 1935 treatment of randomization was far more sophisticated in terms of its philosophical grounding than anything that preceded it, Fisher was aware of (and drew from) previous experiments testing psychic ability.

Notable examples include C.S. Peirce's 1885 test of Fechner's theory of sensation. In Peirce's experiment a deck of cards was used to determine whether the subject would experience one condition or the other. Similarly, in an experiment conducted by Charles Richet 1884, a deck of cards was used to supply a random image to be sent from one person to another as a test of telepathy. Yet for Hacking, the culmination of the methodology is Stanford professor, John Coover's use of randomization in 1912. Coover used the roll of a die (randomization) to determine whether the card would be viewed before or after the recipient made a guess (control), and the recipient was unaware if the person assigned to telepathically transmit the card to them was in fact doing so (blinding).

Hacking's account underrates the history of medicine in the history of randomization. He claims that while medicine might seem like a plausible place to prospect, "seldom is there much statistical know-how and never, I think, any assignment of patients at random" (p. 431). As Chalmers et al. (2012) have detailed, at least in terms of randomization, Jan Baptist van Helmont proposed drawing lots to test treatments in the 1600s and Benjamin Franklin, Antoine Lavoisier, and others used randomization to test Anton Mesmer's "animal magnetism" in the 1700s. In the 1800s trials were conducted to test homeopathy in which treatments and controls were randomized and administered without the patient knowing which they were taking (Podolsky et al. 2016). A die, specifically, was used in 1902 to allocate treatments in assessing the efficacy of a serum treatment for the plague in India; moreover, Nasserwanji Choksy's extension, published in 1908, included both randomization and statistical analysis.

The Salicylate trial is practically contemporaneous with Coover's work. Although Hewlett fails to discuss the logic behind his method of assigning treatments, the Salicylate trial is a particularly sophisticated and balanced administration regime and clearly goes beyond alternating treatments. Nevertheless, it remains unclear if Hewlett or others on the committee intentionally randomized the treatments or whether randomization was the *de facto* result of their design.

Blinding: Keeping the subject ignorant of whether they are experiencing the genuine article dates at least as far back as the sixteenth century and the administration of sham holy water during exorcisms and most notably with the trails debunking Mesmerism in late 1700s (Podolsky et al. 2016). In the nineteenth century, homeopaths were often at the forefront of using placebos to test the efficacy of treatments and towards the end of the century, skepticism over newly introduced commercial products led to the administration of sham treatments by mainstream researchers (Kaptchuk 1998). Placebos were even used as a methodological control by a pharmaceutical company (Parke-Davis) in the 1920s (Gabriel 2014b).

The need to keep doctors ignorant of the treatment assignment is the second component of the double-blind. A recognition of the need to do so arose in the second half of the nineteenth century as doctors became aware of the need to mitigate "the personal equation" of the researcher. As discussed by Podolsky et al. (2016), Hewlett's trial was at the forefront of blinding as a means of preventing the researcher's bias from influencing observations.

Control: The council distinguished between "haphazard clinical trials" in which a doctor formed an opinion on the basis of experience with a drug, and trials like those conducted by the Committee on Therapeutic Research, which were "carefully controlled." Though the Salicylate trial can be seen as controlled by modern standards, the extension of the term included any case where investigators made an effort to rule out confounders (Marks 1997). By this broader standard, controls reach back millennia, for example, to Persian physician Al-Razi (Rhazes) who evaluated blood-letting by applying it to some and withholding it from others.

Alternative allocation began to be used in the nineteenth century as a means of providing a control group. At least as early as 1816, Alexander Lesasier Hamilton used alternative allocation to test bloodletting and the practice became increasingly common towards the end of the century (Chalmers et al. 2012). Chalmers et al., argue that the problem with alternative allocation lay in doctors' knowledge of what treatment a patient might receive, and thus the ability to influence treatment assignment. This ultimately precipitated the increasing disfavor of alternative allocation as a means of randomizing (e.g. most notably in Bradford Hill's famous streptomycin trial in the late 1940s).

It is worth noting that though in his discussion of the blind test, Sollmann (1917) spoke of alternative allocation, in practice the treatments were randomized rather than allocated alternatively. Chalmers et al. (2012) provide other cases during the early twentieth century where these two terms were used interchangeably, suggesting that the distinction between randomization and alternative allocation may not have been clearly demarcated. Whether Sollmann and/or the CPC drew a clear distinction between the two is an open question.

References

17 Stat. 1872, 322–323.

Adams, S. H. (1907). *The great American fraud*. Chicago: P.F. Collier and Son.

American School of Magnetic Healing and J. H. Kelly vs. J.M. McAnnulty, 187 U.S. 94.

Barker, L. F. (1900, July–August). On the present status of therapy and its future. *Bulletin of the Johns Hopkins Hospital*.

Bastedo, W. A. (1915). Clinical experience with liquid parafin (Liquid Petrolatum): A comparative investigation made under the auspices of the council on pharmacy and chemistry. *JAMA, 64*, 808–809.

Bigelow, J. (1858). *Brief expositions of rational medicine: To which is prefixed the paradise of doctors, a fable*. Phillips, Sampson and company.

Boyle, E. W. (2007). *The boundaries of medicine: Redefining therapeutic orthodoxy in an age of reform*. Santa Barbara: ProQuest.

Boyle, E. W. (2013). *Quack medicine: A history of combating health fraud in twentieth-century America*. Santa Barbara: ABC-CLIO.

Burrow, J. (1963). *AMA: Voice of American medicine*. Baltimore: Johns Hopkins Press.

Burrow, J. G. (1970). The prescription-drug policies of the American Medical Association in the progressive era. In *Safeguarding the public: Historical aspects of medicinal drug control*. Baltimore: Johns Hopkins University Press.

Carpenter, D. (2010). *Reputation and power: Organizational image and pharmaceutical regulation at the FDA*. Princeton: Princeton University Press.

Chalmers, I., Dukan, E., Podolsky, S., & Davey Smith, G. (2012). The advent of fair treatment allocation schedules in clinical trials during the 19th and early 20th centuries. *Journal of the Royal Society of Medicine, 105*(5), 221–227.

Council on Pharmacy and Chemistry (CPC). (1905). Acetanilid mixtures. *JAMA, 44*, 1791.

Council on Pharmacy and Chemistry (CPC). (1906). Vin Mariani. *JAMA, 47*, 1751–1753.

Council on Pharmacy and Chemistry (CPC). (1907). Kutnow's powder: Which is it, a "proprietary" or a "patent medicine"? *JAMA, 49*, 787–788.

Council on Pharmacy and Chemistry (CPC). (1910). Cactus Grandiflorus. *JAMA, 54*, 889.

Council on Pharmacy and Chemistry (CPC). (1913). Sinkina. *JAMA, 61*, 1056.

Council on Pharmacy and Chemistry (CPC). (1914a). Buffalo lithia water: Contains one-fifth as much lithium as the potomac river water. *JAMA, 62*, 834–835.

Council on Pharmacy and Chemistry (CPC). (1914b). The German council on pharmacy and chemistry. *JAMA, 62*, 1271–1273.

Council on Pharmacy and Chemistry (CPC). (1915a). Budwells emulsion of Cod-LIVER oil, Nos. 1 and 2. *JAMA, 64*, 684–685.

Council on Pharmacy and Chemistry (CPC). (1915b). Bell-ans (Pa-pay-ans Bells): Another 'patent medicine' foisted on the public through the medical profession. *JAMA, 64*, 1815–1816.

Council on Pharmacy and Chemistry (CPC). (1915c). At the bar of public opinion. *JAMA, 65*, 1373.

Council on Pharmacy and Chemistry (CPC). (1917). Biniodol. *JAMA, 68*, 650.

Dowling, H. (1973). *Medicines for man: The development, regulation, and use of prescription drugs*. New York: Alfred A. Knopf.

Dykstra, D. (1955). The medical profession and patent and proprietary medicines during the 19th century. *Bulletin of the History of Medicine, 29*, 401–419.

Edmunds, C. W. (1909, February 13). Therapeutic progress. *JAMA*.

Edsall, D. (1910). The work of the council of pharmacy and chemistry. *JAMA, 55*, 1701–1705.

Eggleston, C. (1912). The relative view of the "natural" and the synthetic salicylate: A study of the literature. *JAMA, 2057*.

Gabriel, J. M. (2009). A thing patented is a thing divulged: Francis E. Stewart, George S. Davis, and the legitimization of intellectual property rights in pharmaceutical manufacturing, 1879–1911. *Journal of the History of Medicine and Allied Sciences, 64*(2), 135–172.

Gabriel, J. M. (2014a). *Medical Monopolies*. Chicago: The University of Chicago Press.
Gabriel, J. M. (2014b). The testing of sanocrysin: Science, profit, and innovation in clinical trial design, 1926–1931. *Journal of the History of Medicine and Allied Sciences, 69*, 604–632.
Hacking, I. (1988). Origin of randomization in experimental design. *Isis, 79*, 427–451.
Hart, F. (1956). The postal fraud statutes: Their use and abuse. *Food Drug and Cosmetics Law Journal, 11*, 245–261.
Hatcher, R. (1916). The duty of the medical profession to the council on pharmacy and chemistry. *JAMA, 67*, 1339–1342.
Hewlett, A. W. (1913). Clinical effects of 'natural' and 'synthetic' sodium salicylate. *JAMA, 61*, 319–320.
Hilpert, W. (1913). The purity of commercial sodium salicylate. *JAMA*, 1137–1139.
Hilts, P. (2003). *Protecting America's health: The FDA, business, and one-hundred years of regulation*. New York: Alphred A. Knopf.
Holmes, O. W. (1892). *Medical essays 1842–1882* (Vol. 9). Mifflin: Houghton.
Ioannidis, J. P. (2016). Evidence-based medicine has been hijacked: A report to David Sackett. *Journal of clinical epidemiology, 73*, 82–86.
JAMA. (1900). Secret nostrums and the journal. *JAMA, 34*, 1420.
JAMA. (1903). National bureau of medicines and food. *JAMA, 40*, 1736–1738.
JAMA. (1904). Journal clearing house commission. *JAMA, 42*, 1579–1580.
JAMA. (1907). House of delegates. *JAMA, 48*, 1957–1972.
Kaptchuk, T. (1998). Intentional ignorance: A history of blind assessment and placebo controls in medicine. *Bulletin of the History of Medicine, 72*, 389–433.
Marks, H. (1997). *The progress of experiment: Science and therapeutic reform in the United States 1900–1990*. Cambridge: Cambridge University Press.
Meltzer. (1909). The science of clinical medicine: What it ought to be and the men to uphold it. *JAMA, 53*, 508–512.
Parascandola, J. (1992). *The development of American pharmacology: John J. Able and the shaping of a discipline*. Baltimore: The Johns Hopkins Press.
Podolsky, S. H., Jones, D. S., & Kaptchuk, T. J. (2016). From trials to trials: Blinding, medicine, and honest adjudication. In *Blinding as a solution to bias: Strengthening biomedical science, forensic science, and law* (pp. 45–58). *Amsterdam: Elsevier*.
Puckner, W., & Hilpert, W. (1908). Zyme-oid. *JAMA, 50*, 1706.
Puckner, W., & Warren, L. (1910). Noitol and Anadol. *JAMA, 54*, 1704–1705.
Rothstein, W. (1987). *American medical schools and the practice of medicine: A history*. New York: Oxford University Press.
Simmons, G. (1907). The commercial domination of therapeutics and the movement for reform. *JAMA, 48*, 1645.
Smith, A. (1947). The council of pharmacy and chemistry and the chemical laboratory. In M. Fishbein (Ed.), *A history of the American medical association 1847 to 1947* (pp. 885–886). Philadelphia: WB Saunders.
Sollmann, T. (1908). *The broader aims of the council of pharmacy of the American medical association*. Chicago: AMA.
Sollmann, T. (1916). The therapeutic research committee of the council of pharmacy and chemistry of the American medical association. *JAMA, 67*, 1439–1442.
Sollmann, T. (1917). The crucial test of therapeutic evidence. *JAMA, 69*, 199.
Sollmann, T. (1930). The evaluation of therapeutic remedies in the hospital. *JAMA, 94*, 1279–1280.
Sollmann, T. (1961). Why an annual review of pharmacology? *Annual Review of Pharmacology, 1*(1), 1–7.
Starr, P. (1982). *The social transformation of American Medicine: The rise of a sovereign profession and the making of a vast industry*. New York: Basic Books.
Swann, J. (1988). *Academic Scientists and the Pharmaceutical Industry: Cooperative Research in the Twenty-first Century*. Baltimore: The Johns Hopkins University Press.
Temkin, O. (1964). Historical aspects of drug therapy. In P. Talay (Ed.), *Drugs in our society* (pp. 3–16). Baltimore: Johns Hopkins Press.

The Medical Brief. (1910). The blind infatuation of the A.M.A. leaders. *The Medical Brief, 38*, 565–567.

The Medical Standard. (1911). A Critical Laboratory. *The Medical Standard, 34*, 42.

Waddell, J. A. (1911). A comparative investigation of the effects and toxicity of Sodium Salicylates of natural and synthetic origin. *Archives of Internal Medicine, VIII*(6), 784–805.

Warner, J. H. (1978). 'The nature trusting Hersey': American physicians and the concept of the healing power of nature. *Perspectives in American History, 11*(1977–1978), 291–324.

Warner, J. H. (1991). Ideals of Science and their discontents in Late Nineteenth-Century American Medicine. *Isis, 82*, 454–478.

Warner, J. H. (2003). *Against the spirit of system: The French impulse in nineteenth-century American medicine*. JHU Press.

Young, J. H. (1967). *The medical Messiahs: A social history of health Quackery in 20th century America*. Princeton: Princeton University Press.

Chapter 18
On the Normative Foundations of Pharmaceutical Regulation

David Teira

Abstract I argue that behind the 1962 Food and Drug Administration Act we find a combination of two normative principles: a liberal argument for the protection of pharmaceutical markets (in terms of quality control) and a paternalist argument for the protection of pharmaceutical consumers (in terms of drug safety and efficacy). These normative intuitions go hand in hand with the choice of regulatory testing standards: depending on the values the regulator wants to protect, she will avail herself of different testing methods. I explore two potential justifications for regulatory paternalism, in terms of risk aversion and impartiality. I defend our current regulatory arrangement against socialist and libertarian critiques.

18.1 Why Pharmaceutical Regulation?

When philosophers of science started discussing randomized clinical trials (RCTs), now almost 20 years ago, they picked up a thread originally put forward by John Worrall (2002): wrong methodological ideals have undesirable ethical consequences. Adopting RCTs as the gold standard for testing new medical treatments was, for Worrall, epistemically unjustified (Worrall 2007): there were alternative ways to assess causality that were just as reliable or more. And, moreover, it was morally objectionable to carry out RCTs when other testing methods would provide patients a better chance of receiving the treatment they needed earlier. Here I want to ask the opposite question: what were political values behind the choice of RCTs as a testing standard? And how the methodology of RCTs serves these values? Unlike Worrall and other philosophers, I am not going to discuss the priority of RCTs in evidence-based medicine, but rather the choice of RCTs as our regulatory yardstick for drug safety and efficacy–see, in this same spirit, (Landes et al. 2017).

D. Teira (✉)
Universidad Nacional de Educación, Madrid, Spain
e-mail: dteira@fsof.uned.es

A. LaCaze, B. Osimani (eds.), *Uncertainty in Pharmacology*, Boston Studies in the Philosophy and History of Science 338,
https://doi.org/10.1007/978-3-030-29179-2_18

The legal foundations of this choice were laid out for the first time in the US, in the 1962 Kefauver-Harris amendment to the Food, Drug and Cosmetics Act (from now on, the 1962 FDA Act). A pharmaceutical company seeking the Food and Drug Administration (FDA) approval for the commercialization of a new treatment should submit to the agency "adequate and well-controlled clinical studies" for evidence of efficacy and safety.[1] The FDA developed this principle in further detail: The "Form FD-356", published in the *Federal Register* on February 14, 1963, established (Carpenter and Moore 2007, 355–356):

> An application may be incomplete or may be refused unless it includes substantial evidence consisting of adequate and well-controlled investigations, by experts qualified by scientific training and experience to evaluate the effectiveness of the drug involved, on the basis of which it could fairly and responsibly be concluded that the drug will have the effect it purports or is represented to have under the conditions of use prescribed.

The definition of a well-controlled investigation would not be clarified until 1970, when it was formally defined as two well-controlled clinical trials plus one previous or posterior confirmatory trial The standard view among historians and sociologists is that this set of regulations created the modern clinical trial industry. Evidence-based medicine would only arrive three decades later, taking stock of a mass of trials conducted mainly for regulatory purposes.

We therefore need to make some preliminary distinctions: whereas in Evidence Based Medicine (EBM) RCTs should deliver evidence for physicians prescribing treatments, in regulatory RCTs we are merely testing drugs for some predefined levels of safety and efficacy. Whereas these two goals may coincide, they are conceptually different, as EBM supporters often recall: a regulatory trial only shows that a treatment is safe and effective for a particular use, but physicians are often interested in clinical targets not covered in such trials. As we are going to see, pharmaceutical regulation constrains medical prescription, but only to a point: once a drug enters the market, physicians are free to prescribe it independently of the targets covered in the regulatory trials. EBM supporters and regulators may agree on the epistemic superiority of RCTs, but regulators claim this superiority for just one decision (market entrance), whereas EBM supporters ideally want RCT evidence to ground every prescription.

---

[1] I should warn the reader that I do not presuppose an a priori definition of *safety* and *efficacy*: I rather track the public understanding of both concepts in the United States, namely through regulatory decisions. In the policy-making process, medical reformers operationally defined these concepts in terms of the test a treatment should pass in order to be considered safe and effective. As we will see below, prior to 1962, safety was linked to success in a number of laboratory tests; from 1962 onwards, safety and efficacy meant success of a treatment in a randomized clinical trial. Of course, we can adopt tighter definitions of both concepts that would question the adequacy of the testing standard: e.g., randomized clinical trials are notoriously underpowered to detect uncommon side effects, which might be sometimes fatal. For the sake of my analysis, vague as they might seem, it is sufficient to with the operational definitions developed by the medical reformers who crafted American pharmaceutical regulation.

Secondly, philosophers of science have cared about ethical conflicts caused by the trial design within the realm of research ethics: how shall we protect the participants in a trial? Regulators care instead about protecting patients outside trials: namely, the future consumers of experimental drugs. There are potential conflicts (and trade-offs) between these goals, but I will not cover these here (Levine 1986). My first goal here is making explicit the values presiding over current pharmaceutical regulation, taking the United States (US) as a case study. I am going to argue that the 1962 FDA Act combines two normative principles: a liberal case for the regulatory protection of pharmaceutical markets and a paternalist appraisal of patient safety.[2] The liberal view had been already implemented in the two major regulatory landmarks in US legislation, the FDA Acts of 1906 and 1938. The paternalist view was incorporated in 1962 and has been gradually renegotiated ever since (up to the twenty-first Century Cures Act[3]). *Of course, these two principles are just idealizations*: in the actual policy-making process, they were combined with each other in various degrees and mixed with many other intuitions. Yet, I think they provide the strongest normative defense of our current regulatory system and allow us to understand the issues at stake in pharmaceutical regulation.

In a nutshell, the liberal argument appeals to the necessity of quality control regulation in order to avoid fraud and litigation. If we want markets to drive the drug development process, a regulatory agency should perform the role of quality safeguard. The paternalist argument goes a step further and presumes that patients should be prevented from accessing some potentially dangerous compounds for their own safety. I am going to discuss the justification of this paternalist stance, distinguishing two different foundations: *risk aversion* and *bias*. The *risk aversion* argument maintains that consumers want a regulatory agency to prevent them from accessing treatments with serious adverse effects for fear of making bad decisions;[4] the *bias* argument, that patients are risk averse (as assumed in the first argument) and, in addition, that they cannot trust physicians to provide an unbiased estimate of the risks of new treatments, given the pervasive conflicts of interest in pharmaceutical markets. Hence, patients defer to an impartial regulator to constrain the prescribing options of physicians. I am going to contend that the risk aversion

---

[2]Throughout the paper, I will use the following terms as follows. *Pharmaceutical liberalism* will defend the superiority of markets to organize the production and distribution of drugs. *Pharmaceutical socialists* will be those who defend instead the superiority of a State-owned pharmaceutical industry. *Pharmaceutical paternalists* will constrain the freedom of patients to consume drugs, whereas *pharmaceutical libertarians* will oppose any constriction in this regard. Pharmaceutical libertarians will often rely on *liberal* (pro-market) arguments.

[3]Signed by President Barack Obama on December 13, 2016, the twenty-first Century Cures Act aims at increasing and speeding the production of new medical treatments. Its Title II Subtitle D ("Modern Trial Design and Evidence Development") invites the FDA to issue guidance that addresses using alternative statistical methods in clinical trials and in the development and review of drugs. The current normative consensus on pharmaceutical regulation discussed in this paper will surely change as a consequence.

[4]What counts as a *serious* adverse effect historically depends on a number of contextual circumstances that I will not discuss here.

argument is difficult to sustain *per se*, but that paternalism is still defensible on the grounds of regulatory impartiality.

My second claim is that the regulatory testing methods depend on the values the regulator wants to protect. RCTs are a regulatory yardstick in plain accordance with paternalism: as we will see, it is probably the best tool for conducting *pre-market* tests for pharmaceutical consumers who do not want to bear the risks of experimental treatment. RCTs allow the regulator to test these treatments on a (proportionally, small) sample of patients, before authorizing its use in the general population. Other methods, like observational studies may require a larger number of patients to reach the same conclusions. Under the assumption that pharmaceutical consumers are not risk averse, and leaving other methodological concerns aside, a less paternalist regulator may use these latter methods for either pre-market or post-market testing.

Finally, I am going to defend our current blend of liberal and paternalist views against socialist and a libertarian critiques. The socialist view is that paternalist regulators cannot resist the pressure of markets. The libertarian view is that markets can provide better information about drugs than any paternalist regulator. Both objections show that the choice of our drug testing standards depends not only on the values we are pursuing, but on the social world in which we want those testing standards to work. I will discuss, on the one hand, to what extent pharmaceutical markets are dispensable; and, on the other hand, whether markets can bring us, on its own, a consensual standard on drug safety and efficacy.

18.2 Liberal Arguments for Regulation

Let me first present what I take to be *the liberal case for regulation*. During the last 150 years, there have been different regulatory regimes for pharmaceutical markets all over the world. From the standpoint of economics, we can explain the emergence of pharmaceutical regulation from the standpoint of the interests of both drug consumers and producers. Drug tests can help in protecting pharmaceutical markets in, at least, two ways. On the one hand, they can check for quality and prevent a particular type of market failure (*adverse selection*). On the other hand, they can check for safety, reducing litigation costs for companies in case of unwanted secondary effects. In both ways, regulatory intervention would foster innovative companies in finding better drugs for the greater benefit of the consumers. Let us spell this out how in more detail, with some historical illustrations.

Regulation solves a problem pharmaceutical consumers will inevitably face in the market: by their own means they cannot ascertain the quality of a drug, either by simple inspection of their appearance (shape, size, smell…) or by their price. Depending on the circumstances of the patient, the natural rate of variability of their effects (positive or negative) prevents a reliable assessment on the basis of individual experience alone. When buyers or sellers cannot directly determine the quality of a good or service, *adverse selection* (Wilson 2008) can lead to the elimination of all trade in a market: putting it very simply, if consumers are misled by the existence of cheap (but bad quality) drugs, they will be reluctant to pay the more expensive price requested for good quality compounds and the producers of these latter may end up leaving the market. Regulatory agencies may be potentially justified for the sake of economic efficiency, as a remedy for this type of market failure; they certify a given level of pharmaceutical quality.

We could see adverse selection at work in the US market for "proprietary medicines" at the turn of the twentieth century. Pharmaceutical manufacturers sold compounds with undisclosed ingredients and big therapeutic claims, even if most of them were later shown not "to have significant marginal clinical curative power" (Carpenter and Sin 2007, 156).[5] Given the potential for fraud, consumers and reformers within the US administration were campaigning for an official assessment of the quality of food and, to a lesser extent, drugs.[6] In 1902 the Biologics Control Act, enforced by the Public Health Service of the Treasury Department, regulated the sale of "serums, viruses, toxins and analogous products". It set a minimal degree on control on *biologics*: laboratory experiments on the properties of each compound were the official method employed to verify whether they had any "therapeutic or prophylactic value". The guiding principle was that "evidence of usefulness satisfactory to an impartial investigator must be secured" before recommending a license (McCOY 1920, 1553). It was easier though to establish lack of utility in a laboratory test (for lack of active principles) than actual effects in clinical trials, for which there was no consensual design. In other words, the 1902 Act could only prevent adverse selection when it was possible to establish that a compound had no substance with known therapeutic properties.

This approach was strengthened 4 years later, in 1906, when the Pure Food and Drugs Act was passed in Congress. The 1906 Act included provisos about adulterated and misbranded drugs[7]: the Bureau of Chemistry (in the Department of Agriculture) was authorized to seize drugs from the market if the claims featuring in the label (and only there) were untrue. However, the courts would only consider it a fraud if there was evidence that the manufacturer knew these claims to be false. And the testimony of the Bureau's experts about the means to ascertain it was not considered compelling enough to prove whether a manufacturer lied or was simply biased in favor of his product (Marks 1997, 74–75).

From a legal standpoint, the asymmetry of information about drugs between consumers and producers[8] had already been assumed in American tort law during

---

[5]The fight against secret remedies in twentieth century America is explored in (Young 1992). The reference work on the evolution of the FDA is (Carpenter 2010).

[6]See (Okun 1986) about the forerunners of the reform and (Anderson 1958) for an analysis of the campaigns of Harvey Washington Wiley, Chief Chemist in the United States Department of Agriculture and the leading campaigner for the 1906 act.

[7]The debate on the 1906 Act is analyzed in (Young 1989). See also (Marks 1997, 73–74).

[8]Adverse selection presupposes an asymmetry of information: variations in the quality of individual goods can be observed by only one side of the market. In principle, the manufacturer knows better the ingredients of a compound and their properties than any buyer.

the nineteenth century, with a view to protecting consumers from defective pharmaceutical products. Unlike other manufactured goods, purchasers of drugs were entitled to sue the producer in case of tort, instead of the immediate seller. The assumption was that neither the retailer nor the consumer could know the risks or benefits of a drug, but the manufacturer should anticipate them. At the same time, strict liability in tort was imposed on manufacturers: an injured plaintiff did not need to prove intent, negligence or fault, as long as it could be proven that the drug caused the injury. American courts have traditionally favored the customary medical use of drugs in cases of tort claims. Hence, for pharmaceutical manufacturers innovation was potentially costly. Having an independent third party assessment of the safety of drugs, could protect the industry from litigation and foster drug development.

This is a potential (*post hoc*) interpretation of the incorporation of safety tests in the 1938 Federal Food, Drug, and Cosmetic Act (Vardon 2003).[9] Drug producers were now required to demonstrate that their compounds were "safe for use under the conditions prescribed, recommended or suggested in the proposed labelling thereof", before it was commercially distributed. The FDA created a system of expert consultants to decide on the therapeutic merit of the submitted drugs, under the utilitarian principle that a drug was safe if its "proposed use would benefit patients more than it harmed them" (Marks 1997, 72). Again, the applications were screened through laboratory analysis and animal experimentation. If there were still doubts about the toxicity of the drug, clinical evidence was obtained from well controlled trials conducted by a group of trusted experts. Despite an increased emphasis on the need for a sufficient number of cases, if the evidence was not conclusive the last word was for the most experienced expert in each case. This is how the principle of pre-market review established in the 1938 Act was implemented. In order to avoid liability claims, the industry agreed with the FDA that it would not be compelled to provide directions for consumers' use in the labels of its products.

Hence, the first normative pillar of pharmaceutical regulation is based on a conditional: if we want pharmaceutical markets to drive the process of drug development and commercialization, there are economic arguments for regulatory quality controls or safety checks. The antecedent of the conditional is what makes this argument for regulation *liberal*: as we will see below, not everybody would want markets to guide pharmaceutical research. But, to the extent that markets are actually playing that role, there are arguments for having it regulated for its own protection.[10]

---

[9]The passage of the 1938 Act is discussed in (Jackson 1970) and (Carpenter 2010, 73–117).

[10]Julian Reiss (personal communication) objects that my articulation of the liberal argument presupposes that regulation protects pharmaceutical markets better against adverse selection than more market-oriented solutions, such as expert advice on drugs or privately sponsored assessments. As a matter of principle, we cannot rule out these solutions, but it is worth considering why in our past history none of them has ever worked, despite their apparent simplicity. For instance, it is worth considering why the American Medical Association (AMA) dismantled its pioneering Seal of Acceptance program precisely at a time (the 1950s) in which companies were bringing to the

18.3 Regulatory Paternalism and Risk Aversion

Let me now present the paternalist argument for pharmaceutical regulation: regulatory paternalism occurs when the treatments available for physicians to prescribe are constrained not just in terms of quality, as in the previous argument, but also in terms of safety and efficacy. Physicians (and patients) are denied access to certain treatments because they are not considered safe or effective enough, independently of the stakeholders' views. This form of regulatory paternalism, I will argue, underlies the adoption of RCTs as the testing in the 1962 FDA Act. But, as we are going to see, regulatory paternalism was never explicitly justified: it was tacitly grounded on the collective risk aversion to adverse effects triggered by the Thalidomide tragedy. However, on its own, risk aversion is not a solid foundation for regulatory paternalism: we are going to see as well how it has been gradually eroded over the last four decades.

According to (Marks 1997, 155), the road to the 1962 FDA Act started in the 1950s. A minority of therapeutic reformers fought a "shadow war" marshaling arguments to persuade "a largely silent medical community" of the necessity of adopting RCTs to evaluate the therapeutic merit of drugs. The 1950s saw a boom in industrial drug production (some were *wonder drugs*, e.g. antibiotics, but many were just combinations of already available compounds) and, simultaneously, in pharmaceutical advertising that caused much confusion among practitioners about the therapeutic merit of each product (Marks 1997; Podolsky 2015). In this context the idea of testing drugs not just for safety but for *efficacy* arose, and it quickly reached the FDA.

No precise definition of safety or efficacy was given, but it was generally linked with a demand for more strict tests, often expressed in public by FDA officers.[11] Safety had been shown so far by all sorts of laboratory tests and trials. Now, safety had to be observed in RCTs statistically powered to detect a given clinical effects: if these latter could be achieved without serious adverse effects, the new treatment could be considered safe. It was this severe approach to testing which allowed the FDA officer in charge of the Thalidomide application, Frances Kelsey, to resist pressure from Merrell, the American manufacturer, to approve the drug

market many great drugs. According to Podolsky (2015, 31–33) the contributing causes were the combined pressure of the own AMA marketing branch, the lack of resources for assessing every drug submitted and litigation against AMA for failed assessments. Precisely the problems we have been discussing regarding the FDA. What institutional setup would allow a private assessment system to overcome these obstacles?

[11]"As of the summer of 1969, the Administration had not yet codified its interpretation of the Food, Drug and Cosmetic Act's requirement for "substantial evidence" of effectiveness. The main criteria for effectiveness had instead been elaborated informally, often in speeches by Joseph Sadusk, Frances Kelsey, and Commissioner James Goddard in the early and mid-1960s. [...]The May 1970 rules defined "well controlled studies" in flexible but rigorous terms, explicitly excluding "isolated case reports, random experience, and reports lacking the details which permit scientific evaluation." (Carpenter 2010, 354).

for marketing in the U.S. (Carpenter 2010, 217–297). For Kelsey, the evidence about safety submitted by Merrell was insufficient, methodologically defective and at odds with the literature and the expert opinion on the drug. Hence, she resisted approval, gradually increasing the burden of proof, at a point in which Thalidomide was liberally used outside the United States. What initially seemed unjustified stubbornness on the part of Kelsey came later to define the FDA standard regulatory outlook, once the reports of Thalidomide-caused phocomelia reached the US. Defenders of a more severe pharmaceutical regulation in Congress found in the Thalidomide scandal the occasion they needed to pass what would become the 1962 Kefauver-Harris amendment to the Food, Drug and Cosmetics Act. It required from the applicant "adequate and well-controlled clinical studies" for evidence of efficacy and safety. Without them, new treatments would not be allowed in the market.

The physician's decision as to how much a patient could benefit from a given drug was now externally constrained by these new testing standards: whereas in the pre-1962 conception of safety, there was room for clinical judgment in the assessment of drugs, after 1962 prescription was only possible if there was a previous RCT certifying together safety and efficacy. If the FDA considered a compound unsafe or ineffective, it would not be available for prescription, even if the physician and the patient were willing to take their risks. Regulation goes here well beyond the liberal protection of pharmaceutical markets: it adopts a *strong* paternalistic stance regarding pharmaceutical consumers.[12] Of course, medical prescription is, by default, paternalist: the patient is usually not in a position to properly decide on her own treatments. The benevolent supervision of a physician (under the Hippocratic Oath) has been traditionally considered a good enough basis for prescription. The 1962 FDA Act went far beyond and *constrained the number of therapeutic choices available to the physician in terms of efficacy, without any consent from the patient and for the sake of her own health.*

This is what I call *strong regulatory paternalism* and it has not been subjected to much philosophical discussion, because Kelsey's argument by example has been considered almost self-evident by pharmaceutical consumers. Kelsey's paternalistic caution went perhaps beyond the spirit of the 1938 Act, but she clearly avoided adverse effects that no pregnant mother would have desired for her baby. As Dan Carpenter puts it, throughout the following decades "the legacy of thalidomide was a powerful one, and every time a new drug safety issue arose, the memory of the drug and its disfigured victims was evoked" (Carpenter 2010, 300–1). So far, the

[12] As one of my reviewers observed, it would be possible to appeal to the liberal argument for introducing the 1962 FDA Act rather than (or in addition to) the paternalistic arguments. Indeed, the Act could have been justified in terms of the superior protection it offered against adverse selection. But the actual arguments of the Kefauver-Harris hinged on consumer protection understood in a paternalistic manner. The FDA could have tested the drugs and deliver the assessment to the consumers for them to make the decision about taking potentially risky compounds. But this was a route never taken.

US public has not massively complained about an excess of regulatory safety, but sometimes the opposite.[13]

The collective fear of adverse effects has tacitly sustained the regulatory paternalism of the FDA. But, as such, there is nothing intrinsically moral about this fear. As a vocal minority of libertarian critics soon claimed, not every patient wants the same levels of regulatory protection in every circumstance (Carpenter 2010, 300–345). For instance, Louis Lasagna, a John Hopkins pharmacologist who had been an early FDA advocate, became a staunch critic in the 1970s. Together with William Wardell, Lasagna argued that the testing standards of the FDA were sacrificing the welfare of today's patients for the protection of future patients (Wardell and Lasagna 1975). Unlike the tests required in the 1938 Act, RCTs are long and costly experiments that delay the introduction of new treatments in the market. The 1962 FDA Act denied access to experimental treatments to patients who may benefit from them and are willing to bear the risks. For Wardell and Lasagna this was immoral. They advocated instead a return to a weak form of medical paternalism about treatments, such that "if a respectable minority of professional opinion believes in the utility of a drug, then it ought to be available for use by those who believe in it" (Wardell and Lasagna 1975, 38).

In other words, if physicians and patients were willing to bear the risks of a new treatment, when no conclusive evidence was available about its effects, no regulatory authority should prevent them from trying it. Wardell and Lasagna supported their arguments on a libertarian appraisal of the Declaration of Helsinki on human experimentation (article 37th), revised in the same year they published their book to affirm that "concern for the interests of the subject must always prevail over the interests of science and society."

18.4 RCTS and the Justification of Regulatory Paternalism

The 1962 FDA Act combined then the liberal argument for market protection already implemented in 1906 and 1938 with a paternalist argument for patient protection. Whereas the liberal argument requires a weak concept of safety (based on quality control), the paternalist argument strengthens safety subordinating it to efficacy, as shown in RCTs.[14] I am going to contend now that there is indeed a tight link going through *methodology* between the normative principles behind

[13]The construction of the FDA regulatory reputation went far beyond Thalidomide though: see the fifth chapter of (Carpenter 2010) for a full account.

[14]The view that the degree of safety of a treatment must be viewed in light of its therapeutic value emerged at the FDA during the 1950s. So far, the FDA had acted as a liberal regulator that checked for quality and safety, but left to the prescribing physician the assessment of a treatment's therapeutic value. The emerging view wanted to conduct a centralized assessment using RCTs as a standard (Carpenter 2010, 150-ff). The justification for this was a superior degree of protection for the patient, provided by a paternalistic regulator.

pharmaceutical regulation and the drug testing standards adopted. Take *sample size*, for instance: if we assume pharmaceutical consumers to be risk averse, paternalist regulators will want testing methods that minimize the number of patients exposed to the risks of experimental treatments. If we assume instead that pharmaceutical consumers are willing to take their risks, leaving other considerations aside, less paternalist regulators can try methods that require bigger numbers of patients.

In the regulatory development of the 1962 FDA Act, the safety and efficacy of a drug was operationally defined in terms of success in a RCT. The virtue of RCTs in protecting consumers lies in their purported internal validity in grasping the causal effects on a treatment. In a clinical trial it is possible to control for many more factors than, e.g., in an observational study: in principle, this should allow trialists to reach their conclusions with fewer subjects and less biases.[15] In other words, testing a treatment on a comparatively small sample of consumers will allow the regulator to protect a much bigger population. This is precisely what a paternalist regulator wants: since the population of patients is assumed to be risk averse, RCTs minimize the number of patients who need to be exposed to experimental treatments in order to make a decision.

If the assumption is removed and a big enough number of patients are willing to bear the risks, the range of methodological options available for the regulator grows. In a large observational study, the variance of the estimator decreases. Its reliability, as compared to RCTs, will depend on how well the observational design controls for biases.[16] This is the methodological rationale of the regulatory approach advocated by Wardell and Lasagna, for whom the FDA should return to its 1938 functions and just check for drug quality. Physicians should test new drugs with patients who, in their judgment, may benefit from the drug and consent to the risks. With an adequate reporting system, the FDA could detect early serious adverse effects and withdraw compounds, if necessary.

It is crucial to bear in mind this connection between political values and testing methods, because, I contend, the changes in the latter signal implicit shifts in the former. As a matter of fact, the FDA was not completely insensitive to the anti-paternalist argument of its libertarian critics and developed a compassionate use policy, giving access to experimental treatments on a case-by-case basis. But the revolt of AIDS patients against the AZT trials in the late 1980s forced the FDA to articulate a more systematic advanced access policy (Epstein 1996). Organized around the networks created in the 1970s gay movement, Act Up gathered AIDS patients in active opposition to regulatory paternalism, demanding access to new drugs in the name of their individual rights to bear the risks (Epstein 1996, 222). As Larry Kramer, one of their leaders, put it, "AIDS sufferers, who have nothing to lose, are more than willing to be guinea pigs" (*ibid.*).

---

[15] See (Senn 2008) for a discussion in terms of the bias-variance trade-off.

[16] Philosophers of science often argue that the ability of RCTs to control for biases is overestimated (see, e.g., (Borgerson 2009), I take sides here with (Senn 2008), who claims the opposite.

In response the FDA articulated its accelerated approval program, in which new treatments were examined in shorter trials, using surrogate endpoints as predictors of clinical efficacy. For instance, treating a disease like cancer until its actual remission may take years, delaying the regulatory decision accordingly. RCTs can be designed targeting instead a biological marker that obtains earlier after treatment and nonetheless reliably predicts the treatment actual endpoint (survival or death) (Teira 2017). This methodological shift shows, in my view, a compromise between regulatory paternalism and individual rights, in which the former is relaxed *a bit*. Using a surrogate outcome instead of the actual treatment endpoint decreases the levels of efficacy for pharmaceutical consumers: RCTs in which we observe a prediction of clinical efficacy will fail more often than trials in which the actual clinical efficacy is observed. Therefore, the regulator drawing on the former may go wrong more often in her gatekeeping decisions, exposing future patients to bigger risks.

18.5 Regulatory Paternalism and Impartiality

Let me take stock of the argument so far: I have argued that the 1962 FDA Act combines liberal and paternalist ingredients about drug safety and efficacy. These normative intuitions go hand in hand with the testing methods of the regulatory agency. The paternalism of the 1962 FDA Act is implicitly justified in terms of the risk aversion of pharmaceutical consumers. But I think that this justification is far from satisfactory. On the one hand, as a matter of principle, not every patient has the same level of risk aversion in every circumstance, so there is no *a priori* consensual level of regulatory protection. On the other hand, throughout the last few decades, the FDA has *de facto* accepted that there is room for different levels of consumer protection, gradually modifiying its testing methods in a way that relaxes the 1962 Act paternalism.

Why not jump ahead in this slippery slope and let every patient choose what risks to bear if in need of an experimental treatment? I am going to present now an alternative defense of regulatory paternalism, based on the following intuition: in order to take one's chances with an experimental treatment, we want impartial information about the risks we are bearing. If there is uncertainty, we want a fair presentation of our chances. Pharmaceutical markets on its own cannot reliably provide this impartial estimate, since the commercial interest at stake systematically bias the information presented. Hence, it is rational for pharmaceutical consumers to defer to a regulator for the pre-market assessment of new drugs, for lack of a better evidential source. Let me spell this out.

The paternalist and anti-paternalist arguments presented above have a common presupposition: in the own words of Wardell and Lasagna, "any system of weighing evidence will ultimately stand or fall on the honesty and wisdom of the individuals making the judgment" (Wardell and Lasagna 1975, 38). A regulator deciding on the safety and efficacy of a drug on the basis of RCTs and a physician prescribing an untested drug to a patient should do it upon their best clinical judgment. However, historians have shown that the 1962 FDA Act emerged in a context of growing mistrust in the marketplace, where clinical judgment was under a constant threat of distortion caused by pharmaceutical advertisement to physicians. The same Louis Lasagna only a decade earlier argued against the "sleazy advertising" practices of the pharmaceutical industry, trying to persuade physicians with "badly scissored quotes", "pharmaceutical numbers racket", "detail men" visits and so forth (Lasagna 1959, 460–461).[17] Physicians and patients, on their own, cannot discern clinical facts from marketing claims.

There is here is a potential justification for regulatory paternalism in terms of a Ulysses contract (Spellecy 2003): given the pervasive biases of physicians and patients in judging over treatments, and in particular their defective appraisal of the risks they entail, it is rational for them to transfer to a qualified third party (the pharmaceutical regulator) the power to test every new treatment before they can have access to it. Furthermore, if patients cannot trust physicians to prescribe new treatments upon this evidence, it is rational for them to have an impartial regulator to constrain the prescribing options of physicians according to a pre-established level of safety.[18]

Whereas the previous argument was about the willingness of patients to bear the risks of new treatments, this second case is about their chances to obtain unbiased information about those risks from the prescribing physician. And, indeed, behind the 1962 Act we find a growing concern about the partiality of physicians, exposed as they were to all sorts of misleading evidence originating in the marketing departments of the industry.[19] In this regard, the adoption of RCTs as a regulatory yardstick was often defended in terms of the warrants of impartiality they incorporated: methodological devices such as randomization, blinding, and a pre-established interpretation rule (*p*-values) prevented the stakeholders' interests from interfering on the trial result (Teira 2013a). Assuming this protection was effective, RCTs would provide unbiased evidence regarding the safety and efficacy of new treatments, at least as compared with the uncontrolled clinical judgment

[17]For an analysis of how the fight against pharmaceutical marketing drove the adoption of RCTs in the United States see, most recently, (Podolsky 2015).

[18]Julian Reiss rightly objects again that a regulator may not be the only alternative and that less interventionist alternatives are conceivable. However, we have had experience of some of these approaches under our current regulatory regime and their degree of success is sobering: see, for instance, (Jain 2007).

[19]"To get an approximate idea of the current situation in all specialties, the author reviewed the summaries of original investigations which appeared in the *1959–60 Year Book of Drug Therapy*. Out of 394 summaries which gave adequate information about the plan of the study, 225 (57%) related to reports without any explicit comparison with the results of another treatment in similar patients. Every section of the *Year Book* contained reports which ignored the hard fact of unpredictability and simply described the course followed by patients while they were being given a drug. All of this evidence suggests that a great part of the so-called clinical evaluations of new drugs is unscientific, lacks adequate provisions to eliminate bias, and cannot be objectively judged" (Sheps 1961, 651).

of physicians exposed to pharmaceutical advertising. Strong regulatory paternalism would be just the constriction of the prescribing choices of physicians in terms of an impartial assessment of risks.

For this defense of paternalism to hold, we need the pharmaceutical regulator to be actually impartial. After all, if there was no such impartiality, why deferring on a paternalist agency that is no less biased than your physician of choice? Despite the frequent critics against the FDA for being accommodating with the demands of the industry, I concur with Dan Carpenter that so far it has reasonably accomplished its regulatory mission with reasonable fairness.[20] However, no methodology or deliberative procedure provides a definite warrant of impartiality. Bennett Holman aptly describes this process as an ongoing *arms race* between the industry and the regulator in which commercial interests try to by-pass every debiasing method implemented (Holman 2015). Ultimately, any regulator can be *captured*. Regulatory capture occurs when, given a definition of the public interest that the agency should protect (e.g. guaranteeing that only safe and effective drugs get access to markets), there is evidence of the industry acting on the agency and evidence about regulatory actions shifting away from the public interest toward industry interest (Carpenter 2014).

Capture is a threat for *both* the liberal and the paternalist arguments for pharmaceutical regulation. Capture can happen both to institutions checking for quality (liberal safety) or for efficacy (paternalist safety). There are two opposite answers to this threat. On the one hand, there is the socialist approach: if ultimately no regulator can stand the pressure of pharmaceutical markets, let us make these latter shrink as much as we can. On the other hand, there is the libertarian approach: since no regulator is better at estimating risks than the prescribing physician, let markets handle drug assessment. I will examine each of these two arguments in the following sections in order to support further my defense of the *status quo*. I will contend first, against the socialist objection, that we need the levels of innovation brought about by pharmaceutical markets. If we foster competition in these latter, not only there will be more innovation, but the risk of capture will also diminish. Therefore, there is still a chance for an impartial regulatory agency such as the FDA. I will then argue, against the libertarian, that on their own, markets are unlikely to yield the impartial assessment of new drugs that pharmaceutical consumers, as of today, seem to require.

[20]In my view –and I follow Carpenter, once more, here–, both regulatory RCTs and the FDA itself have been reasonably effective so far at fulfilling their main mission: keeping out of the market unsafe compounds. The main evidence for this claim is the number of treatments withdrawn from the market for serious adverse effects (Carpenter et al. 2008). That is, the sort of Thalidomide scenarios that the 1962 FDA Act was aimed at preventing. In this regard, if either the evidence or the committees had been biased, the harm has not big enough to threaten the regulatory authority of the FDA.

18.6 A Socialist View of Pharmaceutical Regulation

According to the most critical voices about the pharmaceutical industry, pharmaceutical capitalism, as we know it, has already captured the FDA and the only possible correction is to minimize the role of markets in the development and distribution of drugs. The Danish physician Peter Gøtzsche exemplifies this position (Gøtzsche 2013): being one of the leading experts in the meta-statistical analysis of RCT biases and a former pharmaceutical insider, Gøtzsche's arguments provide qualified evidence for the widespread view that Big Pharma are intrinsically bad.

For Gøtzsche, agencies like the FDA cannot properly fulfil their mission of testing for safety: according to his estimates, the incorrect use of drugs kills about 100,000 patients per year in the US alone, making it the fourth aggregate cause of death (after heart conditions, cancer and chronic lower respiratory diseases). For these estimates, Gøtzsche relies partly on actual death reports, but also on meta-analyses: if a treatment causes x deaths for every 100 patients treated, it is enough to multiply per the number of patients receiving the treatment in the country to make up a figure. In his view, the failure of the FDA at gatekeeping for safety is plain *corrosive* capture by the pharmaceutical industry: although the legal mandate is there, the FDA is impotent to enforce it. On the one hand, because it relies on industry sponsored trials which are designed in the best interest of the sponsor.[21] Its outcomes are presented in a completely unmanageable format in which actual effects are easy to hide. On the other hand, the FDA officers discourage scientists at the agency to pursue evidence of harm, accept fraudulent data and even react slowly to the evidence of lethal harm.

Leaving aside his estimates about the deathly effects of drugs and sticking to FDA statistics, the numbers are equally appalling: "No less than 51% of drugs have label changes because of major safety issues discovered after marketing; 20% of drugs get new black box warnings; and more than 1 in 20 are withdrawn from the market" (Gøtzsche 2013). It could actually be worse, since the FDA simply neglects the problem of polypharmacy (what happens when different treatments are administered simultaneously), and we have not enough evidence about the safety of combined treatments that have been approved separately.

In Gøtzsche's view, the ultimate cause of this sorry situation is pharmaceutical capitalism: "Pharmaceutical research is perverted by market incentives", he claims. "In 2012, the top 50 companies sold $610 billion in human prescription pharmaceuticals. I have little doubt that we could easily save 95% of this, which are annual savings of $580 billion" (Gøtzsche 2013). Again, if we take these figures at face value, pharmaceutical markets would seem almost dispensable. And this is indeed Gotzche's proposed reform:

[21] For instance, in trials with subjective outcomes, it has been shown that using an assessor who is not blind to the treatments received by patients overestimates the treatment effect. Yet, blinding is not implemented in a systematic manner: see (Hróbjartsson et al. 2012).

- State-owned drug companies and need-driven research
- Publicly conducted trials and publicly available data
- Tougher approval criteria and better labelling
- No need for drug marketing

Summing up, if capitalism is distorting the impartial judgment of regulatory agencies of safety, let us simply minimize the role of pharmaceutical markets in both drug development and commercialization. For the sake of the argument, let us assume that Gøtzsche's estimates are correct: the question is whether nationalizing the pharmaceutical industry would indeed provide us with safer and more effective drugs. Pharmaceutical regulation is, indeed, not an end in itself: it is only meaningful when the industry is capable of producing innovative drugs. In this regard, it is worth recalling an old libertarian point: the USSR, where state-owned companies developed drugs based on public needs, was never pharmaceutically innovative (Jack and Mason 1987). Moreover, we owe to capitalist incentives the development of the first generation of truly effective compounds (e.g, aspirin, anesthesia), despite the medical objections against pharmaceutical markets. According to Joseph Gabriel (2014), in the nineteenth century US, the medical profession considered "ethical" those drug manufacturers that somehow tested their products before commercialization, disclosed the composition of their remedies, abided by purity standards set by the US pharmacopeia, and cultivated their reputation for quality rather than trademarks and advertising, without any patent protection. However, the professional consensus among physicians on these "ethical" standards collapsed by the last quarter of the nineteenth century. In part this was driven by the emergence of companies with strong foundations in chemistry (rather than botanical medicine) and the new "synthetic" compounds. These medicines were clearly more effective than the traditional pharmacopeia, but were nonetheless, "unethically" patented and advertised.[22]

In other words, state-owned companies can fail to deliver innovation (as in the USSR), whereas for-profit firms, despite their "unethical" standards, can innovate on their own initiative. As we saw before, regulation, in its liberal version, aims to protect innovative pharmaceutical companies. The question is: *what is there in markets that ultimately causes the capture of the regulator*, flooding markets with *me-too* drugs and unsafe compounds. This is a question I cannot settle here, but there are good conceptual grounds to defend that the problem is not pharmaceutical capitalism. Let us borrow here an argument by Michele Boldrin and David K. Levine, two economists for whom the *monopolies* created by patents on drugs would explain both lack of pharmaceutical innovation and overinvestment in advertising. In this view, *market competition* would provide the best remedy (Boldrin and Levine 2008).

[22]In the last quarter of the nineteenth century, a majority of the medical profession thought unethical to patent drugs: not disclosing their composition and exploiting them commercially for private benefit (Gabriel 2014).

In a very simple textbook scenario (ignoring competitors and imitators), there is a crucial difference between the incentives of an inventor to innovate in a competitive and a monopolistic regime. Under a number of very simple assumptions about demand and production, it can be shown that in a competitive industry, the inventor's only incentives to innovate are the profits that any cost-reducing invention would yield. When such inventions are protected by patents, other firms cannot use it to lower their costs and compete on prices. Hence, for the monopolistic inventor there is no need to innovate further to obtain the same profit. This profit becomes a *rent*.

According to Boldrin and Levine, patents provide such *rents* to the pharmaceutical industry. Hence, the industry would lack incentives to compete on cost-cutting innovations: *it is actually cheaper to invest in advertising*. Technically this is known as *non-price competition*. For instance, me-too drugs just change the characteristics of a compound without actual innovation and invest instead in advertising the changes as if they were substantial. Only 238 out of the 1035 drugs approved by the FDA between 1989 and 2000 contained new active principles and were given priority ratings on the basis of their clinical performance (Boldrin and Levine 2008, 231).

Summing up, there are conceptual grounds to think that the dysfunctions in pharmaceutical markets leading to regulatory capture can be alleviated if we foster, rather than minimize, competition. Ultimately, this could be achieved with the abolition of pharmaceutical patents, for which Boldrin and Levine have produced powerful theoretical arguments. Without patents, pharmaceutical companies would be forced to invest in developing new treatments and conduct trials that showed actual efficacy. For competitive companies, there would be no need of capturing the regulator, since competition brings out quality drugs. The regulator would protect these companies from rogue competitors producing fake drugs (as in the liberal argument).

Of course, this stylized case is far from conclusive, but it shows, at least, that there is no conceptual contradiction between the two normative principles that guide pharmaceutical regulation. Liberal markets do not necessarily lead to the capture of regulatory agencies. As a matter of fact, Boldrin and Levine's argument about the power of markets is so powerful that it might actually pose the opposite question: if market competition alone can deliver actual innovation, why would we need a regulator at all?

18.7 A Libertarian View of Pharmaceutical Regulation

Julian Reiss has precisely made this radically libertarian case, although also on epistemic grounds: if we question the assumption that RCTs provide the gold standard for testing treatments, regulation becomes a matter of expert judgment (Reiss forthcoming). And, for Reiss, there are no good reasons to believe that the FDA's experts are any better than those of industry. Fostering market competition would provide incentives for the industry to develop treatments of credible quality.

As to the risks entailed by these treatments, the only responsibility of the industry would be to convey them properly to the consumers, so that, in case of adverse effects, tort law would help to adjudicate guilt in court, if necessary.

We might wonder though what a proper signal for credible quality is. In this regard, Reiss goes for *contextualism.* There are different methods for grasping causal effects and each one of them may yield valid conclusions depending on the circumstances. Hence, each pharmaceutical consumer should receive enough education to discern for herself about the quality of the available evidence. Regulatory agencies would be, in this view, dispensable. Yet, I think that both the liberal and the paternalist arguments presented above are still defensible, because there is something that competitive markets, on their own, are unlikely to bring about: evidence based on shared standards of proof.

For a start, no amount of statistical education for consumers is going to make up for the lack of evidence. A serious problem we currently have with industry-sponsored RCTs is that these tests do not address the clinical questions that might be most relevant for actual prescription, but rather those that either would earn the drugs regulatory approval or give them an edge on me-too competitors (Goldacre 2013). It is, perhaps, conceivable that market competition could align the interests of patients, physicians and pharmaceutical producers in such a way that one single test would be good enough for them all. But the economics of science shows that for one such consensus to exist the conflicting parties should agree first on a set of rules defining what counts as a good test (Zamora 2002). In particular, what makes the test impartial regarding the different interest at stake (Teira 2013b). Without this constitutional agreement about methods, pharmaceutical markets will remain in an epistemic "state of nature" with self-interested agents contesting each other's claims about the efficacy of drugs whenever it does not fit their particular interests.

In actual practice, can we expect market competition to bring us somewhat close to this methodological consensus? There is evidence to think that dispensing with patents, good as it may be for innovation, may not help us to reach one such agreement: as Jeremy Greene has recently shown, in the US market for generics: trials are still controversial. On the one hand, this is because it is not so simple to agree on the concept of bioequivalence between drugs with the same active principle:

> [T]wo pills with the same amount of the same therapeutic ingredient can cause different effects on the human body if they dissolve at different times in the stomach, if their active principles appear at different rates in the bloodstream, or if their binders, fillers, dyes, and shellac coatings influence their action in the human body in different ways (Greene 2014, 93–94)

On the other hand, assessing the comparative effectiveness of drugs is equally open to interpretation. Citing Greene again, we need to agree on these three core questions:

> First, how could one measure the comparative effectiveness of drugs for a given therapeutic class? Second, how could one evaluate the comparative safety of these drugs? Third, when two or more drugs were found to have similar safety and efficacy in a general population, were there any subgroups for whom that similarity did not apply? (Greene 2014, 238)

So far, we have not seen a consensual response emerge in generic markets. As a matter of fact, we see a growing number of public institutions dedicated to therapeutic comparisons (e.g, in the UK, the National Institute for Health and Clinical and Excellence) precisely in order to provide this kind of evidence as a sort of *public good*.

Without an agreement on standards of proof, there are reasons to expect that markets alone will not correct any asymmetry of information about drugs. We may expect adverse selection to kick in massively – as it happens today in pharmaceutical marketplaces where regulation is not enforced, as is the case of Nigeria (Peterson).[23] And we should not expect litigation alone to correct it. E.g., if a patient sues a company for incorrect advertising of the risks a treatment involves: on the basis of which evidence is the court to adjudicate the claim? Given the costs of running a clinical trial, in its most modest estimates, the financial asymmetry between companies and consumers will just reinforce the asymmetry of information.[24] Therefore the liberal argument for regulation would still hold: if we want markets to guide the drug development process, there should be a regulatory check for quality.

And the same logic supports the paternalist case for regulation: without a shared standard of proof, patients should blindly trust the clinical judgment of their physicians to choose treatments.[25] This sort of trust is not epistemically indefensible as such, but it presupposes no conflict of interest on the part of the prescribing physician. Can we expect pure market competition to increase or minimize these conflicts? Again, this will depend on our expectations on pharmaceutical advertisement. Can we expect market competition to make it more faithful to the actual properties of treatments (according to a particular yardstick of efficacy)? Will it contain the sort of mistrust in the advertising marketplace that brought about the FDA in the 1950s? My own guess is based on the fact that we have been so far unable to reach a consensus on what is rational to prescribe neither for patented drugs (see, e.g., the exemplary discussion of antibiotics in (Podolsky 2015)) nor for generics (Greene 2014). For patients, it may be rational to have a social convention on what counts as safe and effective for most and constrain the

[23]Kristin Peterson's etnography shows how the lack of enforcement of the pharmaceutical regulation in Nigeria (based on a British system) led to the proliferation of all sort of fake and defective drugs in West Africa. For Peterson, this should be mostly blamed on the neoliberal policies implemented in Nigeria. For our purposes, her case study illustrates instead to what extent minimally regulated markets (without checks for safety and efficacy) can generate a credible signalling system about drug quality. Of course, this particular case does not prove that such signalling systems may not emerge. It just illustrates once more that, at least to the extent of my knowledge, we have not seen these systems emerge in our recent pharmaceutical history.

[24]Even if countries like the UK there are public subsidies for hiring private forensic experts, the costs of RCTs are in a totally different order of magnitude –in the latest guideline "unusually large" forensic costs are those over £5000.

[25]On the epistemic problem of choosing expert advice when statistical evidence is involved, see (González De Prado Salas and Teira 2015).

physicians' choices. This is the sort of Ulysses contract enforced today by the FDA, a basic form of paternalism with a number of established exceptions.[26]

18.8 Concluding Remarks

Let me summarize my arguments to close. I have argued that behind the 1962 FDA Act we find a combination of two normative principles: a liberal argument for the protection of pharmaceutical markets (in terms of quality control) and a paternalist argument for the protection of pharmaceutical consumers (in terms of drug safety and efficacy). These normative intuitions go hand in hand with the choice of regulatory testing standards: depending on the values the regulator wants to protect, she will avail herself of different testing methods.

The risk aversion of pharmaceutical consumers provided the original justification of the paternalism of the 1962 FDA Act. RCTs should screen off unsafe and useless compounds in order to prevent Thalidomide-like tragedies. However, as libertarian critics soon denounced, not every patient requires the same levels of consumer protection in every circumstance. I have developed a different justification for regulatory paternalism in terms of the impartiality of the drug assessment. If a patient wants to take her chances with an experimental treatment, she needs a fair estimation of the risks involved. Against the libertarians, I have argued that markets alone are not likely to provide this impartial assessment. But we need pharmaceutical markets, nonetheless, against the socialist critique, and if we foster proper competition in these markets, we will see that regulatory agencies like the FDA will thrive for everybody's benefit. As Dan Carpenter has often argued, the FDA is quite unique as a government agency, both in terms of its mission and the reputation it has achieved (Carpenter 2010). I think the current arrangement deserves continuity since we do not have solid grounds to think that the alternatives will fare better.

Acknowledgements I am grateful to the editors, Bennett Holman and Julian Reiss for their detailed comments on a previous draft. My research was funded by the grant FFI2014-57258-P The paper was written during my stay at the Instituto de Investigaciones Filosóficas (UNAM, Mexico) with a PREI grant that I gratefully acknowledge. I thank Atocha Aliseda and various audiences at UNAM for their insightful discussion.

[26] Apart from the accelerated approval track discussed above, the FDA also fast tracks breakthrough therapies, improving on current standards of treatment, and grants priority reviews, which gives approval on an application within 6 months.

References

Anderson, O. (1958). *The health of a nation: Harvey W. Wiley and the fight for pure food*. Chicago: University of Chicago Press.

Boldrin, M., & Levine, D. K. (2008). *Against intellectual monopoly*. New York: Cambridge University Press.

Borgerson, K. (2009). Valuing evidence Bias and the evidence hierarchy of evidence-based medicine. *Perspectives in Biology and Medicine, 52*(2), 218–233.

Carpenter, D. P. (2010). *Reputation and power: Organizational image and pharmaceutical regulation at the FDA*. Princeton: Princeton University Press.

Carpenter, D. P. (2014). Corrosive capture? The dueling forces of autonomy and industry influence in FDA pharmaceutical regulation. In D. Carpenter & D. A. Moss (Eds.), *Preventing regulatory capture: Special interest influence and how to limit it* (pp. 152–171). New York: Cambridge University Press.

Carpenter, D., & Moore, C. (2007). Robust action and the strategic use of ambiguity in a bureaucratic cohort: FDA scientists and the investigational new drug regulations of 1963. In S. Skowronek & M. Glassman (Eds.), *Formative acts* (pp. 340–362). Philadelphia: University of Pennsylvania press.

Carpenter, D., & Sin, G. (2007). Policy tragedy and the emergence of economic regulation: The food, drug and cosmetic act of 1938. *Studies in American Political Development, 21*(2), 149–180.

Carpenter, D., Zucker, E. J., & Avorn, J. (2008). Drug-review deadlines and safety problems. *New England Journal of Medicine, 358*(13), 1354–1361.

Epstein, S. (1996). *Impure science. Aids and the politics of knowledge*. Berkeley/Los Angeles: University of California Press.

Gabriel, J. M. (2014). *Medical monopoly: Intellectual property rights and the origins of the modern pharmaceutical industry, synthesis*. Chicago: The University of Chicago Press.

Goldacre, B. (2013). *Bad pharma: How drug companies mislead doctors and harm patients*. (First American ed.). New York: Faber and Faber, Inc., an affiliate of Farrar, Straus and Giroux.

González De Prado Salas, J., & Teira, D. (2015). Choosing expert statistical advice: Practical costs and epistemic justification. *Episteme, 12*(1), 117–129.

Gøtzsche, P. C. (2013). *Deadly medicines and organised crime: How big pharma has corrupted healthcare*. London: Radcliffe Publishing.

Greene, J. A. (2014). *Generic: The unbranding of modern medicine*. Baltimore: Johns Hopkins University Press.

Holman, B. (2015). The fundamental antagonism: Science and commerce in medical epistemology. PhD, University California Irvine.

Hróbjartsson, A., Thomsen, A. S. S., Emanuelsson, F., Tendal, B., Hilden, J., Boutron, I., Ravaud, P., & Brorson, S. (2012). Observer bias in randomised clinical trials with binary outcomes: Systematic review of trials with both blinded and non-blinded outcome assessors. *BMJ, 344*, e111.

Jack, D. B., & Mason, N. P. (1987). The pharmaceutical industry in the U.S.S.R. *Journal of Clinical Pharmacy and Therapeutics, 12*(6), 401–407.

Jackson, C. O. (1970). *Food and drug legislation in the New Deal*. Princeton: Princeton University Press.

Jain, S. (2007). *Understanding physician-pharmaceutical industry interactions*. Cambridge/New York: Cambridge University Press.

Landes, J., Osimani, B., & Poellinger, R. (2017). Epistemology of causal inference in pharmacology. *European Journal for Philosophy of Science, 8*, 1–47. https://doi.org/10.1007/s13194-017-0169-1.

Lasagna, L. (1959). Gripesmanship: A positive approach. *Journal of Chronic Diseases, 10*, 459–468.

Levine, R. J. (1986). *Ethics and regulation of clinical research* (2nd ed.). Baltimore: Urban & Schwarzenberg.
Marks, H. M. (1997). *The progress of experiment. Science and therapeutic reform in the United States, 1900–1990*. New York: Cambridge University Press.
McCoy, G. W. (1920). Official methods of control of remedial agents for human use. *Journal of the American Medical Association, 74*(23), 1553–1555.
Okun, M. (1986). *Fair play in the marketplace: The first battle for pure food and drugs*. Dekalb: Northern Illinois University Press.
Peterson, K. (2014). *Speculative markets: Drug circuits and derivative life in Nigeria, experimental futures: Technological lives, scientific arts, anthropological voices*. Durham/London: Duke University Press.
Podolsky, S. H. (2015). *The antibiotic era: Reform, resistance, and the pursuit of a rational therapeutics*. Baltimore: Johns Hopkins University Press.
Reiss, J. (2017). Meanwhile, why not biomedical capitalism? In K. Elliott & D. Steel (Eds.), *Current controversies in science and values* (pp. 161–175). Routledge: New York.
Senn, S. (2008). Lessons from TGN1412 and TARGET: Implications for observational studies and meta-analysis. *Pharmaceutical Statistics, 7*(4), 294–301.
Sheps, M. C. (1961). The clinical value of drugs: Sources of evidence. *American Journal of Public Health & the Nation's Health, 51*(5), 647–654.
Spellecy, R. (2003). Reviving Ulysses contracts. *Kennedy Institute of Ethics Journal, 13*(4), 373–392.
Teira, D. (2013a). On the impartiality of early British clinical trials. *Studies in History and Philosophy of Biological and Biomedical Sciences, 44*(3), 412–418. https://doi.org/10.1016/j.shpsc.2013.05.003.
Teira, D. (2013b). A contractarian solution to the experimenter's regress. *Philosophy of Science, 80*(5), 709–720.
Teira, D. (2017). Testing oncological treatments in the era of personalized medicine. In G. Boniolo & M. Nathan (Eds.), *Philosophy of molecular medicine* (pp. 236–251). London: Routledge.
Vardon, P. J. (2003). The constitutional political economy of the new drug approval process: Between reason and success. Economics, George Mason University.
Wardell, W. M., & Lasagna, L. (1975). *Regulation and drug development*. Washington D. C.: American Enterprise Institute for Public Policy Research.
Wilson, C. (2008). Adverse selection. In S. N. Durlauf & L. E. Blume (Eds.), *The new Palgrave dictionary of economics*. Basingstoke: Palgrave Macmillan.
Worrall, J. (2002). What evidence in evidence-based medicine? *Philosophy of Science, 69*(3 Supplement), S316–S330.
Worrall, J. (2007). Why there's no cause to randomize. *British Journal for the Philosophy of Science, S 58*(3), 451–488.
Young, J. H. (1989). *Pure food: Securing the federal food and drugs act of 1906*. Princeton: Princeton University Press.
Young, J. H. (1992). *American health quackery: Collected essays*. Princeton: Princeton University Press.
Zamora, J. (2002). Scientific inference and the pursuit of fame: A contractarian approach. *Philosophy of Science, 69*, 300–323.

Chapter 19
After Disclosure

Miriam Solomon

Abstract Industry funding of research is the greatest known systematic threat to the objectivity of medical research. This paper clarifies the nature and scope of industry funding bias and attempts to quantify it. It reviews four kinds of remedy for industry bias suggested so far: disclosure, standards and regulation, steps towards independence for all clinical research, and case by case assessments, finding most of them helpful but not sufficient. The paper proposes two possible further interventions to reduce the effects of industry bias: qualitative and quantitative discounting of industry results.

19.1 Introduction

Industry funding of research is the greatest known systematic threat to the objectivity of medical research. This is a strong claim, based mostly on the work of medical researchers who have systematically compared the results of industry-sponsored and not industry-sponsored research (Bekelman et al. 2003, 454–465; Flacco et al. 2015, 811–820; Kjaergard and Als-Nielsen 2002, 249; Lexchin et al. 2003, 1167–1170; Lundh et al. 2017, MR000033; Schott et al. 2010, 279–285). It is also based on my reflection that even when non-traditional, more nuanced accounts of objectivity are considered (such as Solomon 2001), the epistemic effect of industry sponsorship is still negative, because it always skews the results in the same direction viz. to serve the interests of the sponsors.

My thanks to John R. Clarke for discussions about quantitative discounting. I also thank Adam LaCaze and Barbara Osimani for their helpful comments on an earlier draft of this paper.

M. Solomon (✉)
Temple University, Philadelphia, PA, USA
e-mail: msolomon@temple.edu

A. LaCaze, B. Osimani (eds.), *Uncertainty in Pharmacology*, Boston Studies in the Philosophy and History of Science 338,
https://doi.org/10.1007/978-3-030-29179-2_19

Industry now funds more than half of biomedical research in the USA; public funding (such as the NIH) supports about a third; the rest comes mostly from private not-for-profit sources (Dorsey et al. 2010, 137–143). Most industrial medical research is research into pharmaceuticals, and I will focus on this, noting that the situation with medical device research is thought to be similar (Lundh et al. 2017, MR000033).

The task of this paper is to clarify the nature and scope of industry funding bias, to attempt to quantify it, to evaluate the remedies for the bias suggested so far, and to propose two possible further interventions to reduce the effects of industry bias.

19.2 The Nature and Scope of Industry Funding Bias

Research that is funded by for-profit industry is substantially more likely to yield results that are favorable to the industry than research that is funded independently (whether with public or private non-profit money). "Industry bias," "funding bias," and "pharmaceutical company bias" are widely used terms to describe this systematic skewing of results.

Marcia Angell, the former editor of the *New England Journal of Medicine*, was one of the first to sound the alarm about pharmaceutical company bias, and her 2004 *The Truth about Drug Çompanies: how they deceive us and what to do about it* is just one of several books that have come out over the past 15 years on this topic. Others include Sheldon Krimsky's *Science in the Private Interest: Has the lure of profits corrupted biomedical research?* (2003), David Resnik's *The Price of Truth: how money affects the norms of science* (2007), Ben Goldacre's *Bad Pharma: how drug companies mislead doctors and harm patients* (2012), David Healy's *Pharmageddon* (2012), and Peter Gotzsche's *Deadly Medicines and Organized Crime* (2013). Major medical journal editors—e.g. Marcia Angell (NEJM), Richard Smith (BMJ), and Catherine De Angelis (JAMA) have been particularly prominent critics of pharmaceutical company bias. Richard Smith even titled an essay in *PLoS Medicine* "Medical journals are an extension of the marketing arm of pharmaceutical companies" (Smith 2005, e138). The tone of most of these books and essays is one of moral outrage, framed as the pursuit of money getting in the way of the pursuit of truth and, even more egregious, the health and safety of patients.

At first, the main concern was the incidence of overt research misconduct, such as fabrication of data and concealment of adverse outcomes. That is, industry bias was thought to be the result of intentional dishonesty and deception. There were some prominent cases involving misconduct, such as fabricated and misrepresented studies of rofecoxib (Vioxx), and the concealment of adverse effects for risperidone (Risperdal). But most cases of industry bias do not appear to involve such overt misconduct. Nor is it the case that, generally speaking, industry research has worse or sloppier methodology than independent research, at least as measured by our standard measures of methodological quality (such as CONSORT standards, ICMJE standards and Jadad scores). Several studies have found that industry research is of a methodological quality equal to or greater than independent research (Bekelman et al. 2003, 454–465; Kjaergard and Als-Nielsen 2002, 249; Lexchin et al. 2003, 1167–1170; Lundh et al. 2017, MR000033). The problem is not (or not primarily) "bad science" that results from sloppy methodology and/or fabricated data; even with an apparent commitment to honest and high quality research, biases still operate.

The so-called "replicability crisis"—the failure of clinical trials to produce the same (or statistically similar) results when repeated—(see, for example, Ioannidis 2005, e124) is attributed at least in part to pharmaceutical company bias. (Pharmaceutical company bias is not the only cause of the "replicability crisis," which also occurs in fields such as psychology.[1])

At least some causes of industry bias are well known. These are selective publication of positive results, multiple publication of positive results, weak control arm dosages, excessive use of placebo controls (rather than controls using the best current treatment), selective reporting of favored endpoints, and interpretative spin. Some of this bias operates through explicit decisions and strategic choices, and some through implicit mechanisms such as confirmation bias.

19.3 The Magnitude of Industry Bias

In an absolute sense, industry company bias is modest. No grand cures of formerly intractable diseases are being claimed. Rather, claims are typically run of the mill, of the kind "more effective than current treatment" or "more effective than placebo" or "just as effective with fewer side effects." *The problem is not the absolute size of the bias but the fact that the bias produces results—small benefits that just reach the level of statistical and clinical significance–that are of a comparable magnitude to the effectiveness of most of our medical interventions.* Most pharmaceuticals have modest and variable effects; this is why we need large, double-masked, randomized controlled trials to discern their effects from those of placebo and noise. (Indeed, if effects were large and consistent, large double-masked randomized controlled trials would not be possible, because the results would unmask the intervention.)

There have been a number of attempts to quantify the likelihood of pharmaceutical company bias by comparing the results of studies having industry sponsorship with studies funded independently of industry. Bekelman et al. (2003, 454–465) aggregated the results of 37 studies of financial conflicts of interest before 2002, finding an odds ratio of 3.6 (confidence interval 2.63–4.91). Lexchin et al. (2003, 1167–1170) did a systematic review of 30 studies before 2002, finding an odds ratio of 4.05 (confidence interval 2.98–5.51). Flacco et. al. (Flacco et al. 2015,

[1]This means that there will be more biases to correct for even when (or if) industry bias is taken care of. For example, publication bias and time-to-publication bias reduce and slow the publication of negative results even in independent research.

811–820) analyzed a random sample of trials indexed in 2011 and found an odds ratio of 2.8 (confidence interval 1.6–4.7). Most recently, Lund, Lexchin et al. (2017, MR000033) (an update of Lundh et al. (2012, MR000033)) did a Cochrane review of all papers published before February 2015, finding an odds ratio of 2.05 (confidence interval 1.66–2.52) for results and 2.69 (confidence interval 2.04–3.54) for conclusions. (All the studies have 95% confidence intervals.)

In this context, the odds ratio is $\frac{\text{p(success/pharma funding)}/1 - \text{p(success/pharma funding)}}{\text{p(success/no pharma funding)}/1 - \text{p(success/no pharma funding)}}$

The authors of the Cochrane study (Lundh et al. 2017, MR000033) note that they find less chance of pharmaceutical company bias than previous studies, but argue that this is unlikely to be because pharmaceutical industry bias has decreased. Instead, they believe that their study is the most rigorous. For the purposes of this paper, it is not necessary to agree with them about this interpretation of their results. (I think it is possible that pharmaceutical industry bias has decreased in response to increased oversight.) We can proceed by noting that all the measures of bias have been substantial, and previous measures of the amount of bias have been higher.

One caveat should be mentioned.[2] It is possible that trials funded by industry are, in fact, more likely to be effective than trials funded by universities or governments (so the actual odds ratio may be greater than 1). Perhaps, for example, industry is more likely to fund trials at a later stage of the development of a pharmaceutical, when it has already shown some efficacy. Lundh and Lexchin et al. (2017, MR000033) deny that this is the case based on their familiarity with the data, and I go with their assessment while noting that a formal study of this question would provide a more definitive answer.

19.4 Proposed Remedies

A number of remedies for pharmaceutical industry bias have been implemented; several have been suggested but not yet implemented. They are all worth discussing.

19.4.1 Disclosure

Disclosure was the first remedy proposed for industry bias. Journals started requiring such disclosure in the 1980s. The American Medical Association (AMA) made the following statement in 1990:

> Perhaps the best mechanism available to assuage public (and professional) doubts about the propriety of a research arrangement is full disclosure...The disclosures should be made to the medical center where the research is conducted, to organizations that are funding the

[2]I am grateful to Adam LaCaze for suggesting this.

research and to journals that publish the results of the research. (Anonymous 1990, 2790–2793)

The International Committee of Medical Journal Editors (ICMJE) (which issues recommendations that are accepted by more than 600 medical journals) has a standard form for such disclosures. The current disclosure form requires full disclosure of the sources of research funding, as well as any other financial relationships of the author or author's institution with industry. The idea behind disclosure is that the readers of a journal article will have information about conflicts of interest that they can take into account when evaluating the article.

Why does disclosure seem like an obvious remedy to address conflicts of interest? The answer to this question is in fact not obvious. If we look at the language of the AMA statement above, it seems more concerned with *ethical propriety* than with correcting bias. It makes it sound as though the main problem is *non-disclosure* of the funding, which is viewed as a deception that undermines trust in research results. But the main problem—from an epistemic point of view–is that industry funding (whether disclosed or not) biases research outcomes.

The AMA statement does not make explicit (perhaps it was not thought to be necessary) that when industry funding is disclosed, readers of a journal article can adjust their confidence in the results appropriately. The journal article with a proper disclosure does not mislead—it just leaves the work of discounting to the reader.

As a moral intervention, disclosure works to prevent deception and restore trust, at least for most readers. The reader is told what they think they most need (and ought) to know. As an epistemic intervention, however, *it is only a first step*. How much should we discount[3] industry funded studies? And do we, in fact, do the discounting? The current answer to both these questions is disappointing.

To my knowledge, there have been no discussions of exactly how much discounting there should be for industry funded studies, either individually (the results of each trial) or in meta-analyses (when the results of trials—some funded by industry and some funded independently—are combined). The work is entirely left to the reader of the journal article. For individual trials, the reader has information regarding trial funding, but for meta-analyses, information about trial funding is currently left out of the analysis and therefore inaccessible to the reader. In meta-analysis, a randomized controlled double masked trial funded by a pharmaceutical company is almost always[4] ranked as higher quality evidence than an independent high quality observational study.

While readers often claim that they discount the results of industry funded studies, recent evidence suggests that they do not discount them *at all* (never mind discounting them by the right amount, whatever that is). A 2010 study (Silverman et al. 2010, 265–270) on hypothetical willingness to prescribe a new drug based on

[3]This use of the term "discount" should not be confused with a economics/decision science of the term to describe discounting future utilities (temporal discounting).

[4]In the GRADE system of rating, a high quality observational trial can count as highest quality of evidence. But most ranking systems allow only RCTs as best evidence.

the evidence and a disclosure statement suggests that physicians do not discount at all for conflict of interest.[5]

There is also a finding that disclosure may have "perverse effects" on both the researcher/author and the reader. Loewenstein et al. (2012, 669–670) find that researchers sometimes exaggerate their findings in advance to counteract the expected discounting after disclosure, and feel morally justified in so doing. Moreover, they find that disclosure can create a "burden of disclosure" effect, in which readers feel obliged to trust what they read even more when the authors have "been honest" with them by disclosing competing interests.[6]

In summary, disclosure of competing interests is a widely implemented regulation designed to reduce industry bias. It may suffice ethically (in that it increases transparency), but it does not suffice epistemically (in that the disclosure does not reliably produce discounting by the reader). We still need to figure out how much we ought to discount industry-funded research, and we still need to implement this discounting while avoiding (or correcting for) "perverse effects" of disclosure. Disclosure is a step, not a solution.

19.4.2 Further Standards and Regulations for Research

Disclosure is not the only procedural recommendation for addressing industry bias. Over the last thirty years of experience with industry and other sources of bias, the research community has developed common standards for the proper conduct and publication of clinical trials. For example, the recommendations of the International Committee of Medical Journal Editors (ICMJE)[7] have been adopted by over 600 medical journals. They include disclosure of conflicts of interest, specific criteria for authorship, trial registration, accessibility of study data, peer review, editorial freedom, and constraints on duplicate publication, sponsorship of special issues, and relationships to advertisers. Other recommendations address methodology, for example the CONSORT Statement[8] (endorsed by many medical journals) gives standard recommendations for conduct of randomized controlled trials and the Jadad scale (Oxford quality scoring system) (Jadad et al. 1996, 1–12) is a widely used quantitative measure of trial quality. The standards include requirements for trial registration, statement of initial objectives/hypotheses, measurement of attrition rate, documentation of any changes to the conduct of the trial, and a requirement for the full trial protocol to be accessible. The *Journal of the American Medical*

[5]In this study, the physicans were practicing obstetrician-gynecologists. It is possible that more research oriented readers would discount.

[6]These perverse effects were mostly found in disclosures between physicians and patients. It is not known how much they affect the authors and readers of medical research literature.

[7]See http://www.icmje.org/icmje-recommendations.pdf, accessed September 5, 2016.

[8]See http://www.consort-statement.org/consort-2010, accessed September 5, 2016

Association (JAMA) takes these requirements even further, insisting on independent statistical analysis, with both trial protocol and raw data posted on websites.

These standards and regulations are most important. They are designed to reduce or eliminate many sources of bias, including industry bias. They have been developed over the past thirty years in response to specific, known, sources of error. Without them, we would be in a much worse position when assessing the evidence for a clinical intervention, and there would be many more ways to compromise research quality without discovery or consequences.

However, as mentioned above, our current measures of trial quality find that industry funded research is of equal or higher quality than non-industry funded research, even while industry bias is still present. The Lundh, Lexchin et al. study (2017, MR000033), for example, finds that as measured by four common standards (sequence generation, allocation concealment, followup, and selective outcome reporting), industry funded research has the same *risk* of bias (and a lower risk of bias from the fifth common standard, blinding) while having a higher *actual* bias in outcome assessments (interpreting the outcome differences between industry and non-industry funded research as due to bias). This is thought to be because industry has been mindful of the need to be perceived as producing high quality research, and quick to direct its substantial resources to comply with publicized standards.

This finding suggests that standards and regulations are not, *in practice*, strong enough—to safeguard against the industry bias that currently operates. In some cases, for example selective publication of positive results, this is because of lack of enforcement of standards of reporting of negative results. Trial registries (at least, in the USA) are often left incomplete when results are negative. Moreover, will anyone notice if there is selective reporting of favored endpoints?[9] In other cases, for example the use of weak control arm dosages that make the product look better than it should, it is because standards have not yet been devised or cannot be devised to avoid the techniques of the industry trade. Especially in those areas in which there is room for "scientific judgment," it is going to be difficult to operationalize regulation of research. These areas include choice of control dosages, choice to use placebos rather than active controls, inclusion and exclusion criteria, and interpretative spin.

If the standards and regulations do not do the job—or are not implemented well enough to do so—perhaps biasing strategies can be caught in peer review? Richard Smith (2005, e138) quite reasonably doubts that this is the case:

> . . . there are many ways to hugely increase the chance of producing favourable results, and there are many hired guns who will think up new ways and stay one jump ahead of peer reviewers

---

[9]Adam LaCaze has brought the website http://compare-trials.org to my attention. It is run by a team of researchers at the Center for Evidence-Based Medicine, Oxford University, headed by Ben Goldacre. It tracks unreported and switched outcomes. Hopefully, this will reduce selective reporting.

Bennett Holman[10] refers to the situation as an "asymmetrical arms race" in which one side is driven by truth and the other side by money. It is a situation rather like that of computer virus prevention—with the virus creators always at least one step ahead of the anti-viral software engineers. If that is the case, then some pessimism about the elimination of industry bias is justified.

But it is also reasonable to view the situation as one in which regulations and standards are already reducing industry bias, and in which improvements are still possible. The Lundh, Lexchin et. al. Cochrane review (Lundh et al., MR000033) finds considerably less bias more than ten years after previous measurements. This may be, as Lundh, Lexchin et al. say, because the Cochrane review has a more rigorous methodology than other studies and the amount of bias has not changed. Or it may (also) be because the magnitude of publication bias has decreased due to regulation. Over time, if we keep measuring industry bias at regular intervals, the trend (or the cause of the trend) may become clearer.

19.4.3 Independent Sponsorship of Clinical Research

The most commonly suggested remedy for industry bias is to propose that most or all of clinical research be carried out independently of industry funding. Usually, such proposals are accompanied by the practical recommendation that clinical research be carried out by governmental organization(s) that can be funded through a tax on the industries that develop the interventions.

This proposed remedy has been suggested from the beginning of awareness of the magnitude of industry bias to the present, for example by Marcia Angell (2005), Sheldon Krimsky (2003), Sergio Sismondo (2008, 1909–1914), and Peter Gotzsche (2013). Sometimes slightly weaker suggestions are put forward, for example Lundh et al. (2012, MR000033) call for making data from independent trials a requirement for drug and device approval, and for increasing funds for independent drug and device trials.

These suggestions address the problem of industry bias by proposing to remove it. It is a somewhat drastic suggestion in terms of regulation and institutional reorganization, but preserves an ideal that (much, most, or all of) research should be *free of* conflicts of interest. Going forward, it proposes that medical research should be both epistemically and ethically unsullied, at least by conflicts of interest.

These suggestions come up frequently, yet no steps have been taken to implement them, at least in the USA. They would take a good deal of bipartisan political will to implement, and such will is in short supply at the moment. Even if the suggestions were implemented, the revolving door between industry and government means that the same individuals will spend time in both industry and government, making it likely that their interests cannot be kept "pure."

[10]Informal communication.

As with the other suggestions for reducing industry bias, partial implementation of the idea is likely to make a difference. Increasing independent sponsorship of clinical research is valuable, even if the ideal of all clinical research being independent cannot be realized. Certainly, this suggestion can be part of the response to the problem of industry bias.

19.4.4 Case by Case Assessments

David Resnik and Kevin Elliott (2013, 184–205) have recently proposed a nuanced approach to assessing the impact of financial conflicts of interest in medical research. They recommend against discounting all research that has financial conflicts of interest simply on the grounds that such conflicts can produce biased research. Instead, they suggest considering five factors that vary from case to case: whether researchers, sponsors, or institutions have a significant financial stake in the outcome, whether the financial interests coincide with the epistemic goals, whether the researchers, sponsors, or institutions have a history of biased research, the ease of manipulation of the research to achieve financial goals, and the presence of oversight mechanisms (such as prior registration of clinical trials) that can minimize bias. For example, an industry that has a strong record of replicable research and effective interventions will not need to have its current research discounted. And an industry with a history of biased research that has not pre-registered its current research will deserve to have its research discounted.

There is a lot to be said for this proposal. It avoids the perhaps over-generalizing conclusion that all industry funded research is biased. Instead, it says that such research *may* be biased and needs to be investigated for bias, but may turn out to be unbiased. This is more just; it does not tar one industry with the sins of another. Also, it gives each industry an incentive to build a reputation for reliable research.

The downside to this proposal is that doing a five factor assessment for each piece of research is quite labor intense, and requires information that may not be easily available. It may not be feasible to do it for each study, although perhaps it is feasible to do it for the occasional study or a particular company. Resnik and Elliott also do not say who should do the assessment, and at what point—before or after publication?

The five factors considered cover known sources of industry bias. However, there may be other sources of industry bias that we do not know about, and if that is so, Resnik and Elliott's proposal will not do the necessary work. It is also likely that the five factors will need to be supplemented with additional criteria as some industries develop new strategies to maximize the chance of favorable outcomes. As with guidelines and regulations, there is the possibility of an arms race.

My main concern with Resnik and Elliot's proposal is that, in their attempt to assess each company on its own merits (or demerits), they lose sight of the fact that the *overall* impact of industry funding on research is in the direction of biasing the results towards positive conclusions. We even know the magnitude of the bias: an odds ratio somewhere in the range of 2 to 4. In the absence of a five factor assessment, it is most reasonable to treat industry research as biased—not as innocent until proven guilty. Of course, this has the downside of doing an injustice to less biased research, and perhaps removing an incentive for industries to improve the reliability of their research.

19.5 Proposed Interventions

At this time, I believe that there is a need to flag all medical research funded by industry, because the biases are so persistent. *I propose calling industry data "I-data" and industry results "I-results."* This tag should stay through any systematic review and/or meta-analysis, so that industry bias is taken into account when aggregating the data.

My first suggestion is for a qualitative discounting of I-data and I-results: they should not be included as "level 1 evidence" in evidence hierarchies, because they are much more likely to be biased than independent research. Only independent RCTs belong in level 1 evidence. Perhaps industry trials should be included in "level 2 evidence" (along with observational trials, which are good but not quite as reliable as RCTs[11]).

My second suggestion is for a quantitative discounting: a quantitative penalty to correct for bias (not to eliminate it). The idea is to *adjust the results using the inverse of the most recent odds ratio measurement of industry bias.*[12] For example, suppose that an industry sponsored randomized controlled trial of a new drug reports a result that the drug makes relief of pain three times as likely as placebo. And suppose that the most recent odds ratio measurement of industry bias is Lundh, Lexchin et al.'s (2017, MR000033) 2.05. Then, the adjusted results will be 3 x 1/2.05 = 1.46, a much more modest result. This adjusted odds ratio might be more appropriate for use in meta-analyses in which the results of both industry and independent research are combined.

This would provide an incentive for pharmaceutical companies (at least, those in it for the long term) to collaborate to reduce bias. Alternatively, particular companies (also, those in it for the long run) could strive to establish independent records of reliable research, receiving less discounting of I-results in return.

There is a possibility that such an intervention will backfire–in the same way as disclosure may have backfired–by encouraging companies who feel penalized by discounting to feel licensed to exaggerate their results. Another unwelcome outcome would be the removal of an incentive for companies to improve the reliability of their

[11]The recent GRADE system is an exception that includes high quality observational trials in level 1.

[12]My thanks to John R. Clarke for walking me through this calculation of quantitative discounting.

research. Perhaps the only way to prevent these outcomes is to encourage companies to produce more reliable research by rewarding such efforts with less discounting.

19.6 Conclusions

I wish there were a simple *and* feasible remedy for industry bias. Because, in the current political climate, there is not, I am glad that so much work has already been done to reduce industry bias through guidelines and oversight. My suggestions are intended to take current strategies for minimizing the effects of industry bias a step or two further.

References

Angell, M. (2005). *The truth about the drug companies: How they deceive us and what to do about it* (Rev a updated). New York: Random House Trade Paperbacks.

Anonymous. (1990). Conflicts of Interest in Medical Center/Industry Research Relationships. Council on Scientific Affairs and Council on Ethical and Judicial Affairs. *Jama, 263*(20), 2790–2793.

Bekelman, J. E., Li, Y., & Gross, C. P. (2003). Scope and impact of financial conflicts of interest in biomedical research: A systematic review. *JAMA: The Journal of the American Medical Association, 289*(4), 454–465.

Dorsey, E. R., de Roulet, J., Thompson, J. P., Reminick, J. I., Thai, A., White-Stellato, Z., Beck, C. A., George, B. P., & Moses, H., 3rd. (2010). Funding of US biomedical research, 2003–2008. *Jama, 303*(2), 137–143.

Flacco, M. E., Manzoli, L., Boccia, S., Capasso, L., Aleksovska, K., Rosso, A., Scaioli, G., et al. (2015). Head-to-head randomized trials are mostly industry Sponsored and almost always favor the industry sponsor. *Journal of Clinical Epidemiology, 68*(7), 811–820.

Gıtzsche, P. C. (2013). *Deadly medicines and organised crime: How big pharma has corrupted healthcare*. London: Radcliffe Publishing.

Ioannidis, J. P. (2005). Why most published research findings are false. *PLoS Medicine, 2*(8), e124.

Jadad, A. R., Moore, R. A., Carroll, D., Jenkinson, C., Reynolds, D. J., Gavaghan, D. J., & McQuay, H. J. (1996). Assessing the quality of reports of randomized clinical trials: Is blinding necessary? *Controlled Clinical Trials, 17*(1), 1–12.

Kjaergard, L. L., & Als-Nielsen, B. (2002). Association between competing interests and authors' conclusions: Epidemiological study of randomised clinical trials published in the BMJ. *BMJ (Clinical Research Ed.), 325*(7358), 249.

Krimsky, S. (2003). *Science in the private interest: Has the lure of profits corrupted biomedical research?* Lanham: Rowman & Littlefield Publishers.

Lexchin, J., Bero, L. A., Djulbegovic, B., & Clark, O. (2003). Pharmaceutical industry sponsorship and research outcome and quality: Systematic review. *BMJ (Clinical Research Ed.), 326*(7400), 1167–1170.

Loewenstein, G., Sah, S., & Cain, D. M. (2012). The unintended consequences of conflict of interest disclosure. *Jama, 307*(7), 669–670.

Lundh, A., Sismondo, S., Lexchin, J., Busuioc, O. A., & Bero, L. (2012). Industry sponsorship and research outcome. *The Cochrane Database of Systematic Reviews, 12*, MR000033.

Lundh, A., Lexchin, J., Mintzes, B., Schroll, J. B., & Bero, L. (2017). Industry sponsorship and research outcome. *The Cochrane Database of Systematic Reviews, 2*, MR000033.

Resnik, D. B., & Elliott, K. C. (2013). Taking financial relationships into account when assessing research. *Accountability in Research, 20*(3), 184–205.

Schott, G., Pachl, H., Limbach, U., Gundert-Remy, U., Ludwig, W. D., & Lieb, K. (2010). The financing of drug trials by pharmaceutical companies and its consequences. Part 1: A qualitative, systematic review of the literature on possible influences on the findings, protocols, and quality of drug trials. *Deutsches Arzteblatt International, 107*(16), 279–285.

Silverman, G. K., Loewenstein, G. F., Anderson, B. L., Ubel, P. A., Zinberg, S., & Schulkin, J. (2010). Failure to discount for conflict of interest when evaluating medical literature: A randomised trial of physicians. *Journal of Medical Ethics, 36*(5), 265–270.

Sismondo, S. (2008). How pharmaceutical industry funding affects trial outcomes: Causal structures and responses. *Social Science & Medicine (1982), 66*(9), 1909–1914.

Smith, R. (2005). Medical journals are an extension of the marketing arm of pharmaceutical companies. *PLoS Medicine, 2*(5), e138.

Solomon, M. (2001). *Social empiricism*. Cambridge, MA: MIT Press.

Chapter 20
Sex, Drugs, and How to Deal with Criticism: The Case of Flibanserin

Anke Bueter and Saana Jukola

Abstract In 2015, the US Food and Drug Administration made the controversial decision to approve flibanserin as the first drug against problems of low female sexual desire. This approval has encountered many criticisms; in particular, it has been condemned as an exemplary instance of medicalization and disease-mongering, as lacking sufficient evidential support, and as targeting a highly problematic diagnosis (Hypoactive Sexual Desire Disorder). In this paper, we review these complaints and show that many of the issues in the flibanserin case stem from a failure to properly deal with criticism. We argue that a principle of *uptake of criticism* can help preventing bias in research and drug regulation and can be secured by institutional measures.

20.1 Introduction

There is a new pill for women, and it is pink: In August 2015, the US Food and Drug Administration (FDA) approved flibanserin (brand name *ADDYI*) as a treatment for *Hypoactive Sexual Desire Disorder* (HSDD) in premenopausal women. In this paper, we analyze the FDA's decision and the process leading towards it in terms of procedural and institutional legitimacy. In particular, we argue that this case fails with regard to a principle of *uptake of criticism,* and propose measures to promote the realization of this principle.

The flibanserin case relates to a number of heatedly discussed issues of current medical research and health care. It illustrates that the worries about the detrimental effects of the commercialization of research, pathologization of normal aspects of

---

A. Bueter (✉)
Leibniz University Hannover, Hannover, Germany
e-mail: anke.bueter@philos.uni-hannover.de

S. Jukola
Bielefeld University, Bielefeld, Germany
e-mail: sjukola@uni-bielefeld.de

A. LaCaze, B. Osimani (eds.), *Uncertainty in Pharmacology*, Boston Studies in the Philosophy and History of Science 338,
https://doi.org/10.1007/978-3-030-29179-2_20

human life, and sexism in medicine should be taken seriously. As such, the process leading up to the approval of flibanserin is an excellent case study for showing how philosophical analysis can be not only theoretically interesting but also have potential for being socially relevant. It demonstrates how the epistemic quality and the validity of results from clinical trials, as well as the objectivity or adequacy of the used methods, can be inseparable from the social context of medical practice and research. Moreover, it is an example of how the institutional conditions of medical research and decision-making can have an impact on what is accepted as effective treatment. We will show that analyzing this case by using tools offered by social epistemology makes it possible to offer concrete proposals for how to improve drug approval processes.

While some have celebrated the approval of flibanserin as a milestone for women suffering from low sexual desire, there also have been numerous critiques, both in the medical literature (e.g., Jaspers et al. 2016; Woloshin and Schwartz 2016) and popular media (e.g., Tavernise 2016). Prominently, the case has been condemned as a problematic instance of medicalization and commercialization of research. *Female Sexual Dysfunction* (FSD) in general as well as HSDD in particular have been dismissed as textbook examples of "disease-mongering" (Jutel 2010; Moynihan 2005; Moynihan and Mintzes 2010; Tiefer 2006): as "trying to convince essentially well people that they are sick, or slightly sick people that they are very sick" (Payer 1992, 5). Claims of disease-mongering are usually evoked when pharmaceutical companies are accused of ruthlessly maximizing profit.

In order to avoid the slightly polemical term of disease-mongering, we will call the relevant kind of biologically-oriented medicalization with commercial goals *pharmaceutical medicalization* in the following. Pharmaceutical medicalization of HSDD starts with the hunt for a female Viagra at the end of the 1990s. The precise nature of the alleged biological dysfunction changes along with the pharmaceutical solution en vogue. When Pfizer was trying to establish Viagra as a suitable option for women, FSD was described as a problem of "clitoral erectile dysfunction" or "insufficiency of vaginal engorgement "(Park et al. 1997). However, no convincing correlations were found between levels of desire or arousal and blood stream to the genitals. The next unsuccessful attempt was to recast the problem as an insufficiency of testosterone, when Proctor & Gamble worked on its testosterone patch "Intrinsa" (Simon et al. 2005). Next in line comes flibanserin along with hypotheses about HSDD being caused by an imbalance of neurotransmitters (Clayton 2010).

Pharmaceutical companies have undoubtedly invested a lot of effort to create a female equivalent to Pfizer's bestseller Viagra, as well as a market to sell it; for example, by funding most research in the area, reporting very high prevalence rates of the target condition, or sponsoring numerous Continuing Medical Education measures (Meixel et al. 2016; Moynihan and Mintzes 2010; Tiefer 2006). However, criticizing the approval of flibanserin primarily in terms of medicalization is problematic, as it hinges on HSDD not being a real disease (but rather, e.g., a matter of problematic socio-cultural norms or relationships) or not being severe enough to merit treatment.

As plausible as that may be, it presumes that there is a matter of the fact regarding whether a certain condition is a disease. Yet, the distinction between "disease" (or "disorder") and "health" is notoriously hard to draw and highly contentious. The basic issue is whether it is possible to define the concept of disease in a value-free manner by explicating it in terms of dysfunction (*naturalism*), or whether it simply expresses value-judgments about what kind of conditions and behaviors are undesirable *(normativism)*. *Hybrid accounts* try to reconcile descriptive and normative aspects of our ascriptions of somatic diseases or mental disorders. As this is a complex and ongoing debate, we will stay agnostic concerning the correct definition of "disorder" in this paper.[1]

We therefore focus on finding institutional instead of conceptual safeguards against unnecessary pharmaceutical medicalization. Such an institutional focus has also played a big role in discussions regarding the commercialization of science. It is well known that the financial interests of funders often introduce bias into the research and that industry-funded studies tend to produce more favorable results than publicly funded ones (e.g., Lundh et al. 2017). Many have argued that at least part of the solution lies in institutional measures such as the independent implementation of clinical trials, disclosure of conflicts of interests, or the registration of all clinical trials and publication of their results (e.g., De Angelis et al. 2004; Sismondo and Doucet 2010; Young et al. 2008).

We shall argue that such institutional instruments should be expanded with regard to pharmaceutical medicalization. For example, concerning the case of flibanserin's approval, Woloshin and Schwartz (2016) contend that an essential shortcoming of the approved re-application was its failure to adequately respond to earlier FDA recommendations, such as the implementation of a new efficacy trial with less restrictive entry criteria (see Sect. 20.2). They argue that such a failure may be overcome by adhering to FDA recommendations on major study design elements mandatory for (re-)submissions. As they show in Woloshin et al. (2014), it is currently often the case that applicants ignore such advice by the FDA, yet are ultimately successful in attaining approval. Thus, the incentive for making qualitative improvements to the study design, especially if it is costly or threatens the attainment of significant results, is rather low.

We support Woloshin and Schwartz's argument that it should not be possible to simply ignore former FDA recommendations and want to generalize this point in terms of a principle of *uptake of criticism.* The idea of a properly critical discourse as a cornerstone or even condition of trustworthy science has a long tradition in philosophy and sociology of science (e.g., Merton 1942; Popper 1959; Longino 1990). Essentially, it expresses the idea that engaging with criticism and other points of view is a means of protection against blind spots and bias. Firstly, this requires

[1]Nevertheless, it is highly unlikely that HSDD is a purely biological problem and its evaluation as harmful and treatworthy unaffected by commercial interests as well as value-laden assumptions about what "normal" female sexuality looks like, or should look like (cf. also Sect. 20.3 on the definition of HSDD as a diagnostic category).

opportunities to express and discuss different opinions; secondly, such criticisms also need to be engaged with if the discourse is to promote the trustworthiness of science. Longino further explicates this as a requirement that members of the scientific community "pay attention to the critical discussion taking place and that the assumptions that govern their group activities remain logically sensitive to it" (Longino 1990, 78). We posit that in order to fulfill this requirement of *uptake of criticism*, one might (a) implement the critics' proposals in one's research, or (b) if one is not convinced by it, point out the reasons for this dissent. It is thus not compulsory to essentially change one's position – but it is necessary to engage with counterarguments in a critical discussion (instead of, for example, ignoring other positions or just repeating one's own arguments).

We will argue in the following that the flibanserin case exhibits a lack of such uptake of criticism at several points, and sketch how such problems might be counteracted by institutional measures. This not only pertains to the failure of the drug manufacturer (*Sprout Pharmaceuticals*) to respond to the FDA, but also to how Sprout Pharmaceuticals and the FDA reacted to the DSM's change of diagnostic criteria, the dealing with dissent inside the FDA, or the engagement with criticism in public hearings organized by the FDA.

Before continuing a caveat is in order. Because we want to analyze the case from a new angle, we have decided to criticize the commercial aspects of the drug development and approval process only in passing. Also, one might state that it is not justified to use the principle of uptake of criticism in this case because it was developed for evaluating practices in purely scientific contexts. However, even though the principles of product development and so-called pure science differ to a certain degree (see, e.g., Wilholt 2006), if we want drug regulation to be based on trustworthy scientific evidence, it is justified to evaluate the practices of commercial players on the basis of the principle of uptake of criticism as well.

The structure of the paper is as follows: In Sect. 20.2, we discuss how flibanserin's effectiveness and safety was established. First, we give a brief account of the studies that formed the basis for the approval, and then offer a critical evaluation of the studies with the focus on the uptake of criticism in the process. In Sect. 20.3, we discuss the diagnosis of HSDD and the reasons for its deletion from the DSM-5, as well as the industry's and FDA's reactions to this. Section 20.4 focuses on the role of advocacy in weighing the risks and benefits of flibanserin. In Sect. 20.5, conclusions are drawn.

20.2 Establishing the Clinical Evidence on Flibanserin

The FDA memo summarizing reviews of Addyi states, "there is internal agreement on the facts, but not on whether the demonstrated benefits outweigh the known risks" (Joffe 2015, 23). In this section, we argue that the principle of the uptake of criticism was violated in the process through which these facts on flibanserin's effectiveness and risks were established.

20.2.1 The Approval of Flibanserin

Flibanserin was originally developed as a treatment for depression by the German company Boehringer Ingelheim but did not demonstrate efficacy in trials. In the studies, however, it outperformed placebo with respect to how the participants answered the question "How strong is your sex drive?" Consequently, flibanserin was seen as a potential treatment for HSDD (Gellad et al. 2015).

In 2009, Boehringer Ingelheim applied for an FDA approval of flibanserin to treat HSDD for the first time. The application was based on two efficacy trials (Derogatis et al. 2012; Thorp et al. 2012) which had two co-primary endpoints: (1) the number of satisfying sexual events (SSEs) and (2) sexual desire. The number of SSEs was measured by patient responses to the questions "Did you have a sexual event?" and "Was it satisfying to you?"[2] Sexual desire was measured by using electronic diaries in which the subjects reported the highest intensity of sexual desire on a daily basis. In order to be considered more efficacious than placebo, both endpoints needed to be met. In these two trials, the first primary endpoint, the number of SSEs, had median increase between 0.5 and 1 SSE per month from baseline compared to placebo, and, thus, did meet the threshold of efficacy. However, the drug did not improve sexual desire in the co-primary endpoint. Alarmingly, the studies also indicated that using the drug together with alcohol or other medication could have serious side effects. Consequently, the FDA rejected the application on both efficacy and safety criteria. It asked for a third efficacy trial with less strict participant inclusion criteria and new studies on the interactions between flibanserin and alcohol (Woloshin and Schwartz 2016). After this first rejection, Boehringer Ingelheim decided to stop developing flibanserin on the basis of the regulatory feedback and sold the rights to the drug to Sprout Pharmaceuticals ('Boehringer Pulls the Plug on "Pink Viagra"' 2010).

The second try at getting approval for flibanserin was made in 2013. This application included interaction studies required by the FDA and a new efficacy trial (Katz et al. 2013), in which a small benefit was demonstrated; this time, the increase in SSEs was 0.5 per month. Instead of electronic daily diaries, sexual desire was now measured by using the desire subscale of Female Sexual Function Index (FSFI), which had been used as a secondary outcome measure in the previous studies. With this new primary outcome measure, there was a significant improvement in sexual desire. Sprout Pharmaceuticals had suggested to make this change in measuring instruments for the co-primary endpoint already for the first two efficacy studies after these trials had been concluded and the results were known, but the FDA had not agreed (FDA briefing document 2015, Clinical Background Document p. 20). One reason for the decline was that the content validity of the desire scale of FSFI had raised concerns (FDA briefing document, 2015, Introductory Memorandum p. 4).

[2] Sexual intercourse, oral sex, masturbation, and genital stimulation by the partner counted as sexual events.

The interaction studies included in the second application indicated that adverse events might occur when flibanserin is used together with alcohol or certain other drugs. For example, in a study on the interaction between flibanserin and fluconazole, an antifungal medication that is used for treating, for example, vaginal infections, 3 of 15 women "developed symptomatic hypotension" and one became unresponsive, after which the trial was terminated early. The study on the interaction between flibanserin and alcohol was performed on 25 subjects, 23 of whom were men, and showed that using the drug together with alcohol increased hypotension. This second submission resulted in rejection: The FDA stated that the small benefits of the drug did not outweigh the risks (Woloshin and Schwartz 2016).

In 2015, Sprout Pharmaceuticals submitted the third application for approval. This time, a new safety study examining potential impairment while driving and new pharmacokinetic studies were included in the application (Woloshin and Schwartz 2016). No new data on efficacy were presented. The approval was thus based on the outcomes of the three randomized controlled efficacy trials that were also included in the previous, rejected applications. According to the FDA's responder analysis, in these phase 3 trials, "approximately 8% to 13% of women were at least "much improved" on at least 1 of the primary outcomes" after the placebo effect had been accounted for (Gellad et al. 2015, E1). In August 2015, the FDA approved flibanserin for the treatment of HSDD based on this application.

It is worth mentioning that the first systematic review and meta-analysis of flibanserin studies was published after the approval. Jaspers et al. (2016) analyzed data from all eight trials on flibanserin: the three main studies on which the FDA decision was based, two additional published and three unpublished studies. According to their analysis (p. 5), the benefit of flibanserin is "minimal" and the quality of evidence was very low both for efficacy and safety outcomes.

20.2.2 Methodological and Conceptual Issues of the Flibanserin Studies

As Gellad et al. (2015) remark, there are several noteworthy issues in the studies on flibanserin. Here we shall discuss two of them in particular: concerns related to measurement of efficacy and to extrapolation of the study results.

Philosophers of medicine have tended to put their critical focus on study design and pay less attention to the role that measurement plays in establishing efficacy and effectiveness of treatments (McClimans 2013). However, the validity of measuring instruments is relevant to the validity of study results, and inadequate use of measuring instruments can contribute to false estimations of treatments' effectiveness (Stegenga 2015). In the case of patient-reported outcome measures, the content validity is particularly important: the instrument should measure the concept of interest, and different understandings concerning the questions and their meaning may result in measurement error (McClimans 2013).

The patient-reported outcome measure used in flibanserin studies, the desire domain of FSFI, includes the following two questions: (1) "Over the past 4 weeks, how often did you feel sexual desire or interest?", and (2) "Over the past 4 weeks, how would you rate your level (degree) of sexual desire of interest?" The questionnaire defines sexual desire as "a feeling that includes wanting to have a sexual experience, feeling receptive to a partner's sexual initiation, and thinking or fantasizing about having sex." One could ask: does FSFI measure what it is supposed to measure? According to Forbes et al. (2014), measuring desire with the FSFI is problematic, mainly for conceptual reasons: They doubt that the two questions could capture the complexity of sexual desire. They also criticize this definition for not taking into account evidence suggesting that female sexual desire may be more often triggered than spontaneous (ibid; see also Sect. 20.3). In other words, the concept validity of the instrument that was used for measuring the efficacy of flibanserin has been questioned by pointing out that it is based on a debatable theory of female sexuality.

Moreover, treating low sexual desire as a biological phenomenon that takes place at the level of an individual and not as a consequence of interpersonal dynamics could have resulted in disregarding the possibility that the patients' improvement in the studies was partly caused by factors not related to the drug. In the studies, the efficacy of flibanserin was not apparent before at least 8 weeks of usage. Brotto (2015) speculates that this time period may be needed for overcoming the interpersonal and personal difficulties that are partly responsible for low sexual desire. Furthermore, it is not clear how well the studies were blinded (Jaspers et al. 2016), and thus the placebo and placebo by proxy response in the studies may have manifested itself in complicated ways (Bradford 2012): As the side effects of flibanserin are relatively strong, some women in the trials may have been able to guess that they are receiving the treatment instead of a placebo, which, in turn, would present a threat to the trials' validity (See, Howick 2011, ch. 6 and 7). Testing whether the blinding was successful would have been rather simple: By asking the participants after the study whether they believed they had received the drug or placebo and then comparing the result with an expected chance distribution.[3]

The FDA Memorandum on Study Endpoints (FDA Briefing Document, Memorandum on Study Endpoints, pp. 6–7) mentions another concern related to using FSFI as a measuring instrument: Because sexual desire is defined as a feeling that includes wanting to have a sexual experience, feeling receptive to a partner's sexual initiation, and thinking or fantasizing about having sex, it is not clear "what is driving any change identified on the assessment (e.g., receptivity, sexual fantasies, and/or initiating sex)." It is possible that a change in the frequency of sexual fantasies could increase the score, even if none of the other components improved. As the memorandum states, it is unclear whether this would be a clinically meaningful benefit. Among other problematic aspects of using the FSFI scale are that the respondents are expected to reliably remember their sexual desire

---

[3] We thank an anonymous reviewer for pointing this out.

levels over the past 4 weeks and that the scale is based on the assumption that higher level of sexual desire is always better.[4]

A noteworthy point concerning the measuring instruments in flibanserin studies is that the first two studies (DeRogatis et al. 2012; Thorp et al. 2012) did not use FSFI for measuring sexual desire; in these trials, the primary desire end point was measured by diary reports on the most intense desire. Moreover, as was mentioned in the previous subsection, Sprout Pharmaceuticals had suggested changing the measure for the co-primary endpoint also for the first two studies after their results were known (FDA briefing document, Clinical Background Document). Post hoc changes in the primary outcome measures are generally considered to be bad scientific practice as they can be used as means of acquiring preferred results. The choice of a measuring instrument is one of the numerous decisions scientists have to make in the course of conducting studies. These choices are not neutral but can influence the results that are acquired and it is important to be willing to question the reasons for particular choices (McClimans and Browne 2011).

Because of the specific trial history, it does not seem far-fetched to assume that the change in the measuring instrument was motivated by the desire to make the drug seem as efficacious as possible. Also, it is relevant to point out that the critical voices raising questions on the use of the instrument have been disregarded in the process of flibanserin's approval. The FDA (FDA briefing document, Memorandum on Study Endpoints) recommended using the eDiary daily measure instead of the FSFI scale, and according to Gellad et al. (2015), despite concerns raised about whether FSFI is optimal for measuring desire, the switch in primary endpoints or the validity of FSFI were not in the focus of the advisory committee. Here, both the manufacturer of the drug and the members of the advisory committee seem to have failed at taking seriously the importance of criticism and responding to it: The critical comments concerning the use of measuring instruments and the theoretical background assumptions that underlie them have been bypassed. This violates the principle of uptake of criticism.

Furthermore, the flibanserin studies have been criticized for their weak extrapolability (e.g., Brotto 2015; Gellad et al. 2015)): In clinical practice, flibanserin is likely to be used by a population that differs from the studies' population. A striking issue regarding extrapolation and the evaluation of possible risks in this case is that the safety of using flibanserin together with alcohol was tested in a trial in which 23 of 25 participants were male. Sprout stated that the reason for this was trouble recruiting moderate female drinkers (FDA briefing document, 2015, Clinical Background document; see also Sect. 20.4). It seems reasonable to state that in this case there is a significant difference between the trial subjects and the target patients, which may undermine the applicability of the results.

Presumably, the drug will be used by patients who do not satisfy the criteria for HSDD and who have other conditions or medications that increase the likelihood of adverse reactions. Especially one of the inclusion criteria, namely, that the

[4]We thank an anonymous reviewer for pointing this out.

women who took part in the studies had agreed to participate in sexual activities monthly, is likely to make the study population different from the typical users with HSDD in the clinical context. Moreover, the exclusion criteria were strict regarding comorbidities and additional medication. In a letter sent to the manufacturer after the first failed application for approval, the FDA required that the third efficacy trial would have to have less restrictive entry criteria for the subjects. The first two trials had excluded subjects with depression, anxiety and other comorbidities, or not in stable, communicative, heterosexual relationships. Despite these requests, the third conducted study still had equally restrictive entry criteria ("Fixed 100 mg Every Evening of Flibanserin vs Placebo in Premenopausal Women With Hypoactive Sexual Desire Disorder"; Woloshin and Schwartz 2016). This is another demonstration of the failure to respond to criticism.

The process of producing evidence on flibanserin, exhibits practices that have been criticized in literature describing the questionable methods used by pharmaceutical industry (e.g., Sismondo 2008). Also, the producer of the drug has not been sensitive to the critical feedback by the FDA either by adjusting their methodology or justifying their chosen ways of conducting research, which violates the principle of the uptake of criticism. This, in turn, gives reasons to doubt whether the factual basis for evaluating the risks and benefits of flibanserin really is as solid as Joffe (2015) implies.

20.3 Assessing the Diagnosis

Next we discuss how the potential user group for flibanserin was established. HSDD has recently been deleted from the DSM-5 (APA 2013), which has led to considerable opposition from industry-linked researchers. In the DSM-IV (APA 1994), it was defined by the monothetic criterion of "persistent or recurrently deficient or absent sexual fantasies or desire for sexual activity," supplemented by "the disturbance causes marked distress or interpersonal difficulties." Additionally, the diagnostic criteria excluded cases that could be accounted for by other medical conditions, mental disorders, or substance-abuse. From the DSM-III onwards, the classification of sexual dysfunctions has relied on a particular model of normal sexual functioning: Masters and Johnson's (1966) linear model of the human sexual response cycle, complemented by a preliminary phase of desire (Brotto 2010). The presumed chronology of (satisfying) sexual events starts with desire, followed by arousal, orgasm, and resolution. The DSM matched this model by dividing sexual dysfunctions into "Sexual Desire Disorders" (e.g., HSDD), "Sexual Arousal Disorders" (e.g., *Female Sexual Arousal Disorder*, FSAD, which was mainly defined as lack of genital lubrication-swelling response), and "Orgasmic Disorders." A fourth category was problems with pain during intercourse.

The diagnosis of HSDD was thus already given when the race for drugs against female sexual dysfunctions began in the late 1990s; it was not simply made up by pharmaceutical companies (notwithstanding, diagnoses of female sexual dysfunctions have a long and troublesome history impacted by gender norms and commercial interests; cf. Bueter 2017; Moynihan 2005). What still had to be established, however, was its prevalence. Repeatedly mentioned in the debate on HSDD and flibanserin is the misleading number of 43% of women suffering from some kind of sexual dysfunction. This rate is taken from Laumann et al. (1999)[5]: 43% of women in this survey reported having any kind of sexual problem at some point in the last 12 months, including issues of lubrication, difficulties to climax, or anxiety over performance. Problems of low desire were most common among these complaints and said to affect 27%–32% of women. There are other epidemiological studies that report similarly high prevalence rates of low desire, but there are also huge differences in results, ranging from 7% to around 40% of women being affected. This problematic variability stems from differences in entry criteria as well as methods (e.g., questionnaires, face-to-face interview, telephone-interview), diagnostic tools, cut-off points, and time-periods considered (see Brotto 2010 for an overview).

Another important point is that these rates are in any case not equivalent to the prevalence of HSDD, since the diagnosis requires not only the presence of low desire, but also that this causes marked distress *and* is not better accounted for by relationship issues or other medical conditions. If the presence of distress is determined, the prevalence range drops markedly. For example, an often-quoted study from Shifren et al. (2008) found that 34% of women reported issues of low desire, but only 10% reported low desire and distress. Risk factors for low desire and distress were found in generally poor health, depression, age, surgical menopause, or urinary incontinence. Thus, these 10% are also not equivalent to an HSDD diagnosis, which would exclude such cases. Again, other studies found prevalence rates of low desire and distress ranging from 7% to 26% (Brotto 2010).

To sum up, there are problems with establishing the prevalence rates of HSDD. Different studies report diverse results, which are based on different methods, study populations, and diagnostic criteria. Illuminating in this respect is a study by McCabe and Goldhammer (2013), which assesses a population of 741 women on the basis of different diagnostic criteria (and varying degrees of stringency in their operationalization), such as the DSM-IV criteria or scores on the Female Sexual Desire Questionnaire. The respective rates of HSDD for this sample range from 3% to 31% depending on the criteria used. In addition, the high prevalence numbers often reported neglect the requirement that low sexual desire is actually perceived as distressing by the concerned women and ignore other factors that can affect levels of desire and distress, such as medical conditions or relationship issues. These problems undermine claims of a large number of women with unmet medical needs.

Furthermore, there have been issues concerning the conceptual and empirical adequacy of the DSM-IV criteria for HSDD, which have ultimately led to its controversial deletion from the DSM-5. Following recommendations of the DSM-

[5]Two of the authors worked as consultants for Pfizer and other pharma companies; the conflicts of interest were "inadvertently omitted" in the original paper, but disclosed later on.

5 Workgroup on Sexual and Gender Identity Disorders[6] (Brotto 2010; Graham 2010), both HSDD and FSAD are now combined into the diagnosis of *Female Sexual Interest/Arousal Disorder* (FSIAD). The workgroup's recommendations were guided by the intention to avoid an inflation of HSDD diagnoses and to "raise the bar" (Brotto et al. 2011, 223) for disorder ascriptions. This intention was grounded on the extremely high prevalence rates reported by some studies (see above), making high rates of false positives likely. Moreover, the described variety of reported prevalence rates was taken to indicate a need for further explication of the diagnostic criteria, e.g., by defining the vague criterion of "persistent or recurrent" as requiring a minimum symptom duration of 6 months.

A more fundamental critique concerned the linear model of the human sexual response cycle and its problematic distinction between desire and arousal. As mentioned briefly in subsection 20.2.2., this distinction has been challenged by alternative accounts (such as the circular model of desire, Basson 2000) that consider all desire as triggered and point out that desire often develops *during,* not necessarily *before* sexual activity. As Brotto (2010) highlights, there is no definite empirical support for one or the other model. Rather, studies show considerable diversity as to which account of sexual response women subscribe to. In addition, the DSM's operationalization of a lack of desire as absence of sexual fantasies seems problematic in the light of studies suggesting that such fantasies do not play an important role in many women's (healthy) desire/arousal. Thus, basing a diagnosis of HSDD on such an absence, as well as on a lack of desire for sexual activity *before* such activity, threatens to overpathologize women whose sexual response does not fit the DSM model, without thereby being dysfunctional (ibid. 224–227).

The workgroup therefore recommended a polythetic approach that would be able to better accommodate diversity in functional sexual response and overcome the problematic distinction between desire and arousal by combining HSDD and FSAD. Possible symptoms of the new category of FSIAD include (1) an absent/reduced interest in sexual activity; (2) absent/reduced sexual thoughts or fantasies; (3) absent/reduced initiation of sexual activity and being typically unreceptive to partner's initiatives; (4) absent/reduced excitement or pleasure during sexual activity in 75%–100% of sexual encounters; (5) absent/reduced interest/arousal in response to sexual cues; and (6) absent/reduced genital or non-genital sensations during activity in 75%–100% of encounters. These symptoms have to cause clinically significant distress in the individual and not be better explained by other health issues or (relationship) stressors (APA 2013, 433 ff).

The original proposal demanded 4 out of these 6 symptoms for a diagnosis; the actual DSM-5 category, however, requires only 3 out of 6. A possible reason for this retreat is that while it was suggested that the proposed criteria would be tested for reliability and validity in field trials, this did not happen due to limited resources and other funding priorities of the APA. This is unfortunate since it reinforces

[6]In contrast to the heavy industry links of a lot of other works on HSDD, the DSM-5 adopted a policy that not only made conflicts of interests transparent but also limited them (APA 2008).

criticisms of FSIAD as lacking evidence by researchers who have been key figures in the establishment of HSDD's assumed high prevalence and the development of flibanserin:

> Any such changes should be based on confirmed data [...] rather than on theoretical speculation or expert opinion, as in the current proposal. (DeRogatis et al. 2011, 218).

> It seems to us that the creation of FSIAD was based on ideology and personal beliefs, rather than on published scientific and clinical evidence. (Balon and Clayton 2014, 1229).

It is argued that there is no evidence for the reliability, validity, or utility of FSIAD, and that no underlying pathology has been demonstrated (Balon and Clayton 2014). While this is true, and is indeed problematic, it also holds for many other DSM-diagnoses – especially for the former diagnoses of HSDD and FSAD. Moreover, the DSM workgroup's recommendations were based on an extensive review of the data on the prevalence of HSDD as well as the empirical merits of different models of sexual response cycles. Even though this is repeatedly referred to by Brotto and her colleagues (e.g., Brotto et al. 2011; Graham et al. 2014), the critique of FSIAD as lacking *any* evidence keeps being repeated. Meanwhile, the actual evidence for the former diagnostic construct's problems to capture the diversity of "normal" sexual functioning is never even mentioned. A distinctive characteristic of this debate on FSIAD is the failure of one side of it (the one with heavy industry ties) to actually respond to criticism and counter arguments.

Besides the alleged lack of evidence, these critics do, however, formulate further reasons in favor of HSDD. For example, DeRogatis et al. (2011) claim that there are data that show HSDD and FSAD to have distinct symptom patterns (which should therefore be differentiated). They refer to treatment studies by DeRogatis and others (not published at the time) as well as to baseline characteristics from patients of the flibanserin trials that support this. However, such a result of different symptom patterns should not be surprising, since the entry criteria for the respective study populations were based on the diagnoses of HSDD and FSAD. To recruit a study population on the basis of two sets of (symptom-based) criteria, and then conclude that the overall sample exhibits two distinct symptom profiles, suffers from circularity and does not prove the existence of two different underlying dysfunctions (see also Brotto et al. 2011).

Another recurrent critique of FSIAD concerns potential harms of the earlier diagnoses' deletion. It is worried that former patients will not meet the new diagnostic criteria and therefore be excluded from treatment:

> The establishment of the FSIAD diagnosis has a potential to inflict harm (e.g., by excluding women from a DSM-5 diagnosis who currently meet the DSM-IV criteria for HSDD or FSAD). It is also unclear what is going to happen with regards to treatments established for these disorders and not indicated for treatment of FSIAD. [...] It is not clear whether insurance companies will cover the cost of such treatments in women diagnosed with FSIAD. (Balon and Clayton 2014, 1229; see also Graham et al. 2014 for a response)

As with the evidence indicating problems with the former diagnostic criteria, the opponents of FSIAD – while raising the important issue of potential harm – never mention worries and arguments concerning risks of over-diagnosis and – treatment.

Without presuming any absolute standard, the question of how to balance risks of under- versus over-diagnosis always remains. We will not argue to have the one correct answer here. However, regarding how to strike an appropriate balance in such situations of considerable uncertainty, characteristics of the scientific debate are important. The DSM workgroup (and many others) have expressed concerns about high rates of false positives on the basis of an empirical and conceptual critique of the Masters and Johnson model, the extraordinarily high prevalence rates reported by some studies, the variability of reported prevalence rates, and vagueness of the diagnostic criteria. They have also repeatedly reacted to critiques such as FSIAD lacking evidential support. Researchers expressing worries about false negatives and a potential exclusion of women from treatment, on the other side, have neither given any good reason as to why a high rate of false negatives seems likely, nor reacted to problems with the diagnoses of HSDD and FSAD, or addressed the concerns of over-diagnosis raised by the questions regarding prevalence. In conclusion, while FSIAD has problems (such as the lack of field trials) one can hardly say that a large extent of unmet medical need of women suffering from HSDD has been established. On the contrary, accusations of pharmaceutical medicalization have quite a substantial backing.

Here, we see a problematic lack of uptake of criticism on two levels. The first pertains to the discussion about the deletion of HSDD and its replacement by FSIAD. The DSM workgroup has provided an opportunity for critical discussion by publishing reviews and proposals before their actual implementation, and it is, of course, legitimate for industry-linked researchers to seize this opportunity. However, they fail in this discussion to actually deal with critiques regarding their own work and do not address the issues that have been raised. Instead, they keep repeating rather polemical claims about the workgroup's proposal being ideology-driven. In order to create a more fruitful debate, one simple measure, which could be implemented by using existing systems of peer review, would be to make it an explicit requirement for publication to respond to formerly published work that directly criticizes the positions brought forward.

Secondly, considering this taxonomic change based on substantial criticism of the diagnosis of HSDD, it is problematic that the FDA approved flibanserin for treating the condition. While the FDA summary review acknowledges the deletion of HSDD from the DSM, it seems to assume that FSIAD is simply a combination of HSDD and FSAD (Joffe 2015, 4). This misses the central idea that motivated the change, however – to avoid false positives promoted by the old diagnoses, for example, by the introduction of more explicit criteria. Maybe FSIAD did not go far enough in this respect, and probably there is a good deal of overlap between the respectively defined populations. The extent of this overlap, however, still has to be established empirically. Both Sprouts's reapplication and the FDA thus should have paid more attention to the change of the relevant diagnostic construct; for example, Sprout could be asked to demonstrate to what degree their study populations meet the stricter criteria of FSIAD. In any case, in deciding on the approval of drugs for certain indications, the FDA should be sensitive towards the current definition and classification of these indications. This does not imply that they have to follow the DSM or other nosologies blindly – instead, they should critically review nosological changes and give reasons for deviating from the contemporary status quo in classification.

20.4 Risk-Benefit Assessment and the Role of Advocacy

Having discussed the questionable quality of evidence for flibanserin's efficacy and safety as well as problems with the diagnostic category it targets, we will now take a closer look at the FDA's assessment of the drug's risk-benefit-profile. The first and second applications for flibanserin's approval were rejected because of the questionable efficacy, diagnostic tool, switch in endpoints, and safety of the drug, leading the FDA to conclude that the risks do not outweigh the benefits, and the 2015 application did not contain much more data or studies to overcome these concerns (Woloshin and Schwartz 2016). However, this third application was successful, thus expressing a change in the evaluation of the risk-benefit-ratio.

It is important to make clear at this point that the decision to approve the drug was actually far from unanimous: Of the Advisory Committee, 18 to 6 voted for it, under the condition of further regulations such as a ban on direct-to-consumer marketing and clear labeling of risks. On the other side, both the Clinical Pharmacology Review team and the Clinical Review team were against approval, on grounds of the known methodological difficulties and because the risk-benefit ratio was considered unfavorable (Joffe 2015, 22 f.). Yet, these votes were overruled by the division director, who argued that flibanserin does provide an improvement for some women (about 10% of patients).[7]

> The question comes down to whether we withhold flibanserin treatment from everyone in the United States – including those who would have meaningful benefit – or whether there is a reasonable approach to allow flibanserin to be marketed, so that it is available to patients who need it. (ibid. 23 f.)

In order to establish that there is a need for a treatment of low desire that outweighs its risks, both Boehringer Ingelheim and Sprout Pharmaceuticals have invested considerably into increasing the condition's publicity. For example, Boehringer Ingelheim sponsored numerous CME courses, television programs, and internet awareness campaigns such as "sexandahealthieryou.org" and "Sex Brain Body: Make the Connection", providing information as well as online tools for a "sexual satisfaction check-up". Similarly, Sprout has been involved as a sponsor of the "Even the Score" campaign that lobbied for the approval of flibanserin by reframing it as an issue of gender equality and feminism. While it is common for patient advocacy groups to receive funding from pharmaceutical companies, these

[7] We see an additional issue regarding the uptake of criticism here, namely, how the FDA deals with dissent inside it's committees and boards. However, it is beyond the scope of this paper to analyze this in more detail.

campaigns are exceptional in blurring the line between advocacy and marketing. Not only are they repeatedly referring to experts that make an appearance as such in all the different PR efforts, but these experts are also the very same researchers directly involved in the industry's hunt for a female Viagra. Moreover, the information given to the public is highly misleading. For example, the Even the Score website unanimously states that 43% of women have a sexual dysfunction (referring to Laumann et al. 1999), the biggest part of which (10%) suffer from HSDD (referring to Shifren et al. 2008, see above). Moreover, it claims that before the approval of Addyi, there have been 26 treatment options for men with sexual dysfunctions, but none for women. This is interpreted as an outcome of gender bias: "For decades, men's sexual dysfunction has been prioritized over women's" (http://eventhescore.org/the-problem/, retrieved 05/27/16).

Both the 2010 rejection and the 2015 approval were preceded by FDA Advisory Committee Hearings, which invited the public to contribute viewpoints and opinions on the matter. These opportunities were seized by representatives of the above campaigns in order to defend women's right to choose themselves what risks to take. It might thus be argued that the approval of FDA is actually an outcome of it responding to criticism by advocates and patients. Yet, we hold that the situation is more complicated than that.

Including members of the public in the critical discussion informing decision-making in drug regulation can be justified: Regarding issues such as whether there is an unmet medical need and what kinds of risk the associated suffering merits, the requirement of "uptake of criticism" can plausibly be extended to include patient voices. However, it then becomes a decisive question whose voices are actually heard, how much weight is given to them, and who is considered to represent certain viewpoints. The claims of Even the Score did not pass unchallenged on the 2015 FDA meeting. A decade before the first public hearing, Leonore Tiefer had already started the "NewView" campaign that fights against pharmaceutical medicalization of female sexuality and questioned not only the scientific soundness of Sprouts clinical trials and the HSDD diagnosis, but also their commitment to feminism (www.newviewcampaign.org, Tiefer 2006).

Although it is impossible to break down a broad issue such as feminism to a few central theses and there is considerable variety among feminists, Even the Score's appeal to gender equality as an argument for the approval of flibanserin is pretty far-fetched. Feminist activism regarding women's health issues has a long tradition. Of great historical importance is the women's health movement of the 1970s, which began to fight against pathologizing female sexuality and reproduction (e.g., childbirth or menopause). Feminism in health care is thus usually intricately linked to a critique of medicalization, understood as a means to control and exploit women's bodies (e.g., Dreifus 1978; Ehrenreich and English 1978). Another cornerstone of feminism has always been the resistance to biological reductionism regarding the complex issues of sex and gender. Portraying the declaration of an (allegedly very common) sexual experience of women as a biological pathology to be treated with pharmaceuticals as feminist is, therefore, quite a stretch.

Another important strand of the feminist women's health movement was a critique of ignoring sex and gender differences in non-reproductive health. Basically, this is an issue of extrapolating data from exclusively male study populations to all adults. Due to an increased inclusion of women in clinical trials, it became clear that this ignorance has led to dangerous overgeneralizations in many areas (e.g., heart diseases, auto-immune diseases, cancer), and that many drugs, tested exclusively on men, brought less benefits and more side-effects for women (cf. Bueter 2017 on the women's health movement). Against this background, it seems questionable to invoke feminism as a rhetoric strategy, but to perform the required study on interactions with alcohol on a population of 23 men and 2 women.

Tiefer and her colleagues have also pointed out that Even the Score's claim of the ratio of 26:1 for treatments of male/female sexual dysfunction is deceptive. There are no treatments for low libido in men, but several drugs (plus generics) that target blood flows. They have also argued that feminism is not about the quantity of consumer choices, and that Even the Score's critique of the FDA's rejection of flibanserin as paternalistic fails to address that such consumer choices happen in a certain cultural environment with certain social norms regarding sexuality and normality (see www.newviewcampaign.org).

Thus, while an integration of public contributions to risk-benefit-assessments seems in principle appropriate, the requirements of a diversity of perspectives and uptake of criticism should be extended to these contributions, too. For example, this could mean that pro and contra advocates are required to deal with counter-arguments (instead of just repeating their claims, such as a gender bias of the FDA). Especially since there were two consequent public hearings, this would have been an excellent opportunity to actually engage participants in a critical discussion.

In any case, the arguments of feminist advocates associated with the New View campaign against flibanserin make it difficult to interpret the FDA's approval as a result of uptake of public criticism. In the summary review of that decision, Joffe states that this approval was not a reaction to accusations of gender inequality. It is, however, unclear what led to the decision to approve the drug instead, given that the application did not include new evidence on the efficacy of the drug. One reason given for the decision was that there are no other pharmaceutical solutions for HSDD. However, as we have seen, even the number of 10% potentially profiting patients is subject to several sources of uncertainty. Even if it were better established, one might just read the situation the other way around, with 90% of patients not experiencing any meaningful improvement, but possibly severe side-effects. Moreover, using the lack of alternative drugs as an argument for approving a drug with a questionable risk-benefit profile seems acceptable only if one assumes that the treatment for HSDD cannot be other than pharmaceutical. It in fact ignores that while there are no alternative drugs available, psychological interventions do seem to be helpful in some patients suffering from low desire (Frühauf et al. 2013). This point has been repeatedly brought up by the New View campaign, yet never really got addressed by either Sprout or the FDA. This is another example of failing to respond to criticism.

20.5 Conclusion

In this paper, we have shown that the flibanserin case presents multiple violations of a principle that is traditionally taken to be one of the conditions for trustworthy science, namely, the uptake of criticism: Firstly, the manufacturer of the drug and researchers associated with it failed to respond to critical remarks challenging theoretical assumptions underlying the methods that were used for establishing the prevalence of HSDD and the effectiveness of the drug needed for treating it. Secondly, the FDA did not require that its recommendations concerning how the studies on flibanserin should be conducted would actually be followed. In addition, the public hearings, which aimed at integrating voices from involved lay groups, were exploited by a group tightly associated with the manufacturer, and, consequently, failed at serving the purpose of offering a forum for genuine critical interaction.

In order to better fulfill the principle of the uptake of criticism, at least the following institutional measures should be considered: After a rejected application, reacting to the FDA's recommendations concerning, for example, the use of measuring instruments, inclusion criteria for study subjects, and other issues related to study design, should be a prerequisite for submitting a reapplication. Both the FDA and industry need to take into account changes in the relevant disease taxonomy when new drugs are submitted for approval. Involving the public in the assessment of risks and benefits of new treatments is a way of integrating different perspectives into the process and making criticism possible. Yet, the FDA needs to be aware of the possibility that the public hearings are being used for marketing purposes. Because of this, transparency concerning the financial and other ties of advocacy groups is desirable. Moreover, the participants should actually engage in responding to arguments made by other discussants. Providing such an institutional framework and incentives for critical discourse has the potential to work as a safeguard against bias not only in research, but also in drug regulation.

To conclude, we hope that our analysis of the approval of flibanserin shows how a detailed case study can both illustrate problems that apply to medical research and drug regulation in general, and help to develop suggestions for institutional improvements. These suggestions concerning the uptake of criticism connect the case to approaches developed in the social epistemology of science, helping to apply and further explicate these conceptual resources.

References

American Psychiatric Association (APA). (1994). *Diagnostic and statistical manual of mental disorders* (DSM-IV) (4th ed.). Washington, DC: APA.

APA. (2008). *Principles for DSM-5 Task Force and Work Group Members*. http://web.archive.org/web/20120404114422/http://www.psych.org/MainMenu/Research/DSMIV/DSMV/BOTPrinciples.aspx. Assessed 8 Aug 2016.

APA. (2013). *Diagnostic and statistical manual of mental disorders* (DSM-5) (5th ed.). Washington, DC: APA.

Balon, R., & Clayton, A. H. (2014). Female sexual interest/arousal disorder: A diagnosis out of thin air. *Archives of Sexual Behavior, 43*, 1227–1229.

Basson, R. (2000). The female sexual response: A different model. *Journal of Sex and Marital Therapy, 26*, 51–64.

Boehringer Pulls the Plug on "Pink Viagra". (2010, October 8). https://www.reuters.com/article/us-boehringer-flibanserin/boehringer-pulls-the-plug-on-pink-viagra-idUSTRE6970TN20101008. Accessed 13 Sept 2017.

Bradford, A. (2012). Listening to placebo in clinical trials for female sexual dysfunction. *The Journal of Sexual Medicine, 10*, 451–459.

Brotto, L. A. (2010). The DSM diagnostic criteria for hypoactive sexual desire disorder in women. *Archives of Sexual Behavior, 39*, 221–239.

Brotto, L. A., et al. (2011). Should sexual desire and arousal disorders in women be merged? A response to DeRogatis, Clayton, Rosen, Sand, and Pyke (2010). *Archives of Sexual Behavior, 40*, 221–225.

Brotto, L. (2015). Flibanserin. *Archives of Sexual Behavior, 44(8)*, 2103–2105.

Bueter, A. (2017). Androcentrism, feminism, and pluralism in medicine. *Topoi, 36*, 521–530.

Clayton, A. H. (2010). The pathophysiology of hypoactive sexual desire disorder in women. *International Journal of Gynecology & Obstetrics, 110*, 7–11.

De Angelis, C., Drazen, J., Frizelle, F., Haug, C., Hoey, J., Horton, R., Kotzin, S., Laine, C., Marusic, A., Overbeke, A., Schroeder, T., Sox, H., & Van Der Weyden, M. (2004). Clinical trial registration. A statement from the International Committee of Medical Journal Editors. *New England Journal of Medicine, 35*, 1250–1251.

DeRogatis, L. R., et al. (2011). Should sexual desire and arousal disorders in women be merged? *Archives of Sexual Behavior, 40*, 217–219.

DeRogatis, L., Komer, L., Katz, M., & et al. VIOLET Trial investigators. (2012). Treatment of hypoactive sexual desire disorder in premenopausal women: Efficacy of flibanserin in the VIOLET study. *The Journal of Sexual Medicine, 9*(4), 1074–1085.

Dreifus, C. (Ed.). (1978). *Seizing our bodies: The politics of women's health*. New York: Vintage Books.

Ehrenreich, B., & English, D. (1978). *For her own good: 150 years of the experts' advice to women*. New York: Anchor Books.

FDA briefing document. (2015). http://www.fda.gov/downloads/AdvisoryCommittees/CommitteesMeetingMaterials/Drugs/DrugSafetyandRiskManagementAdvisoryCommittee/UCM449088.pdf FDA. Accessed 5 June 2016.

Fixed 100 mg Every Evening of Flibanserin vs Placebo in Premenopausal Women with Hypoactive Sexual Desire Disorder. ClinicalTrials.gov. https://clinicaltrials.gov/ct2/show/NCT00996164. Accessed 4 Aug 2016.

Forbes, M., Baillie, A., & Schniering, C. (2014). Critical flaws in the female sexual dysfunction index and the international index of erectile dysfunction. *The Journal of Sexual Research, 57*, 485–491.

Frühauf, S., et al. (2013). Efficacy of psychological interventions for sexual dysfunction: A systematic review and meta-analysis. *Archives of Sexual Behavior, 42*, 915–933.

Gellad, W., Flynn, K., & Alexander, G. (2015). Evaluation of Flibanserin. Science and advocacy at the FDA. *JAMA, 314*(9), 869–870.

Graham, C. A. (2010). The DSM diagnostic criteria for female sexual arousal disorder. *Archives of Sexual Behavior, 39*, 240–255.

Graham, C. A., Brotto, L. A., & Zucker, K. J. (2014). Response to Balon and Clayton (2014): Female sexual interest/arousal disorder is a diagnosis more on firm ground than thin air. *Archives of Sexual Behavior, 43*, 1231–1234.

Howick, J. (2011). *Philosophy of evidence-based medicine*. Oxford: BMJ Books/Wiley-Blackwell.

Jaspers, L., Feys, F., Bramer, W., Franco, O., Leusink, P., & Laan, E. (2016). Efficacy and safety of Flibanserin for the treatment of hypoactive sexual desire disorder in women. A systematic review and meta-analysis. *JAMA Internal Medicine, 176*(4), 453–562.

Joffe, H. (2015). *Division director review*. http://www.accessdata.fda.gov/drugsatfda_docs/nda/2015/022526Orig1s000SumRedt.pdf. Accessed 6 June 2016.

Jutel, A. (2010). Framing disease: The example of female hypoactive sexual desire disorder. *Social Science & Medicine, 70*, 1084–1090.

Katz, M., Derogatis, L., Ackerman, R., & et al. Begonia Trial Investigators. (2013). Efficacy of flibanserin in women with hypoactive sexual desire disorder: results from the BEGONIA trial. *The Journal of Sexual Research, 10*(7), 1807–1815.

Laumann, E. O., Paik, A., & Rosen, R. C. (1999). Sexual dysfunction in the United States: Prevalence and predictors. *JAMA, 281*, 537–544.

Longino, H. (1990). *Science as social knowledge: Values and objectivity in scientific inquiry*. Princeton: Princeton University Press.

Lundh, A., Lexchin, J., Mintzes, B., Schroll, J. B., & Bero, L. (2017). Industry sponsorship and research outcome. *Cochrane Database of Systematic Reviews, 2*, MR000033. https://doi.org/10.1002/14651858.MR000033.pub3.

Masters, W. H., & Johnson, V. E. (1966). *Human sexual response*. Boston: Little, Brown.

McCabe, M. P., & Goldhammer, D. L. (2013). Prevalence of women's sexual desire problems: What criteria do we use? *Archives of Sexual Behavior, 42*, 1073–1078.

McClimans, L. (2013). The role of measurement in establishing evidence. *The Journal of Medicine and Philosophy, 38*(5), 520–538.

McClimans, L., & Browne, J. (2011). Choosing a patient-reported outcome measure. *Theoretical Medicine and Bioethics, 32*, 47–60.

Meixel, A., Yanchar, E., & Fugh-Berman, A. (2016). Hypoactive sexual desire disorder: Inventing a disease to sell low libido. *Journal of Medical Ethics, 41*, 859–862.

Merton, R. K. (1942). *The sociology of science: Theoretical and empirical investigations*. Chicago: University of Chicago Press.

Moynihan, R. (2005). The marketing of a disease: Female sexual dysfunction. *British Medical Journal, 330*, 192–194.

Moynihan, R., & Mintzes, B. (2010). *Sex, lies, + pharmaceuticals: How drug companies plan to profit from female sexual dysfunction*. Vancouver: Greystone Books.

Park, K., Goldstein, I., Andry, C., Siroky, M. B., Krane, R. J., et al. (1997). Vasculogenic female sexual dysfunction: The hemodynamic basis for vaginal engorgement insufficiency and clitoral erectile insufficiency. *International Journal of Impotence Research, 9*, 27–38.

Payer, L. (1992). *Disease-mongers: How doctors, drug companies, and insurers are making you feel sick*. New York: Wiley.

Popper, K. R. (1959). *The logic of scientific discovery*. London: Routledge.

Shifren, J. L., et al. (2008). Sexual problems and distress in United States women: Prevalence and correlates. *Obstetrics & Gynecology, 112*, 970–978.

Simon, J., et al. (2005). Testosterone patch increases sexual activity and desire in surgically menopausal women with hypoactive sexual desire disorder. *The Journal of Clinical Endocrinology & Metabolism, 90*, 5226–5233.

Sismondo, S. (2008). How pharmaceutical industry funding affects trial outcomes: Causal structures and responses. *Social Science & Medicine, 66*(9), 1909–1914.

Sismondo, S., & Doucet, M. (2010). Publication ethics and the ghost management of medical publications. *Bioethics, 24*(6), 273–283.

Stegenga, J. (2015). Measuring effectiveness. *Studies in History and Philosophy of Biological and Biomedical Sciences., 54*, 62–71.

Tavernise, S (2016) 'Female Viagra' Only Modestly increase sexual satisfaction, study finds. *The New York Times*. February 29.

Thorp, J., Simon, J., Dattani, D., & et al. DAISY Trial Investigators. (2012). Treatment of hypoactive sexual desire disorder in premenopausal women: Efficacy of flibanserin in the DAISY study. *Journal of Sexual Medicine, 9*(3), 793–804.

Tiefer, L. (2006). Female sexual dysfunction: A case study of disease mongering and activist resistance. *PLoS Medicine, 3*, e178.

Wilholt, T. (2006). Design rules: Industrial research and epistemic merit. *Philosophy of Science, 73*(1), 66–89.

Woloshin, S., Schwartz, L. M., Frankel, B., & Faerber, A. (2014). US food and drug administration and design of drug approval studies. *JAMA, 312*, 2163–2165.

Woloshin, S., & Schwartz, L. M. (2016). US food and drug administration approval of flibanserin: Even the score does not add up. *JAMA Internal Medicine, 176*, 439–442.

Young, N., Ioannidis, J., & Al-Ubaydli, O. (2008). Why current publication practices may distort science. *PLoS Medicine, 5*(10), e201. https://doi.org/10.1371/journal.pmed.0050201.

CPSIA information can be obtained
at www.ICGtesting.com
Printed in the USA
LVHW021213270721
693809LV00003B/293

9 783030 291785